Aut[illegible] Electrical and Electronic Systems

Classroom Manual

5th Edition

Chek-Chart

John F. Kershaw, Ed.D.
Revision Author

James D. Halderman
Series Advisor

Upper Saddle River, New Jersey
Columbus, Ohio

Library of Congress Cataloging-in-Publication Data

Automotive electrical and electronic systems / Chek-Chart: John F. Kershaw, revision author; James D. Halderman, series advisor—5th ed.
p. cm.
Includes index.
Contents: 1. Classroom manual— 2. Shop manual.
ISBN 0-13-049884-X
1. Automobiles—Electric equipment. 2. Automobiles—Electronic equipment. 3. Automobiles—Electronic equipment—Maintenance and repair. 4. Automobiles—Electric equipment—Maintenance and repair. I. Kershaw, John F. II. Halderman, James D., III. Chek-Chart Publications (Firm)

TL272.A786 2005
629.2'7—dc22 2004044646

Executive Editor: Ed Francis
Editorial Assistant: Jennifer Day
Production Editor: Stephen C. Robb
Production Supervision: Brenda Averkamp, Carlisle Publishers Services
Design Coordinator: Diane Y. Ernsberger
Cover Designer: Jeff Vanik
Cover photo: Super Stock
Production Manager: Matt Ottenweller
Marketing Manager: Mark Marsden

This book was set in Times by Carlisle Communications, Ltd. It was printed and bound by Courier Kendallville, Inc. The cover was printed by Phoenix Color Corp.

Portion of materials contained herein have been reprinted with permission of General Motors Corporation, Service and Parts Operations. License Agreement #0310805.

Pearson Education Ltd.
Pearson Education Singapore Pte. Ltd.
Pearson Education Canada, Ltd.
Pearson Education—Japan
Pearson Education Australia Pty. Limited
Pearson Education North Asia Ltd.
Pearson Educación de Mexico, S.A. de C.V.
Pearson Education Malaysia Pte. Ltd.

10 9 8 7 6 5 4 3 2 1
ISBN 0-13-049884-X

Introduction

Automotive Electrical and Electronic Systems is part of the Chek-Chart Series in Automotive Technology, which also includes:

- *Automatic Transmissions and Transaxles*
- *Automotive Brake Systems*
- *Automotive Heating, Ventilation, and Air Conditioning*
- *Automotive Manual Drive Train and Rear Axle*
- *Automotive Steering, Suspension, and Wheel Alignment*
- *Automotive Engine Repair and Rebuilding*
- *Engine Performance, Diagnosis, and Tune-Up*
- *Fuel Systems and Emission Controls.*

Each book in the Chek-Chart series aims to help instructors teach students to become competent and knowledgeable professional automotive technicians. The texts are the core of a learning system that leads a student from basic theories to actual hands-on experience.

The entire series is job-oriented, designed for students who intend to work in the automotive service profession. Knowledge gained from these books and the instructors enables students to get and keep jobs in the automotive repair industry. Learning the material and techniques in these volumes is a giant leap toward a satisfying, rewarding career.

Like other titles in the Chek-Chart series, *Automotive Electrical and Electronic Systems* consists of a *Classroom Manual* and a *Shop Manual.* The two-volume approach provides an effective presentation of the descriptive information and study lessons, along with representative testing, repair, and overhaul procedures. The manuals are to be used together; the descriptive material in the *Classroom Manual* reinforces the application material in the *Shop Manual.*

Each manual is divided into 15 stand-alone chapters. The *Classroom Manual* and the *Shop Manual* effectively complement each other. For example, *Classroom Manual* Chapter 7, "Automotive Battery Operation," is a companion chapter to *Shop Manual* Chapter 7, "Battery Diagnosis and Testing." When application information in the *Shop Manual* needs content explanation, a cross-reference will direct readers to a corresponding area of the *Classroom Manual.* Similarly, cross-references in the *Classroom Manual* link content to the *Shop Manual.* The complete, readable, and well-thought-out presentation of the chapters will facilitate instructors' effort. Students, in turn, will benefit from the many learning aids included, as well as from the thoroughness of the presentation.

Since 1929, the Chek-Chart Series in Automotive Technology has provided vehicle specification, training, and repair information to the professional automotive service field. *Automotive Electrical and Electronic Systems* has been a key component of the series for many years. Relying on the comprehensive Chek-Chart data bank, John F. Kershaw, Ed.D., researched and extensively rewrote the material in previous editions to develop this fifth edition. Series Advisor James D. Halderman contributed significantly to this effort.

Because of the comprehensive material, the hundreds of high-quality illustrations, and the inclusion of the latest automotive technology, instructors and students alike find that these books maintain their value over the years. Chek-Chart publications form the core of many master technicians' professional libraries.

How to Use This Book

WHY ARE THERE TWO MANUALS?

Unless you are familiar with the other books in this series, *Automotive Electrical and Electronic Systems* is unlike any other textbook you have used before. It is actually two books, the *Classroom Manual* and the *Shop Manual*. They have different purposes and should be used together.

The *Classroom Manual* teaches what a technician needs to know about electrical and electronic theory, systems, and components. The *Classroom Manual* is valuable in class and at home, both for study and for reference. The text and illustrations can be used for years hence to refresh your memory about the basics of automotive electrical and electronic systems and also about related topics in automotive history, physics, mathematics, and technology. This 5th edition text is based upon detailed learning objectives, which are listed in the beginning of each chapter.

The *Shop Manual* teaches test procedures, troubleshooting techniques, and how to repair the systems and components introduced in the *Classroom Manual*. The *Shop Manual* provides the practical, hands-on information required for working on automotive electrical and electronic systems. Use the two manuals together to understand fully how the systems work and how to make repairs when something is not working. This 5th edition text is based upon the 2002 NATEF (National Automotive Technicians Education Foundation) Tasks, which are listed in the beginning of each chapter. The 5th edition *Shop Manual* contains Job Sheet assessments that cover the 56 tasks in the NATEF 2002 A6 Electrical/Electronics repair area.

WHAT IS IN THESE MANUALS?

The following key features of the *Classroom Manual* make it easier to learn and remember the material:

- Each chapter is based on detailed learning objectives, which are listed in the beginning of each chapter.
- Each chapter is divided into self-contained sections for easier understanding and review. This organization clearly shows which parts make up which systems and how various parts or systems that perform the same task differ or are the same.
- Most parts and processes are fully illustrated with drawings or photographs. Important topics appear in several different ways, to make sure other aspects of them are seen.
- A list of Key Terms begins each chapter. These terms are printed in **boldface** type in the text and defined in the Glossary at the end of the manual. Use these words to build the vocabulary needed to understand the text.
- Review Questions are included for each chapter. Use them to test your knowledge.
- Every chapter has a brief summary at the end to help you review for exams.
- Brief but informative sidebars augment the technical information and present "real world" aspects of the subject matter.

The *Shop Manual* has detailed instructions on test, service, and overhaul procedures for modern electrical and electronic systems and their components. These are easy to understand and often include step-by-step explanations of the procedure. The *Shop Manual* contains:

- ASE/NATEF tasks, which are listed in the beginning of each chapter and form the framework for the chapter's content
- A list of Key Terms at the beginning of each chapter (These terms are printed in boldface type where first used in the text.)
- Helpful information on the use and maintenance of shop tools and test equipment
- Safety precautions
- Clear illustrations and diagrams to help you locate trouble spots while learning to read service literature
- Test procedures and troubleshooting hints that help you work better and faster
- Repair tips used by professionals, presented clearly and accurately
- A sample test at the back of the manual that is similar to those given for Automotive Service

Excellence (ASE) certification (Use this test to help you study and prepare when you are ready to be certified as an electrical and electronics expert.)

WHERE SHOULD I BEGIN?

If you already know something about automotive electrical and electronic systems and how to repair them, this book is a helpful review. If you are just starting in automotive repair, then this book provides a solid foundation on which to develop professional-level skills.

Your instructor has designed a course that builds on what you already know and effectively uses the available facilities and equipment. You may be asked to read certain chapters of these manuals out of order. That's fine. The important thing is to really understand each subject before moving on to the next.

Study the Key Terms in boldface type and use the review questions to help understand the material. When reading the *Classroom Manual,* be sure to refer to the *Shop Manual* to relate the descriptive text to the service procedures. When working on actual vehicle systems and components, look to the *Classroom Manual* to keep the basic information fresh in your mind. Working on such a complicated piece of equipment as a modern automobile is not easy. Use the information in the *Classroom Manual,* the procedures in the *Shop Manual,* and the knowledge of your instructor to guide you.

The *Shop Manual* is a good book for work, not just a good workbook. Keep it on hand while actually working on a vehicle. It will lie flat on the workbench and under the chassis, and it is designed to withstand quite a bit of rough handling.

When you perform actual test and repair procedures, you need a complete and accurate source of manufacturer specifications and procedures for the specific vehicle. As the source for these specifications, most automotive repair shops have the annual service information (on paper, CD, or Internet formats) from the vehicle manufacturer or an independent guide.

Acknowledgments

The publisher sincerely thanks the following vehicle manufacturers, industry suppliers, and organizations for supplying information and illustrations used in the Chek-Chart Series in Automotive Technology.

Allen Testproducts
American Isuzu Motors, Inc.
Automotive Electronic Services
Bear Manufacturing Company
Borg-Warner Corporation
DaimlerChrysler Corporation
Delphi Corporation
Fluke Corporation
Fram Corporation
General Motors Corporation
Honda Motor Company, Ltd.
Jaguar Cars, Inc.
Marquette Manufacturing Company
Mazda Motor Corporation
Mercedes-Benz USA, Inc.
Mitsubishi Motor Sales of America, Inc.
Nissan North America, Inc.
The Prestolite Company
Robert Bosch Corporation
Saab Cars USA, Inc.
Snab-on Tools Corporation
Toyota Motor Sales, U.S.A., Inc.
Vetronix Corporation
Volkswagen of America
Volvo Cars of North America

The comments, suggestions, and assistance of the following reviewers were invaluable: Rick Escalambre, Skyline College, San Bruno, CA, and Eugene Wilson, Mesa Community College, Mesa, AZ.

The publisher also thanks Series Advisor James D. Halderman.

Contents

HAZ-MAT and Safety

LEARNING OBJECTIVES

Upon completion and review of this chapter, you should be able to:

- Define the Occupational Safety and Health Act and describe the Mission of the Occupational Safety and Health Administration.
- Identify hazardous waste materials in accordance with state and federal regulations and follow proper safety precautions while handling hazardous waste materials.
- Explain the term *Material Safety Data Sheet (MSDS).*
- Define the steps required to safely handle and store automotive fuels.
- Describe four different types of fires, and identify the type of fire extinguisher required for each type of fire.
- Explain the steps needed for electrical equipment safety.
- Explain how to safely use hand and power tools and compressed air equipment
- Explain how to safely operate a hydraulic lift and a hydraulic jack, and when and how to use jack stands.
- Explain how to operate a vehicle safely in the automotive shop.

KEY TERMS

Combustible Materials
Corrosive
Eyewash Fountain
Fire Extinguishers
Hazard Communication Standard
Hazardous Waste Materials
Ignitable
Lift Pads
Lockout/Tagout
Material Safety Data Sheet (MSDS)
Occupational Safety and Health Administration (OSHA)
Radioactive
Reactive
Right-to-know Laws
Safety Glasses
Safety Stands
Solvents
Toxic
Used Oil

INTRODUCTION

The safe handling of hazardous waste material is extremely important in the automotive shop. The improper handling of hazardous material affects us all, not just those in the shop. Shop personnel must be familiar with their rights regarding hazardous waste disposal. Right-to-know laws explain these rights. Also, shop personnel must be familiar with hazardous materials in the automotive shop, and the proper disposal methods for these materials according to state and federal regulations.

The knowledge of safety precautions prevents serious personal injury and expensive property damage. Technicians who work with electrical and electronic systems must be familiar with all types of safety, including electrical safety, fuel handling, fires, tools, hydraulic jacks, and safe vehicle operation in the shop. The first step in providing a safe shop is learning about all types of safety precautions. However, the second, and most important step in this process is applying our knowledge of safety precautions while working in the shop, which will be covered in Chapter 1 of the *Shop Manual.* When shop employees have a careless attitude toward safety, accidents are more likely to occur. All shop personnel must develop a serious attitude toward safety in the shop. The result of this serious attitude is that automotive technicians will learn and adopt all shop safety rules.

OCCUPATIONAL SAFETY AND HEALTH ACT

The U.S. Congress passed the Occupational Safety and Health Act of 1970 with the following purposes:

> To assure safe and healthful working conditions for working men and women; by authorizing enforcement of the standards developed under the Act; by assisting and encouraging the States in their efforts to assure safe and healthful working conditions; by providing for research, information, education, and training in the field of occupational safety and health; and for other purposes.

The OSH Act also authorized enforcement of the standards developed under the Act by the **Occupational Safety and Health Administration (OSHA).** Since approximately 25 percent of workers are exposed to health and safety hazards on the job, the OSHA standards are necessary to monitor, control, and educate workers regarding health and safety in the workplace.

In Canada, the same function is performed by the Workplace Hazardous Materials Information System (WHMIS), so employees and employers residing in Canada need to become familiar with WHMIS.

For more information about workplace safety, see the section on "Lifting and Carrying Heavy Objects" in Chapter 1 of the *Shop Manual.*

HAZARDOUS WASTE

Hazardous waste materials are chemicals or components that the shop no longer needs, which pose a danger to the environment and people if they are disposed of in ordinary garbage cans or sewers. However, one should note that no material is considered hazardous waste until the shop has finished using it and is ready to dispose of it. The Environmental Protection Agency (EPA) publishes a list of hazardous materials that is included in the Code of Federal Regulations. The EPA considers waste hazardous if it is included on the EPA list of hazardous materials or it has one or more of these characteristics:

- **Reactive:** Any material that reacts violently with water or another chemical is considered hazardous. When exposed to low-pH (acid) solutions, if a material releases cyanide gas, hydrogen sulfide gas, or similar gases, it is considered hazardous.
- **Corrosive:** If a material burns the skin, or dissolves metals and other materials, a technician should consider it hazardous.

CAUTION: When handling hazardous waste material, one must always wear the proper protective clothing and equipment detailed in the right-to-know laws; this includes respirator equipment. All recommended procedures must be followed accurately. Personal injury may result from improper clothing, equipment, and procedures when handling hazardous materials.

- **Toxic:** Materials are hazardous if they leak one or more of eight different heavy metals in concentrations greater than 100 times the primary drinking water standard.
- **Ignitable:** A liquid is hazardous if it has a flash point below 1400°F (600°C) and a solid is hazardous if it ignites spontaneously.
- **Radioactive:** Any substance that emits measurable levels of radiation is hazardous. If individuals bring containers of highly radioactive substances into the shop environment, qualified personnel with the appropriate equipment must test them.

Improper Disposal Methods

Never, under any circumstances, use these methods to dispose of hazardous waste material:

- Pouring hazardous wastes on weeds to kill them
- Pouring hazardous wastes on gravel streets to prevent dust
- Throwing hazardous wastes in a Dumpster
- Disposing of hazardous wastes anywhere but an approved disposal site
- Pouring hazardous wastes down sewers, toilets, sinks, or floor drains

WARNING: Hazardous waste disposal laws include serious penalties for anyone responsible for breaking these laws.

Resource Conservation and Recovery Act (RCRA)

Federal and state laws control the disposal of hazardous waste materials. Every shop employee must be familiar with these laws. Hazardous waste disposal laws include the *Resource Conservation and Recovery Act (RCRA).* This law states that hazardous material users are responsible for hazardous materials from the time they become a waste until the proper waste disposal is completed. Many shops hire an independent hazardous waste hauler to dispose of hazardous waste material. The shop owner or manager should have a written contract with the hazardous waste hauler. However, the user must store hazardous waste material properly and safely, and be responsible for the transportation of this material until it arrives at an approved hazardous waste disposal site and is processed according to the law. Rather than have hazardous waste material hauled to an approved hazardous waste disposal site, a shop may choose to recycle the material in the shop. The RCRA controls the following types of automotive waste:

- Paint and body repair products waste
- Solvents for parts and equipment cleaning
- Batteries and battery acid
- Mild acids used for metal cleaning and preparation
- Waste oil and engine coolant (antifreeze)
- Air conditioning refrigerants
- Engine oil filters

Material Safety Data Sheets (MSDS)

The **right-to-know laws** state that employees have a right to know if the materials they use at work are hazardous. The right-to-know laws started with the **Hazard Communication Standard** published by OSHA in 1983. Originally, this document was intended for chemical companies and manufacturers that required employees to handle hazardous materials in their work situation. Meanwhile, the federal courts have decided to apply these laws to all companies, including automotive service shops. Under the right-to-know laws, the employer has three responsibilities regarding the handling of hazardous materials by their employees. First, all employees must be trained about the types of hazardous materials they will encounter in the work place. The employees must be informed about their rights under legislation regarding the handling of hazardous materials.

All hazardous materials must be properly labeled, and information about each hazardous material must be posted on **material safety data sheets (MSDS)** available from the manufacturer as shown in Figure 1-1. In Canada, the WHMIS also uses MSDSs.

The employer has a responsibility to place MSDSs where they are easily accessible by all employees. The MSDSs provide the following information about the hazardous material: chemical name, physical characteristics, protective

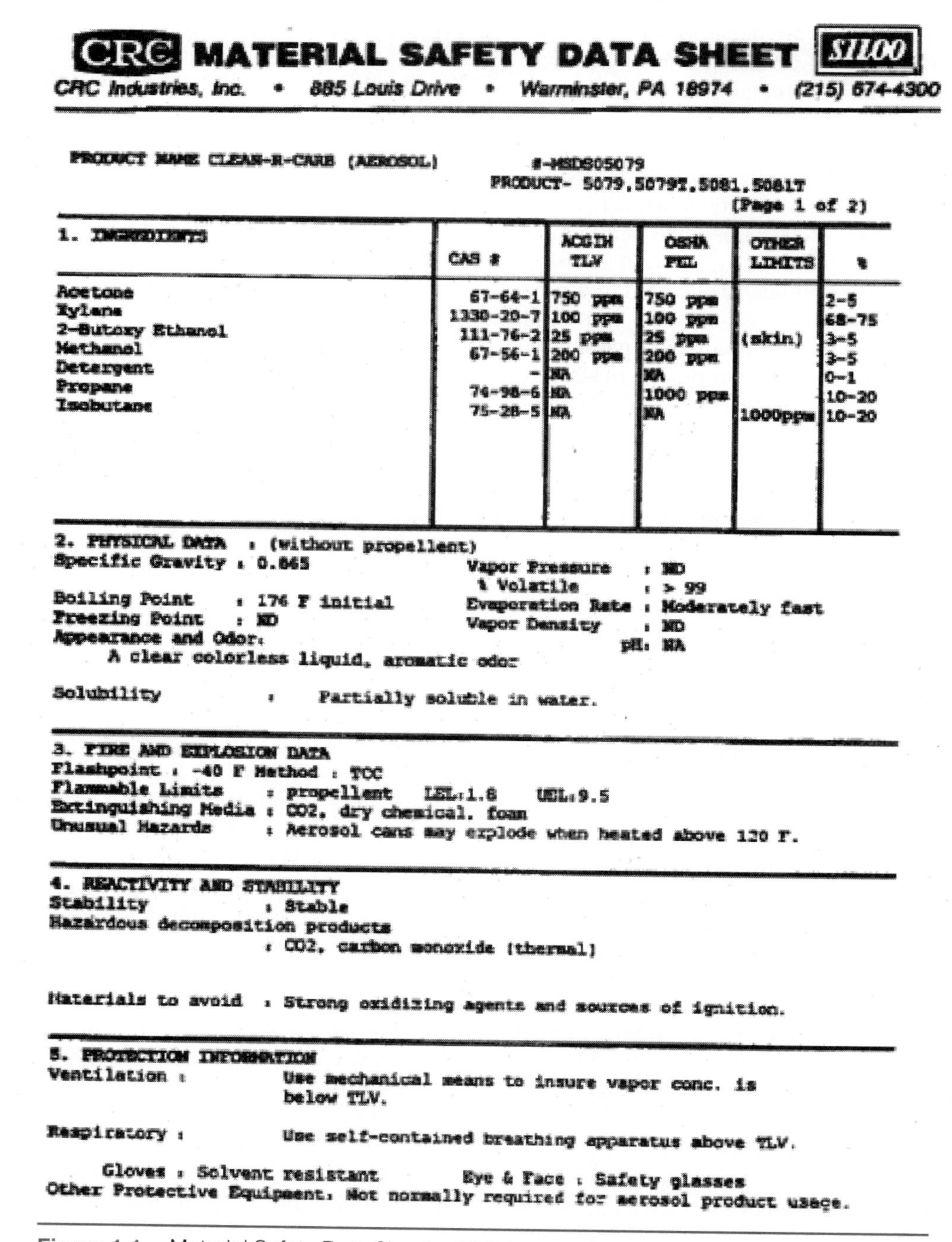

CRC MATERIAL SAFETY DATA SHEET SILICO

CRC Industries, Inc. • 885 Louis Drive • Warminster, PA 18974 • (215) 674-4300

PRODUCT NAME CLEAN-R-CARB (AEROSOL) #-MSDS05079
PRODUCT- 5079,5079T,5081,5081T
(Page 1 of 2)

1. INGREDIENTS	CAS #	ACGIH TLV	OSHA PEL	OTHER LIMITS	%
Acetone	67-64-1	750 ppm	750 ppm		2-5
Xylene	1330-20-7	100 ppm	100 ppm		68-75
2-Butoxy Ethanol	111-76-2	25 ppm	25 ppm	(skin)	3-5
Methanol	67-56-1	200 ppm	200 ppm		3-5
Detergent	-	NA	NA		0-1
Propane	74-98-6	NA	1000 ppm		10-20
Isobutane	75-28-5	NA	NA	1000ppm	10-20

2. PHYSICAL DATA : (without propellent)
Specific Gravity : 0.865
Boiling Point : 176 F initial
Freezing Point : ND
Appearance and Odor:
A clear colorless liquid, aromatic odor
Vapor Pressure : ND
% Volatile : > 99
Evaporation Rate : Moderately fast
Vapor Density : ND
pH: NA

Solubility : Partially soluble in water.

3. FIRE AND EXPLOSION DATA
Flashpoint : -40 F Method : TCC
Flammable Limits : propellent LEL:1.8 UEL:9.5
Extinguishing Media : CO2, dry chemical, foam
Unusual Hazards : Aerosol cans may explode when heated above 120 F.

4. REACTIVITY AND STABILITY
Stability : Stable
Hazardous decomposition products
: CO2, carbon monoxide (thermal)

Materials to avoid : Strong oxidizing agents and sources of ignition.

5. PROTECTION INFORMATION
Ventilation : Use mechanical means to insure vapor conc. is below TLV.

Respiratory : Use self-contained breathing apparatus above TLV.

Gloves : Solvent resistant Eye & Face : Safety glasses
Other Protective Equipment: Not normally required for aerosol product usage.

Figure 1-1. Material Safety Data Sheets. (CRC Industries, Inc.)

handling equipment, explosion/fire hazards, incompatible materials, health hazards, medical conditions aggravated by exposure, emergency and first-aid procedures, safe handling, and spill/leak procedures.

Second, the employer has a responsibility to make sure that all hazardous materials are properly labeled. The label information must include health, fire, and reactivity hazards posed by the material, and the protective equipment necessary to handle the material. The manufacturer must supply all warning and precautionary information about hazardous materials and this information must be read and understood by the employee before handling the material.

Third, employers are responsible for maintaining permanent files regarding hazardous materials. These files must include information on hazardous materials in the shop; proof of employee training programs; and information about accidents such as spills or leaks of hazardous materials. The employer's files must also include proof that employees' requests for hazardous material information such as MSDSs have been met. The employer must maintain a general right-to-know compliance procedure manual.

Used Oil

Simply put, **used oil** is exactly that—any petroleum-based or synthetic oil that has been used (Figure 1-2). During normal use, impurities such as dirt, metal scrapings, water, or chemicals can get mixed in with the oil, so that in time the oil no longer performs well. Eventually, this used oil must be replaced with virgin or re-refined oil to do the job at hand. The EPA's used oil management standards include a three-pronged approach to determine if a substance meets the definition of used oil. To meet the EPA's definition of used oil, a substance must meet each of the following three criteria:

- **Origin:** The first criterion for identifying used oil is based on the origin of the oil. Used oil must have been refined from crude oil or made from synthetic materials. Animal and vegetable oils are excluded from the EPA's definition of used oil.
- **Use:** The second criterion is based on whether and how the oil is used. Oils used as lubricants, hydraulic fluids, heat transfer fluids, buoyants, and for other similar purposes are considered used oil. Oil such as the bottom clean-out waste from virgin fuel-oil storage tanks or virgin fuel oil recovered from a spill do not meet the EPA's definition of used oil because these oils have never been "used." The EPA's definition also excludes products used as cleaning agents or solely for their solvent properties, as well as certain petroleum-derived products like antifreeze and kerosene.
- **Contaminants:** The third criterion is based on whether the oil is contaminated with physical or chemical impurities. In other words, to meet the EPA's definition, used oil must become contaminated as a result of being used. This aspect of the EPA's definition includes residues and contaminants generated from handling, storing, and processing used oil. Physical contaminants could include metal shavings, sawdust, or dirt. Chemical contaminants could include solvents, halogens, or salt water.

Figure 1-2. Changing engine oil.

The following are *not* used oil:

- Waste oil that is bottom clean-out waste from virgin fuel storage tanks, virgin fuel oil spill cleanups, or other oil wastes that have not actually been used.
- Products such as antifreeze and kerosene.
- Vegetable and animal oils, even when used as lubricants.
- Petroleum distillates used as solvents.

Improper Handling of Used Oil

The following information provides facts concerning the irresponsible management of used oil:

- The release of only one gallon of used oil (a typical oil change) can make a million gallons of fresh water undrinkable. This amount of fresh water is enough to satisfy the water requirements of 50 people for a full year!
- If used oil is dumped down the drain and enters a sewage treatment plant, concentrations as small as 50–100 ppm (parts per million) in the wastewater can foul sewage treatment processes.

- A film of used oil on top of the water's surface blocks sunlight and prevents oxygen absorption, which reduces plant and animal life in a body of water.

Contamination

Never mix a listed hazardous waste, gasoline, wastewater, halogenated solvents, antifreeze, or an unknown waste material with used oil. Adding any of these substances will cause the used oil to become classified as hazardous waste.

Disposal of Used Oil

Once oil has been used, it can be collected, recycled, and used over and over again. An estimated 380 million gallons of used oil are recycled each year. Recycled used oil can sometimes be used again for the same job or can take on a completely different task. For example, used motor oil can be re-refined and sold at the store as motor oil or processed for furnace fuel oil. Aluminum rolling oils also can be filtered on site and used over again. After collecting used oil in an appropriate container (such as a 55-gallon steel drum) the material must be disposed of in one of two ways:

- Shipped off site for recycling
- Burned in an on-site or off-site EPA-approved heater for energy recovery

Used Oil Storage

- Used oil must be stored in compliance with existing underground storage tank (UST) or above ground storage tank (AGST) standards, or kept in separate containers such as those shown in Figure 1-3. *Containers* are portable receptacles, such as a 55-gallon steel drum.
- Keep used oil storage drums in good condition. This means that they should be covered, secured from vandals, properly labeled, and maintained in compliance with local fire codes. Frequent inspections for leaks, corrosion, spillage, and so on are an essential part of container maintenance. Never store used oil in anything other than tanks and storage containers. Used oil may also be stored in units that are permitted to store regulated hazardous waste.

Figure 1-3. Used AGST oil storage container. (Eagle Manufacturing Company)

Solvent Handling

Solvents are used to clean parts, tools, and other items essential in vehicle maintenance and collision repair operations. Perhaps the most popular of these are naphtha petroleum distillates and additives found in carburetor cleaners. Mineral spirits and kerosene are also commonly used solvents.

Solvent Hazardous and Regulatory Status

Most solvents are classified as hazardous wastes. Other characteristics of solvents include the following:

- Solvents with flash points below 140°F are considered flammable and are federally regulated like gasoline by the Department of Transportation (DOT).
- Solvents and oils with flash points above 140°F are considered combustible and are regulated like motor oil by the DOT.

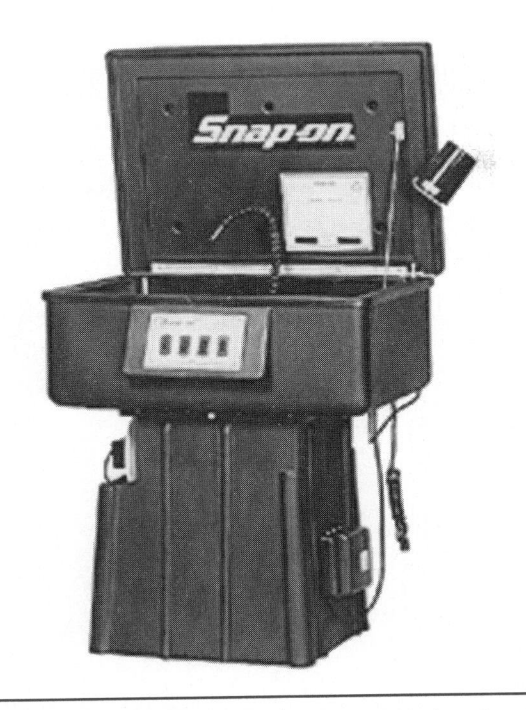

Figure 1-4. Small parts cleaner. (Reproduced under license from Snap-on Technologies, Inc.)

It is the responsibility of the repair shop to determine if their spent solvent is hazardous waste. Waste solvents that are considered hazardous waste have a flash point below 140°F (60°C). Hot water or aqueous parts cleaners may be used to avoid disposing of spent solvent as hazardous waste. Solvent-type parts cleaners with filters are available to greatly extend solvent life and reduce spent solvent disposal costs (Figure 1-4). Solvent reclaimers are available to clean and restore the solvent so it lasts indefinitely.

Used Solvents

Used or spent solvents are liquid materials that have been generated as waste and may contain xylene, methanol, ethyl ether, and methyl isobutyl ketone (MIBK). These materials must be stored in OSHA-approved safety containers with the lids or caps closed tightly. These storage receptacles must show no signs of leaks or significant damage due to dents or rust. In addition, the containers must be stored in a protected area equipped with secondary containment or a spill protector, such as a spill pallet. Additional requirements include the following:

- Containers should be clearly labeled "Hazardous Waste," along with the date the material was first placed into the storage receptacle.
- Labeling is not required for solvents being used in a parts washer.
- Used solvents will not be counted toward a facilities monthly output of hazardous waste if the vendor under contract removes the material.

Used solvents may be disposed of by recycling with a local vendor such as SafetyKleen to have them removed according to specific terms in the vendor agreement.

LEAD-ACID BATTERY WASTE

About 70 million spent lead-acid batteries are generated each year in the United States alone. Lead is classified as a toxic metal and the acid used in lead-acid batteries is highly corrosive. The vast majority (95–98 percent) of these batteries are recycled through lead reclamation operations and secondary lead smelters for use in the manufacture of new batteries.

Battery Hazardous and Regulatory Status

Used lead-acid batteries must be reclaimed or recycled in order to be exempt from hazardous waste regulations. Leaking batteries must be stored and transported as hazardous waste. Some states have more strict regulations, which require special handling procedures and transportation. According to the Battery Council International, 22 states currently base their used battery laws on a model regulation that includes the following provisions:

- Prohibits lead-acid battery disposal in landfills or incinerators
- Requires batteries to be delivered to a battery retailer, wholesaler, recycling center, or lead smelter
- Requires all retailers of automotive batteries to post a sign that displays the universal recycling symbol and indicates the retailer's specific requirements for accepting used batteries

CAUTION: Battery electrolyte contains sulfuric acid, which is a very corrosive substance capable of causing serious personal injury, such as skin burns and eye damage. In addition, the battery plates contain lead, which is highly poisonous. For this reason, disposing of batteries improperly can cause environmental contamination and lead to severe health problems.

Battery Handling and Storage

Batteries, whether new or used, should be kept indoors if possible. The storage location should be an area specifically designated for battery storage and must be well ventilated (to the outside). If outdoor storage is the only alternative, a sheltered and secured area with acid-resistant secondary containment is strongly recommended. It is also advisable that acid-resistant secondary containment be used for indoor storage. In addition, batteries should be placed on acid-resistant pallets and never stacked!

Battery Disposal

Used battery disposal can be handled as follows:

- By purchasing batteries from companies that pick up and transport used batteries to EPA-approved recycling facilities
- By returning defective batteries and batteries under warranty to the supplier in accordance with established policies and warranty credit terms.

FUEL SAFETY

Gasoline is a very explosive liquid! The expanding vapors that come from gasoline are extremely dangerous. These vapors are present even in cold temperatures. Vapors formed in gasoline tanks on many cars and trucks are controlled, but vapors from gasoline storage may escape from the can, resulting in a hazardous situation. Therefore, one must place gasoline storage containers in a well-ventilated space.

Although diesel fuel is not a volatile as gasoline, the same basic rules apply to diesel fuel and gasoline storage, as follows:

- Approved gasoline storage cans have a flash-arresting screen at the outlet (Figure 1-5). These screens prevent external ignition sources from igniting the gasoline within the can when someone pours the gasoline or diesel fuel.

Figure 1-5. Approved fuel-storage can. (Eagle Manufacturing Company)

- Technicians must always use red-painted, approved gasoline containers to allow for proper hazardous substance identification.
- Do not fill gasoline containers completely full. Always leave the level of gasoline at least one inch from the top of the container. This action allows expansion of the gasoline at higher temperatures. If gasoline containers are completely full, the expansion forces gasoline from the can and creates a dangerous spill.
- If gasoline or diesel fuel containers must be stored, place them in a designated storage locker or facility.
- If a gasoline container must be transported, be sure it is secured against upsets.
- Do not store a partially filled gasoline container for long periods, because it may give off vapors and produce a potential danger. Never leave gasoline containers open except while filling the container, or pouring gasoline from the container.
- Do not prime an engine with gasoline while cranking the engine.
- Never use gasoline as a cleaning agent.

- Always connect a ground strap to containers when filling or transferring fuel or other flammable products from one container to another, in order to prevent static electricity, which could result in explosion and fire.

FIRE SAFETY

Observing fire safety rules may prevent a fire in the shop. Following the proper fire safety rules and procedures may also make the difference between extinguishing a fire with minimum damage and having a fire get out of control, causing very expensive damage. Follow these fire safety rules and procedures in the shop:

- Familiarize yourself with the location and operation of all shop fire extinguishers.
- If a fire extinguisher is used, report it to management so that the extinguisher will be recharged.
- Do not use any type of open-flame heater to heat the work area.
- Do not turn on the ignition switch or crank the engine with a fuel line disconnected.
- Store all **combustible materials** such as gasoline, paint, and oily rags in approved safety containers.
- Clean up gasoline, diesel fuel, oil, or grease spills immediately.
- Always wear clean shop clothes. Do not wear oil-soaked clothes.
- Do not allow sparks and flames near batteries.
- Welding tanks must be securely fastened in an upright position.
- Do not block doors, stairways, or exits.
- Do not smoke when working on vehicles.
- Do not smoke, or create sparks, near flammable materials or liquids.
- Store combustible shop supplies, such as paint, in a closed steel cabinet.
- Gasoline and diesel fuel must be kept in approved safety containers.
- If a fuel tank is removed from a vehicle, do not drag the tank on the shop floor.
- Call the fire department as soon as a fire begins, then attempt to extinguish the fire.
- Familiarize yourself with different types of fires and fire extinguishers, and know the type of extinguisher to use on each fire.
- Know the approved fire escape route from your classroom or shop to the outside of the building.
- If a fire occurs, do not open doors or windows. This action creates an extra draft that makes the fire worse.
- Do not put water on a gasoline fire, because the water will make the fire worse.
- If possible, stand six to ten feet from the fire and aim the fire extinguisher nozzle at the base of the fire with a sweeping action.
- If a fire produces a lot of smoke in the room, remain close to the floor to obtain oxygen and avoid breathing smoke.
- If the fire is too hot or the smoke makes breathing difficult, get out of the building.
- Do not re-enter a burning building.
- Keep solvent containers covered except when pouring from one container to another. When flammable liquids are transferred from bulk storage, the bulk container should be grounded to a permanent shop fixture such as a metal pipe. During this transfer process, ground the bulk container to the portable container (Figure 1-6). These ground wires prevent the buildup of a static electric charge, which could result in a spark and disastrous explosion. Always discard, or clean, empty solvent containers, because fumes in these containers are a fire hazard.

Figure 1-6. Solvent container. (Eagle Manufacturing Company)

FIRE EXTINGUISHERS

Fire extinguishers (Figure 1-7) are among the most important pieces of safety equipment. All shop personnel must know the location of the fire extinguishers in the shop. If you have to waste time looking for an extinguisher after a fire starts, the fire could get out of control before you get the extinguisher into operation. Fire extinguishers should be located where they are easily accessible at all times. Everyone working in the shop must know how to operate the fire extinguishers. A decal on each fire extinguisher identifies the type of chemical in the extinguisher and provides operating information (Figure 1-8). Shop personnel should be familiar with the following types of fires and fire extinguishers:

- **Class A fires** are those involving ordinary combustible materials such as paper, wood, clothing, and textiles. Multipurpose dry chemical extinguishers are used on these fires.
- **Class B fires** involve the burning of flammable liquids such as gasoline, diesel fuel, oil, paint, solvents, and greases. These fires may be extinguished with multipurpose dry chemical extinguishers. In the past, fire extinguishers containing halogen, or halon, were used to extinguish class B fires. The chemicals in this type of extinguisher attach to the hydrogen, hydroxide, and oxygen molecules to stop the combustion process almost instantly. However, the resultant gases from the use of halogen-type extinguishers were very toxic and harmful to the operator of the extinguisher. Halon fire extinguishers are now illegal to manufacture because of chlorofluorocarbon (CFC) regulations.
- **Class C Fires** involve the burning of electrical equipment such as wires, motors, and switches. These fires may be extinguished with multipurpose dry chemical extinguishers.
- **Class D Fires** involve the combustion of metal chips, turnings, and shavings. Dry chemical extinguishers are the only type of extinguisher recommended for these fires.

Figure 1-7. Fire extinguisher symbol.

ELECTRICAL EQUIPMENT SAFETY

Electrical components have a number of safety concerns for technicians to be aware of in their daily use. The major concerns are as follows:

- Technicians must replace or repair frayed cords on electrical equipment such as shop lights, drills, grinders, wheel aligners, wheel balancers, overhead electric hoists, and cleaning equipment.
- All electric cords from lights and electric equipment must have a ground connection. The ground connector is the round terminal in a three-prong electrical plug. Do not use a two-prong adapter to plug in a three-prong electrical cord. Three-prong electrical outlets should be mandatory in all shops.
- Do not leave electrical equipment running and unattended.

HAND TOOL SAFETY

Improper use and care of hand tools causes many shop accidents. One needs to follow these tool safety steps:

- Maintain tools in good condition and keep them clean. Worn tools may slip and result in hand injury. When using a hammer with a loose head, the head may fly off and cause personal injury or vehicle damage. Your hand may slip off a greasy tool, caus-

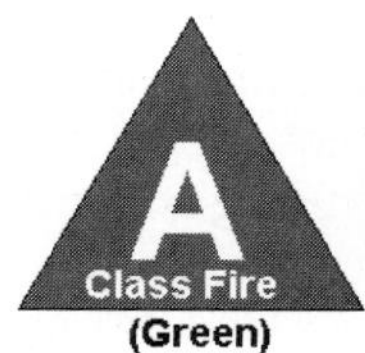

(Green)

Class A Fire Extinguisher: This fire extinguisher is used on common combustibles by soaking and cooling the fibers to prevent re-ignition. It uses pressurized water, foam or multipurpose (ABC-rated*) dry chemical extinguishers. Carbon Dioxide or normal (BC-rated*) dry chemical extinguishers are not to be used on class fires

(Red)

Class B Fire Extinguisher: This fire extinguisher is used on flammable liquids, greases or gases. They work by removing the oxygen and preventing any vapors from reaching a new ignition source or starting the chemical chain reaction. It uses foam, carbon dioxide, normal (BC-rated*) dry chemical, multipurpose (ABC-rated*) dry chemical, and Halon mediums

(Blue)

Class C Fire Extinguisher: using an extinguishing medium not capable of conducting electricity uses this fire extinguisher to extinguish energized electrical equipment. Carbon dioxide, ordinary (BC-rated*) dry chemical, multipurpose (ABC-rated*), and Halon fire extinguishers may be used to fight Class C fires.**

(Yellow)

Class D Fires involve the combustion of metal chips, turnings, and shavings. These metals include: magnesium, titanium, potassium and sodium as well as pyrophoric organometallic reagents such as alkyllithiums, Grignards and diethylzinc. These materials burn at high temperatures and will react violently with water, air, and/or other chemicals. Handle with care!! Dry chemical or Metal/Sand extinguishers are the only types of extinguisher recommended for these fires. They simply smother the fire.

Figure 1-8. Fire classes table.

ing some part of your body to hit the vehicle; for example, your head may hit the vehicle hood.

- Using the wrong tool for the job may cause damage to the tool, the fastener, or your hand, if the tool slips. If you use a screwdriver as a chisel or pry bar, the blade may shatter, causing serious personal injury.
- Use sharp pointed tools with caution. Always check your pockets before sitting on the vehicle seat. A screwdriver, punch, or chisel in the back pocket may put an expensive tear in the upholstery. Do not lean over fenders with sharp tools in your pockets.
- Tool tips that are intended to be sharp should be kept in a sharp condition. Sharp tools, such as chisels, will do the job faster with less effort.

For more information about hand tool safety; see the section on "Housekeeping" in Chapter 1 in the *Shop Manual.*

POWER TOOL SAFETY

Power tools use electricity, shop air, or hydraulic pressure as a power source. Careless operation of power tools may cause personal injury and vehicle damage. Follow these steps for safe power tool operation:

- Do not operate power tools with frayed power cords.
- Be sure the power tool cord has a proper ground connection.
- Do not stand on a wet floor while operating an electric power tool.
- Always unplug an electric power tool before servicing the tool.
- Do not leave a power tool running and unattended.
- When using a power tool on small parts, do not hold the part in your hand. The part must be secured in a bench vise, or with locking pliers.

- Do not use a power tool on a job where the maximum capacity of the tool is exceeded.
- Be sure that all power tools are in good condition, and always operate these tools according to the tool manufacturer's recommended procedure.
- Make sure all protective shields and guards are in position.
- Maintain proper body balance while using a power tool.
- Always wear safety glasses or a face shield.
- Wear ear protection.
- Follow the equipment manufacturer's recommended maintenance schedule for all shop equipment.
- Never operate a power tool unless you are familiar with the tool manufacturer's recommended operating procedure. Serious accidents occur from improper operating procedures.
- Always make sure that the wheels are securely attached and in good condition on the electric grinder.
- Keep fingers and clothing away from grinding and buffing wheels. When grinding, or buffing a small part, it should be held with a pair of locking pliers.
- Always make sure the sanding or buffing disc is securely attached to the sander pad.
- Special heavy-duty sockets must be used on impact wrenches. If ordinary sockets are used on an impact wrench, they may break and cause serious personal injury.
- Never operate an air chisel unless the tool is securely connected to the chisel with the proper retaining device.
- Never direct a blast of air from an air gun against any part of the body. If air penetrates the skin and enters the bloodstream, it can cause very serious health problems and even death.

COMPRESSED AIR EQUIPMENT SAFETY

The shop air supply contains high-pressure air in the shop compressor and air lines. Serious injury or property damage may result from careless operation of compressed air equipment. When these steps are followed, safety is improved regarding compressed air equipment:

- Safety glasses or a face shield should be worn for all shop tasks, including those involving the use of compressed air equipment.
- Wear ear protection when using compressed air equipment.
- Always maintain air hoses and fittings in good condition. If an end suddenly blows off an air hose, the hose will whip around, which could cause personal injury.
- Do not direct compressed air against the skin, This air may penetrate the skin, especially through small cuts or scratches. If air penetrates the skin and enters the bloodstream, it can be fatal or can cause serious health complications.
- Use only air-gun nozzles approved by OSHA.
- Do not use an air gun to blow off clothing or hair.
- Do not clean the workbench or floor with compressed air. This action may blow very small parts against your skin or into your eye. Small parts blown by compressed air may cause vehicle damage. For example, if the vehicle in the next stall has the air cleaner removed, a small part may go into the air intake. When the engine is started this part will likely be pulled into a cylinder by engine vacuum, and the part will penetrate through the top of the engine piston.
- Never spin bearings with compressed air, because the bearing will rotate at extremely high speed. Under this condition the bearing may be damaged or disintegrate, causing personal injury.
- All pneumatic tools must be operated according to the manufacturer's recommended operating procedure.
- Follow the equipment manufacturer's recommended maintenance schedule for all compressed air equipment.

VEHICLE OPERATION

When driving a customer's vehicle, one must observe certain shop driving rules and precautions to prevent accidents and maintain good customer relations.

1. Before driving a vehicle, always make sure the brakes are operating and fasten the safety belt.

2. Check to be sure there are no people or objects under the vehicle before you start the engine.
3. If the vehicle is parked on a lift, be sure the lift is fully down and the lift arms or components are not contacting the chassis.
4. If the vehicle contains a diesel engine equipped with glow plugs, make sure the glow plugs have cycled off before starting the engine.
5. Check to see if there are any objects directly in front or behind, the vehicle before driving away.
6. Always drive slowly in the shop, and watch carefully for personnel and other moving vehicles.
7. Make sure the shop door is up high enough to so there is plenty of clearance between the top of the vehicle and the door.
8. Watch the shop door to be certain that it is not coming down as you attempt to drive under the door.
9. If a road test is necessary, obey all traffic laws, and never drive in a reckless manner.
10. Do not squeal tires when accelerating, or turning corners.

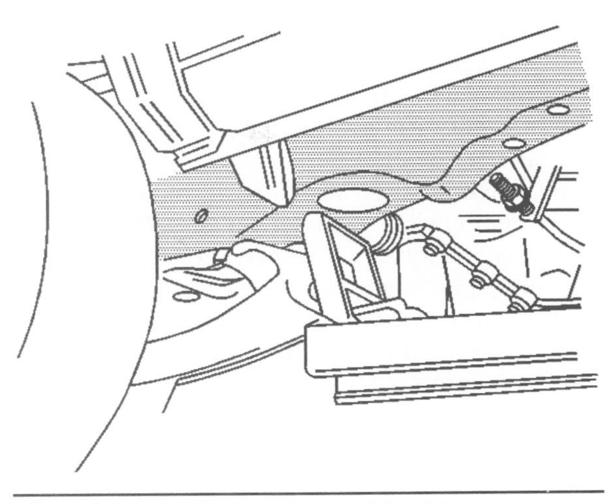

Figure 1-9. Vehicle lift. (GM Service and Parts Operations)

LIFT SAFETY

Figure 1-9 shows a typical aboveground electric-hydraulic automotive lift. Special lift safety precautions and procedures must be followed when raising a vehicle on a lift. Follow these steps for lift safety:

- Always be sure the lift is completely lowered before driving a vehicle on or off the lift.
- Be sure the **lift pads** on the lift are contacting the vehicle manufacturer's recommended lift points (Figure 1-10) shown in the service manual. If the proper lift points are not used, components under the vehicle such as brake lines or body parts may be damaged. Failure to use the recommended lift points may cause the vehicle to slip off the lift, resulting in severe vehicle damage and personal injury.
- Before a vehicle is raised or lowered, close the doors and hood.
- After the vehicle has been lifted a short distance off the floor, stop the lift and check the contact between the lift pads and the vehicle chassis to be sure the lift pads are still on the recommended lift points.
- When a vehicle is raised on a lift, be sure the safety mechanism is in place to prevent the lift from dropping accidentally.

CAUTION: When you raise a vehicle on a lift, it must be raised high enough to allow engagement of the lift locking mechanism. If the lock mechanism is not engaged, the lift may drop suddenly, causing personal injury or vehicle damage.

- Do not hit, or run over, lift arms and adapters when driving a vehicle on, or off, the lift. Have a co-worker guide you when driving a vehicle onto the lift. Do not stand in front of a lift with the vehicle coming towards you.
- Before lowering a vehicle on a lift, always make sure there are no objects, tools, or people under the vehicle.
- Do not rock a vehicle on a lift during a service job.

- While a vehicle is raised on a lift, removal of some heavy components may cause imbalance on the lift.
- Do not raise a vehicle on a lift with people in the vehicle.
- When raising vans on a lift, remember they are higher than passenger cars. Be sure you have adequate clearance between the top of the vehicle and the shop ceiling.
- Do not raise a four-wheel-drive truck with a frame contact lift, because it may damage axle joints.

HYDRAULIC JACK AND SAFETY STAND SAFETY

Never work under a car or truck unless **safety stands** (Figure 1-11) are placed securely under the vehicle chassis, and the vehicle is resting on these stands. Before lifting a vehicle with a floor jack, be sure that the jack lift pad is positioned securely under a recommended lift point on the vehicle.

CAUTION: Always make sure the safety stand weight capacity rating exceeds the vehicle weight that is lowered onto the stands. If the weight placed on the safety stands exceeds the safety stand weight capacity, the safety stands may collapse, resulting in personal injury or vehicle damage.

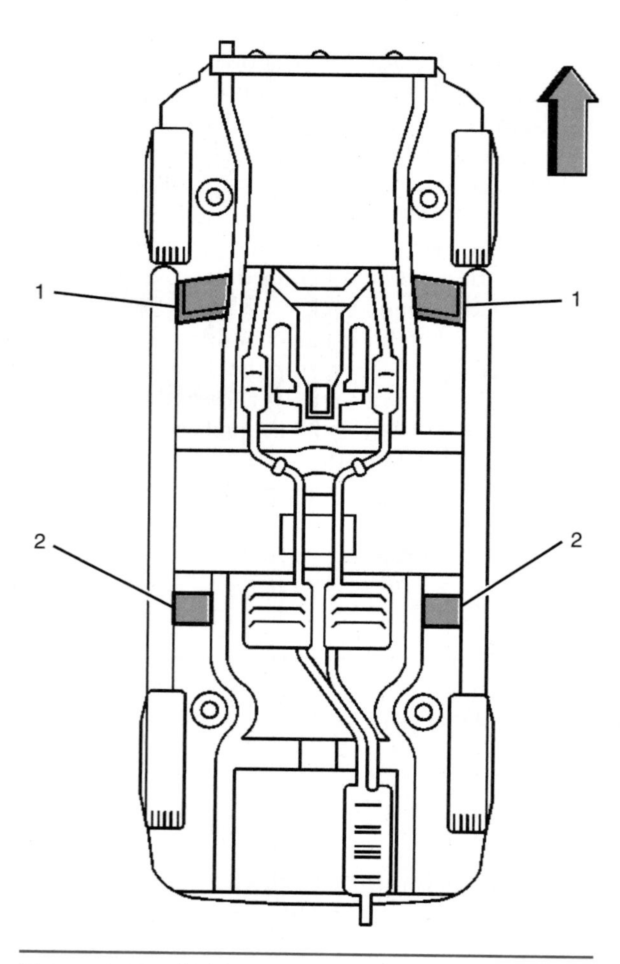

Figure 1-10. Vehicle lift pads. (GM Service and Parts Operations)

Accidents involving the use of floor jacks and safety stands may be avoided if these safety precautions are followed:

- Do not the front end of a vehicle with the jack placed under a radiator support it; may cause severe damage to the radiator and the support.
- Position the safety stands under a strong chassis member such as the frame or the axle housing. The safety stands must contact the vehicle manufacturer's recommended lift points.

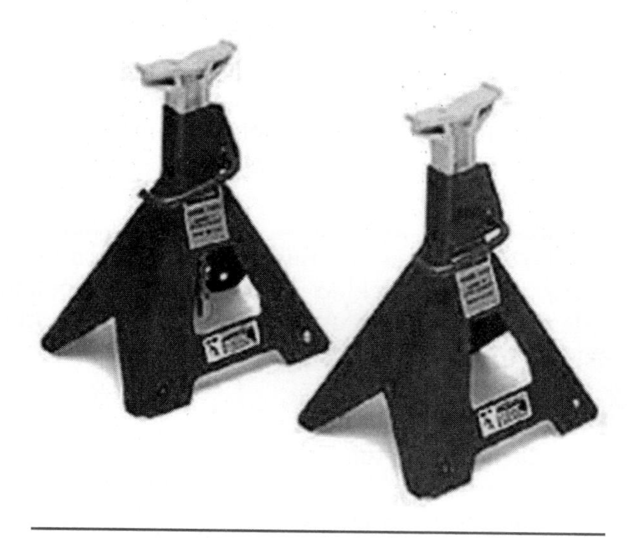

Figure 1-11. Safety stands. (Reproduced under license from Snap-on Technologies, Inc.)

- Since the floor jack is on wheels, the vehicle and jack tend to move as the vehicle is lowered from a floor jack onto jack stands. Always be sure the safety stands remain under the chassis member during this operation, and be sure the safety stands do not tip. All the safety stand legs must remain in contact with the shop floor.
- When the vehicle is lowered from the floor jack onto safety stands, remove the floor jack from under the vehicle. Never leave a jack handle sticking out from under a vehicle. Someone may trip over the handle and injure him- or herself.

EYE INJURIES

Causes of Eye Injuries

Eye injuries may occur in various ways in an automotive shop. Some of the common eye accidents are as follows:

- Thermal burns from excessive heat
- Irradiation burns from excessive light, such as from an arc welder
- Chemical burns from strong liquids such as battery electrolyte
- Foreign material in the eye
- Penetration of the eye by a sharp object
- A blow from a blunt object
- Wearing safety glasses and observing shop safety rules will prevent most eye accidents.

Eyewash Fountains

If a chemical gets in your eyes it must be washed out immediately to prevent a chemical burn. An **eyewash fountain** is the most effective way to wash the eyes. An eyewash fountain is similar to a drinking water fountain, but the eyewash fountain has water jets placed throughout the fountain top. Every shop should be equipped with some eyewash facility (Figure 1-12). Be sure you know the location of the eyewash fountain in the shop.

Figure 1-12. Eyewash fountain. (Haws Corporation)

Safety Glasses and Face Shields

The mandatory use of eye protection with safety glasses or a face shield is one of the most important safety rules in a shop. **Safety glasses** protect the eyes (Figure 1-13); face shields protect the face (Figure 1-14). When using an electric grinder or buffer, you must wear safety glasses or a face shield. Many shop insurance policies require the use of eye protection in the shop. Some automotive technicians have been blinded in one, or both, eyes because they did not bother to wear safety glasses. All safety glasses should provide some type of side protection (Figure 1-13). When selecting a pair of safety glasses they should feel comfortable on your face. If they are uncomfortable, you may tend to take them off, leaving the eyes unprotected. A technician should wear a face shield when handling hazardous chemicals or when using an electric grinder or buffer (Figure 1-14).

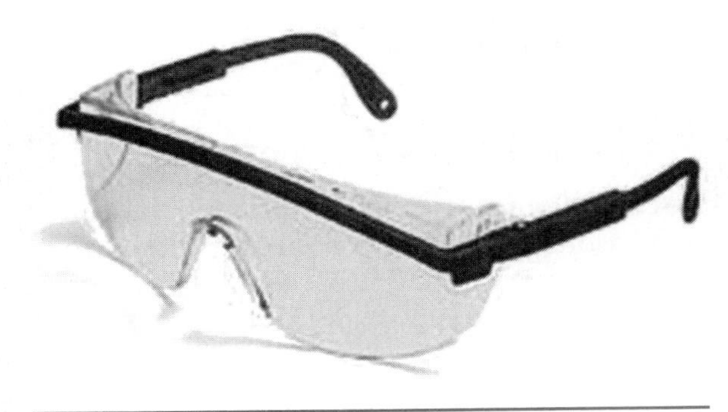

Figure 1-13. Safety glasses. (Reproduced under license from Snap-on Technologies, Inc.)

Figure 1-14. Face shield. (Reproduced under license from Snap-on Technologies, Inc.)

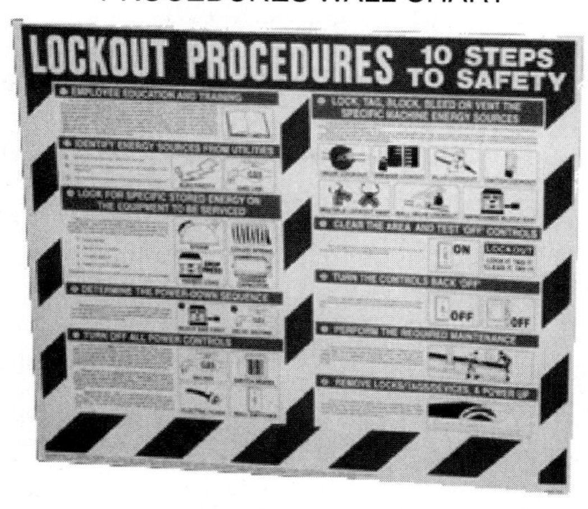

Figure 1-16. Lockout/tagout. (© Lab Safety Supply, Inc., Janesville, WI. Reproduced with permission.)

Figure 1-15. First-aid kit. (Swift First Aid)

FIRST-AID KITS

First-aid kits should be clearly identified and conveniently located (Figure 1-15). These kits contain items such as bandages and ointment required for minor cuts. All shop personnel must be familiar with the location of first-aid kits. At least one of the shop personnel should have basic first-aid training, and this person should be in charge of administering first aid and keeping first-aid kits filled.

LOCKOUT/TAGOUT

Lockout/tagout is an OSHA procedure required in all workplaces. This procedure is designed to prevent electrical equipment from being started while being repaired or maintained. With lockout/tagout, all of the necessary switches must be opened, locked out, and tagged (Figure 1-16). Any person that may be involved with a piece of equipment must be trained in this procedure.

SUMMARY

The United States Occupational Safety and Health Act of 1970 assures safe and healthful working conditions and authorizes enforcement of safety standards by the Occupational Safety and Health Administration (OSHA). Hazardous waste materials are chemicals or components that are no longer needed in the shop. They may pose a danger to people or the environment if they are disposed of in ordinary garbage cans or sewers. Hazardous materials fall into the following five categories: reactive, corrosive, toxic, ignitable, and radioactive.

The Resource Conservation and Recovery Act (RCRA) law states that hazardous material users are responsible for hazardous materials from the time they become a waste until the proper waste disposal is completed. All hazardous material information must be posted on material safety data sheets (MSDSs), which are available from the manufacturer. MSDSs provide critical information about the material that the technician must have access to.

Used oil is any petroleum-based or synthetic oil that has been used. If impurities such as gasoline, wastewater, halogenated solvents, antifreeze, or unknown waste are mixed in with the oil, in time they can become hazardous waste. After collecting used oil in an appropriate container, the material must be disposed and stored using an EPA (Environmental Protection Agency)-approved process.

Solvents are used to clean parts, tools, and other items essential in vehicle maintenance and collision repair operations and most of them are classified as hazardous wastes. It is the responsibility of the repair shop to determine if their spent solvent is hazardous waste, if so, must be stored in OSHA-approved safety containers with the lids or caps closed tightly.

The expanding vapors that come from gasoline are extremely dangerous. Therefore, one must place gasoline storage containers in a well-ventilated space. Although diesel fuel is not a volatile as gasoline, the same basic rules apply to diesel fuel and gasoline storage.

Following the proper fire safety rules and procedures may also make the difference between extinguishing a fire with minimum damage and having a fire get out of control, causing very expensive damage. All shop personnel must know the location of the fire extinguishers in the shop and should be familiar with the four classifications of fires (A-B-C-D) and the types of fire extinguisher used to put them out.

Electrical components have a number of safety concerns, such as frayed cords and faulty ground connections. Improper use and care of hand tools causes many shop accidents. Power tools use electricity, shop air, or hydraulic pressure as a power source. Careless operation of power tools may cause personal injury and vehicle damage.

The shop air supply contains high-pressure air in the shop compressor and air lines. Serious injury or property damage may result from careless operation of compressed-air equipment. Safety glasses, or a face shield, should be worn for all shop tasks, including those tasks involving the use of compressed-air equipment.

When driving a customer's vehicle, you must observe certain shop driving rules and precautions to prevent accidents and maintain good customer relations. When you raise a vehicle on a lift, it must be raised high enough to allow engagement of the lift locking mechanism. If the lock mechanism is not engaged, the lift may drop suddenly causing personal injury or vehicle damage. Be sure the lift pads on the lift are contacting the vehicle manufacturer's recommended lift points, as shown in the service manual. Never work under a car or truck unless safety stands are placed securely under the vehicle chassis and the vehicle is resting on stands.

Eye injuries may occur in various ways in an automotive shop. If a chemical gets in your eyes it must be washed out immediately to prevent a chemical burn. An eyewash fountain is the most effective way to wash the eyes. To avoid these injuries, the mandatory use of eye protection, with safety glasses or a face shield, is one of the most important safety rules in an automotive shop. First-aid kits should be clearly identified and conveniently located. These kits contain such items as bandages and ointment required for minor cuts. All shop personnel must be familiar with the location of first-aid kits. Lockout/tagout is an OSHA procedure designed to prevent electrical equipment from being started while being repaired or maintained.

Review Questions

1. Technician A says breathing asbestos dust may cause high blood pressure. Technician B says oily rags should be stored in an OSHA-approved container to avoid a fire hazard. Who is right?
   a. A only
   b. B only
   c. Both A and B
   d. Neither A nor B

2. Two technicians are discussing shop hazards. Technician A says high-pressure air from an air gun may penetrate the skin. Technician B says air in the blood stream may be fatal. Who is right?
   a. A only
   b. B only
   c. Both A and B
   d. Neither A nor B

3. Technician A says hazardous waste never reacts violently with water. Technician B says hazardous-waste haulers are responsible for hazardous waste materials from the time they become waste until the proper waste disposal is completed. Who is right?
   a. A only
   b. B only
   c. Both A and B
   d. Neither A nor B

4. When discussing electrical, gasoline, and diesel fuel safety, which of the following statements is right?
   a. A two-prong adapter may be used to plug in a three-prong electrical cord.
   b. Electric cords from lights and electrical equipment do not require a ground connection.
   c. When stored in the shop, gasoline or diesel fuel containers should be completely full.
   d. Gasoline containers must be secured when they are transported.

5. The subject of shop fire fighting is being discussed. Technician A says that Class B fires are extinguishable using water. Technician B says that Class D fires are extinguishable using a dry chemical extinguisher. Who is right?
   a. A only
   b. B only
   c. Both A and B
   d. Neither A nor B

6. While discussing hazardous waste disposal, Technician A says the right-to-know laws require employers to train employees regarding hazardous waste materials. Technician B says the right-to-know laws do not require employers to inform employees of their rights regarding hazardous waste materials. Who is right?
   a. A only
   b. B only
   c. Both A and B
   d. Neither A nor B

7. While discussing hazardous materials, Technician A says a solid that ignites spontaneously is considered a hazardous material. Technician B says a liquid with a flash point above 140°F (60°C) is considered a hazardous material. Who is right?
   a. A only
   b. B only
   c. Both A and B
   d. Neither A nor B

8. Two technicians are discussing hazardous waste disposal: Technician A says certain types of hazardous waste react violently with water. Technician B says hazardous waste users are responsible for hazardous waste materials from the time they become waste until the proper waste disposal is completed. Who is right?
   a. A only
   b. B only
   c. Both A and B
   d. Neither A nor B

9. When handling hazardous materials in an automotive repair shop, which of these items is correct?
   a. Air conditioning refrigerants are not considered hazardous waste materials.
   b. Hazardous waste materials must not be recycled in the shop.
   c. MSDSs contain information on handling hazardous materials.
   d. Used engine coolant cannot be recycled.

10. Technicians are discussing shop layout and safety. Technician A says some shops have

a tool room where special tools are located. Technician B says each piece of safety equipment must be in an accessible location. Who is right?
a. A only
b. B only
c. Both A and B
d. Neither A nor B

11. While discussing shop rules, Technician A says loose, long hair in the shop presents no problem. Technician B says breathing carbon monoxide may cause headaches. Who is correct?
a. A only
b. B only
c. Both A and B
d. Neither A nor B

12. Congress passed the Occupational Safety and Health ACT in 1970. Technician A says that the purpose of this act was to ensure a safe and healthful work environment by enforcing safety standards. Technician B says that more than 50% of the workers are exposed to health and safety hazards on the job. Who is correct?
a. A only
b. B only
c. Both A and B
d. Neither A nor B

13. While discussing fire safety, Technician A says that gasoline containers are painted blue for proper identification. Technician B says that you never fill a gasoline container to the top. Who is right?
a. A only
b. B only
c. Both A and B
d. Neither A nor B

14. Whie discussing the importance of fire extinguishers, Technician A says it is not necessary for all shop personnel to know their location. Technician B says that fire extinguishers containing halon or halogen are used to extinguish Class B fires involving flammable liquids. Who is right?
a. A only
b. B only
c. Both A and B
d. Neither A nor B

15. Technician A says that all information about each hazardous material must be posted on a material safety data sheet (MSDS). Technician B says MSDSs provide extensive information, such as chemical name, explosive fire hazards, and so on. Who is right?
a. A only
b. B only
c. Both A and B
d. Neither A nor B

16. While discussing solvent handling, which of the following is correct?
a. Used solvent is not a hazardous waste substance.
b. It is not shop responsibility to determine whether a substance is hazardous.
c. Solvent recyclers may be used to recycle antifreeze.
d. Solvents are hazardous, with a flash point below 140°F.

17. When discussing electrical, gasoline, and diesel fuel safety, which of these statements is right?
a. Two-prong adapter may be used to plug in a three-prong electrical cord.
b. Electric cords from lights and electrical equipment do not require a ground connection.
c. When stored in the shop gasoline or diesel fuel containers should be completely full.
d. Gasoline containers must be secured when they are transported.

18. Technician A says that if a shop generates more than 250 gallons of hazardous waste each month, it is considered a Large Quantity Generator. Technician B says that an EPA hazardous waste ID number is only required for shops that generate more than 300 gallons of hazardous waste per month. Who is right?
a. A only
b. B only
c. Both A and B
d. Neither A nor B

19. All of the following statements concerning outdoor battery storage are true, Except:
a. Batteries may be stacked two high.
b. Batteries should be stored on an acid-resistant pallet.
c. Secondary containment is recommended.
d. The storage area should be secured.

20. Which of these is correct when operating a lift?
    a. When lifting a single-rear-axle truck on a twin-post lift, do not place the rear wheels in the floor depression.
    b. On a frame-contact twin-post lift, position the lift arms under the manufacturer's recommended lift points.
    c. On a triple-post-lift for big trucks, only the front post can be moved forward or rearward.
    d. Some twin-post-lifts have a maximum capacity of 8,000 lb. (3628.80 kg) per post.

21. Two technicians are discussing screwdriver use. Technician A says care must be taken to use a screwdriver whose blade fits snugly into the in the screw. Technician B says the screwdriver could slip out of the screw and stab the technician's hand. Who is right?
    a. A only
    b. B only
    c. Both A and B
    d. Neither A nor B

22. Electrical equipment uses a three-prong plug for which of the following reason?
    a. To power the equipment
    b. To protect the user from electrocution
    c. To protect the equipment from overload
    d. To protect the operator from eye hazards

23. Compressed air is likely to cause a fatality if:
    a. Used to dry wet components
    b. Air pressure falls below 25 psi
    c. Used out of doors in low temperatures
    d. Forced into the blood stream

2

Introduction to Electricity

LEARNING OBJECTIVES

Upon completion and review of this chapter, you should be able to:

- Define electricity, atomic structure, and electron movement and explain atomic theory in relation to battery operation, using like and unlike charges.
- Explain the different sources of electricity.
- Describe the scientists who were instrumental in the development of the different tenets of electrical theory.

KEY TERMS

Amber
Atom
Battery
Current
Electricity
Electrolyte
Electron
Electrostatic Discharge (ESD)
Horsepower
Ion
Matter
Neutron
Nucleus
Photoelectricity
Piezoelectricity
Proton
Static Electricity
Thermocouple
Thermoelectricity
Valence
Voltage

INTRODUCTION

This chapter reviews what electricity is in its basic form. We will look at natural forms of electrical energy, and the people who have historically played a part in developing the theories that explain electricity. It is extremely important for automotive technicians to learn all that they can about basic electricity and electronics, because in today's modern automobile, there is a wire or computer connected to just about everything.

Many of us take for granted the sources of electricity. Electricity is a natural form of energy that comes from many sources. For example, electricity is in the atmosphere all around us; a generator just puts it in motion. We cannot create or destroy energy, only change it, yet we can successfully produce electricity and harness it by changing various other forms of natural energy.

WHAT IS ELECTRICITY?

Was electricity invented or discovered? The answer is that electricity was discovered. So who discovered electricity? Ben Franklin? Thomas Edison figured out how to use electricity to make light bulbs, record players, and movies. But neither of these scientist/inventors had anything to do with the discovery of electricity.

Actually, the Greeks discovered electricity. They found that if they took **amber** (a translucent, yellowish resin, derived from fossilized trees) and rubbed it against other materials, it became charged with an unseen force that had the ability to attract other lightweight objects such as feathers, somewhat like a magnet picks up metal objects. In 1600, William Gilbert published a book describing these phenomena. He also discovered that other materials shared the ability to attract, such as sulfur and glass. He used the Latin word "elektron" to describe amber and the word "electrica" for similar substances. Sir Thomas Browne first used the word electricity during the 1600s. Two thousand years after ancient Greece, electricity is all around us. We use it every day. But what exactly is electricity? You can't see it. You can't smell it. You can't touch it; well, you could touch it, but it would probably be a *shocking* experience and could cause serious injury. Next, we will discuss atomic structure to find a more exact definition of electricity.

ATOMIC STRUCTURE

We define **matter** as anything that takes up space and, when subjected to gravity, has weight. There are many different kinds of matter. On Earth, we have classified over a hundred elements. Each element is a type of matter that has certain individual characteristics. Most have been found in nature. Examples of natural elements are copper, iron, gold, and silver. Other elements have been produced only in the laboratory. Every material we know is made up of one or more elements.

Let's say we take a chunk of material—a rock we found in the desert, for example—and begin to divide it into smaller parts. First we divide it in half. Then we test both halves to see if it still has the same characteristics. Next we take one of the halves and divide it into two parts. We test those two parts. By this process, we might discover that the rock contains three different elements. Some of our pieces would have the characteristics of copper, for example. Others would show themselves to be carbon, yet others would be iron.

Atoms

If you could keep dividing the material indefinitely, you would eventually get a piece that only had the characteristics of a single element. At that point, you would have an **atom,** which is the smallest particle into which an element can be divided and still have all the characteristics of that element. An atom is the smallest particle of a chemical element that can take part in a chemical reaction. An atom is so small that it cannot be seen with a conventional microscope, even a very powerful one. An atom is itself made up of smaller particles. You can think of these as universal building blocks. Scientists have discovered many particles in the atom, but for the purpose of explaining electricity, we need to talk about just three: electrons, protons, and neutrons.

All the atoms of any particular element look essentially the same, but the atoms of each element are different from those of another element. All atoms share the same basic structure. At the center of the atom is the **nucleus,** containing **protons** and **neutrons,** as shown in Figure 2-1. Orbiting around the nucleus, in constant motion, are the **electrons.** The structure of the atom resembles planets in orbit around a sun.

The exact number of each of an atom's particles—protons, neutrons, and electrons—depends on which element the atom is from. The simplest atom is that of the element hydrogen. A hydrogen atom (Figure 2-1) contains one proton, one neutron, and one electron. Aluminum, by comparison, has 13 protons, 14 neutrons, and 13 electrons. These particles—protons, neutrons, and electrons—are important to us because they are used to explain electrical charges, voltage, and current. Electrons orbit the nucleus of an atom in

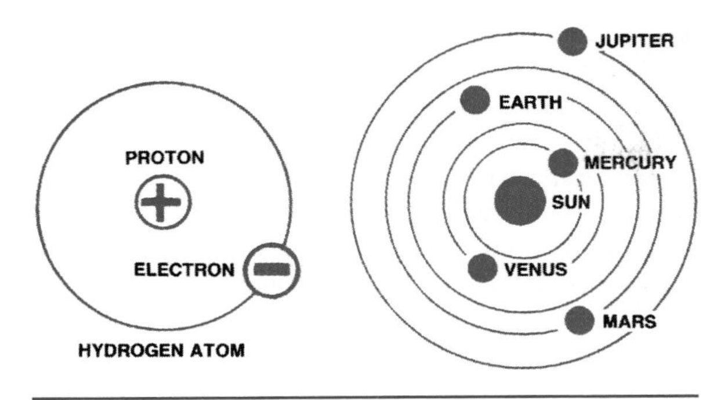

Figure 2-1. In an atom (left), electrons orbit protons in the nucleus in the same way the planets orbit the Sun.

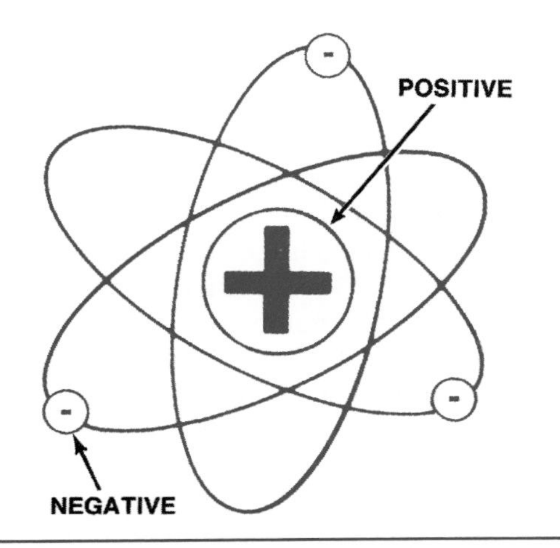

Figure 2-2. The charges within an atom.

a concentric ring known as a *shell.* The nucleus of an atom comprises 99.9% of its mass. The number of protons in the nucleus is the atomic number of any given element and the sum of the neutrons and protons is the atomic mass number.

There are two types of force at work in every atom. Normally, these two forces are in balance. One force comes from electrical charges and the other force, centrifugal force, is generated when an object moves in a circular path.

Electrical Charges

Neutrons have no charge, but *electrons* have a negative electrical charge. *Protons* carry a positive electrical charge (Figure 2-2). Opposite electrical charges always attract one another; so particles or objects with opposite charges tend to move toward each other unless something opposes the attraction. Like electrical charges always repel; particles and objects with like charges tend to move away from each other unless the repelling force is opposed.

Figure 2-3. Unlike and like charges of a magnet.

In its normal state, an atom has the same number of electrons as it does protons. This means the atom is electrically neutral or balanced because there are exactly as many negative charges as there are positive charges. Inside each atom, negatively charged electrons are attracted to positively charged protons, just like the north and south poles of a magnet, as shown in Figure 2-3. Ordinarily, electrons remain in orbit because the centrifugal force exactly opposes the electrical charge attraction. It is possible for an atom to lose or gain electrons. If an atom loses one electron, the total number of protons would be one greater than the total number of electrons. As a result, the atom would have more positive than negative charges. Instead of being electrically neutral, the atom itself would become positively charged.

All electrons and protons are alike. The number of protons associated with the nucleus of an atom identifies it as a specific element. Electrons have 0.0005 of the mass of a proton. Under normal conditions, electrons are bound to the positively charged nuclei of atoms by the attraction between opposite electrical charges.

An atom is held together because of the electrical tendency of unlike charges attracting and like charges repelling each other. Positively charged protons hold the negatively charged electrons in their orbital shells. This occurs because like electrical charges repel each other; the electrons do not collide. All matter is composed of atoms and an electrical charge is a component of all atoms, so all matter is electrical, in essence. The phenomenon we describe as *electricity* concerns the behavior of atoms that have become for whatever reason, unbalanced or electrified.

Electric Potential—Voltage

We noted that a balance (Figure 2-4) between centrifugal force and the attraction of opposing charges keeps electrons in their orbits. If anything upsets that balance, one or more electrons may leave orbit to become free electrons. When a number of free electrons gather in one location, a charge of electricity builds up. This charge may also be called a difference in electric "potential". This difference in electric potential is more commonly known as **voltage** and can be compared to a difference in pressure that makes water flow. When this potential or pressure causes a number of electrons to move in a single direction, the effect is current flow. So the definition of **current** is the flow of electrons. Any atom may possess more or fewer electrons than protons. Such an unbalanced atom would be described as negatively (an excess of electrons) or positively (a net deficit of electrons) charged and known as an ion (Figure 2-5). An **ion** is an atom that has gained or lost an electron. Ions try to regain their balance of equal protons and electrons by exchanging electrons with nearby atoms. This is known as the flow of electric current or **electricity.** For more information about voltage and current, see the section on "Electrical Units of Measurement" in Chapter 3 of the *Shop Manual.*

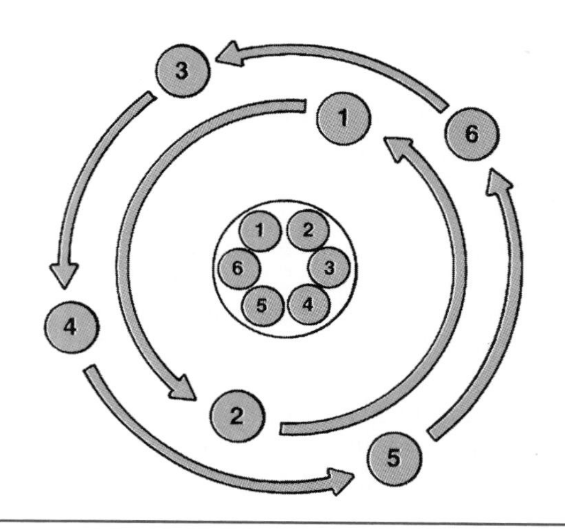

Figure 2-4. A balanced atom.

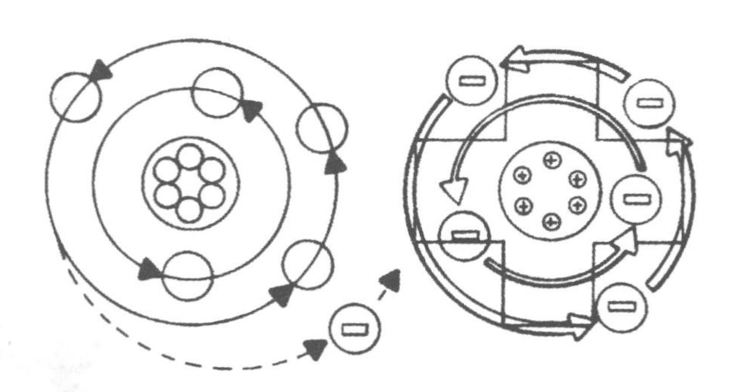

Figure 2-5. An unbalanced atom.

Valence

The concentric orbital paths, or shells, of an atom proceed outward from the nucleus. The electrons in the shells closest to the nucleus of the atom are held most tightly while those in the outermost shell are held more loosely. The simplest element, hydrogen, has a single shell containing one electron. The most complex atoms may have seven shells. The maximum number of electrons that can occupy shells one through seven are, in sequence: 2, 8, 18, 32, 50, 72, 98. The heaviest elements in their normal states have only the first four shells fully occupied with electrons; the outer three shells are only partially occupied. The outermost shell in any atom is known as its **valence** ring. The number of electrons in the valence ring will dictate some basic characteristics of an element.

The chemical properties of atoms are defined by how the shells are occupied with electrons. An atom of the element helium whose atomic number is 2 has a full inner shell. An atom of the element neon with an atomic number of 10 has both a full first and second shell (2 and 8): its second shell is its valence ring (Figure 2-6). Other more complex atoms that have eight electrons in their outermost shell, even though this shell might not be full, will resemble neon in terms of their chemical inertness. **Valence** represents the ability to combine.

Remember that an ion is any atom with either a surplus or deficit of electrons. Free electrons can rest on a surface or travel through matter (or a vacuum) at close to the speed of light. Electrons resting on a surface will cause it to be negatively charged.

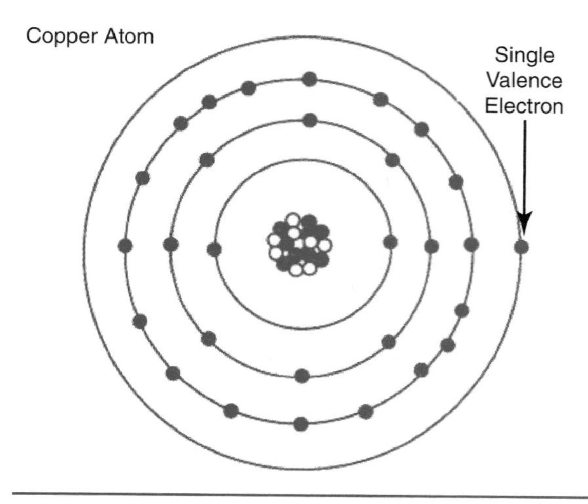

Figure 2-6. Valence ring.

Because the electrons are not moving, that surface is described as having a negative static electrical charge. The extent of the charge is measured in voltage or charge differential. A stream of moving electrons is known as an electrical current. For instance, if a group of positive ions passes in close proximity to electrons resting on a surface, they will attract the electrons by causing them to fill the "holes" left by the missing electrons in the positive ions. Current flow is measured in amperes: one ampere equals 6.28×10^{18} electrons (1 coulomb) passing a given point per second.

SOURCES OF ELECTRICITY

Lightning

Benjamin Franklin (1706–1790) proved the electrical nature of thunderstorms in his famous kite experiment, established the terms *positive* and *negative,* and formulated the conventional theory of current flow in a circuit. Franklin was trying to prove that the positive and negative electron distribution in the clouds produced the static electricity that causes lightning, as shown in Figure 2-7. Natural negatively charged particles will produce lightning when they find a path negative to positive.

■ Benjamin Franklin's Theory:

When the science of electricity was still young, the men who studied it were able to use electricity without really understanding why and how it worked. In the early 1700s, Benjamin Franklin, the American printer, inventor, writer, and politician, brought his famed common sense to the problem. Although he was not the first to think that electricity and lightning were the same, he was the first to prove it. He also thought that electricity was like a fluid in a pipe that flowed from one terminal to the other. He named the electrical terminals *positive* and *negative* and suggested that current moved from the positive terminal to the negative terminal. Benjamin Franklin created what we now call the *Conventional Theory of Current Flow.*

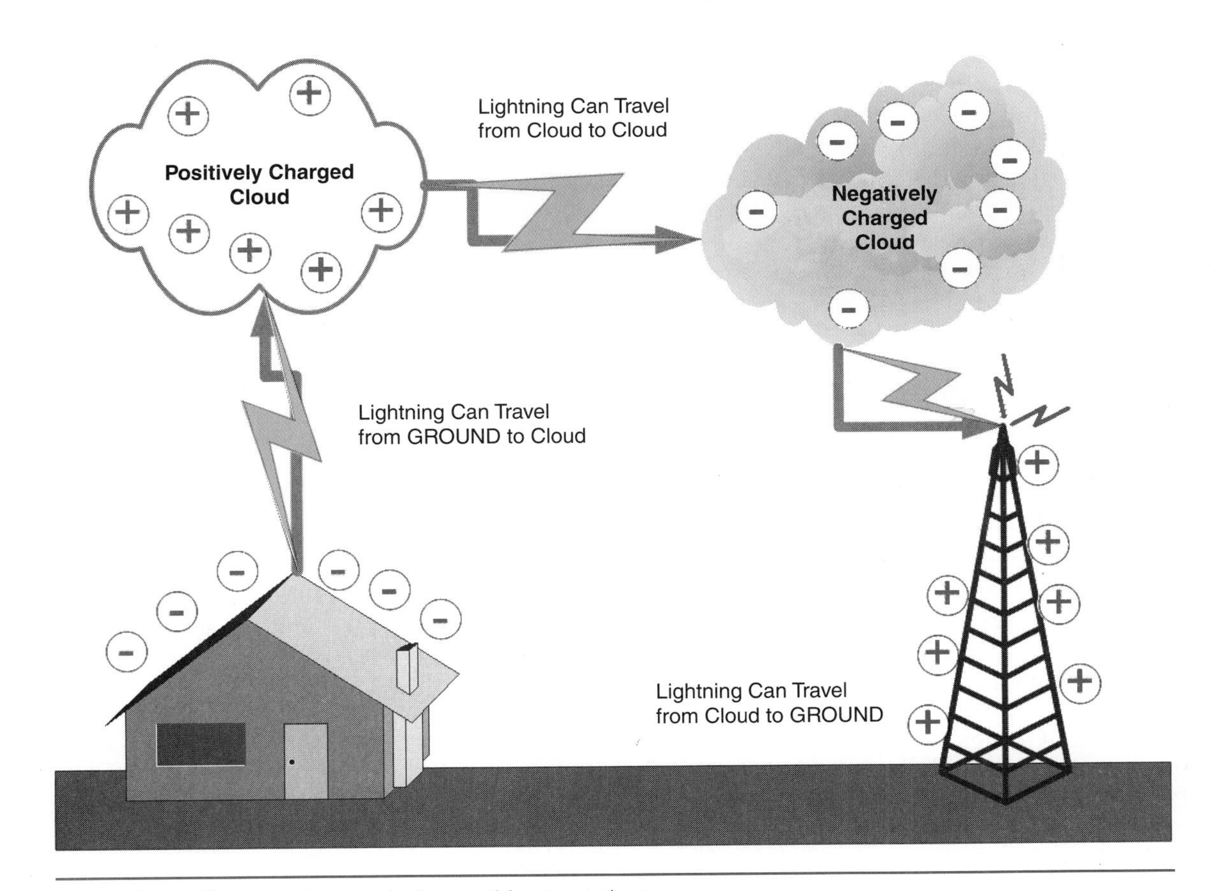

Figure 2-7. Electron charges in the earth's atmosphere.

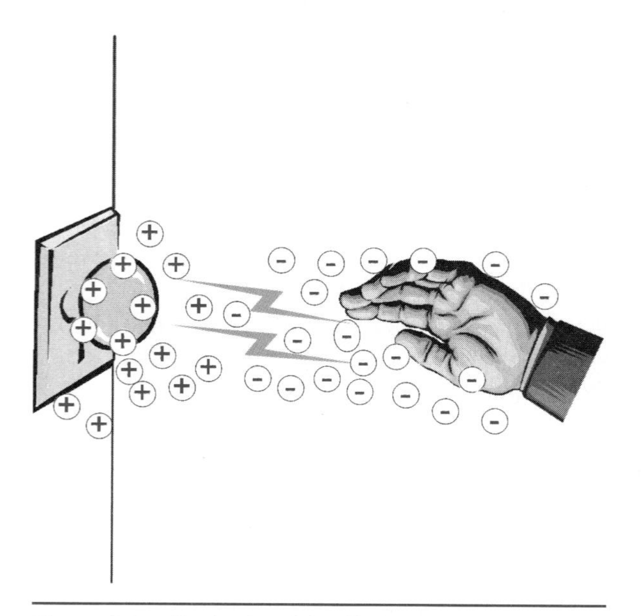
Figure 2-8. Static electricity discharge to metal object.

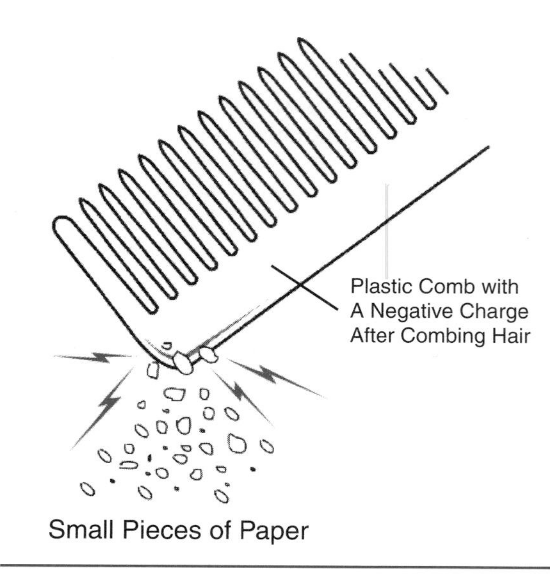

Figure 2-9. Static electricity discharge attraction.

Static Electricity

A number of factors, such as friction, heat, light, and chemical reactions, can "steal" electrons (natural negatively charged particles) from a surface; when this occurs, the surface becomes positively charged. Providing positive ions remain at rest, the surface will have a positive static electrical charge differential, which is known as static electricity. **Static electricity** is electricity at rest or without any motion. Every time a person walks across a carpet, electrons or negatively charged particles are "stolen" from the carpet surface and this has an electrifying effect (*electrification*) on both the substance from which electrons are stolen (the carpet) and the moving body that performs the theft. When the moving body has accumulated a sufficient charge differential (measured in voltage), the excess electrons will be discharged when they touch a metal object such as a doorknob (Figure 2-8) through an arc and balance the charge. Electrification results in both attractive and repulsive forces. In electricity, like charges repel and unlike charges attract.

When you run a plastic comb through someone's hair the electrons will be stolen by the comb, giving it a negative charge. This will subsequently attract small pieces of paper, as shown in Figure 2-9. The experiment will always work better on a dry day because electrons can easily travel through humid air and the accumulated charge will dissipate rapidly. Two balloons rubbed on a woolen fiber will both acquire a negative charge and will therefore tend to repel each other.

Static electricity can also be referred to as frictional electricity because it results from the contact of two surfaces. Chemical bonds are formed when any surfaces contact and if the atoms on one surface tend to hold electrons more tightly, the result is the theft of electrons. Such contact produces a charge imbalance by pulling electrons of one surface from that of the other; as electrons are pulled away from a surface, the result is an excess of electrons (negative charge) and a deficit in the other (positive charge). The extent of the charge differential is, of course, measured in voltage. While the surfaces with opposite charges remain separate, the charge differential will exist. When the two polarities of charge are united, the charge imbalance will be canceled. Static electricity is an everyday phenomenon, as described in the examples in the opening to this chapter, and it involves voltages of 1,000 volts to 50,000 volts. An automotive technician should always use a static grounding strap when working with static-sensitive electronic devices such as PCMs and ECMs.

Electrostatic Field

The attraction between opposing electrical charges does not require contact between the objects involved, as shown in Figure 2-10. This

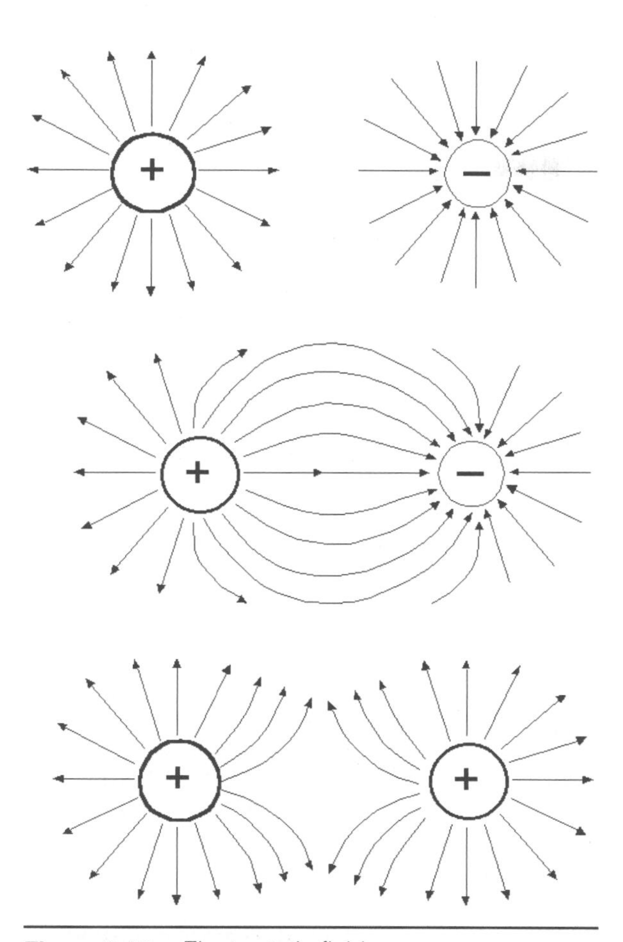

Figure 2-10. Electrostatic field.

is so because invisible lines of force exist around a charged object. Taken all together, these lines of force make up an electrostatic field. Such fields are strongest very close to the charged object and get weaker as they extend away from the object.

Electrostatic Discharge (ESD)

An electrostatic charge can build up on the surface of your body. If you touch something, your charge can be discharged to the other surface. This is called **electrostatic discharge (ESD).** Figure 2-11 shows what the ESD symbol looks like. The symbol tells you that the component is a solid-state component. Some service manuals use the words "solid-state" instead of the ESD symbol. Look for these indicators and take the suggested ESD precautions when you work on sensitive components. We will cover this subject in detail in Chapter 10 of this manual.

Figure 2-11. ESD symbol.

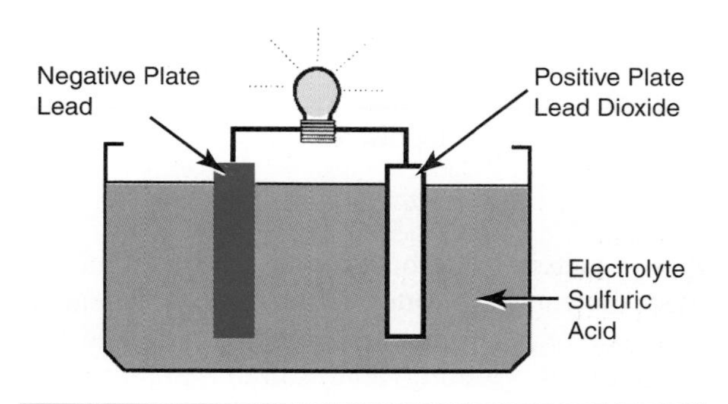

Figure 2-12. Automotive battery operation.

Chemical Source

A **battery** (Figure 2-12) is a means of producing direct current (DC) from a chemical reaction. In the lead-acid battery, a difference in electrical force potential is created by the chemical interaction of lead and lead dioxide submerged in a sulfuric acid electrolyte. An **electrolyte** is a chemical solution that usually includes water and other compounds that conduct electricity. In the case of automotive battery, the solution is water and sulfuric acid.

When the battery is connected into a completed electrical circuit, current begins to flow from the battery. This current is produced by chemical reactions between the active materials in the two kinds of plates and the sulfuric acid in the electrolyte (Figure 2-12). The lead dioxide in the positive plate is a compound of lead and oxygen. Sulfuric acid is a compound of hydrogen and the sulfate radical. During discharge, oxygen in the positive

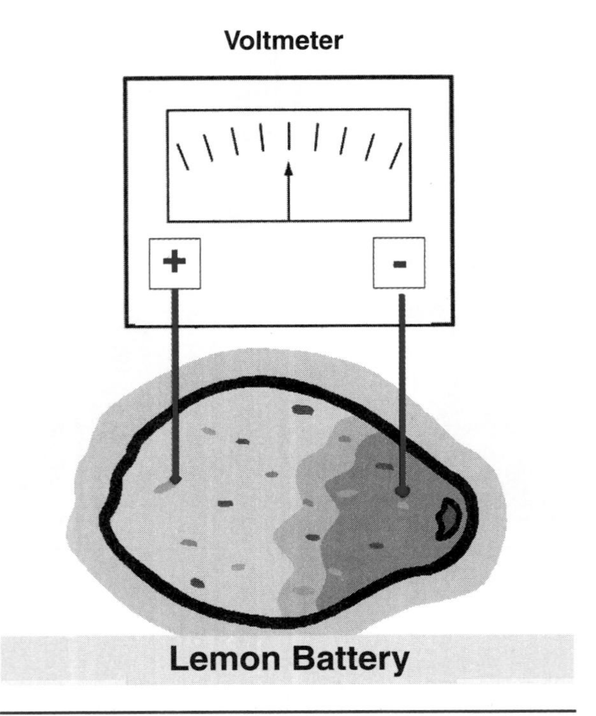

Figure 2-13. Lemon powered battery.

active material combines with hydrogen in the electrolyte to form water. At the same time, lead in the positive active material combines with the sulfate radical, forming lead sulfate. Figure 2-13 shows a very simplified version of a battery powered by a lemon. The availability and amount of electrical energy that can be produced in this manner is limited by the active area and weight of the materials in the plates and by the quantity of sulfuric acid in the electrolyte. After most of the available active materials have reacted, the battery can produce little or no additional energy, and the battery is then discharged.

Thermoelectricity

Applying heat to the connection point of two dissimilar metals can create electron flow (electricity), which is known as **thermoelectricity** (Figure 2-14). This affect was discovered by a German scientist named *Seebeck* and is known as the *Seebeck Effect.* Seebeck called this device a **thermocouple,** which is a small device that gives off a low voltage when two dissimilar metals are heated. The pyrometer is a thermocouple used to measure exhaust gas temperature that consists of two dissimilar metals (iron and constantin, a copper-tin alloy) joined at the "hot" end and connected to a sensitive voltmeter at the gauge end. As temperature increases at the hot end, the reading will increase at the millivoltmeter on the display gauge.

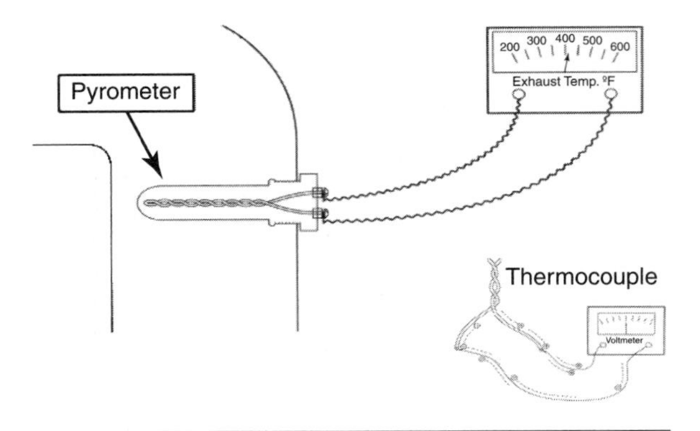

Figure 2-14. Pyrometer thermocouple.

Photoelectricity

Light is composed of particles called photons that are pure energy and contain no mass. However, when sunlight contacts certain materials, such as selenium and cesium, electron flow is stimulated and is called **photoelectricity** (Figure 2-15). Photoelectric cells are used as sensors to control headlight beams and automatic daylight/night mirrors. Solar energy is light energy from the sun that is gathered in a photovoltaic solar cell.

Piezoelectricity

Some crystals, notably quartz or barium titanate, become electrified when subjected to direct pressure, the potential difference increasing with pressure increase. A change in the potential of electrons between the positive and negative terminal creates electricity know as **piezoelectricity.** The term comes from the Greek word "piezo," which means pressure. Figure 2-16 shows that when these materials, quartz or barium titanate, undergo physical stress or vibration, a small oscillating voltage is produced.

Piezoelectric sensors are used as detonation sensors on electronically controlled engines. The typical knock sensor (Figure 2-17) produces about 300 millivolts of electricity and vibrates at a 6,000-hertz (cycles per second) frequency, which is the frequency that the cylinder walls vibrate at during detonation.

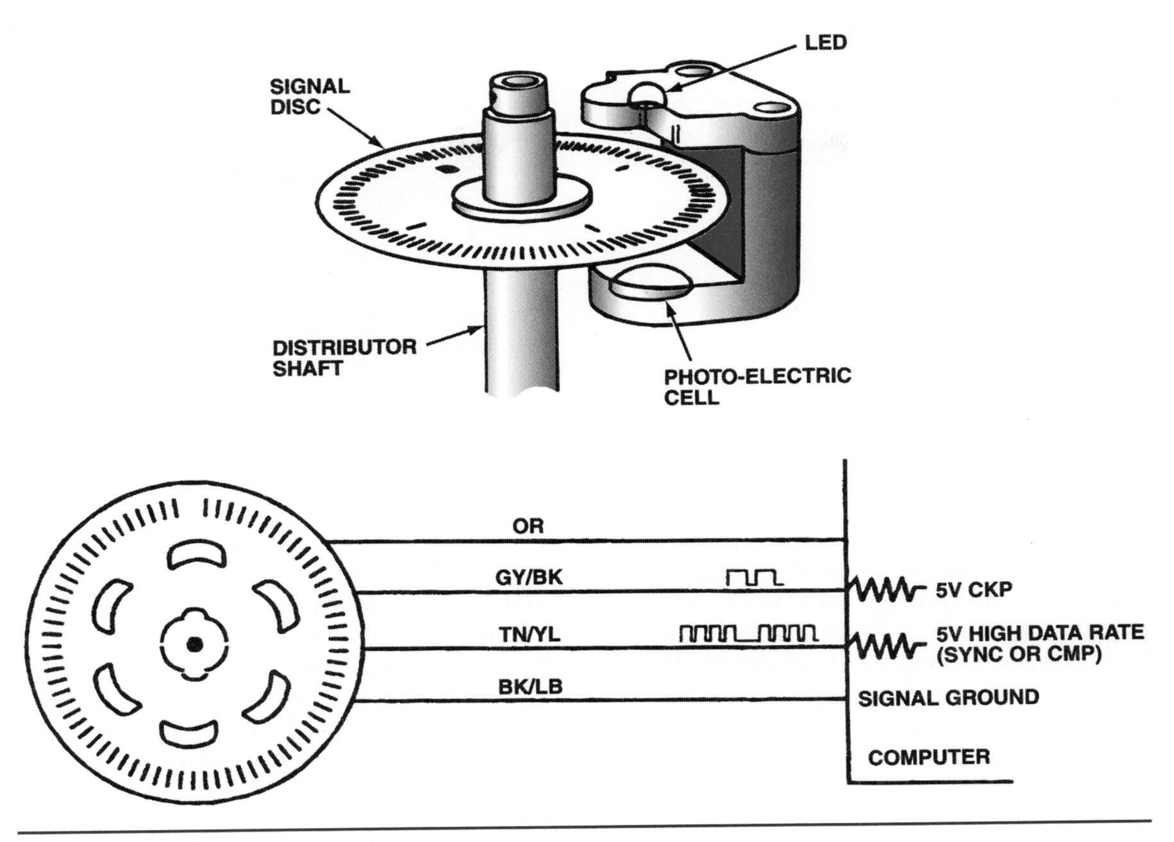

Figure 2-15. Photoelectric cell sensor.

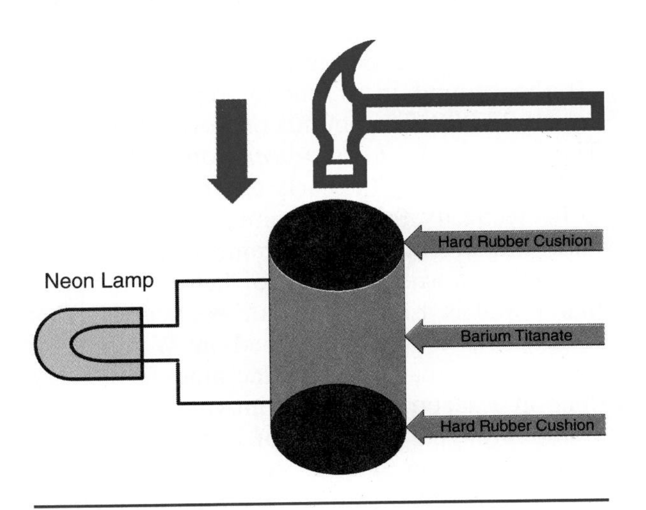

Figure 2-16. Piezoelectric effect.

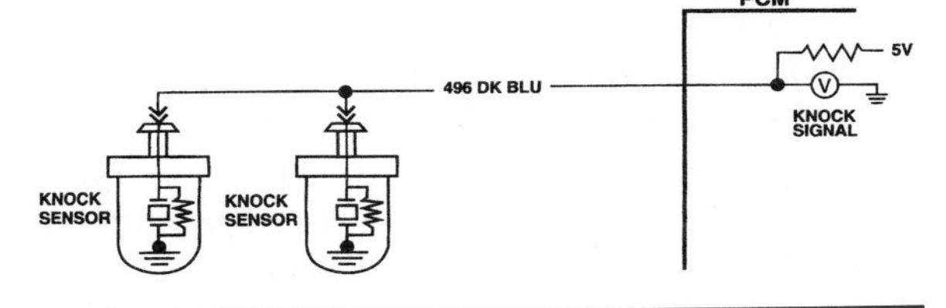

Figure 2-17. Piezoelectric knock sensor. (GM Service and Parts Operations)

■ Squeeze a Rock Get a Volt

In 1880, the French chemists Pierre and Marie Curie discovered the phenomenon of piezoelectricity, which means, "electricity through pressure." They found that when pressure is applied to a crystal of quartz, tourmaline, or Rochelle salt, a voltage is generated between the faces of the crystal. Although the effect is only temporary, while the pressure lasts, it can be maintained by alternating between compression and tension. Piezoelectricity is put to practical use in phonograph pickups and crystal microphones, where mechanical vibrations (sound waves) are converted into varying voltage signals. Similar applications are used in underwater hydrophones and piezoelectric stethoscopes.

Applying a high-frequency alternating voltage to a crystal can create a reverse piezoelectric effect. The crystal then produces mechanical vibrations at the same frequency, which are called ultrasonic sound waves because they are beyond our range of hearing. These ultrasonic vibrations are used, among other things, to detect sonar reflections from submarines and to drill holes in diseased teeth.

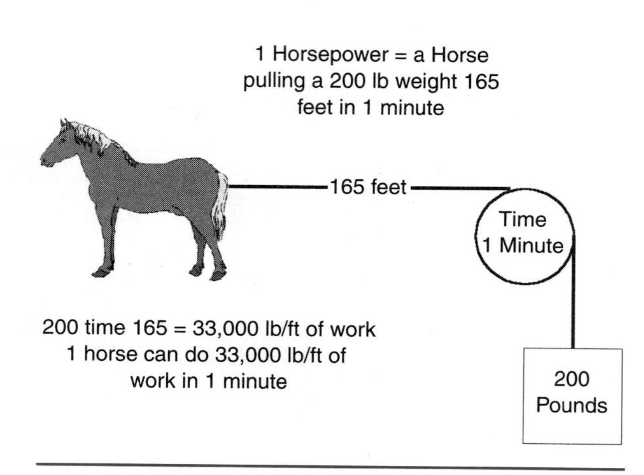

Figure 2-18. Horsepower.

HISTORICAL FIGURES IN ELECTRICITY

In 1767, *Joseph Priestly* established that electrical charges attract with a force inversely proportional to distance. In 1800, *Alessandro Volta* invented the first battery. *Michael Faraday* (1791–1867) opened the doors of the science we now know as electromagnetism when he published his law of induction, which simply states that a magnetic field induces an electromotive force in a moving conductor. *Thomas Edison* (1847–1931) invented the incandescent lamp in 1879, but perhaps even more importantly, built the first central power station and electrical distribution system in New York City in 1881. This provided a means of introducing electrical power into industry and the home.

The discovery of the *electron* by *J.J. Thomson* (1856–1940) in 1897 introduced the science of electronics and quickly resulted in the invention of the diode (1904), the triode (1907), and the transistor (1946). *Andre Marie Ampere* established the importance of the relationship between electricity and magnetism. In 1800, *Alessandro Volta* discovered that if two dissimilar metals were brought in contact with a salt solution, a current would be produced, this invention is now known as the battery. The German physicist *George Simon Ohm* (1787–1854) proved the mathematical relationship between electrical potential (voltage), electrical current flow (measured in amperes) and the resistance to the current flow (measured in ohms: symbol Ω).

Another person who influenced electrical technology was a Scottish inventor named *James Watt* (1736–1819). Of all the inventors who made the Industrial Revolution possible, James Watt was probably the best known. In 1764 he realized the steam engine produced a considerable amount of wasted energy. Over the next several years he developed, patented, and refined several of his steam engines. To express the capacity of his engines, he developed a comparison between what his engine could do and the average work capacity of a normal-size horse. He found a horse could walk 165 feet in one minute pulling a 200-pound weight (165 ft. × 200 lb. = 33,000 ft.-lb. per minute, or 550 foot-pounds of work per second) so he called this work one **horsepower** (Figure 2-18), which is a measure of mechanical power or the rate that work is done. One horsepower is needed to lift 550 pounds 1 foot off the ground in 1 second, one horsepower equals 33,000 foot-pounds of work per minute. The term *brake horsepower* comes from the method of testing the early engines.

In electricity, the metric unit *joule* refers to the base unit of energy measurement. This unit is also called a *watt,* in honor of James Watt, such that 1 watt is a rate of 1 joule per second. This relationship is best described in Watt's Law, which states that 1 watt is the amount of work done in 1 second by 1 volt moving a charge of 1 amp through a resistance of 1 ohm in a circuit. Horsepower can also be expressed in units of electrical power or watts; the simple conversion is 1 horsepower = 746 watts.

These incredible inventors contributed substantial knowledge to the establishment of our current understanding of electricity. As a result, many of their names have become standard household terms, as in a 60-watt light bulb, a 20-amp circuit breaker, a 120-volt wall outlet, and a 60-hertz AC (alternating current) electrical circuit.

SUMMARY

The Greeks discovered the first type of electricity in the form of static electricity when they observed that amber rubbed with fur would attract lightweight objects such as feathers. Static electricity is electricity at rest or without any motion. All matter is composed of atoms and electrical charge is a component of all atoms, so all matter is electrical in essence. An atom is the smallest part of an element that retains all of the properties of that element. All atoms share the same basic structure. At the center of the atom is the nucleus, containing protons, neutrons, and electrons. When an atom is balanced, the number of protons will match the number of electrons and the atom can be described as being in an electrically neutral state. The phenomenon we describe as *electricity* concerns the behavior of atoms that have become, for whatever reason, unbalanced or ionized. Electricity may be defined as the movement of free electrons from one atom to another.

An electrostatic charge can build up on the surface of your body. If you touch something, your charge can be discharged to the other surface, which is called electrostatic discharge (ESD). An automotive technician should always use a static grounding strap when working with static-sensitive electronic devices.

When light contacts certain materials, such as selenium and cesium, electron flow is stimulated and is called photoelectricity. Solar energy is light energy (photons) from the sun that is gathered in a photovoltaic solar cell. A photon is pure energy that contains no mass. Thermoelectricity is electricity produced when two dissimilar metals are heated to generate an electrical voltage. A thermocouple is a small device made of two dissimilar metals that gives off a low voltage when heated. Piezoelectricity is electricity produced when materials such as quartz or barium titanate are placed under pressure. The production of electricity from chemical energy is demonstrated in the lead-acid battery. Electromagnetic induction is the production of electricity when a current is carried through a conductor and a magnetic field is produced.

Andre Marie Ampere established the importance of the relationship between electricity and magnetism. Alessandro Volta discovered that if two dissimilar metals were brought in contact with a salt solution, a current would be produced; this invention is now known as the battery. George Simon Ohm showed a relationship between resistance, current, and voltage in an electrical circuit; he developed what is known as Ohm's Law. James Watt developed a method used to express a unit of electrical power known as Watt's Law.

Review Questions

1. The general name given every substance in the physical universe is which of the following:
   a. Mass
   b. Matter
   c. Compound
   d. Nucleus
2. The smallest part of an element that retains all of its characteristics is which of the following:
   a. Atom
   b. Proton
   c. Compound
   d. Neutron
3. The particles that orbit around the center of an atom are called which of the following:
   a. Electrons
   b. Molecules
   c. Nucleus
   d. Protons
4. An atom that loses or gains one electron is called which of the following:
   a. Balanced
   b. An element
   c. A molecule
   d. An ion
5. Technician A says the battery provides electricity by releasing free electrons. Technician B says the battery stores energy in a chemical form. Who is right?
   a. A only
   b. B only
   c. Both A and B
   d. Neither A nor B
6. Static electricity is being discussed. Technician A says that static electricity is electricity in motion. Technician B says an electrostatic charge can build up on the surface of your body. Who is right?
   a. A only
   b. B only
   c. Both A and B
   d. Neither A nor B
7. What people discovered electricity?
   a. The Italians
   b. The Germans
   c. The Greeks
   d. The Irish
8. Technician A says batteries produce direct current from a chemical reaction. Technician B says that an electrolyte is a chemical solution of water and hydrochloric acid that will conduct electricity. Who is right?
   a. A only
   b. B only
   c. Both A and B
   d. Neither A nor B
9. Two technicians are discussing thermoelectricity. Technician A says applying heat to the connection point of two dissimilar metals can create electron flow (electricity). Technician B says a thermocouple is a small device made of two dissimilar metals that gives off a low voltage when heated. Who is right?
   a. A only
   b. B only
   c. Both A and B
   d. Neither A nor B
10. Technician A says when sunlight contacts certain materials, electron flow is stimulated. Technician B says solar energy is light energy from the moon that is gathered in a photovoltaic solar cell. Who is right?
    a. A only
    b. B only
    c. Both A and B
    d. Neither A nor B
11. Two technicians are discussing how piezoelectricity works. Technician A says it is electricity produced when barium titanate is placed under pressure. Technician B says when no change in the potential of electrons between positive and negative terminal occurs, the barium titanate creates electricity. Who is right?
    a. A only
    b. B only
    c. Both A and B
    d. Neither A nor B
12. James Watt's term *horsepower* is being discussed. Technician A says one horsepower equals 33,000 foot-pounds of work per hour. Technician B says one horsepower would be produced when a horse walked 165 feet in one minute pulling

a 500-pound weight or 165 ft. × 500 lb. = 33,000 ft-lb. Who is right?
a. A only
b. B only
c. Both A and B
d. Neither A nor B

13. Technician A says Andre Marie Ampere established the importance of the relationship between electricity and magnetism. Technician B says Alessandro Volta discovered that if two dissimilar metals were brought in contact with a water solution, a current would be produced. Who is right?
a. A only
b. B only
c. Both A and B
d. Neither A nor B

3

Electrical Fundamentals

LEARNING OBJECTIVES

Upon completion and review of this chapter, you should be able to:

- Explain the terms conductor, insulator, and semiconductor, and differentiate between their functions.
- Identify and explain the basic electrical concepts of resistance, voltage, current, voltage drop, and conductance.
- Define the two theories of current flow (conventional and electron) and explain the difference between DC and AC current.
- Explain the cause-and-effect relationship in Ohm's law between voltage, current, resistance, and voltage drop.
- Define electrical power and its Ohm's law relationship.
- Define capacitance and describe the function of a capacitor in an automotive electrical circuit.

KEY TERMS

Ampere
Capacitance
Capacitor
Circuit
Conductors
Conventional Theory
Current
Ground
Insulators
Ohm
Ohm's Law
Resistance
Resistors
Semiconductors
Voltage
Voltage Drop
Watt

INTRODUCTION

This chapter reviews all of the basic electrical principles required to understand electronics and the automotive electrical/electronic systems in the later chapters. An automotive technician must have a thorough grasp of the basis of electricity and electronics. Electronics has become the single most important subject area and the days when many technicians could avoid working on an electrical circuit through an entire career are long past. This course of electrical study will cover conductors and insulators, characteristics of electricity, the complete electrical circuit, Ohm's Law, and finally capacitance and capacitors.

CONDUCTORS AND INSULATORS

The ease with which an electron moves from one atom to another determines the conductivity of the material. Conductance is the ability of a material to carry electric current. To produce current flow, electrons must move from atom to atom, as shown in Figure 3-1. Materials that readily permit this flow of electrons from atom to atom are classified as conductors. Materials that inhibit or prevent a flow of electrons are classified as insulators. Some examples of conductors are copper, aluminum, gold, silver, iron, and platinum. Some examples of insulators are glass, mica, rubber, and plastic.

- A **conductor** is generally a metallic element that contains fewer than four electrons in its outer shell.
- An **insulator** is a non-metallic substance that contains more than four electrons in the outer shell.
- **Semiconductors** are a group of materials that cannot be classified either as conductors or insulators: they have exactly four electrons in their outer shell. Silicon is an example of a semiconductor.

Conductive metals, even when in an electrically neutral state, contain vast numbers of electrons moving from atom to atom at random. When a battery is placed at either end of a conductor such as copper wire and a complete circuit is formed, electrons are pumped from the more negative terminal to the more positive until either the charge differential ceases to exist or the circuit is opened. The number of electrons does not change.

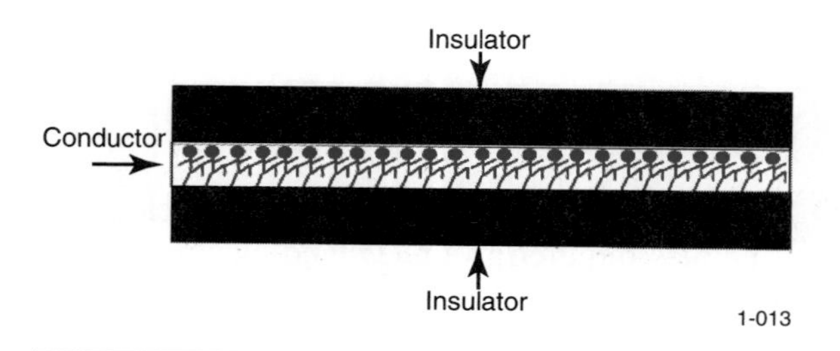

Figure 3-1. Conductors and insulators.

Wires

A wire in a wiring harness is made up of a conductor and an insulator. The metal core of the wire, typically made of copper, is the conductor. The outer jacket (made of plastic or other material) coating the core is the insulator. Under normal circumstances, electrons move a few inches per second. Yet when an electrical potential is applied to one end of a wire, the effect is felt almost immediately at the other end of that wire. This is so because the electrons in the conductor affect one another, much like billiard balls in a line.

CHARACTERISTICS OF ELECTRICITY

Voltage

We have said that a number of electrons gathered in one place effect an electrical charge. We call this charge an electrical potential or "voltage." **Voltage** is measured in volts (V). Since it is used to "move electrons," an externally applied electrical potential is sometimes called an "electromotive force" or EMF. Potential, voltage, and EMF all mean the same thing. You can think of voltage as the electrical "pressure" (Figure 3-2) that drives electron flow or current, similar to water pressure contained in a tank. When voltage is applied to a disconnected length of wire (open circuit), there is no sustained movement of electrons. No current flows in the wire, because current flows only when there is a difference in potential (Figure 3-3). Check the voltage with a meter, and you will find that it is the same at both ends of the wire. (See the section on "DMM Operating Principles" in Chapter 3 of the *Shop Manual*.)

If voltage is applied to one end of a conductor, a different potential must be connected to the other end of that conductor for current to flow. In a typical automotive circuit, the positive terminal of the vehicle battery is the potential at one end of a conductor, and the negative terminal is the potential at the other end (Figure 3-4).

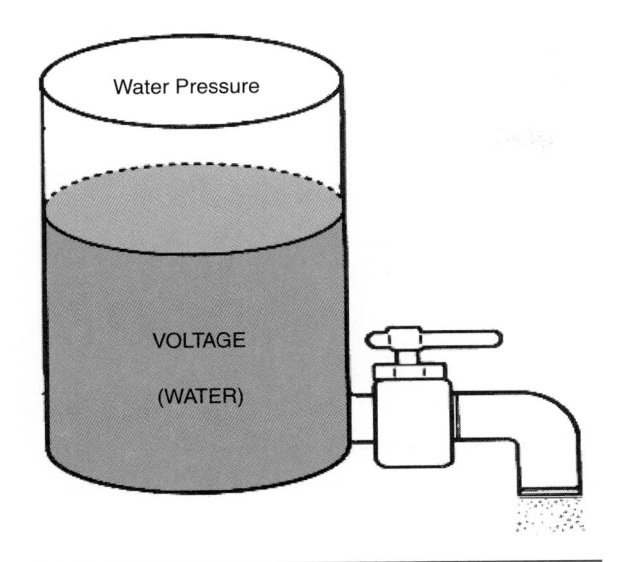

Figure 3-2. Voltage and water pressure.

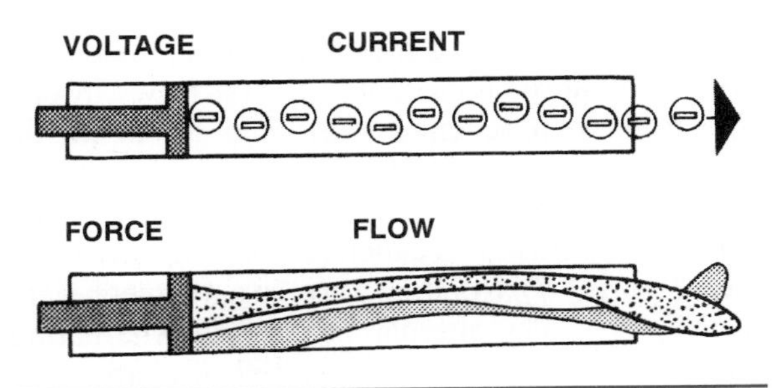

Figure 3-3. Voltage pushes current flow like force pushes water flow.

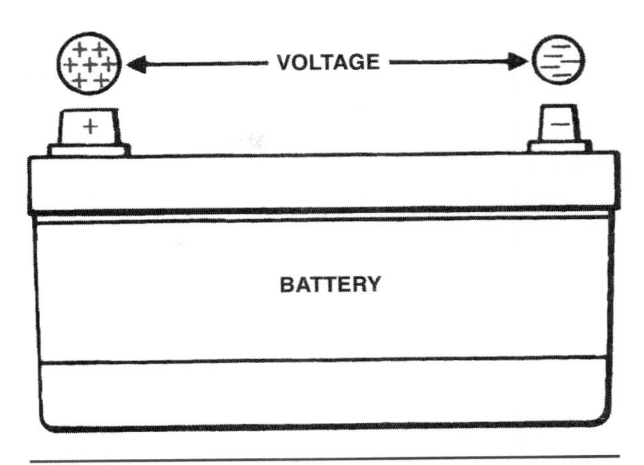

Figure 3-4. Voltage is a potential difference in electromotive force.

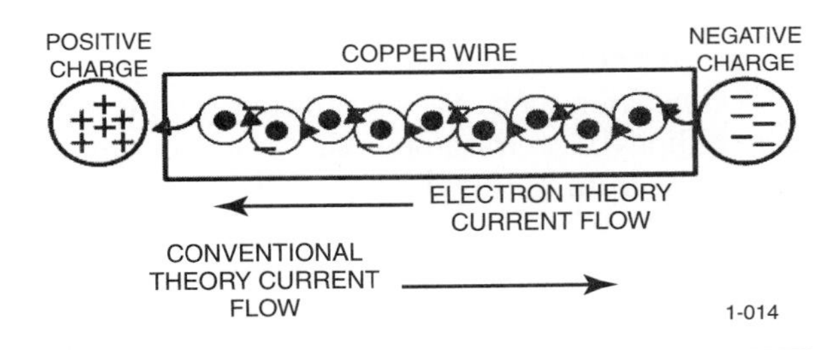

Figure 3-5. Current flow.

Current

The movement of electrons in a circuit is the flow of electricity, or current flow (Figure 3-5), which is measured in **amperes** (A). This unit expresses how many electrons move through a circuit in one second. A current flow of 6.25×10^{18} electrons per second is equal to one ampere. A coulomb is 6.281×10^{18} electrons per second. 1 coulomb/sec=1 ampere.

Current Flow

Current flow will occur only if there is a path and a difference in electrical potential; this difference is known as *charge differential* and is measured in voltage. Charge differential exists when the electrical source has a deficit of electrons and therefore is positively charged. Electrons are negatively charged and unlike charges attract, so electrons flow toward the positive source.

See the section on "DMM Operating Principles" in Chapter 3 of the *Shop Manual.*

Conventional Current Flow and Electron Theory

In automotive service literature, current flow is usually shown as flowing from the positive terminal to the negative terminal. This way of describing current flow is called the **conventional theory.** Another way of describing current flow is called the *electron theory,* and it states that current flows from the negative terminal to the positive terminal. The conventional theory and electron theory are two different ways of describing the same current flow.

Essentially, both theories are correct. The electron theory follows the logic that electrons move from an area of many electrons (negative charge) to one of few electrons (positive charge). However, in describing the behavior of semiconductors, we often describe current as moving from positive to negative. The important thing to know is which theory is being used by the service literature you

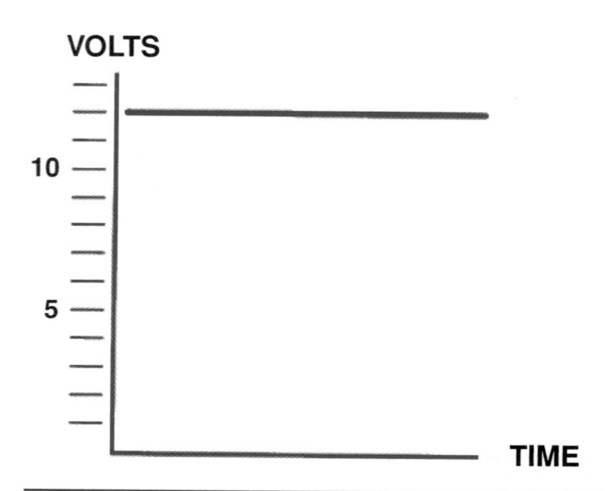

Figure 3-6. Direct current.

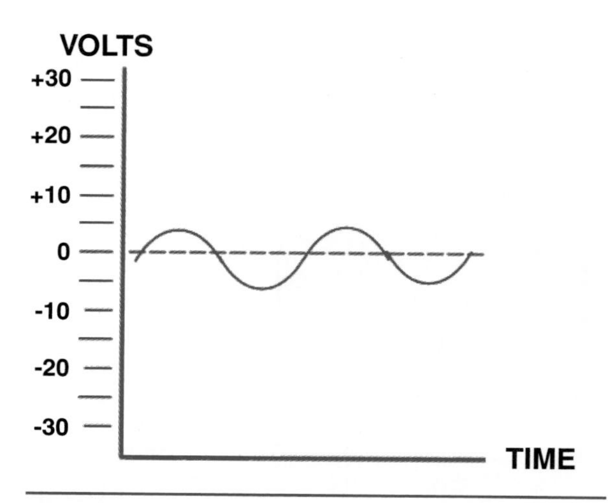

Figure 3-7. Alternating current.

happen to be using; most schematics use conventional current flow theory.

A conductor, such as a piece of copper wire, contains billions of neutral atoms whose electrons move randomly from atom to atom, vibrating at high frequencies. When an external power source such as a battery is connected to the conductor, a deficit of electrons occurs at one end of the conductor and an excess of electrons occurs at the other end, the negative terminal will have the effect of repelling free electrons from the conductor's atoms while the positive terminal will attract free electrons. This results in a flow of electrons through the conductor from the negative charge to the positive charge. The rate of flow will depend on the charge differential (or potential difference/voltage). The charge differential or voltage is a measure of electrical pressure. The role of a battery, for instance, is to act as a sort of electron pump. In a closed electrical circuit, electrons move through a conductor, producing a displacement effect close to the speed of light.

The physical dimensions of a conductor are also a factor. The larger the cross-sectional area (measured by wire gauge size) the more atoms there are over a given sectional area, therefore the more free electrons; therefore, as wire size increases, so does the ability to flow more electrical current through the wire.

Direct Current

When a steady-state electrical potential is applied to a circuit, the resulting current flows in one direction. We call that *direct current* or DC (Figure 3-6). Batteries produce a steady-state, or DC, potential. The advantage of using DC is it can be stored electro-chemically in a battery.

Alternating Current

The electrical potential created by a generator is not steady state, it fluctuates between positive and negative. When such a potential is applied to a circuit, it causes a current that first flows in one direction, then reverses itself and flows in the other direction. In residential electrical systems, this direction reversal happens 60 times a second. This type of current is called *alternating current* or AC (Figure 3-7). Rotating a coil in a magnetic field usually produces alternating current. Alternating current describes a flow of electrical charge that cyclically reverses, due to a reversal in polarity at the voltage source.

Automotive generators produce AC potential. Alternating current is easier to produce in a generator, due to the laws of magnetism, but it is extremely difficult to store. Generators incorporate special circuits that convert the AC to DC before it is used in the vehicle's electrical systems. (Alternating current is also better suited than DC for transmission through power lines.) The frequency at which the current alternates is measured in cycles. A *cycle* is one complete reversal of current from zero though positive to negative and back to the zero point. Frequency is usually measured in cycles per second or hertz.

Resistance

More vehicles can travel on a four-lane superhighway in a given amount of time than on a two-lane country road. A large-diameter pipe can flow more fluid than a small-diameter pipe. A similar characteristic applies to electricity. A large wire can carry

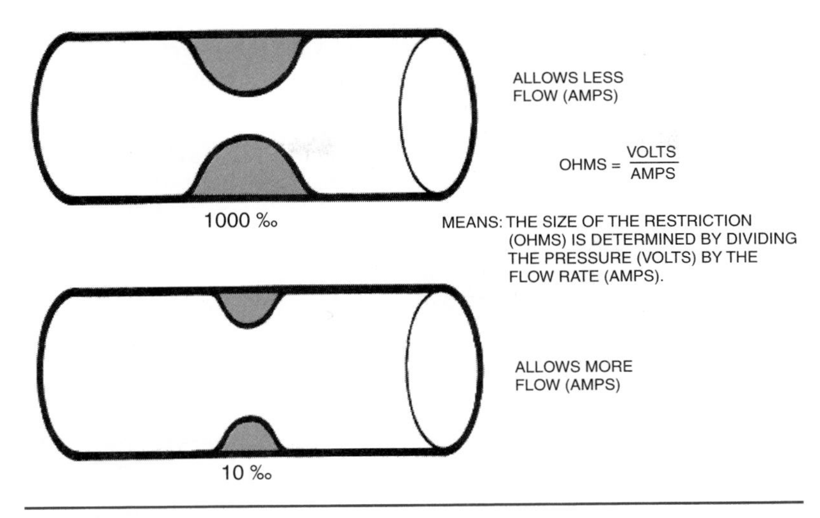

Figure 3-8. Resistance.

more current than a small wire. The reason the large wire carries more current is that it offers less *resistance* to current flow. All materials contain some resistance. For information about measuring resistance, see the section on "DMM Operating Principles" in Chapter 3 of the *Shop Manual.*

Resistance opposes the movement of electrons, or current flow. All electrical devices and wires have some resistance. Materials with very low resistance are called conductors; materials with very high resistance are called insulators. As resistance works to oppose current flow, it changes electrical energy into some other form of energy, such as heat, light, or motion.

Resistance factors (Figure 3-8) determine the resistance of a conductor by a combination of the following:

- **Atomic structure (how many free electrons):** The more free electrons a material has, the less resistance it offers to current flow. Example: copper versus aluminum wire.
- **Length:** The longer a conductor, the higher the resistance.
- **Width (cross-sectional area):** The larger the cross-sectional area of a conductor, the lower the resistance. For example: 12-gauge versus 20-gauge wire.
- **Temperature:** For most materials, the higher the temperature, the higher the resistance. There are a few materials whose resistance goes down as temperature goes up.

The condition of a conductor can also have a large affect on its resistance (Figure 3-9). Broken strands of wire, corrosion, and loose connections can cause the resistance of a conductor to increase.

Wanted and Unwanted Resistance

Resistance is useful in electrical circuits. We use it to produce heat, make light, limit current, and regulate voltage. However, resistance in the wrong place can cause circuit trouble. Sometimes you can predict that high (unwanted) resistance is present by just looking at an electrical connection or component. Expect resistance to be high if the material is discolored or if a connection appears loose. Resistance can also be affected by the physical condition of a conductor. For example, battery terminals are made of lead, ordinarily an excellent conductor. However, when a battery terminal is covered with corrosion, resistance is substantially increased. This makes the terminal a less effective conductor.

Ohms

The basic unit of measurement for resistance is the **ohm.** The symbol for ohms is the Greek letter omega (Ω). If the resistance of a material is high (close to infinite ohms), it is an insulator. If the resistance of a material is low (close to zero ohms), it is a conductor.

Resistors

Resistors are devices used to provide specific values of resistance in electrical circuits. A common type is the carbon-composition resistor, which is available in many specific resistance values. They

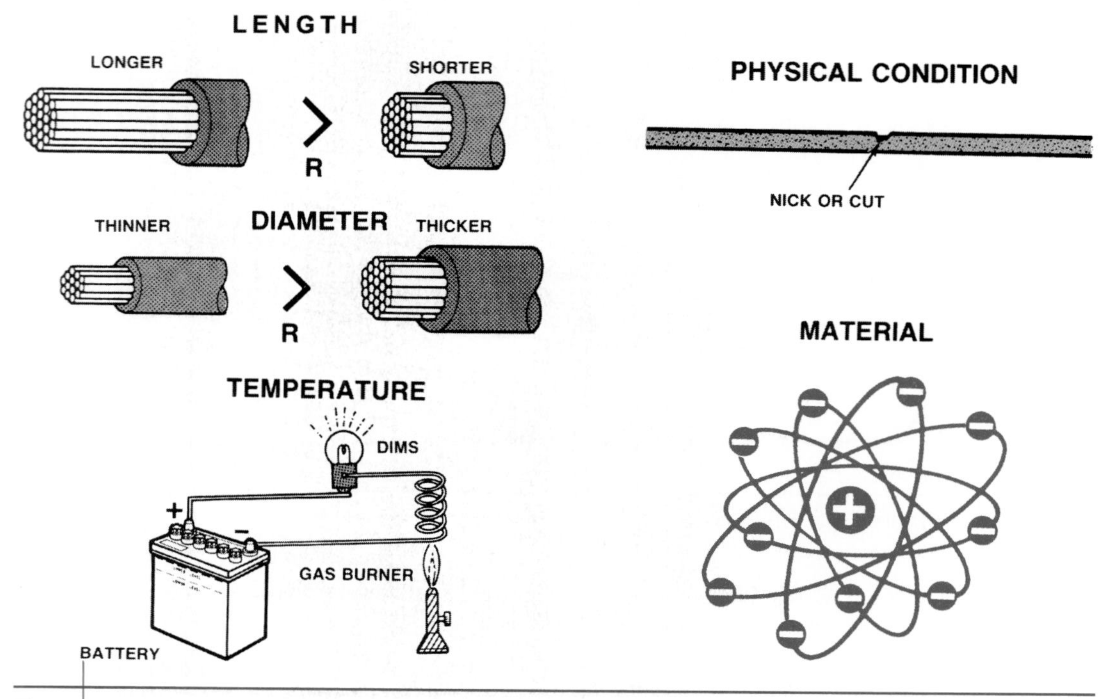

Figure 3-9. Resistance factors.

are generally marked with colored bands that make up a code expressing each resistor's value. The size of the resistor determines how much heat the device can dissipate, and therefore how much power it can handle.

COMPLETE ELECTRICAL CIRCUIT

For current to flow continuously from a voltage supply, such as a battery, there must be a complete circuit, as shown in Figure 3-10. A **circuit** is a path for electric current. Current flows from one end of a circuit to the other when the ends are connected to opposite charges (positive and negative) We usually call these ends *power* and *ground.* Current flows only in a closed or completed circuit. If there is a break somewhere in the circuit, current cannot flow. We usually call a break in a circuit an open. Most protection circuits contain a source of power, conductive material (wires) load, controls, and a ground. These elements are connected to each other with conductors, such as a copper wire. The primary power source in a car or truck is the battery. As long as there is no external connection between the positive and negative sides, there is no flow of electricity. Once an external connection is made between them, the free electrons have a path to flow on, and the entire electrical system is connected between the positive and negative sides.

Power (Voltage Source)

The 12-volt battery is the most common voltage source in automotive circuits. The battery is an electrochemical device; In other words, it uses chemicals to create electricity. It has a positive side and a negative side, separated by plates. Typically, the positive (+) part of the battery is made up of lead peroxide. The negative (−) part of the battery is made up of sponge lead. A pasty chemical called electrolyte is used as a conductor between the positive and negative parts. Electrolyte is made from water and sulfuric acid. So, within the battery there exists an electrical potential. The operation of the alternator continuously replenishes the electrical potential of the battery to prevent it from "draining."

Load

The load is any device in a circuit that uses electricity to do its job. For example, the loads in a circuit could include a motor, a solenoid, a relay, and

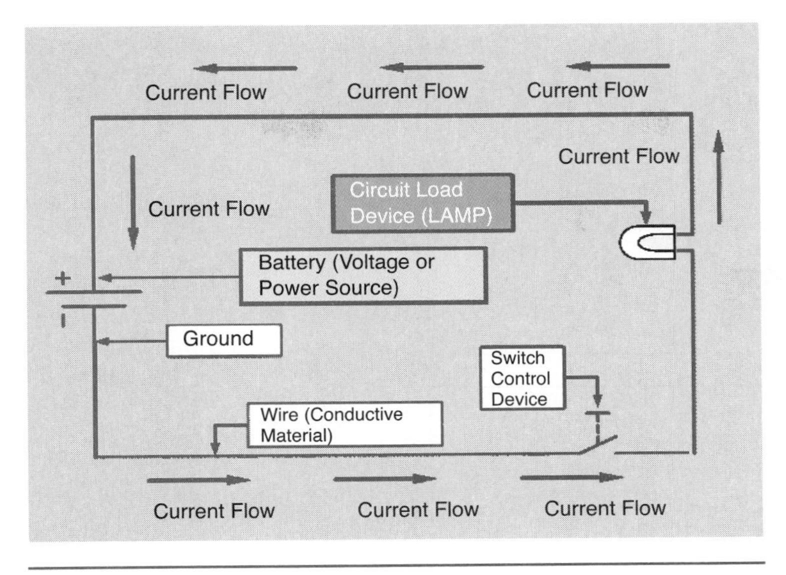

Figure 3-10. The complete electrical circuit.

a light bulb. All loads offer some resistance to current flow.

Controls

Control devices perform many different jobs, such as turning lights on and off, dimming lights, and controlling the speed of motors. Control devices work by completely stopping current flow or by varying the rate of flow. Controls used to stop current flow include switches, relays, and transistors. Controls used to vary the rate include rheostats, transistors, and other solid-state devices. Control devices can be on the positive or negative side of the circuit.

Ground

In a closed circuit, electrons flow from one side of the voltage source, through the circuit, and then return to the other side of the source. We usually call the return side of the source the **ground.** A circuit may be connected to ground with a wire or through the case of a component. When a component is case-grounded, current flows through its metal case to ground. In an automobile, one battery terminal is connected to the vehicle chassis with a conductor. As a result, the chassis is at the same electrical potential as the battery terminal. It can act as the ground for circuits throughout the vehicle. If the chassis didn't act as a ground, the return sides of vehicle circuits would have to reach all the way back to the ground terminal of the battery.

Conductive Material

Conductive materials, such as copper wire with an insulator, readily permit this flow of electrons from atom to atom and connect the elements of the circuit: power source, load, controls, and ground.

Negative or Positive Ground

Circuits can use a negative or a positive ground. Most vehicles today use a negative ground system. In this system, the negative battery terminal is connected to the chassis.

Voltage Drop

In an electrical circuit, resistance behaves somewhat like an orifice restriction in a refrigerant circuit. In a refrigerant circuit, pressure upstream of the orifice is higher than that downstream of the orifice. Similarly, voltage is higher before the resistance than it is after the resistance. The voltage change across the resistance is called a **voltage drop** (Figure 3-11).

Voltage drop is the voltage lost or consumed as current moves through resistance. Voltage is highest where the conductor connects to the voltage source, but decreases slightly as it moves through the conductor. If you measure voltage before it

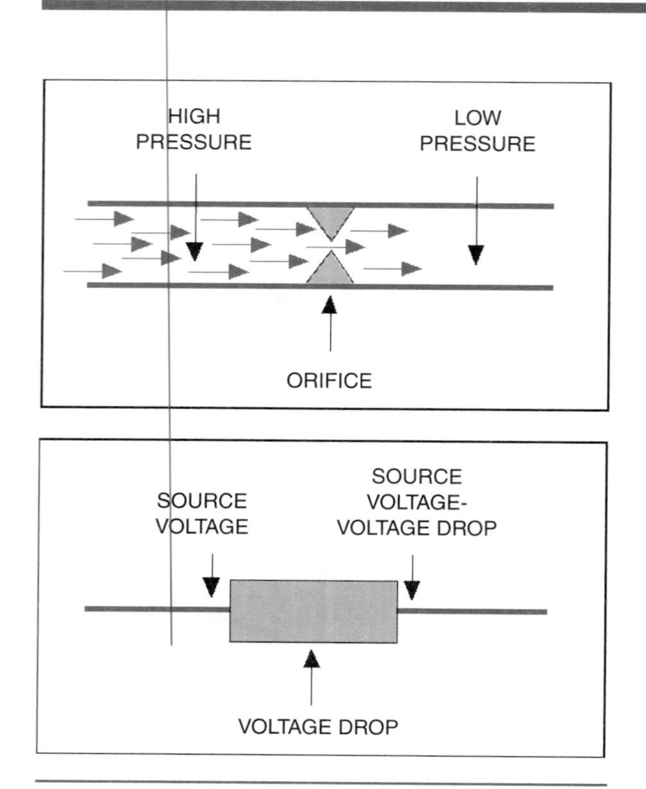

Figure 3-11. Voltage drop. (GM Service and Parts Operations)

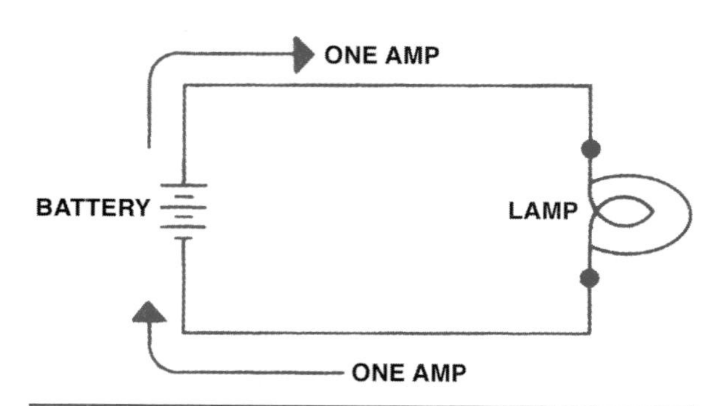

Figure 3-12. If this battery provides 1 volt of pressure, the resistance of the lamp must be 1 ohm.

goes into a conductor and measure it again on the other side of a conductor, you will find that the voltage has decreased. This is voltage drop. When you connect several conductors to each other, the voltage will drop each time the current passes through another conductor.

Voltage drop is the result of a total applied voltage that is not equal at both ends of a single load circuit. Whenever the voltage applied to a device (a load) is less than the source voltage there is a resistance between the two components. The resistance in a circuit opposes the electron flow, with a resulting voltage loss applied to the load. This loss is voltage drop. Kirchhoff's Voltage Law states that the sum of the voltage drops in any circuit will equal the source or applied voltage. For more information about measuring voltage, See the section on "DMM Operating Principles" in Chapter 3 of the *Shop Manual.*

OHM'S LAW

The German physicist George Simon Ohm (1787–1854) proved the mathematical relationship between electrical potential (voltage), electrical current flow (measured in amperes) and the resistance to the current flow (measured in ohms). **Ohm's Law** states that voltage equals current times resistance, and is expressed as E=IR. Ohm's Law is based on the fact that it takes 1 volt of electrical potential to push 1 ampere of current through 1 ohm of resistance, as demonstrated in Figure 3-12.

Ohm's Law Units

When introducing electricity, analogies are often made between electrical circuits and hydraulic circuits; these analogies will be used in the following explanations.

Voltage

Voltage is the electrical pressure or potential difference. Using a hydraulic analogy, voltage in an electrical circuit is similar to water pressure. Just as in the hydraulic circuit, voltage may be present as potential energy in an electrical circuit without any current flow. In the Ohm's Law equation, voltage is represented by the letter *E* for *electromotive force*. One ampere is equal to 6.28×10^{18} electrons (1 columb) passing a given point in an electrical circuit in one second.

Current

Current is the flow of electrons measured in amperes. If the analogy of a hydraulic circuit is used once again, amps can be compared to gpm (gallons per minute). If current flow is to be measured in an electric circuit, the circuit must be electrically active or closed—that is, actively flowing current. In the Ohm's Law equation, current is represented by the letter *I* for *intensity* or amperage. The equation is

$$I = \frac{E}{R}$$

Resistance

Resistance to the flow of electrons through a circuit is measured in *ohms,* whose symbol is Ω. The resistance to the flow of free electrons through a conductor results from the innumerable collisions that occur and, generally, the greater the cross-sectional area of the conductor (wire gauge size), the less resistance to current flow, simply because there are more available free electrons. Once again, using the hydraulic analogy, the resistance to fluid flow through a circuit would be defined by the pipe internal diameter or flow area. In an electrical circuit, resistance generally increases with temperature because of collisions between free electrons and vibrating atoms: as the temperature of a conductor increases, the tendency of the atoms to vibrate also increases and so does the incidence of colliding free electrons. In the Ohm's Law equation, resistance is represented by the letter *R*, for resistance. The equation is

$$R = \frac{E}{I}$$

A complete electrical circuit is an arrangement that permits electrical current to flow, at its simplest, this would require a power source, a load, and a means of connecting supply and return paths to the power source. A circuit is described as closed when current is flowing and open when it is not.

Here is an easy way to remember how to solve for any part of the equation: To use the "solving circle" in Figure 3-13, cover the letter you don't know. The remaining letters give the equation for determining the unknown quantity.

In the circuit in Figure 3-14, the source voltage is unknown. The resistance of the load in the circuit is 2 ohms. The current flow through the circuit is 6 amps. Since the volts are missing, the correct equation to solve for voltage is volts = amps × ohms. If the known units are inserted into the equation, the current can be calculated. Performing the multiplication in the equation yeilds in 12 volts. volts = 6 × 2 as the source voltage in the circuit.

In the circuit in Figure 3-15, the current is unknown. The resistance of the load in the circuit is 2 ohms. The source voltage is 12 volts. Since

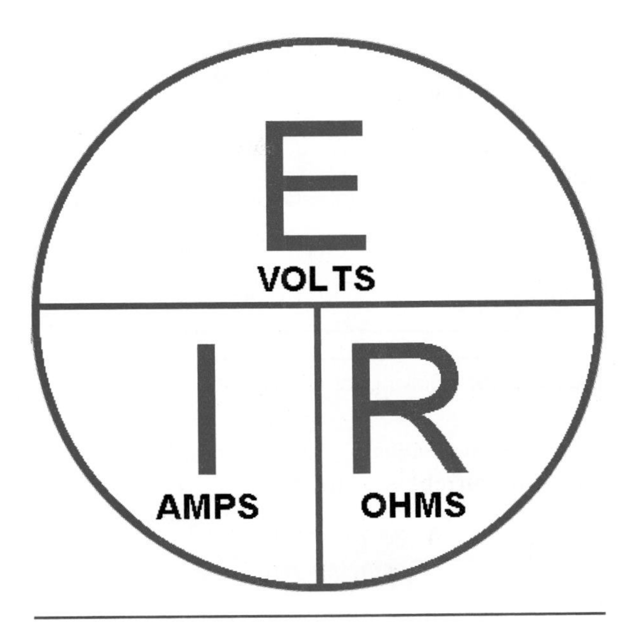

Figure 3-13. Ohm's Law solving circle.

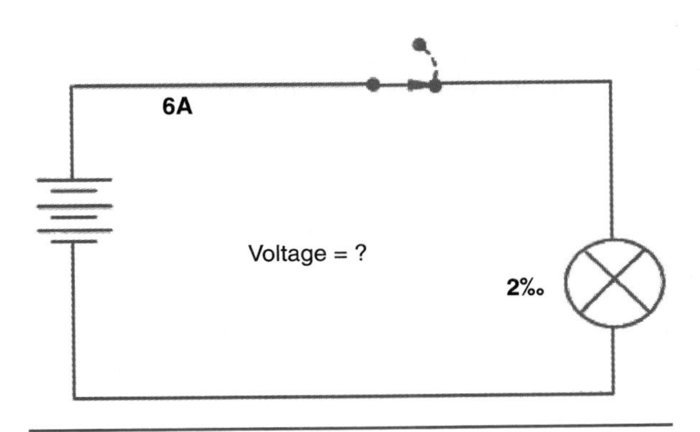

Figure 3-14. Solving for voltage.

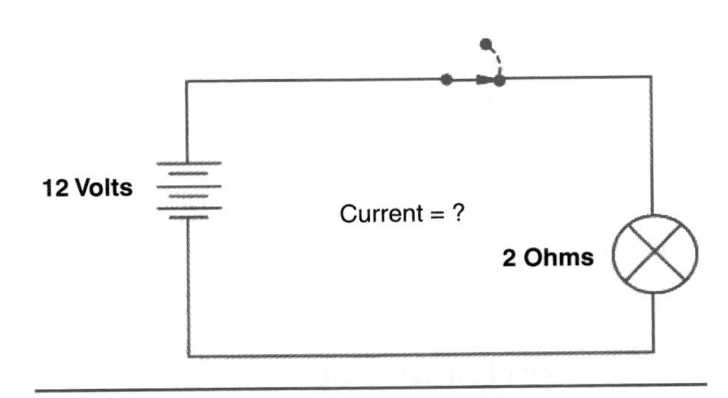

Figure 3-15. Solving for current.

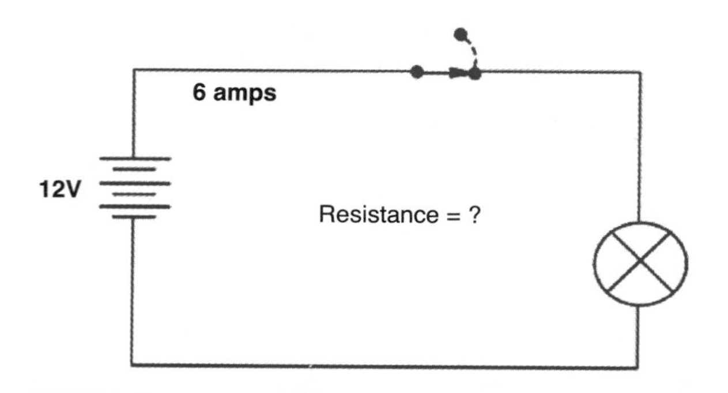

Figure 3-16. Solving for resistance.

the amps are missing, the correct equation to solve for current is as follows:

$$Amps\ (I) = \frac{Volts}{Ohms} = \frac{E}{R} = \frac{12}{2} = 6\ amps$$

If the known units are inserted into the equation, the current can be calculated. Performing the division in the equation yields in 6 amps as the current flow in the circuit.

In the circuit in Figure 3-16, the resistance is unknown. The current flow through the circuit is 6 amps. The source voltage is 12 volts. Since the ohms are missing, the correct equation to solve for resistance is as follows:

$$Ohms\ (R) = \frac{Volts}{Amps} = \frac{E}{I} = \frac{12}{6} = 2\ ohms$$

If the known units are inserted into the equation, the resistance can be calculated. Performing the division in the equation yields 2 ohms as the resistance in the circuit.

Ohm's Law General Rules

Ohm's Law shows that both voltage and resistance affect current. Current never changes on its own—it changes only if voltage or resistance changes. Current cannot change on its own because voltage causes current through a conductor and all conductors have resistance. The amount of current can change only if the voltage or the conductor changes. Ohm's Law says if the voltage in a conductor increases or decreases, the current will increase or decrease. If the resistance in a conductor increases or decreases, the current will decrease or increase. The general Ohm's Law rule, assuming the resistance doesn't change, is as follows:

- As voltage increases, current increases
- As voltage decreases, current decreases

For more information about measuring voltage, resistance, and current, see the section on "DMM Operating" in Chapter 3 of the *Shop Manual.*

Ohm's Law Chart

The table in (Figure 3-17) summarizes the relationship between voltage, resistance, and current. This table can predict the *effect of changes* in voltage and resistance or it can predict the *cause of changes* in current. In addition to showing what happens to current if voltage or resistance changes, the chart also tells you the most likely result if both voltage and resistance change.

If the voltage increases (Column 2), the current increases (Column 1)—*provided the resistance stays the same* (Column 3). If the voltage decreases (Column 2), the current decreases (Column 1)—*provided the resistance stays the same* (Column 3). For example: Solving the three columns mathematically, if 12 volts ÷ 4 ohms = 3 amps, and the voltage increases to 14 volts/4 ohms = 3.5 amps, the current will increase with the resistance staying at 4 ohms. Decrease the voltage to 10 volts: 10 volts/4 ohms = 2.5 amps, or a decrease in current. In both cases, the resistance stays the same.

If the resistance increases (Column 3), the current decreases (Column 1)—*provided the voltage stays the same* (Column 2). If the resistance decreases (Column 3), the current increases (Column 1)—*provided the voltage stays the same* (Column 2). For example: solving the three columns mathematically, 12 volts ÷ 4 ohms = 3 amps; increase resistance to 5 ohms and keep the voltage at 12 volts: 12 volts/5 ohms = 2.4 amps, a decrease in current. Decrease the resistance to 3 ohms and keep the voltage at 12 volts:12 volts/3 ohms = 4 amps, or an increase in current. In both cases, the voltage stays the same.

POWER

Power is the name we give to the rate of work done by any sort of machine. The output of automotive engines is usually expressed in horsepower; so is the output of electric motors. Many electrical devices are rated by how much electrical power they consume, rather than by how much power they produce. Power consumption is expressed in **watts:** 746 watts = 1 horsepower.

	I-Current (Amps)	E-Voltage (Volts)	R-Resistance (Ohms)
IF VOLTAGE INCREASES or DECREASES	Increases ⇑	Increases ⇑	Stays Same
	Increases ⇑	Increases ⇑	Decreases ⇓
	Stays Same	Increases ⇑	Increases ⇑
	Decreases ⇓	Decreases ⇓	Stays Same
	Stays Same	Decreases ⇓	Decreases ⇓
	Decreases ⇓	Decreases ⇓	Increases ⇑
If RESISTANCE INCREASES or DECREASES	Decreases ⇓	Stays Same	Increases ⇑
	Decreases ⇓	Decreases ⇓	Increases ⇑
	Stays Same	Increases ⇑	Increases ⇑
	Increases ⇑	Stays Same	Decreases ⇓
	Stays Same	Decreases ⇓	Decreases ⇓
	Increases ⇑	Increases ⇑	Decreases ⇓
CURRENT will INCREASE or DECREASE	Increases ⇑	Stays Same	Decreases ⇓
	Increases ⇑	Increases ⇑	Stays Same
	Increases ⇑	Increases ⇑	Decreases ⇓
	Decreases ⇓	Stays Same	Increases ⇑
	Decreases ⇓	Decreases ⇓	Stays Same
	Decreases ⇓	Decreases ⇓	Increases ⇑

Figure 3-17. Ohm's Law relationship table.

Power Formula

We describe the relationships among power, voltage, and current with the Power Formula. The basic equation for the Power Formula is as follows:

$$P = I \times E \text{ or watts} = \text{amps} \times \text{volts}$$

Power is the product of current multiplied times voltage. In a circuit, if voltage or current increases, then power increases. If voltage or current decreases, then the power decreases.

The most common applications of ratings in watts are probably light bulbs, resistors, audio speakers, and home appliances. Here's an example of how we determine power in watts: If the total current (I) is equal to 10 amps, and the voltage (E) is equal to 120 volts, then

$$P = 120 \times 10 = 1200 \text{ watts}$$

You can multiply the voltage times the current in any circuit and find how much power is consumed. For example, a typical hair dryer can draw almost 10 amps of current. You know that the voltage in your home is about 120. Multiply these two values and you get 1200 watts.

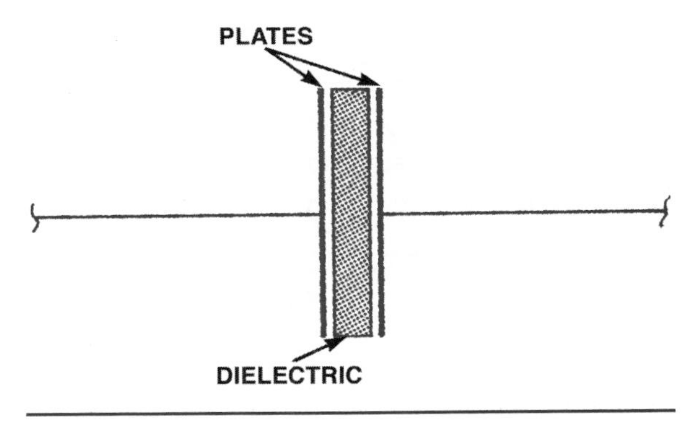

Figure 3-18. A simple capacitor.

CAPACITANCE

Earlier, you learned that a resistor is any device that opposes current flow. A **capacitor** (Figure 3-18) is a device that opposes a change in voltage. The property of opposing voltage change is called **capacitance,** which is also used to describe the electron storage capability of a capacitor. Capacitors are sometimes referred to as condensers because they do the same thing; that is, they store electrons.

There are many uses for capacitors. In automotive electrical systems, capacitors are used to store energy, to make up timer circuits, and as filters. Actual construction methods vary, but a simple capacitor can be made from two plates of conductive material separated by an insulating material called a *dielectric*. Typical dielectric materials are air, mica, paper, plastic, and ceramic. The greater the dielectric properties of the material used in a capacitor, the greater the resistance to voltage leakage.

Energy Storage

When the capacitor (Figure 3-19) is charged to the same potential as the voltage source, current flow stops. The capacitor can then hold its charge when it is disconnected from the voltage source. With the two plates separated by a dielectric, there is nowhere for the electrons to go. The negative plate retains its accumulation of electrons, and the positive plate still has a deficit of electrons. This is how the capacitor stores energy.

When a capacitor is connected to an electrical power source, it is capable of storing electrons from that power source (Figure 3-20). When the capacitor's charge capability is reached, it will cease to accept electrons from the power source. An electrostatic field exists between the capacitor plates and no current flows in the circuit. The charge is retained in the capacitor until the plates are connected to a lower-voltage electrical circuit; at this point, the stored electrons are discharged from the capacitor into the lower-potential (lower-voltage) electrical circuit.

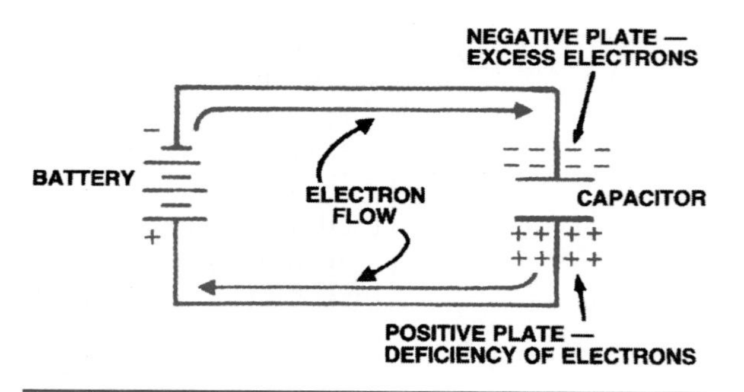

Figure 3-19. As the capacitor is charging, the battery forces electrons through the circuit.

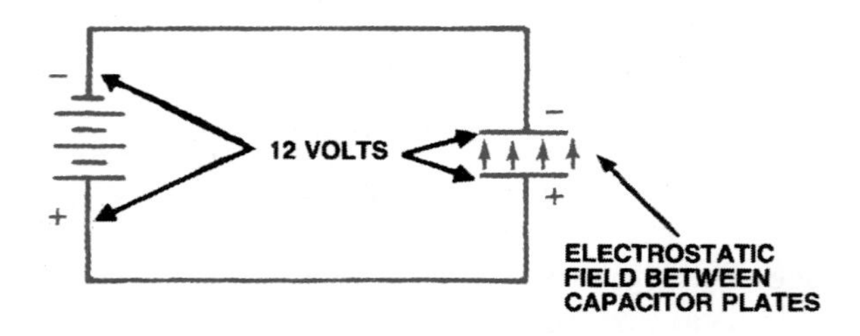

Figure 3-20. When the capacitor is charged, there is equal voltage across the capacitor and the battery. An electrostatic field exists between the capacitor plates. No current flows in the circuit.

Capacitor Discharge

A charged capacitor can deliver its stored energy just like a battery. When capacitors provide electricity, we say they *Discharge*. Used to deliver small currents, a capacitor can power a circuit for a short time (Figure 3-21).

In some circuits, a capacitor can take the place of a battery. Electrons can be stored on the surface of a capacitor plate. If a capacitor is placed in a circuit with a voltage source, current flows in the circuit briefly while the capacitor "charges"; that is, electrons accumulate on the surface of the plate connected to the negative terminal and move away from the plate connected to the positive terminal. Electrons move in this way until the electrical charge of the capacitor is equal to that of the voltage source. How fast this happens depends on several factors, including the voltage applied and the size of the capacitor; generally speaking, it happens very quickly.

The number of electrons in capacitor is identical whether it is in the electrified on neutral states. What changes is the location of the electrons. The

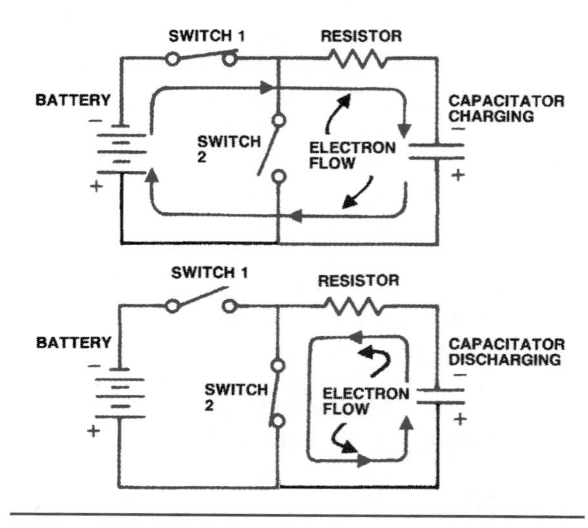

Figure 3-21. The capacitor is charged through one circuit (top) and discharged through another.

electrons in a fully charged capacitor will in time leak through the dielectric until both conductor plates have an equal charge; at this point, the capacitor would be described as being in a discharged condition.

The ability to store electrons or capacitance is measured in farads (named after Michael Faraday [1791–1867], the discoverer of the principle). One farad is the ability to store 6.28×10^{18} electrons at a 1-volt charge differential. In addition to being rated in farads, capacitors are also rated by the maximum voltage that can be handled. When replacing a capacitor, never use a capacitor with a lower voltage rating. Most capacitors have much less capacitance, so they are rated in picofarads (trillionths of a farad) and microfarads (millionths of a farad).

- 1 farad = 1F
- 1 microfarad = 1μF = 0.000001 F
- 1 picofarad = 1pF – 0.000000000001 F

plates and varies the capacitance. In some variable-capacitance type capacitors, the dielectric may be air. One engine manufacturer uses a variable-capacitance type sensor to digitally signal throttle position, which provides an accurate digital signal. Electrolytic capacitors tend to have much higher capacitance ratings than non-electrolytic types and they are polarized, so they must be connected accordingly in a circuit; the dielectric in this type of capacitor is the oxide formed on the aluminum conductor plate.

Capacitors are used extensively in electronic circuits performing the following roles:

- **AC/DC filter:** Steadies a DC voltage wave offset sometimes caused by exposure sunlight.
- **Power supply filter:** Smooths a pulsating voltage supply into a steady DC voltage form.
- **Spike suppressant:** When digital circuits are switched at high speed, transient (very brief) voltage reductions can occur. Capacitors can eliminate these spikes or glitches by compensating for them.
- **Resistor-capacitor circuits (R-C circuits):** Circuits that incorporate a resistor and a capacitor and are used to reshape a voltage wave or pulse pattern from square wave to sawtooth shaping or modify a wave to an alternating pattern.

WARNING: When working on electrical circuits, it should be noted that a capacitor could retain a charge for a considerable time after the circuit current flow has ceased and accidental discharge can result in an electric shock to the handler.

CAUTION: When working on electrical circuits, it should be noted that a capacitor could retain a charge after the circuit current flow has ceased and cause damage to circuit components.

Types of Capacitors

Many capacitors used in electric and electronic circuits are fixed-value capacitors; they are classified by capacitance and voltage ratings (Figure 3-22). Some capacitors have a variable capacity; these may have a combination of fixed plates and moving plates with a shaft that rotates the moving

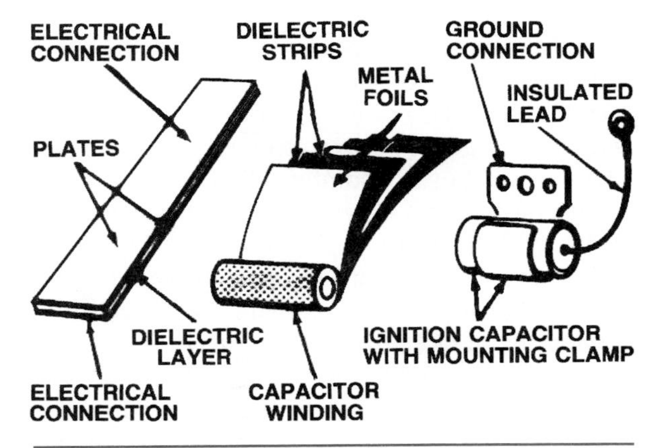

Figure 3-22. Types of capacitors.

Calculating Total Capacitance

The total capacitance of a circuit (Figure 3-23) is dependent on how the capacitors are designed in the circuit. When capacitors are in parallel, total capacitance is determined by the following equation:

$$C_T = C_1 + C_2 + C_3$$

When capacitors are in series, total capacitance is determined by the following equation:

$$C_t = \frac{1}{\frac{1}{C_t} + \frac{1}{C_t}}$$

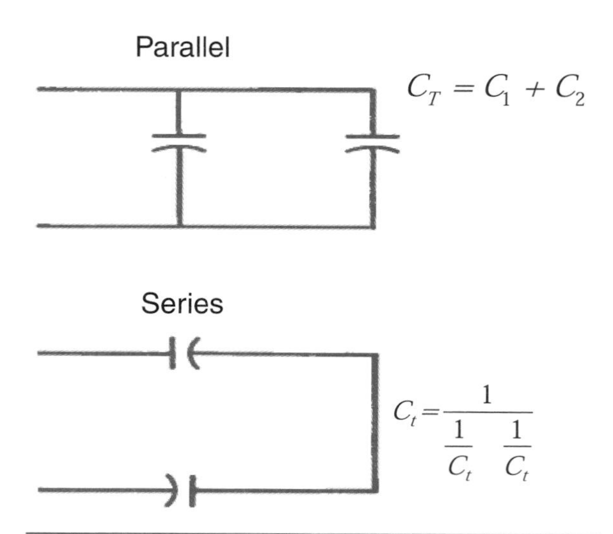

Figure 3-23. Calculating total capacitance here.

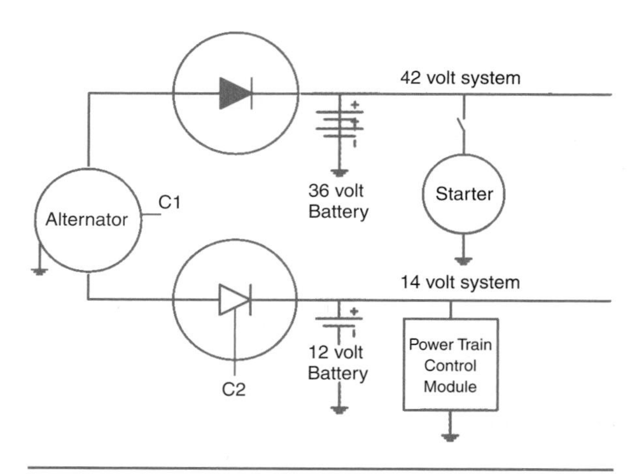

Figure 3-24. Dual 42-volt /14-volt system.

Future Trends in Voltage: The 42-Volt System

The voltage of the automotive electrical system was not always at today's 12 volts. It was raised once from 6 volts to 12 volts around 1955. The reason was that a higher ignition energy was required for the higher-compression V8 engines being introduced, prompting the need for a higher-voltage electrical system. Occurring almost simultaneously with this ignition issue were new automotive features such as radios, higher-power headlamps, and more powerful electric starting motors, all of which were stretching the capabilities of the existing 6-volt system. A pressing need for more reliable ignition drove the implementation of a single higher-energy 12-volt battery (with 14-volt regulation). The transition took about two years.

Today's automobile industry is now faced with a similar situation. Many new electronic features are emerging. Some of them are to meet tighter emission and fuel economy regulations, such as electromechanical (EM) engine valve actuators and convenience items such as in-vehicle information technology systems; some are to satisfy increased desires for safety and comfort with features such as electronic brakes and steering. Many of these cannot be practically introduced using the currently available 14-volt power supply, and a higher-voltage supply would definitely be beneficial to some of the existing automotive components. The comparative complexity of today's 14-volt electrical system and the large number of different components designed to operate at 14 volts make the prospect of changing to any new voltage more difficult. Moreover, there are some specific components, such as light bulbs and low-power electronic control modules, that still require 12- to 14-volt operation.

Before the entire electrical system is transferred to a higher voltage, a dual-voltage electrical system will most likely appear. In a dual-voltage system, a new higher-voltage system is introduced to accommodate desired higher power components and features while simultaneously preserving the present 14-volt system for some interim period and for those components that require the present lower voltage.

42-Volt/14-Volt Electrical System Design

Due to the use of hybrid and mybrid vehicles, it is very likely there will be a dual-voltage electrical architecture before a complete single 42-volt system. In a 42/14-volt dual-voltage system, the governing partitioning philosophy is that high-power loads will generally be allocated to the 42-volt bus, while the low-power electronic loads, including most key-off loads, will be allocated to the 14-volt bus.

Figure 3-24 shows a dual system, including a complicated alternator with two sets of stator windings to deliver power separately to the 42- and 14-volt buses, combined with an integral starter motor. This design is often referred to as an Integrated Starter Alternator (ISA). The dual-stator/starter motor version uses a standard field control (labeled C1) to regulate the 42-volt bus voltage and a phase-controlled converter (C2) to regulate the 14-volt bus voltage. Here again, high-power loads, including the cranking motor, are allocated to the 42-volt bus, while the 14-volt bus supplies the low-power electronic control modules. Separate batteries are used for each bus. There are some difficulties in controlling individual stator windings for optimal outputs in this dual-stator winding architecture.

SUMMARY

A conductor is a metallic element that contains fewer than four electrons in its outer shell. An insulator is a non-metallic substance that contains more than four electrons in the outer shell. Semiconductors are a group of materials that cannot be classified either as conductors or insulators; they have exactly four electrons in their outer shell. Current flow is measured by the number of free electrons passing a given point in an electrical circuit per second. Electrical pressure or charge differential is measured in *volts,* resistance in *ohms* and current in *amperes*. If a hydraulic circuit analogy is used to describe an electrical circuit, *voltage* is equivalent to fluid pressure, *current* to the flow in gpm, and *resistance* to flow restriction.

Capacitors are used to store electrons: this consists of conductor plates separated by a dielectric. Capacitance is measured in farads: capacitors are rated by voltage and by capacitance.

Review Questions

1. Which of the following methods can calculate circuit current?
   a. Multiplying amps times ohms
   b. Dividing ohms by volts
   c. Dividing volts by ohms
   d. Multiplying volts times watts
2. Which of the following methods can calculate circuit resistance?
   a. Dividing amps into volts
   b. Dividing volts into amps
   c. Multiplying volts times amps
   d. Multiplying amps times ohms
3. Which of the following causes voltage drop in a circuit?
   a. Increase in wire size
   b. Increase in resistance
   c. Increase in insulation
   d. Decrease in current
4. Which of the following describes a function of a capacitor in an electrical circuit?
   a. Timing device
   b. Rectification
   c. DC-to-AC conversion
   d. Inductance
5. Which of the following is a measure of electrical current?
   a. Amperes
   b. Ohms
   c. Voltage
   d. Watts
6. A material with many free electrons is a good
   a. Compound
   b. Conductor
   c. Insulator
   d. Semiconductor
7. A material with four electrons in the valence ring is which of these:
   a. Compound
   b. Insulator
   c. Semiconductor
   d. Conductor
8. The conventional theory of current flow says that current flows
   a. Randomly
   b. Positive to negative
   c. Negative to positive
   d. At 60 cycles per second
9. Voltage is which of these items:
   a. Applied to a circuit
   b. Flowing in a circuit
   c. Built into a circuit
   d. Flowing out of a circuit
10. The unit that represents resistance to current flow is which of these items:
    a. Ampere
    b. Volt
    c. Ohm
    d. Watt
11. In automotive systems, voltage is supplied by which of these components:
    a. Alternator
    b. ECM
    c. DC Generator
    d. Voltage Regulator
12. Which of the following characteristic does *not* affect resistance?
    a. Diameter of the conductor
    b. Temperature of the conductor
    c. Atomic structure of the conductor
    d. Direction of current flow in the conductor
13. The resistance in a longer piece of wire is which of these:
    a. Higher
    b. Lower
    c. Unchanged
    d. Higher, then lower
14. According to Ohm's Law, when one volt pushes one ampere of current through a conductor, the resistance is:
    a. Zero
    b. One ohm
    c. One watt
    d. One coulomb
15. The sum of the voltage drops in a circuit equals which of these:
    a. Amperage
    b. Resistance
    c. Source voltage
    d. Shunt circuit voltage
16. Which of the following is a measure of electrical pressure?
    a. Amperes
    b. Ohms
    c. Voltage
    d. Watts

17. Which of the following causes voltage drop in a circuit?
    a. Increase in wire size
    b. Increase in resistance
    c. Increase in insulation
    d. Decrease in current

18. Where E = volts, I = amperes, and R = resistance, Ohm's Law is written as which of these formulas:
    a. $I = E \times R$
    b. $E = I \times R$
    c. $R = E \times I$
    d. $E = 12 \times R$

19. In a closed circuit with a capacitor, current will continue to flow until the voltage charge across the capacitor plates becomes which of these:
    a. Less than the source voltage
    b. Equal to the source voltage
    c. Greater than the source voltage
    d. Equal to the resistance of the plates

20. Capacitors are also called which of these items:
    a. Diodes
    b. Resistors
    c. Condensers
    d. Dielectrics

21. Capacitors are rated in
    a. Microcoulombs
    b. Megawatts
    c. Microfarads
    d. Milliohms

4

Magnetism

LEARNING OBJECTIVES

Upon completion and review of this chapter, you should be able to:

- Define magnetism, electromagnetism, electromagnetic induction, and magnetomotive force.
- Compare the units of magnetism to electricity: magnetic force to current, field density to voltage, and reluctance to resistance.
- Explain the use and operation of automotive circuit components that use electromagnetic induction and magnetism, to include alternators, motors, starters, relays, and solenoids.

KEY TERMS

AC Generator/Alternator
Commutator
Electromagnet
Electromagnetic induction
Electromagnetic Interference (EMI)
Electromagnetism
Left-Hand Rule
Lines of Force
Magnetic Field
Magnetic Field Intensity
Magnetic Flux
Mutual Induction
Relay
Reluctance
Right-Hand Rule
Self-Induction
Transformers

INTRODUCTION

In this chapter you will learn about magnetism, electromagnetism, electromagnetic induction, and magnetomotive force. This knowledge is applied thorough the explanation of the operation of automotive circuit components that use electromagnetic induction and magnetism, such as DC and AC generators (alternators), motors, starters, relays, and solenoids.

MAGNETISM

Historical Facts

Magnetism was first observed in lodestone and the way ferrous (iron-based) metals reacted to it. In the 1600s, *Sir William Gilbert* discovered that the earth was a great magnet with north and south poles. Gilbert shaped a piece of lodestone into a sphere and demonstrated that a small compass placed at any spot on the sphere would always point, as it does on the earth, toward the north pole. Lodestone is also known as magnetite.

Lodestones

If you suspend a bar of lodestone (Figure 4-1) by a string, the same end will always rotate to point toward the earth's north pole. In most materials, the magnetic poles of the composite molecules are arranged randomly, so there is no magnetic force. In certain metals such as iron, nickel, and cobalt, the molecules can be aligned so that their north poles all point in one direction and their south poles point in the opposite direction. The molecules align themselves naturally in a lodestone. Some materials have good magnetic retention, which means that when they are magnetized they retain their molecular alignment. Other materials are capable of maintaining their molecular alignment only when positioned within a magnetic field; when the field is removed, the molecules rearrange themselves randomly and the substance's magnetic properties are lost.

Polarity

All magnetism is essentially electromagnetism, in that it results from the kinetic energy of electrons. Whenever an electric current is flowed through a conductor, a magnetic field is created. When a bar-shaped permanent magnet (Figure 4-2) is cut in two, each piece assumes the magnetic properties of the parent magnet, with individual north and south poles. When a magnet is freely suspended, the poles tend to point toward the north and south magnetic poles of the earth, which led to development of the compass.

Magnetism provides a link between mechanical energy and electricity. By the use of magnetism, an automotive generator converts some of the mechanical power developed by the engine to electromotive potential (EMF). Going the other direction, magnetism allows a starter motor to convert electrical energy from the battery into mechanical power to crank the engine. A magnet can be any object or device that attracts iron, steel, and other magnetic materials. There are three basic types of magnets as follows:

- Natural magnets
- Man-made magnets
- Electromagnets

Reluctance

The term **reluctance** is used to describe *resistance* to the movement of magnetic lines of force. Using permeable (susceptible to penetration) materials within magnetic fields reduces reluctance. The permeability of matter is rated by ascribing a rating of 1 for air, generally considered to be a poor conductor of magnetic lines of force; in contrast, iron would be ascribed a permeability factor of

Figure 4-1. A freely suspended natural magnet (lodestone).

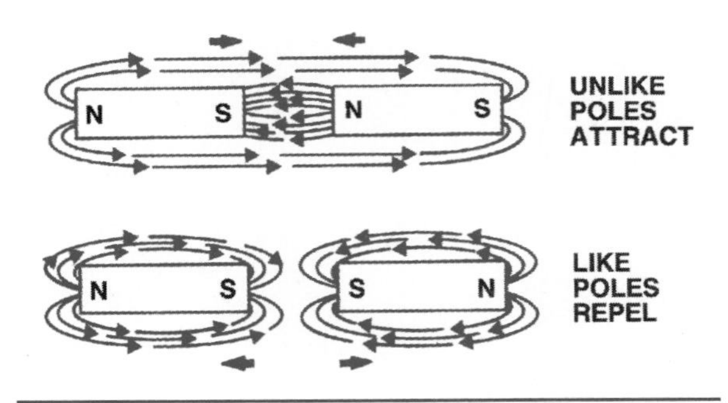

Figure 4-2. Magnetic poles behave like electrically charged particles.

2,000 and certain ferrous alloys may have values exceeding 50,000.

Magnetic Fields and Lines of Force

A **magnetic field** (Figure 4-3) is made up of many invisible **lines of force,** which are also called *lines of flux.* **Magnetic flux** is another term applied to lines of force, which can be compared to *current* in electricity: They come out of one pole and enter the other pole. The flux lines are concentrated at the poles and spread out into the areas between the poles.

Magnetic Field Intensity

The **magnetic field intensity** refers to the magnetic field strength (force) exerted by the magnet and can be compared to *voltage* in electricity. The magnetic field existing in the space around a magnet can be demonstrated if a piece of cardboard is placed over a magnet and iron filings are sprinkled on top of the cardboard. The pattern produced as the filings arrange themselves on the cardboard shows the *flux lines.* A weak magnet has relatively few flux lines; a strong magnet has many. The number of flux lines is sometimes described as *flux density,* as shown in Figure 4-4.

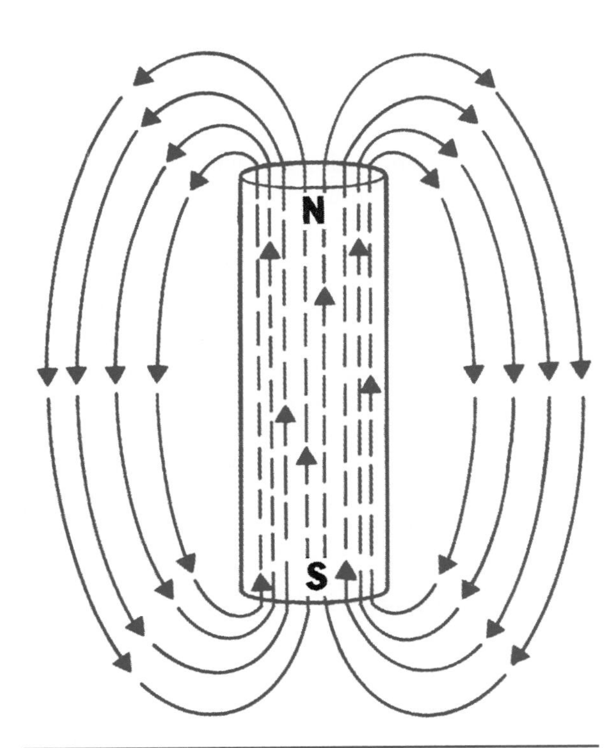

Figure 4-3. Magnetic field/lines of force.

Magnetism Summary

- Flux lines are directional and exit from the magnet's north pole and enter through the south pole.
- The flux density (concentration) indicates the magnetic force. A powerful magnetic field will exhibit a dense flux field, whereas a weak magnetic field will exhibit a low-density flux field.
- The flux density is always greatest at the poles of a magnet.
- Flux lines do not cross each other in a permanent magnet.
- Flux lines facing the same direction attract while flux lines facing opposite directions tend to repel.

Atomic Theory and Magnetism

In an atom, all of the electrons in their orbital shells also spin on their own axes, in much the same way the planets orbit the sun and rotate axially, producing magnetic fields. Because of their axial rotation, each electron can be regarded as a minute permanent magnet. In most atoms, pairs of electrons spinning in opposite directions produce magnetic fields that cancel each other out. An atom of iron has 26 electrons, 22 of which are paired. In the second-from-the-outermost shell,

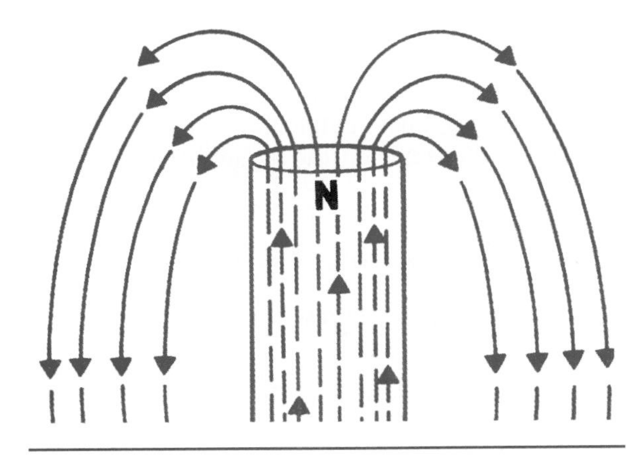

Figure 4-4. Flux density equals the number of lines of force per unit area.

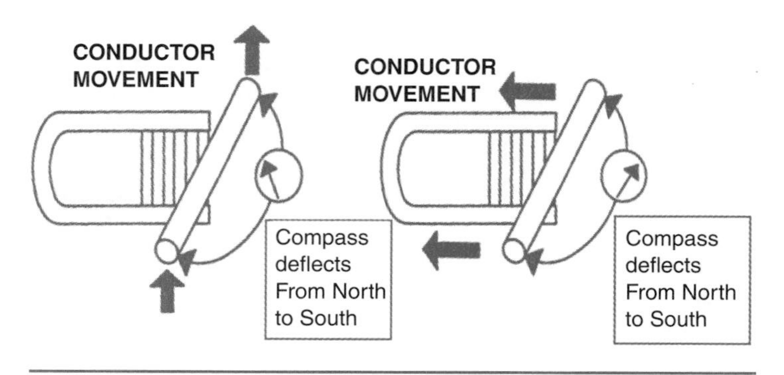

Figure 4-5. Electromagnet.

four of the eight electrons are not paired, meaning that they rotate in the same direction and do not cancel each other out. This fact accounts for the magnetic character of the metal iron.

ELECTROMAGNETISM

In 1820, scientists discovered that current-carrying conductors are surrounded by a magnetic field. Current flowing through a conductor such as copper wire creates a magnetic field surrounding the wire. This effect can be observed by passing a compass lengthwise over a copper wire through which current is flowing, as shown in Figure 4-5. The needle will deflect from its north-south orientation when this occurs. Any magnetic field created by electrical current flow is known as **electromagnetism.**

Straight Conductor

The magnetic field surrounding a straight, current-carrying conductor consists of several concentric cylinders of flux the length of the wire, as in Figure 4-6. The strength of the current determines how many flux lines (cylinders) there will be and how far out they extend from the surface of the wire.

Electromagnetic Field Rules

The following rules apply with electromagnetic fields:

- Magnetic lines of force do not move when the current flowed through a conductor remains constant. When current flowed through the conductor increases, the magnetic lines of force extend further away from the conductor.

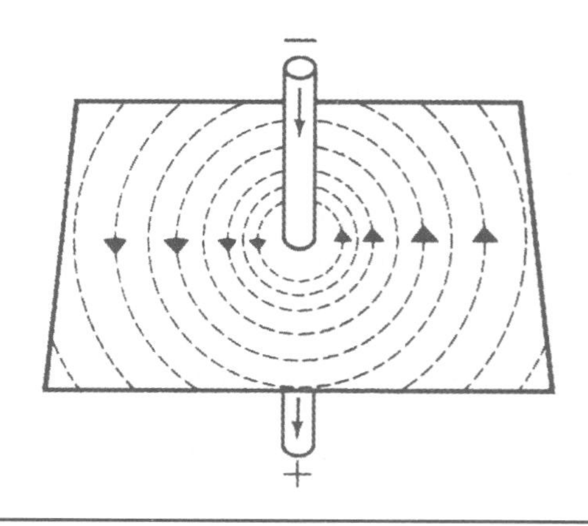

Figure 4-6. A magnetic field surrounds a straight current-carrying conductor.

- The intensity and strength of magnetic lines of force increase proportionally with an increase in current flow through a conductor. They also decrease proportionally with a decrease in current flow through the conductor.

Left-Hand Rule

Magnetic flux cylinders have direction, just as the flux lines surrounding a bar magnet have direction. The **left-hand rule** is a simple way to determine this direction. When you grasp a conductor with your left hand so that your thumb points in the direction of electron flow (− to +) through the conductor, your fingers curl around the wire in the direction of the magnetic flux lines, as shown in Figure 4-7.

Right-Hand Rule

It is important to note at this point that in automotive electricity and magnetism, we use the conventional theory of current (+ to −), so you use the **right-hand rule** to determine the direction of the magnetic flux lines, as shown in Figure 4-8. The right-hand rule is used to denote

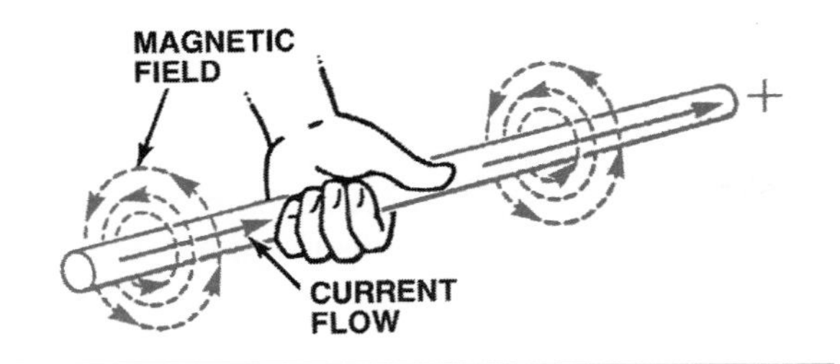

Figure 4-7. Left-hand rule for field direction; used with the electron-flow theory.

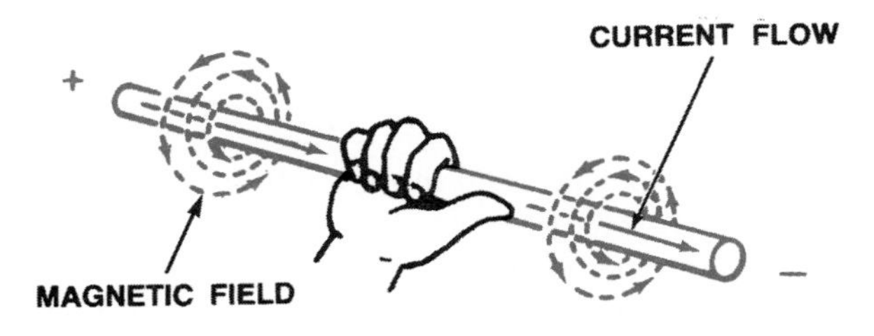

Figure 4-8. Right-hand rule for field direction; used with the conventional flow theory.

the direction of the magnetic lines of force, as follows: The right hand should enclose the wire, with the thumb pointing in the direction of conventional current flow (positive to negative), and the finger tips will then point in the direction of the magnetic lines of force, as shown in Figure 4-8. *For the rest of this chapter, the electron-flow theory (negative to positive) and the left-hand rule are used.*

Field Interaction

The cylinders of flux surrounding current-carrying conductors interact with other magnetic fields. In the following illustrations, the cross symbol (+) indicates current moving inward, or away from you. It represents the tail of an arrow. The dot symbol (.) represents an arrowhead and indicates current moving outward, or toward you (Figure 4-9).

If two conductors carry current in opposite directions, their magnetic fields are also in opposite directions (according to the left-hand rule). If they are placed side by side, Figure 4-10, the opposing flux lines between the conductors create a strong magnetic field. Current-carrying conductors tend to move out of a strong field into a weak field, so the conductors move away from each other (Figure 4-11).

If the two conductors carry current in the same direction, their fields are in the same direction. As seen in Figure 4-12, the flux lines between the two conductors cancel each other out, leaving a very weak field. In Figure 4-13, the conductors are drawn into this weak field; that is, they move closer together.

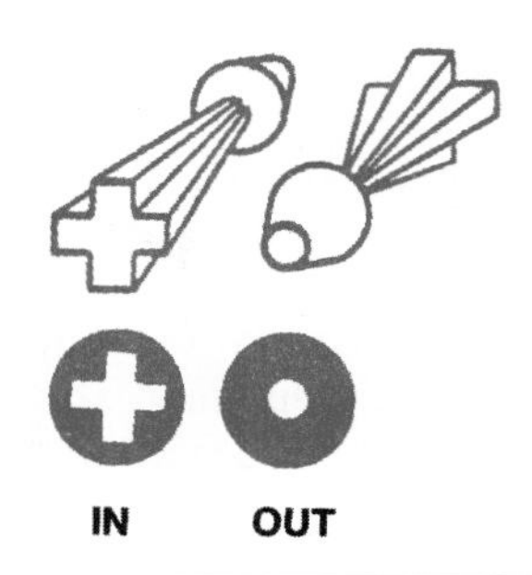

Figure 4-9. Current direction symbols.

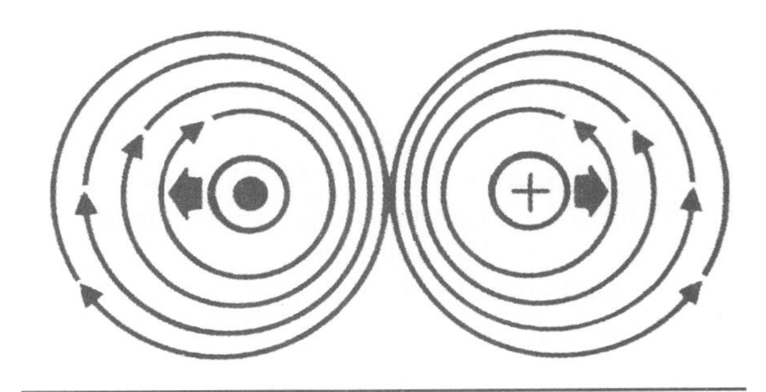

Figure 4-10. Conductors with opposing magnetic fields.

Figure 4-11. Conductors will move apart into weaker fields.

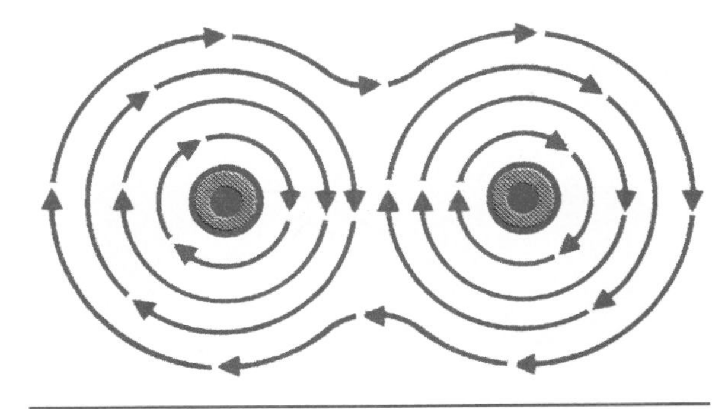

Figure 4-12. Conductors with the same magnetic fields.

Motor Principle

Electric motors, such as automobile starter motors, use field interaction to change electrical energy into mechanical energy (Figure 4-14). If two conductors carrying current in opposite directions are placed between strong north and south poles, the magnetic field of the conductor interacts with the magnetic fields of the poles. The clockwise field of the top conductor adds to the fields of the poles and creates a strong field beneath the conductor. The conductor tries to move up to get out of this strong field. The counterclockwise field of the lower conductor adds to the field of the poles and creates a strong field above the conductor. The conductor tries to move down to get out of this strong field. These forces cause the center of the motor or armature where the conductors are mounted to turn in a clockwise direction. This process is known as magnetic repulsion. For more information about electric motors, see the section on "Diagnostic Strategies" in Chapter 4 of the *Shop Manual*.

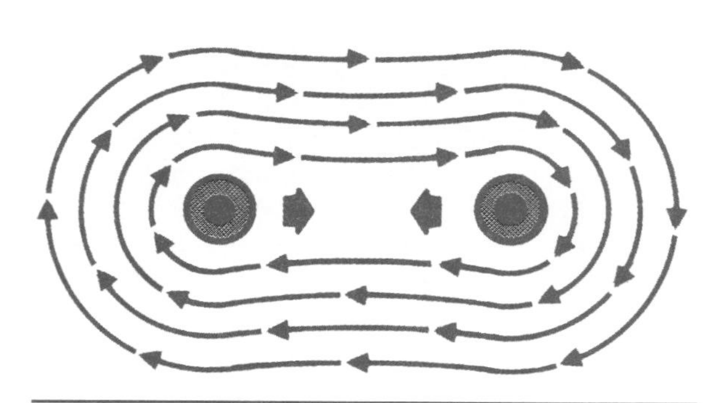

Figure 4-13. Conductors will move together into the weak field.

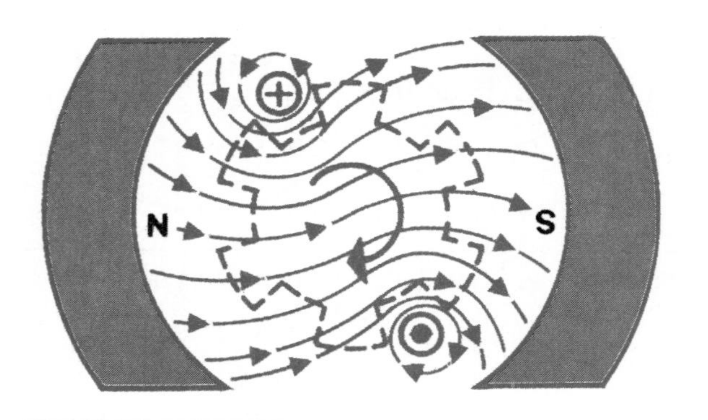

Figure 4-14. Electric motors use field interaction to produce mechanical energy and movement.

Loop Conductor

Bending the wire into a loop can strengthen the field around a straight conductor. As the wire is bent, the fields, which meet in the center of the loop, combine their strengths (Figure 4-15). The left-hand rule also applies to loop conductors.

Coil Conductor

If several loops of wire are made into a coil, the magnetic flux density is further strengthened. Flux lines around a coil are the same as the flux lines around a bar magnet (Figure 4-16). They exit from the north pole and enter at the south pole. Use the left-hand rule to determine the north pole of a coil. If you grasp a coil with your left hand so that your fingers point in the direction of electron flow, your thumb points toward the north pole of the coil, Figure 4-17. Increasing the number of turns in the wire, or increasing the current through the coil, or both, can strengthen the magnetic field of a coil.

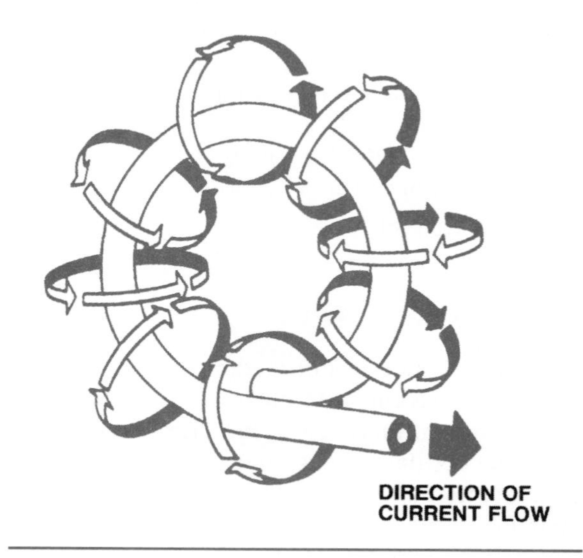

Figure 4-15. Loop conductor. (Delphi Corporation)

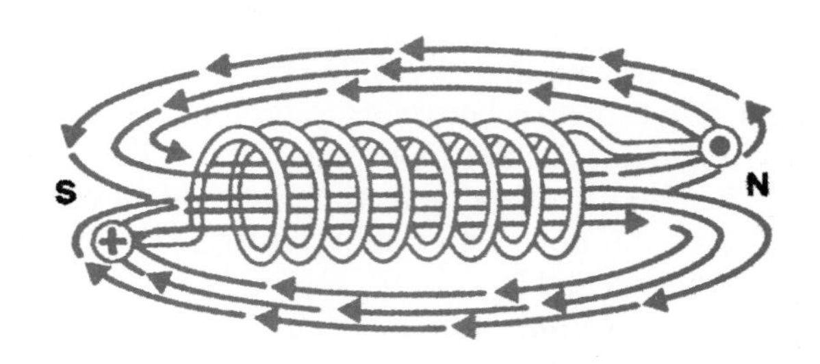

Figure 4-16. Coil conductor.

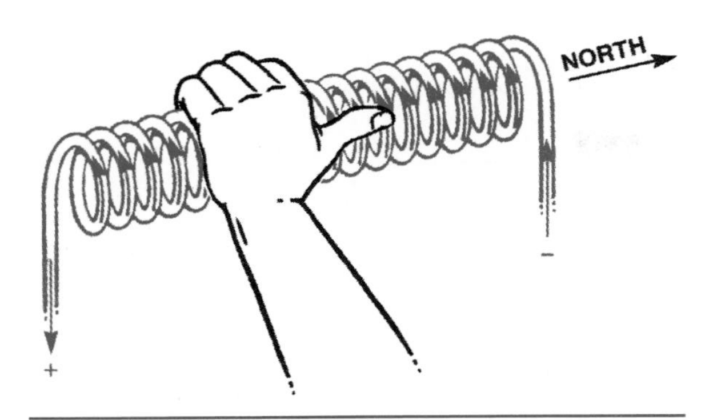

Figure 4-17. Left-hand rule for a coil.

Electromagnets

There is a third way to strengthen the magnetic field surrounding a current-carrying conductor. Because soft iron is very permeable, magnetic flux lines pass through it easily. If a piece of soft iron is placed inside a coiled conductor, the flux lines concentrate in the iron core, as shown in Figure 4-18, rather than pass through the air, which is less permeable. This concentration of force greatly increases the strength of the magnetic field inside the coil. A coils with an iron core is called an **electromagnet.**

Electromagnetic field force is often described as magnetomotive force (mmf). Magnetomotive force (mmf) is determined by the following two factors:

- The amount of current flowed through the conductor.
- The number of turns of wire in a coil.

Magnetomotive force is measured in ampere-turns (At). Ampere-turn factors are the number of windings (complete turns of a wire conductor) and the quantity of current flowed (measured in amperes). If a coil with 100 windings has 1 ampere of current flowed through it, the result will be a magnetic field strength rated at 100 At. A coil with 10 windings could produce an identical magnetic field strength rating with a current flow of 10 amperes. The actual field strength would have to factor in reluctance. In other words, the actual field strength of both the above coils would be increased if the coil windings were to be wrapped around an iron core.

Relays

One common automotive use of electromagnets is in a device called a **relay.** A relay is a control device that allows a small amount of current to

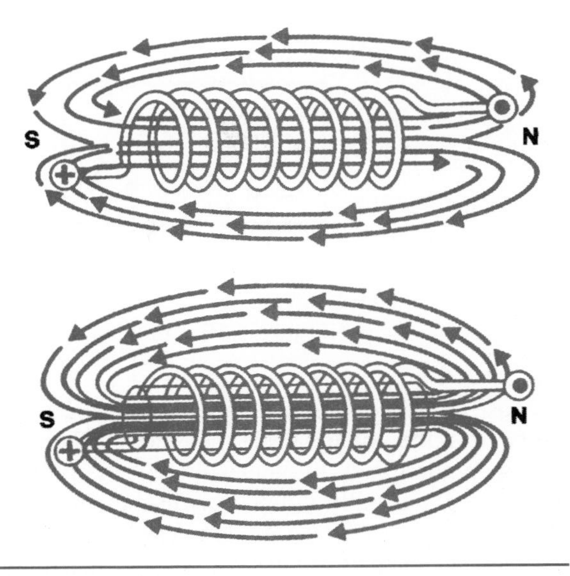

Figure 4-18. Electromagnets.

trigger a large amount of current in another circuit. A simple relay (Figure 4-19) contains an electromagnetic coil in series with a battery and a switch. Near the electromagnet is a movable flat blade, or *armature,* of some material that is attracted by a magnetic field. The armature pivots at one end and is held a small distance away from the electromagnet by a spring (or by the spring steel of the armature itself). A contact point made of a good conductor is attached to the free end of the armature. Another contact point is fixed a small distance away. The two contact points are wired in series with an electrical load and the battery. When the switch is closed, the following occurs:

1. Current travels from the battery through the electromagnet.
2. The magnetic field created by the current attracts the armature, bending it down until the contact points meet.
3. Closing the contacts allows current in the second circuit from the battery to the load.

When the switch is opened, the following occurs:

1. The electromagnet loses its current and its magnetic field.
2. Spring pressure brings the armature back.
3. The opening of the contact points breaks the second circuit.

Relays may also be designed with normally closed contacts that open when current passes through the electromagnet.

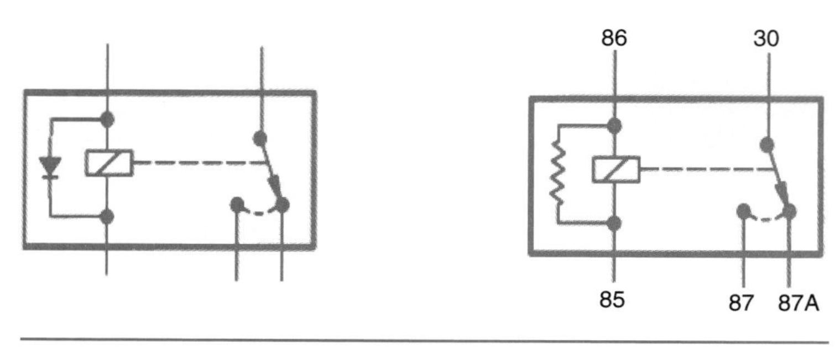

Figure 4-19. Electromagnetic relay.

Most relays contain a device that protects circuitry from the voltage spike that occurs when the coil is de-energized. In older vehicles, the protective device is usually a diode (as in the circuit on the left in Figure 4-19). A diode is a semiconductor device that can be useful in several ways. In a relay, the diode is located in parallel with the coil, where it dissipates the voltage spike. (You'll learn more about how diodes work in a Chapter 10.)

Today many automobile relays include a resistor, rather than a diode, to protect the control circuit (as in the circuit on the right in Figure 4-19). The resistor dissipates the voltage spike in the same way that a diode does. For more information about relays, see the section on "Diagnostic Strategies" in Chapter 4 of the *Shop Manual.*

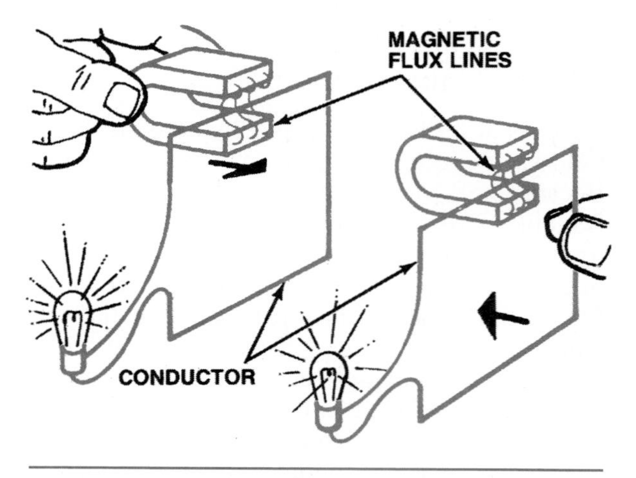

Figure 4-20. Voltage can be induced by the relative motion between a conductor and a magnetic field.

ISO Relays

In Figure 4-19, on the right, the five terminals with specific numbers assigned to them (#85, #86, etc.) show that this relay, like many others now used in vehicles, is an ISO relay. ISO relays, as required by the International Organization for standardization (ISO), are the same size and have the same terminal pattern. They're used in many major-component circuits, and are often located in a vehicle's *underhood junction block* or *power distribution center.* For more information about ISO relays, see the section on "Diagnostic Strategies" in Chapter 4 of the *Shop Manual.*

ELECTROMAGNETIC INDUCTION

Only a decade after the discovery of magnetic fields surrounding current-carrying conductors, more discoveries were made about the relationship between electricity and magnetism. The modem automotive electrical system is based in great part upon the principles of electromagnetic induction discovered in the 1830s. Along with creating a magnetic field with current, it is also possible to create current with a magnetic field. Magnetic flux lines create an electromotive force, or voltage, in a conductor if either the flux lines or the conductor is moving (relative motion). This process is called **electromagnetic induction,** and the resulting electromotive force is called induced voltage (Figure 4-20). If the conductor is in a complete circuit, current exists.

It happens when the flux lines of a magnetic field cut across a wire (or any conductor). It does not matter whether the magnetic field moves or the wire moves. When there is relative motion between the wire and the magnetic field, a voltage is produced in the conductor. The induced voltage causes a current to flow; when the motion stops, the current stops.

Voltage is induced when magnetic flux lines are broken by a conductor (Figure 4-20). This relative motion can be a conductor moving across a magnetic field (as in a DC generator), or a magnetic field moving across a stationary conductor (as in AC generators and ignition coils). In both cases, the induced voltage is caused by relative motion between the conductor and the magnetic flux lines.

Voltage Strength

Induced voltage depends upon magnetic flux lines being broken by a conductor. The strength of the voltage depends upon the rate at which the flux lines are broken. The more flux lines broken per unit of time, the greater the induced voltage. If a single conductor breaks one million flux lines per second, 1 volt is induced.

There are four ways to increase induced voltage, as follows:

- Increase the strength of the magnetic field, so there are more flux lines.
- Increase the number of conductors that are breaking the flux lines.
- Increase the speed of the relative motion between the conductor and the flux lines so that more lines are broken per time unit.
- Increase the angle between the flux lines and the conductor to a maximum of 90 degrees.

No voltage is induced if the conductors move parallel to, and do not break any, flux lines, as shown in Figure 4-21. Maximum voltage is induced if the conductors break flux lines at 90 degrees (Figure 4-22). Induced voltage varies proportionately at angles between 0 and 90 degrees.

We know voltage can be electromagnetically induced, and we can measure it and predict its behavior. Induced voltage creates current. The direction of induced voltage (and the direction in which current moves) is called polarity and depends upon the direction of the flux lines and the direction of relative motion. An induced current moves so that its magnetic field opposes the motion that induced the current. This principle, called Lenz's Law, is based upon Newton's observation that every action has an equal and opposite reaction. The relative motion of a conductor and a magnetic field is opposed by the magnetic field of the current it has induced. This is why induced current can move in either direction, and it is an important factor in the design and operation of voltage sources such as alternators.

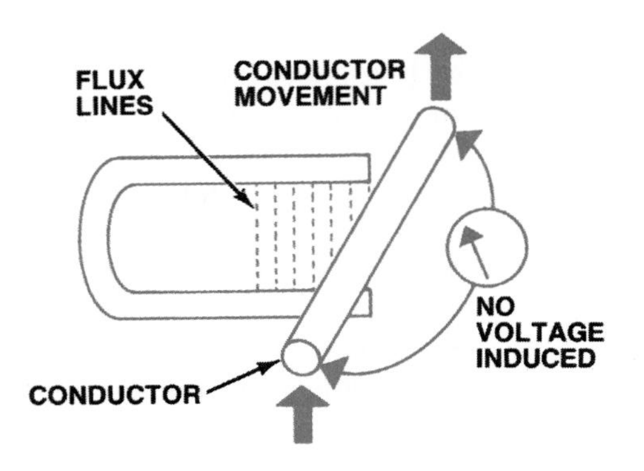

Figure 4-21. No voltage is induced if no flux lines are broken.

DC Generator Principles

The principles of electromagnetic induction are employed in generators for producing DC current. The basic components of a DC generator are shown in Figure 4-23. A framework composed of laminated iron sheets or an other ferromagnetic metal has a coil wound on it to form an electromagnet. When current flows through this coil, magnetic fields are created between the pole pieces, as shown. Permanent magnets could also be employed instead of the electromagnet.

To simplify the initial explanation, a single wire loop is shown between the north and south pole pieces. When this wire loop is turned within the magnetic fields, it cuts the lines of force and a voltage is induced. If there is a complete circuit from the wire loop, current will flow. The wire loop is connected to a split ring known as a **commutator**, and carbon brushes pick off the electric energy as the commutator rotates. Connecting wires from the carbon brushes transfer the energy to the load circuit.

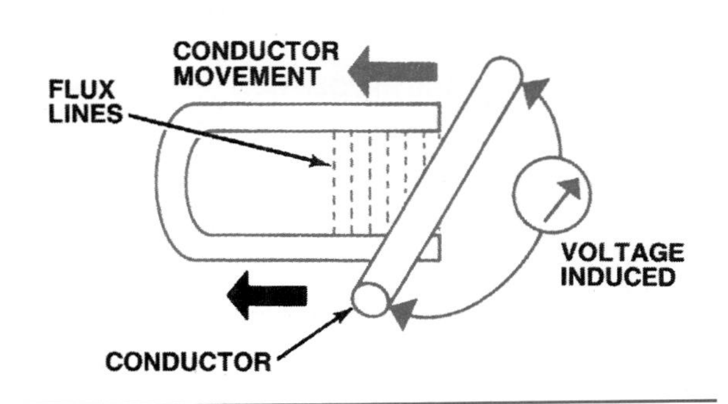

Figure 4-22. Maximum voltage is induced when the flux lines are broken at a 90-degree angle.

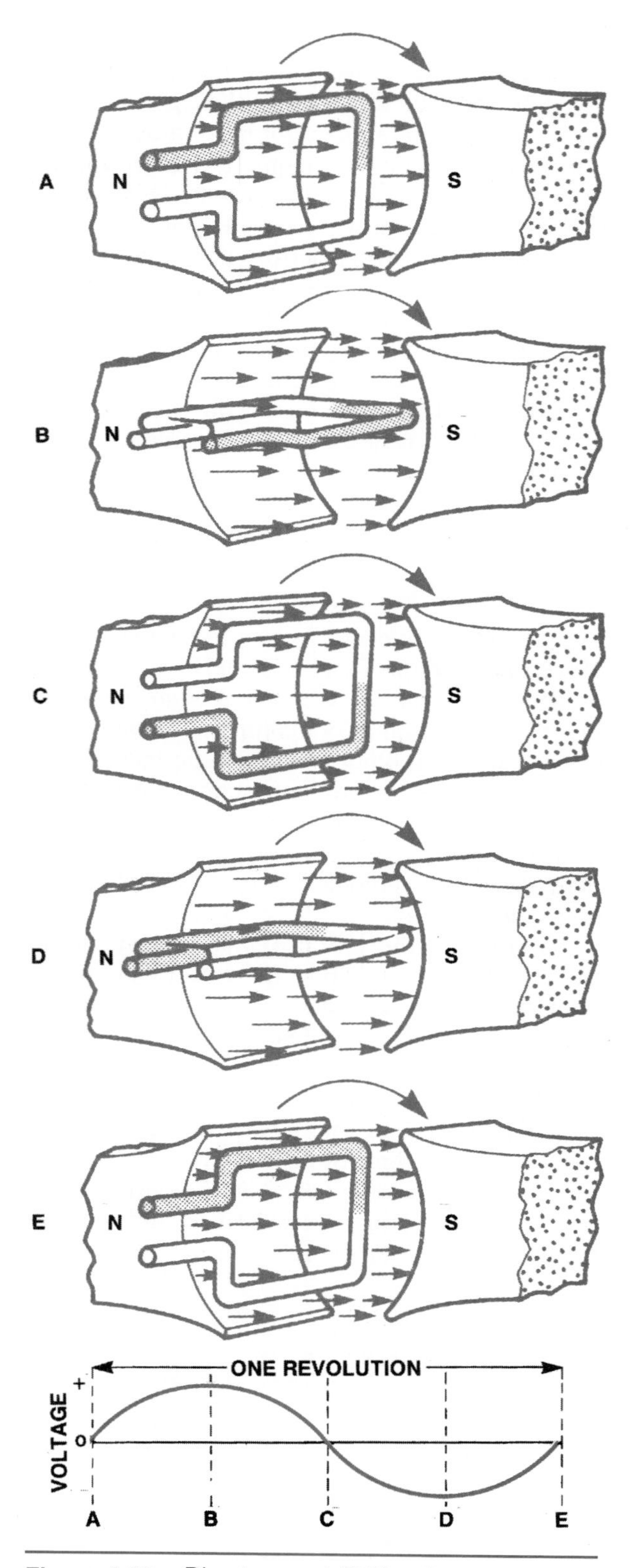

Figure 4-23. Direct current (DC) voltage generated in a rotating loop conductor.

When the wire loop makes a half-turn, the energy generated rises to a maximum level, then drops to zero, as shown in parts B and C of Figure 4-23. As the wire loop completes a full rotation the induced voltage would reverse itself and the current would flow in the opposite direction (AC current) after the initial half-turn. To provide for an output having a single polarity (DC current), a split-ring commutator is used. Thus, for the second half-turn, the carbon brushes engage commutator segments opposite to those over which they slid for the first half-turn, keeping the current in the same direction. The output waveform is not a steady-level DC, but rises and falls to form a pattern referred to as *pulsating DC*. Thus, for a complete 360-degree turn of the wire loop, two waveforms are produced, as shown in Figure 4-23.

The motion of a conductor may induce DC voltage across a stationary magnetic field. This is the principle used to change mechanical energy to electrical energy in a generator. A looped conductor is mechanically moved within the magnetic field created by stationary magnets, as in Figure 4-23. In Figure 4-23A, the voltage is zero because the conductor motion is parallel to the flux lines. As the conductor moves from A to B, the voltage increases because it is cutting across the flux lines. At B, the voltage is at a maximum because the conductor is moving at right angles to the flux lines, breaking the maximum number.

From position B to C, the voltage decreases to zero again because fewer lines are broken. At C, the conductor is again parallel to the flux lines. As the conductor rotates from C to D, voltage increases. However, the induced voltage is in the opposite direction because the conductor is cutting the flux lines in the opposite direction. From position D to E, the cycle begins to repeat. Figure 4-23 shows how voltage is induced in a loop conductor through one complete revolution in a magnetic field. The induced voltage is called alternating current (AC) voltage because it reverses direction every half cycle, as shown on the graph at the bottom of the figure. Because automotive battery voltage is always in one direction, the current it produces always flows in one direction. This is called direct current (DC). Alternating current cannot be used to charge the battery, so the AC must be changed (rectified) to DC. This is done in a generator by the commutator. In a simple, single-loop generator, the commutator would be a split ring of conductive material connected to the ends of the conductor. Brushes of conductive material ride on the surface of the two commutator segments.

Induced current travels from the conductor through the commutator and out through the brushes (Figure 4-24A). At the instant the looped conductor is turned so that the induced current changes direction (Figure 4-24B), the commutator also rotates under the brushes, so that the brushes now contact the opposite commutator segments. Current now flows out of the other half of the commutator, but the same brush is there to receive it. This design is called a brush-rectified, or commutator-rectified, generator, and the output is called pulsating direct current. Actual DC generators have many armature windings and commutator segments. The DC voltages overlap to create an almost continuous DC output.

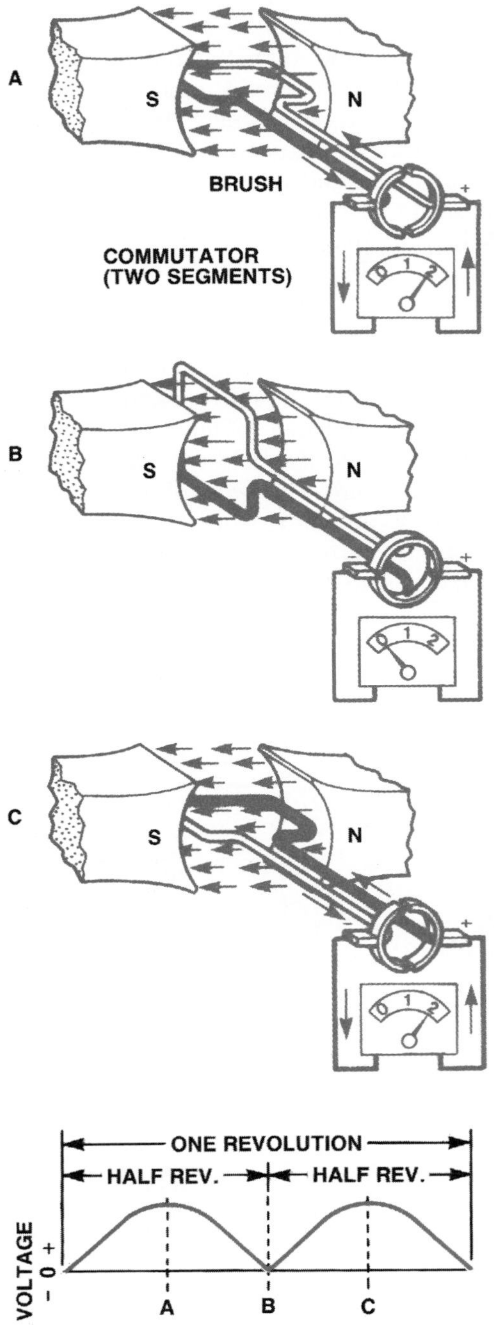

Figure 4-24. The commutator and brushes conduct pulsating direct current from the looped conductor.

AC Generator (Alternator) Principles

Since 1960, virtually all automobiles have used an **AC generator (alternator)**, in which the movement of magnetic lines through a stationary conductor (Figure 4-25) generates voltage. A magnet called a rotor is turned inside a stationary looped conductor called a stator (Figure 4-25). The induced current, like that of a DC generator, is constantly changing its direction. The rotation of the magnetic field causes the stator to be cut by flux lines, first in one direction and then the other. The AC must be rectified to match the battery DC by using diodes, which conduct current in only one direction. This design is called a diode-rectified alternator. We will study diodes in Chapter 10 and alternators in Chapter 8 of this book.

See the section on "Diagnostic Strategies" in Chapter 4 of the *Shop Manual*.

Self-Induction

Up to this point, our examples have depended upon mechanical energy to physically move either the conductor or the magnetic field. Another form of relative motion occurs when a magnetic field is forming or collapsing. When current begins to flow in a coil, the flux lines expand as the magnetic

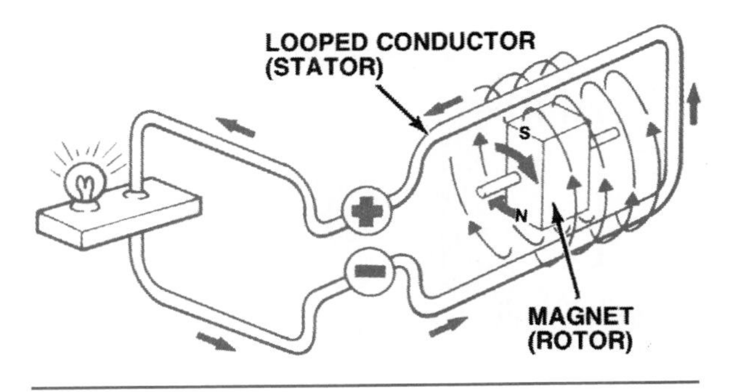

Figure 4-25. A simplified alternator.

field forms and strengthens. As current increases, the flux lines continue to expand, cutting through the wires of the coil and actually inducing another voltage within the same coil, which is known as **self-induction.** Following *Lenz's Law*, this self-induced voltage tends to *oppose* the current that produces it. If the current continues to increase, the second voltage opposes the increase. When the current stabilizes, the counter voltage is no longer induced, because there are no more expanding flux lines (no relative motion). When current to the coil is shut off, the collapsing magnetic flux lines self-induce a voltage in the coil that tries to maintain the original current. The self-induced voltage *opposes* and *slows down* the *decrease* in the original current. The self-induced voltage that opposes the source voltage is called counterelectromotive force (CEMF).

Mutual Induction

When two coils are close together, energy may be transferred from one to the other by magnetic coupling called mutual induction. **Mutual induction** means that the expansion or collapse of the magnetic field around one coil induces a voltage in the second coil. Usually, the two coils are wound on the same iron core. One coil winding is connected to a battery through a switch and is called the primary winding. The other coil winding is connected to an external circuit and is called the secondary winding.

When the switch is open (Figure 4-26A), there is no current in the primary winding. There is no magnetic field and, therefore, no voltage in the secondary winding. When the switch is closed (Figure 4-26B), current is introduced and a magnetic field builds up around both windings. The primary winding thus changes electrical energy from the battery into magnetic energy of the expanding field. As the field expands, it cuts across the secondary winding and induces a voltage in it. A meter connected to the secondary circuit shows current.

When the magnetic field has expanded to its full strength (Figure 4-26C), it remains steady as long as the same amount of current exists. The flux lines have stopped their cutting action. There is no relative motion and no voltage in the secondary winding, as shown on the meter. When the switch is opened (Figure 4-26D), primary current stops and the field collapses. As it does, flux lines cut across the secondary winding, but in the opposite direction. This induces a secondary voltage with current in the opposite direction, as shown on the meter.

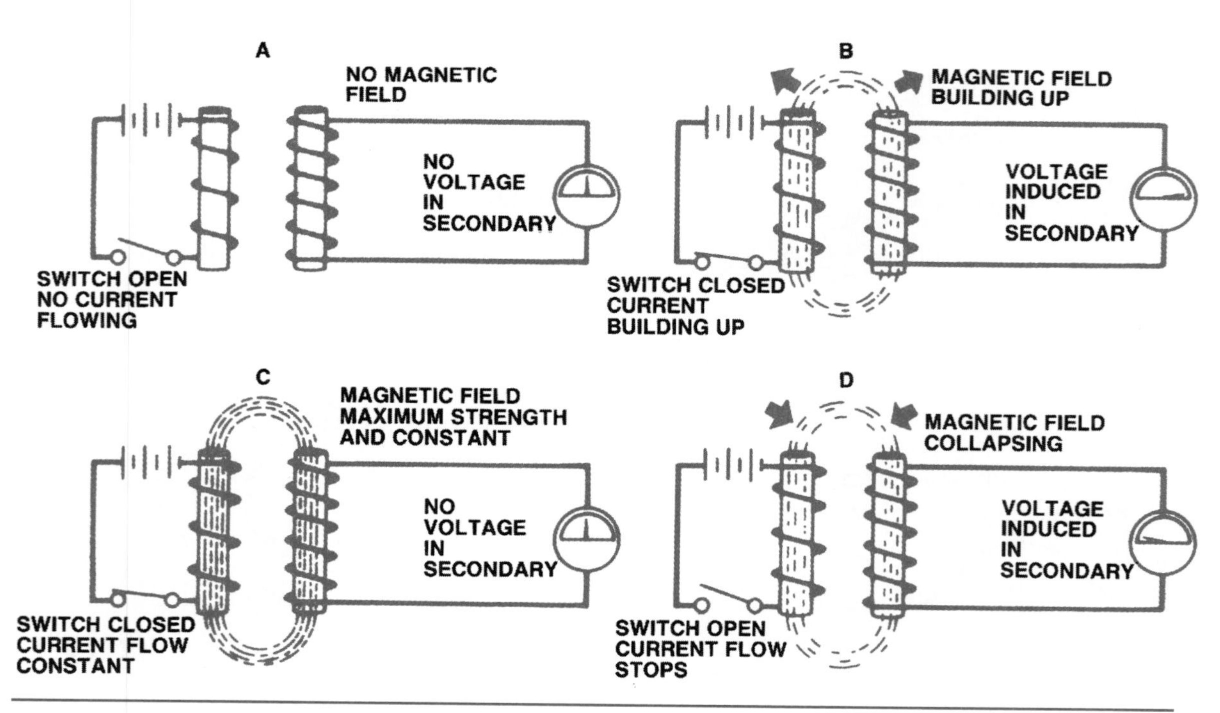

Figure 4-26. Mutual induction.

TRANSFORMERS

Transformers (Figure 4-27) are electrical devices that work on the principle of mutual induction. Transformers are typically constructed of a primary winding (coil), secondary winding (coil) and a common core. When alternating current or pulsating direct current is applied to the primary winding, a voltage is induced in the secondary winding. The induced voltage is the result of the primary winding's magnetic field collapsing. The principle of a transformer is essentially that of flowing current through a primary coil and inducing current flow in a secondary or output coil. Variations on this principle would be coils that are constructed with a movable core, which permits their inductance to be varied.

Transformers can be used to step up or step down the voltage. In a step-up transformer, the voltage in the secondary winding is increased over the voltage in the primary winding, due to the secondary winding having more wire turns than the primary winding. Increasing the voltage through the use of a transformer results in decreased current in the secondary winding. An ignition coil is an example of a step-up transformer operating on pulsating direct current. A transformer that steps down the voltage has more wire turns in the primary winding than in the secondary winding. These transformers produce less voltage in the secondary but produce increased current.

Mutual induction is used in ignition coils (Figure 4-28), which are basically step-up transformers. In an ignition coil, low-voltage primary current induces a very high secondary voltage because of the different number of turns in the primary and secondary windings.

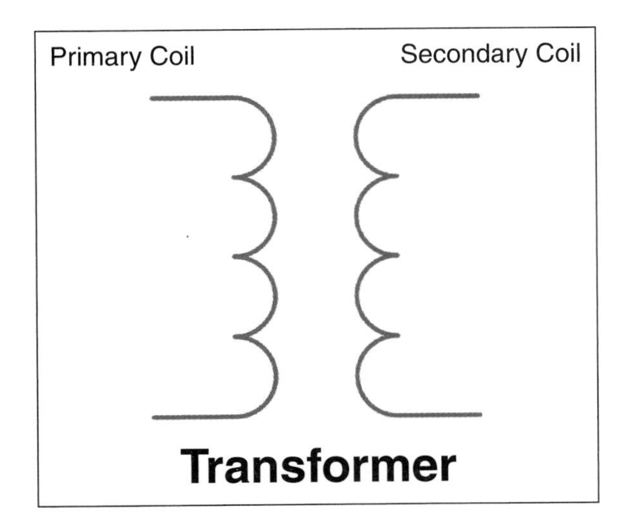

Figure 4-27. Transformer.

ELECTROMAGNETIC INTERFERENCE (EMI) SUPPRESSION

Until the advent of the onboard computer, **electromagnetic interference (EMI)** was not a source of real concern to automotive engineers. The problem was mainly one of radiofrequency interference (RFI), caused primarily by the use of secondary ignition cables containing a low-resistance metal core. These cables produced electrical impulses that interfered with radio and television reception.

Radiofrequency interference was recognized in the 1950s and brought under control by the use of secondary ignition cables containing a high-resistance, nonmetallic core made of carbon, linen, or fiberglass strands impregnated with graphite. In addition, some manufacturers

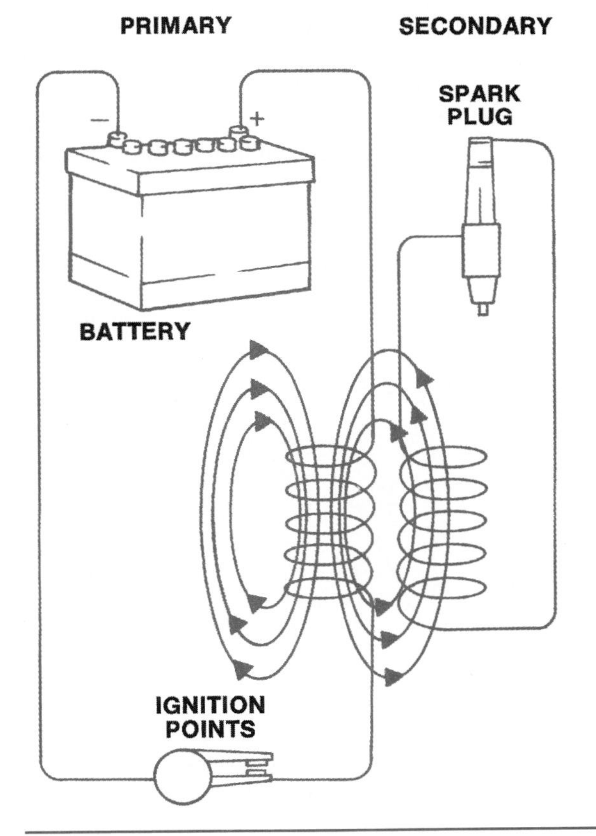

Figure 4-28. Mutual induction in the ignition coil produces voltage across the spark plugs.

even installed a metal shield inside their distributors to further reduce RFI radiation from the breaker points, condensers, and rotors.

As the use of electronic components and systems increased, the problem of electromagnetic interference reappeared with broader implications. The low-power digital integrated circuits now in use are extremely sensitive to EMI signals that were of little or no concern before the late 1970s. For more information about EMI, see the "Logic Probe" section in Chapter 4 of the *Shop Manual.*

Interference Generation and Transmission

Whenever there is current in a conductor, an electromagnetic field is created. When current stops and starts, as in a spark plug cable or a switch that opens and closes, field strength changes. Each time this happens, it creates an electromagnetic signal wave. If it happens rapidly enough, the resulting high-frequency signal waves, or EMI, interfere with radio and television transmission or with other electronic systems such as those under the hood. This is an undesirable side effect of the phenomenon of electromagnetism. Figure 4-29 shows common sources of EMI on an automobile.

Static electric charges caused by friction of the tires with the road, or the friction of engine drive belts contacting their pulleys, also produce EMI. Drive axles, drive shafts, and clutch or brake lining surfaces are other sources of static electric charges.

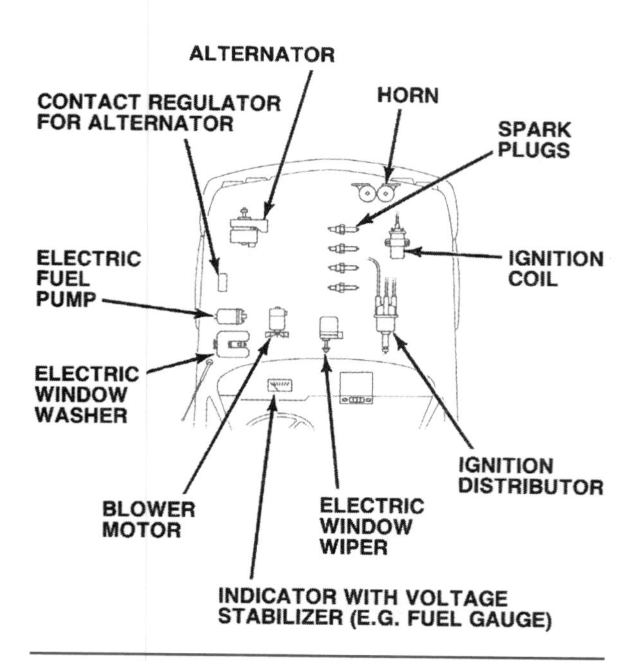

Figure 4-29. Sources of electromagnetic interference (EMI) in an automobile.

There following four ways of transmitting EMI can all be found in an automobile:

- Conductive coupling through circuit conductors (Figure 4-30).
- Capacitive coupling through an electrostatic field between two conductors (Figure 4-31).
- Inductive coupling as the magnetic fields between two conductors form and collapse (Figure 4-32).
- Electromagnetic radiation (Figure 4-33).

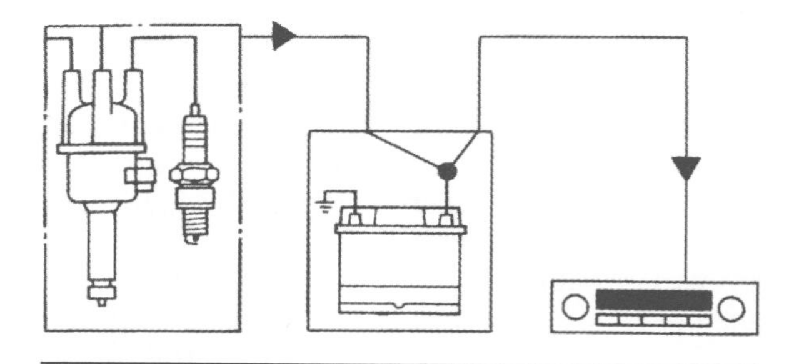

Figure 4-30. Wiring between the source of the interference and the receiver transmits conductive-coupling interference.

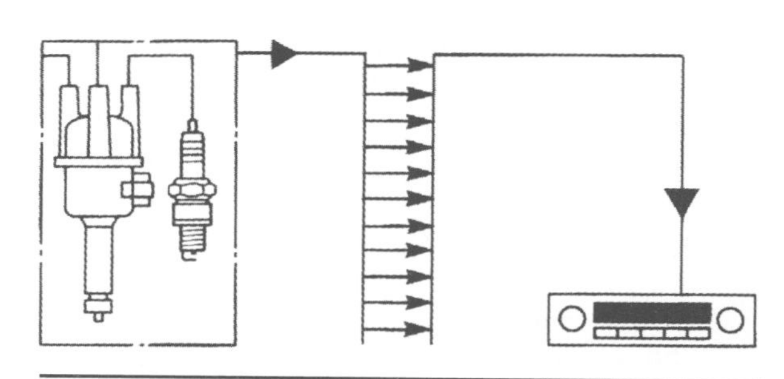

Figure 4-31. Capacitive field between adjacent wiring transmits conductive-coupling interference.

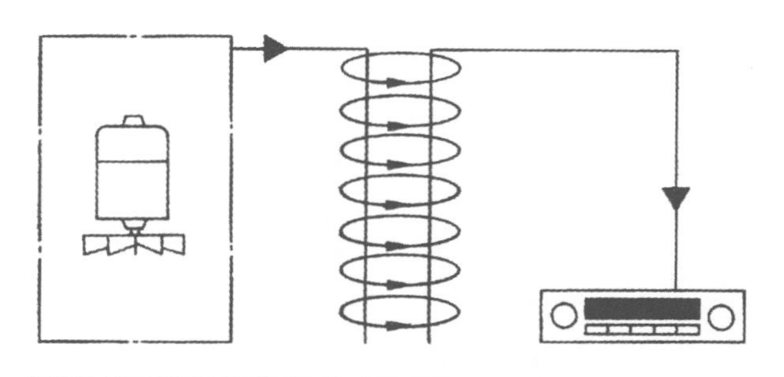

Figure 4-32. Inductive-coupling interference transmitted by an electromagnetic field between adjacent wiring.

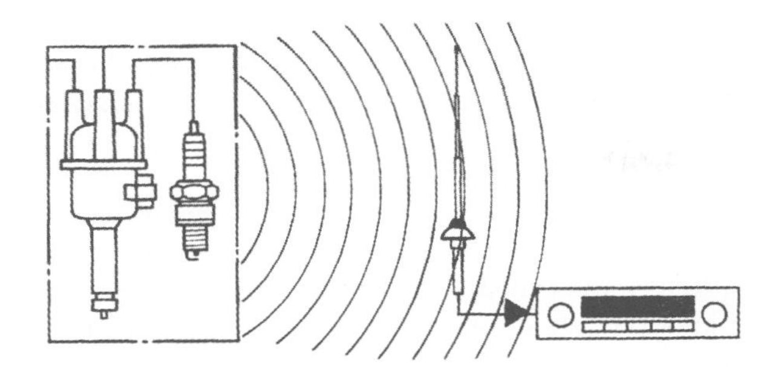

Figure 4-33. Radiation interference: EMI waves travel through the air and are picked up by wiring that acts as a receiving antenna.

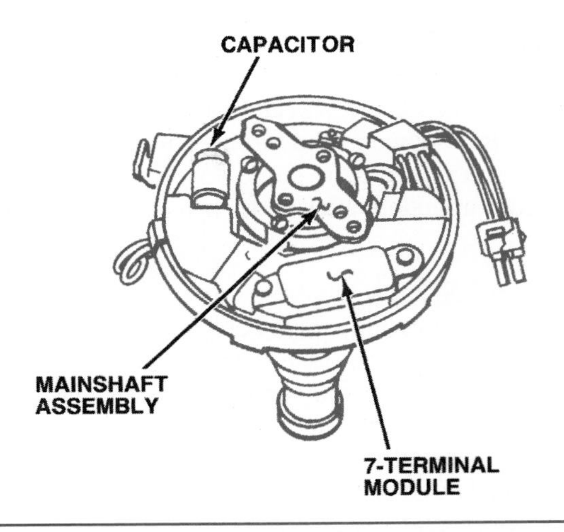

Figure 4-34. The capacitor attached to a GM HEI ignition module protects the module from EMI.

EMI Suppression Devices

Just as there are four methods of EMI transmission, there are four general ways in which EMI is reduced, as follows:

- By the addition of resistance to conductors, which suppresses conductive transmission and radiation
- By the use of capacitors and radio choke coil combinations to reduce capacitive and inductive coupling
- By the use of metal or metalized plastic shielding, which reduces EMI radiation in addition to capacitive and inductive coupling
- By an increased use of ground straps to reduce conductive transmission and radiation by passing the unwanted signals to ground

Resistance Suppression

Adding resistance to a circuit to suppress RFI works only for high-voltage systems (for example, changing the conductive core of ignition cables). The use of resistance to suppress interference in low-voltage circuits creates too much voltage drop and power loss to be efficient.

The only high-voltage system on most vehicles is the ignition secondary circuit. Although this is the greatest single source of EMI, it is also the easiest to control by the use of resistance spark plug cables, resistor spark plugs, and the silicone grease used on the distributor cap and rotor of some electronic ignitions.

Suppression Capacitors and Coils

Capacitors are installed across many circuits and switching points to absorb voltage fluctuations. Among other applications, they are used as follows:

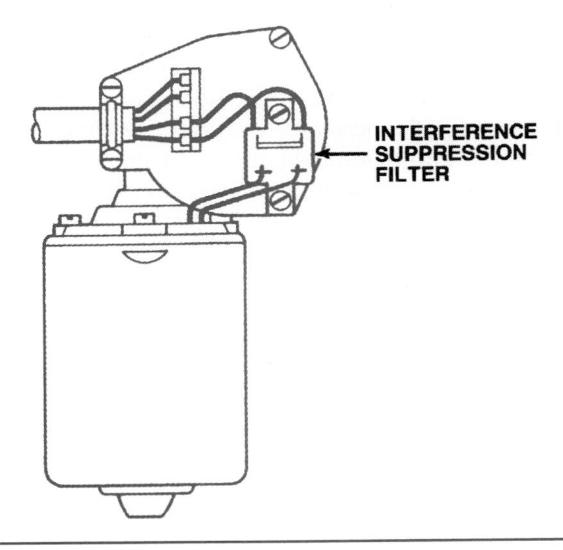

Figure 4-35. Interference-suppression capacitors and choke coils are attached to electric motors, like the Bosch wiper motor shown. (Reprinted by permission of Robert Bosch GmbH)

- Across the primary circuit of some electronic ignition modules (Figure 4-34)
- Across the output terminal of most alternators
- Across the armature circuit of electric motors

Radio choke coils reduce current fluctuations resulting from self-induction. They are often combined with capacitors to act as EMI filter circuits for windshield wiper and electric fuel pump motors (Figure 4-35). Filters may also be incorporated in wiring connectors.

Shielding Metal

Shields, such as the ones used in breaker point distributors, block the waves from components that create RFI signals. The circuits of onboard computers are protected to some degree from external electromagnetic waves by their metal housings.

Ground Straps

Ground or bonding straps between the engine and chassis of an automobile help suppress EMI conduction and radiation by providing a low-resistance circuit ground path. Such suppression ground straps are often installed between rubber-mounted components and body parts, Figure 4-36. On some models ground straps are installed between body parts, such as the hood and a fender panel where no electrical circuit exists, Figure 4-36. In such a case, the strap has no other job than to suppress EMI. Without it, the sheet-metal body and hood could function as a large capacitor. The space between the fender and hood could form an electrostatic field and couple with the computer circuits in the wiring harness routed near the fender panel. For more information about ground straps, see the section on "Diagnostic Strategies" in Chapter 4 of the *Shop Manual.*

EMI Suppression

Interference suppression is now a critical automotive engineering task because the modem automobile has an increased need for EMI suppression.

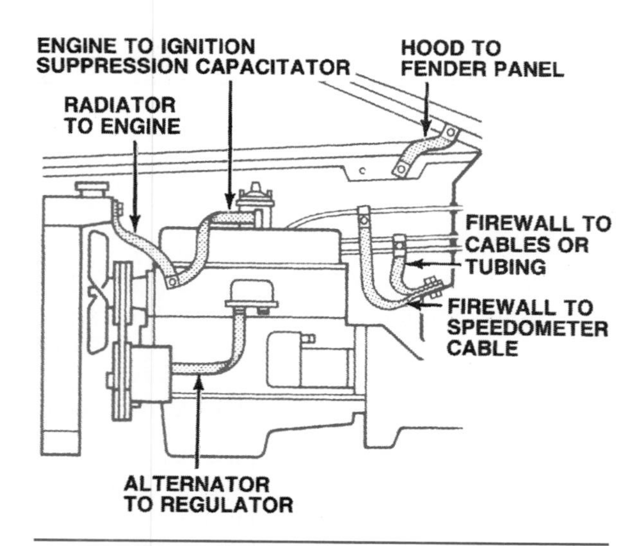

Figure 4-36. Ground straps are installed in many areas of the engine compartment to suppress EMI. (Reprinted by permission of Robert Bosch GmbH)

The increasing use of cellular telephones, as well as onboard computer systems, are only two of the factors that have made interference suppression extremely important.

Even small amounts of EMI can disrupt the operation of an onboard digital computer, which operates on voltage signals of a few millivolts (thousandths of a volt) and milliamperes (thousandths of an ampere) of current. Any of the interference transmission modes discussed earlier are capable of creating false voltage signals and excessive current in the computer systems. False voltage signals disrupt computer operation, while excessive current causes permanent damage to micro-electric circuitry.

As the complexity and number of electronic systems continues to increase, manufacturers are using multiplex wiring systems to reduce the size and number of wiring harnesses, which also reduces EMI. Multiplexing is a method of sending more than one electrical signal over the same channel.

SUMMARY

Electricity can be generated in several ways. The most important way for automotive use is by magnetism. Magnetism is a form of energy caused by the alignment of atoms in certain materials. It is indicated by the ability to attract iron. Some magnetic materials exist in nature; others can be artificially magnetized. The magnetic properties of some metals, such as iron, are due to electron motion within the atomic structure. Reluctance is resistance to the movement of magnetic lines of force: iron cores have permeability and are used to reduce reluctance in electromagnetic fields.

Lines of force, called flux lines, form a magnetic field around a magnet. Flux lines exit the north pole and enter the south pole of a magnet. Magnetic flux lines also surround electrical conductors. As current increases, the magnetic field of a conductor becomes stronger. Voltage can be generated by the interaction of magnetic fields around conductors.

The relative movement of a conductor and a magnetic field generates voltage. This process is called induction. Either the conductor or the magnetic field may be moving. The strength of the induced voltage depends on the strength of the magnetic field, the number of conductors, the speed of the relative motion, and the angle at which the conductors cut the flux lines. Electromagnetic induction is used in generators, alternators, electric motors, and coils. Magnetomotive force (mmf) is a measure of electromagnetic field strength. The unit of measure used for mmf is ampere-turns. The principle of a transformer can be summarized by describing it as flowing current through a primary coil and inducing a current flow in a secondary or output coil.

Electromagnetism can also generate electromagnetic interference (EMI) and radiofrequency interference (RFI). Such interference can disrupt radio and television signals, as well as electronic system signals. Many devices are used to suppress this interference in automotive systems.

Review Questions

1. Which of the following can store energy in the form of an electromagnetic charge?
   a. Thermocouple
   b. Induction coil
   c. Potentiometer
   d. Capacitor
2. Current flows through a heated thermocouple because of the two-way flow of
   a. Electrons between dissimilar materials
   b. The blockage of free electrons between the metals
   c. The one-way transfer of free electrons between the metals
   d. Random electron flow
3. Which of the following forms of generating electricity is *not* widely used in an automobile?
   a. Heat
   b. Pressure
   c. Chemistry
   d. Magnetism
4. The lines of force of a magnet are called
   a. Flux lines
   b. Magnetic polarity
   c. Magnetic lines
   d. Flux density
5. A material through which magnetic force can easily flow has a high
   a. Reluctance
   b. Permeability
   c. Capacitance
   d. Magnetic attraction
6. The left-hand rule says that if you grasp a conductor in your left hand with your thumb pointing in the direction of the electron (– to +) flow,
   a. Your fingers will point in the direction of the magnetic flux lines.
   b. Your fingers will point in the opposite direction of the magnetic flux lines.
   c. Your fingers will point at right angles to the magnetic flux lines.
   d. Your fingers will point at a 45-degree angle to the magnetic flux lines.
7. The left-hand rule is useful to determine
   a. The direction of current flow
   b. The length of the magnetic flux lines
   c. The strength of the magnetic field
   d. Flux density
8. When two parallel conductors carry electrical current in opposite directions, their magnetic fields will
   a. Force them apart
   b. Pull them together
   c. Cancel each other out
   d. Rotate around the conductors in the same direction
9. The motor principle of changing electrical energy into mechanical energy requires
   a. Two semiconductors carrying current in opposite directions
   b. Two semiconductors carrying current in the same direction
   c. Two conductors carrying current in opposite directions
   d. Two conductors carrying current in the same direction
10. Which of the following will *not* increase induced voltage?
    a. Increasing the strength of the magnetic field
    b. Increasing the number of conductors cutting flux lines
    c. Increasing the speed of the relative motion between the conductor and the flux lines
    d. Increasing the angle between the flux lines and the conductor beyond 90 degrees
11. The ________ in an automobile DC generator rectifies AC to DC.
    a. Armature
    b. Commutator
    c. Field coil
    d. Loop conductor
12. In an ignition coil, low-voltage primary current induces a very high secondary voltage because of
    a. The different number of wire turns in the two windings.
    b. An equal number of turns in the two windings
    c. The constant current flow through the primary winding
    d. The bigger wire in the secondary winding
13. To reduce EMI, manufacturers have done all of the following except:
    a. Using low resistance in electrical systems
    b. Installing metal shielding in components
    c. Increasing the use of ground straps
    d. Using capacitors and choke coils

5

Series, Parallel, and Series-Parallel Circuits

LEARNING OBJECTIVES

Upon completion and review of this chapter, you should be able to:

- Define a series circuit.
- Identify the series circuit laws and apply Ohm's Law for voltage, current, and resistance.
- Define a parallel circuit.
- Identify the parallel-circuit laws and apply Ohm's Law for voltage, current, and resistance.
- Define Kirchhoff's Voltage Drop Law and Current Law.
- Define a series-parallel circuit.
- Identify the series and parallel circuit loads of a series-parallel circuit.
- Using Ohm's Law, solve series-parallel circuits for voltage, current, and resistance.

KEY TERMS

Circuit
Kirchhoff's Law of Current
Kirchhoff's Law of Voltage Drops
Parallel Circuit
Series Circuit
Series-Parallel Circuits
Voltage Drop

INTRODUCTION

In this chapter you will apply the individual series and parallel circuit laws learned in previous chapters to a combination circuit consisting of some components connected in series and some in parallel. Series/parallel electrical circuits seem complicated, but are in fact fairly simple to understand if you remember which circuit laws apply to each circuit load component. Most vehicle electrical circuits used today contain several series-parallel circuits, or portions of a series-parallel circuit.

BASIC CIRCUITS

In Chapter 3, we explained that a **circuit** is a path for electric current (Figure 5-1). Current flows from one end of a circuit to the other when the ends are connected to opposite charges (positive

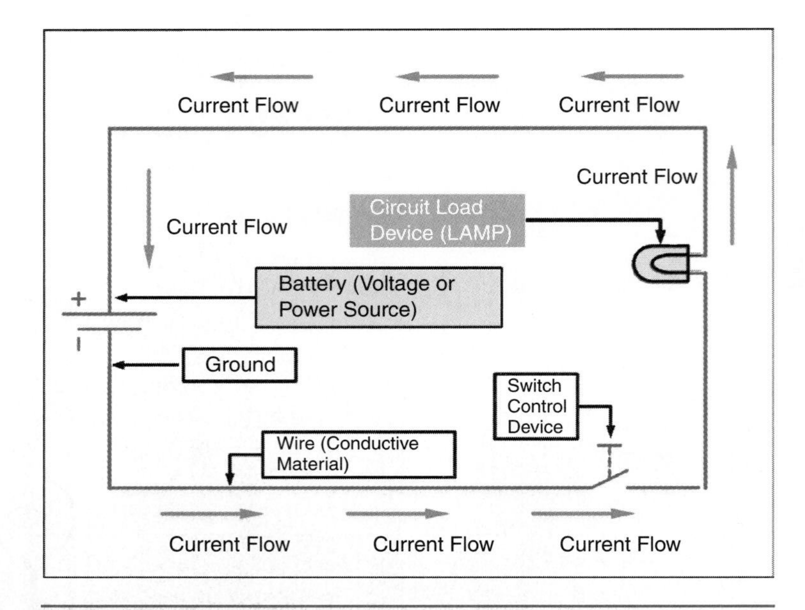

Figure 5-1. Basic circuit.

and negative). We usually call these ends *power* and *ground.* Current flows only in a closed or completed circuit. If there is a break somewhere in the circuit, current cannot flow. We usually call a break in a circuit an *open.* Every automotive circuit contains a source of power, protection, a load, controls, wires (conductive material) and a ground. These elements are connected to each other with conductors.

SERIES CIRCUIT

A **series circuit** (Figure 5-2) is the simplest kind of circuit. Automotive systems almost never include a pure series circuit. A series circuit may have several components such as switches, resistors, and lamps, but they are connected in such a manner that there is only one path for current flow through the circuit. These are characteristics of all series circuits:

- Voltage drops add up to the source voltage.
- There is only one path for current flow.
- The same current flows through every component. In other words, you would get the same current measurement at any point along the circuit.
- Since there is only one path, an open anywhere in the circuit stops current flow.
- Individual resistances add up to the total resistance.

For more information about series circuits, see the section on "Circuit Devices" in Chapter 5 of the *Shop Manual.*

PARALLEL CIRCUIT

A **parallel circuit** is one with multiple paths for current flow, meaning that the components in the circuit are connected so that current flow can flow through a component without having first flowed through other components in the circuit. These are characteristics of all parallel circuits (Figure 5-3):

- There is more than one path for current flow. Each current path is called a branch.
- All of the branches connect to the same positive terminal and the same negative terminal. This means the same voltage is applied to all of the branches.
- Each branch drops the same amount of voltage, regardless of resistance.
- The current flow in each branch can be different depending on the resistance. Total current in the circuit equals the sum of the branch currents.
- The total resistance is *always* less than the smallest resistance in any branch.

We call the top segment of this circuit the *main line* because it's the lead connecting the voltage source

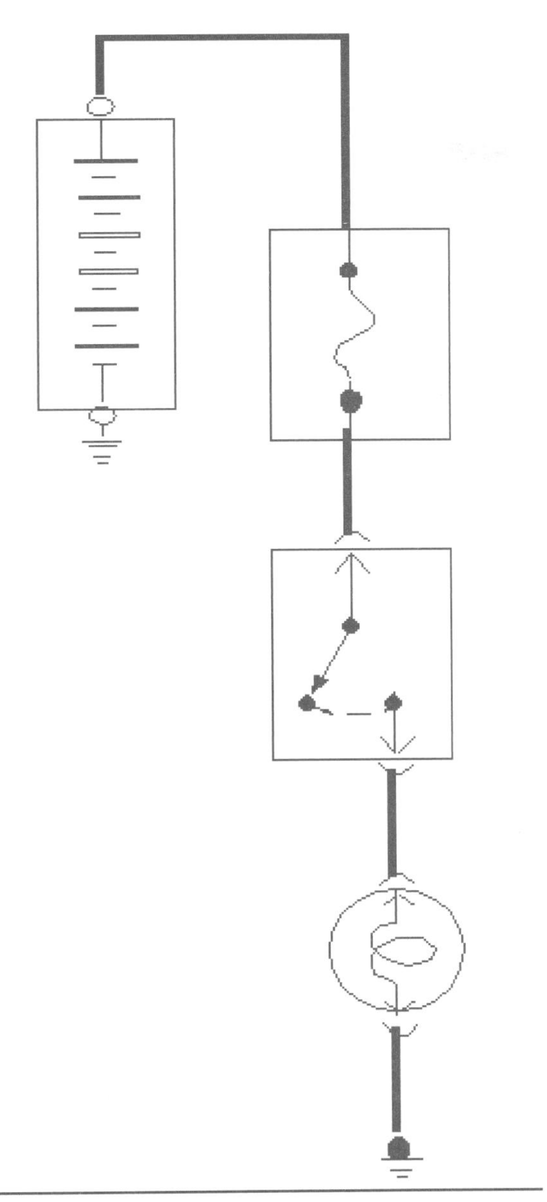

Figure 5-2. Series circuit. (GM Service and Parts Operations)

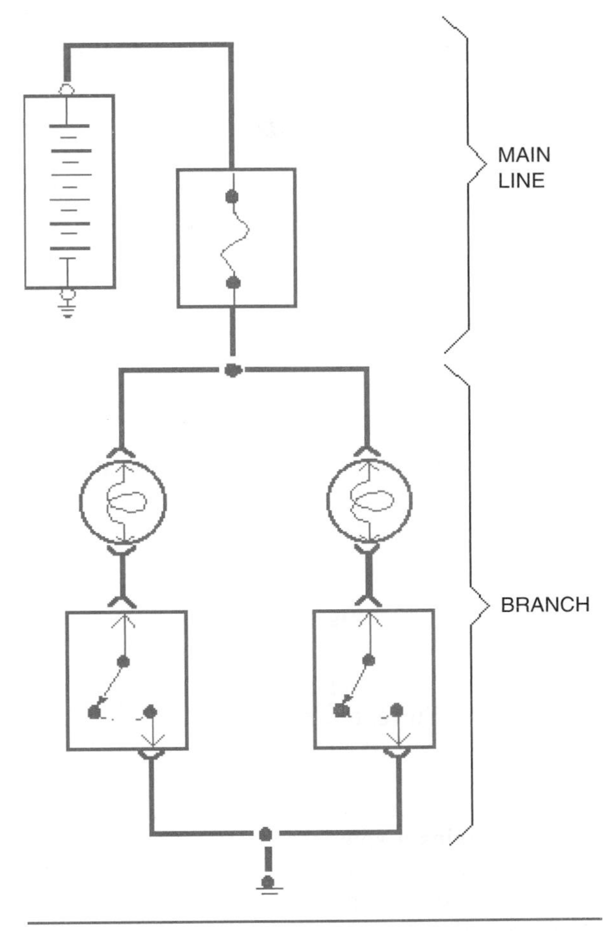

Figure 5-3. Parallel circuit. (GM Service and Parts Operations)

to the other branches. For more information about parallel circuits, see the section on "Circuit Protectors" in Chapter 5 of the *Shop Manual*.

SERIES CIRCUIT VOLTAGE DROPS

As current passes through resistance, energy is converted. It's tempting to say that energy is used up, but that's not strictly accurate. In truth, energy can't be used up; it can only be converted to some other form, such as heat, motion, or light. In any case, the effect of this change in energy is that the voltage before a resistance is greater than the voltage after the resistance. We call this a **voltage drop,** and we usually talk about a voltage drop across a resistance or load. As electricity moves through a resistance or load, there is a change in potential, but the current does not change.

Using a circuit schematic, along with Ohm's Law, we can calculate the voltage drop across a single resistance. The total of all voltage drops in a series circuit (as shown in Figure 5-4), is always equal to the source voltage (also called the applied voltage). This is known as **Kirchhoff's Law of Voltage Drops.**

Kirchhoff's Law Of Voltage Drops (Kirchhoff's 2nd Law) states that voltage will

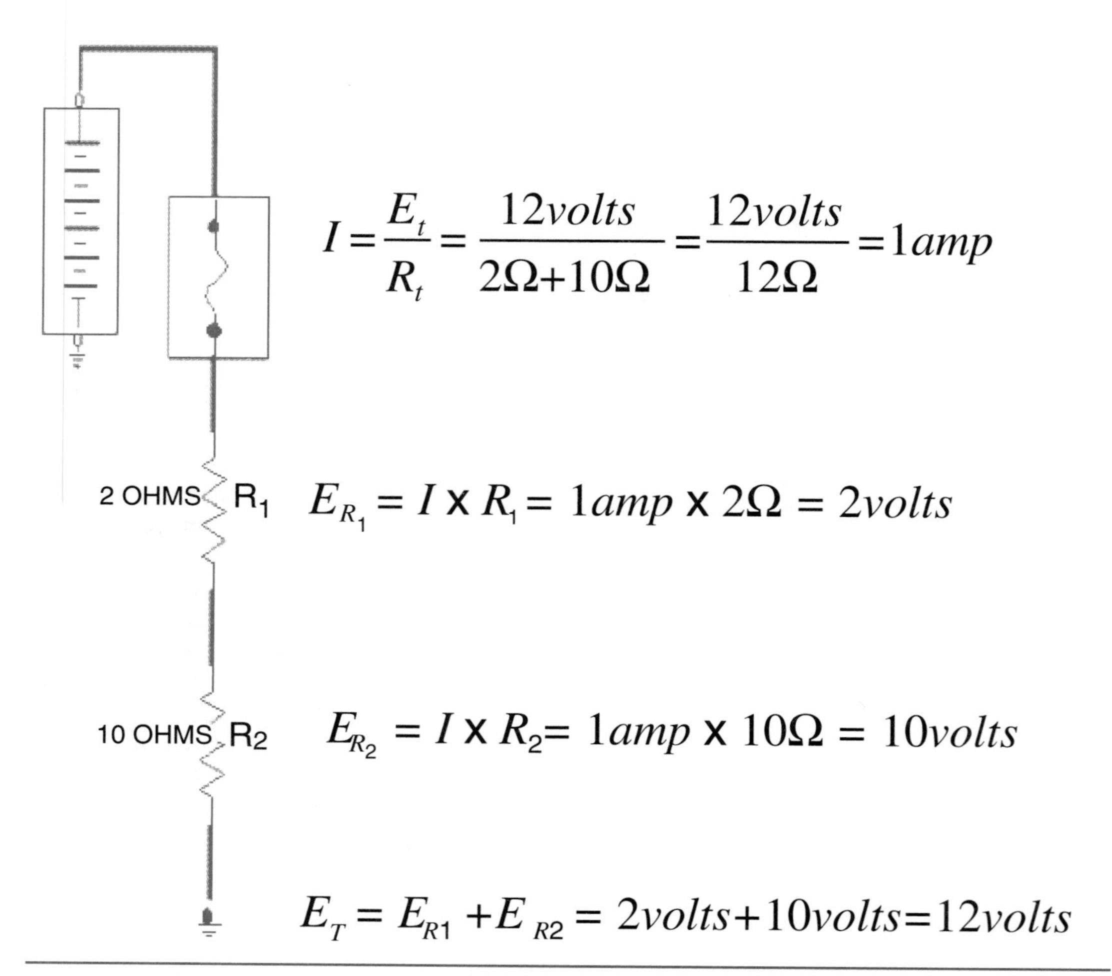

Figure 5-4. Calculating series circuit voltage drops. (GM Service and Parts Operations)

drop in exact proportion to the resistance and that the sum of the voltage drops must equal the voltage applied to the circuit.

Series Circuit Exercise

Exercise Objective: Demonstrate that the sum of the voltage drops for each load in a series circuit is equal to the source voltage.

When you measure any voltage, you measure the *difference* in potential between two points. Components in circuits cause voltage drops. Each voltage drop is a difference in potential between two points—one point before a load, the other point after the load. When you add together all of the voltage drops in a circuit, the total will equal the supply voltage. For more information about applying Kirchhoff's Law of Voltage Drops, see the "Circuit Faults" section in Chapter 5 of the *Shop Manual.*

Assemble a circuit as shown in Figure 5-5. Measure the voltage drop across each of the following circuit sections and record the readings in the spaces provided in the illustration.

Does the sum of the voltage drops equal the applied voltage?________

Why do the voltage drops in a series circuit add up to the source voltage?

Measure the values of the resistors:

R_1:________________________________

R_2:________________________________

R_3:________________________________

R_4:________________________________

PARALLEL CIRCUIT VOLTAGE DROPS

In a *parallel circuit,* each branch drops the same voltage, regardless of the resistance. If the resistance values are not the same, then different amounts of current will flow in each branch (Figure 5-6). According to Kirchhoff's Law of Current, the current that flows through a parallel circuit divides into each path in the circuit: When the current flow in each path is added, the total current will equal the current flow leaving the power source. When calculating the current flow in parallel circuits, each current flow path must be treated as a series circuit or the total resistance of the circuit must be calculated before calculating total current. When performing calculation on a parallel circuit, it should be remembered that more current will always flow through the path with the least resistance.

Kirchhoff's Law of Current (Kirchhoff's 1st Law) states that the current flowing into a junction or point in an electrical circuit must equal the current flowing out.

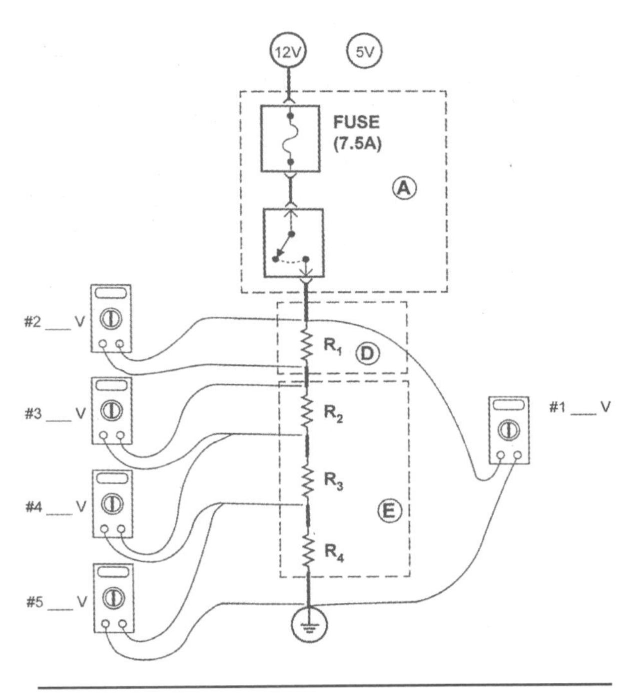

Figure 5-5. Series circuit voltage drop exercise. (GM Service and Parts Operations)

Parallel Circuit Voltage Drops Exercise

Exercise Objective: Demonstrate that all branches of a parallel circuit drop an equal amount of voltage.

In a parallel circuit, each branch drops the same voltage, regardless of the resistance. If the resistance values are not the same, then a different amount of current flows in each branch. Assemble the circuit shown in Figure 5-7. Measure voltage between each of the following pairs of points and record the readings.

Record Measurements:

1 to 2 ______________________ volts

3 to 4 ______________________ volts

5 to 6 ______________________ volts

7 to 8 ______________________ volts

Are the voltage drops in the three branches equal?________

Does the sum of the 3-4, 5-6, and 7-8 voltage drops equal the supply voltage?________ Should it?________

Why do you think all branches of a parallel circuit drop an equal amount of voltage?

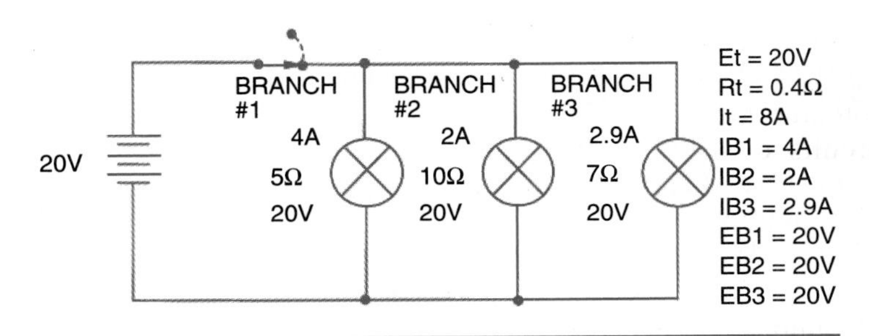

Figure 5-6. Parallel circuit voltage drops.

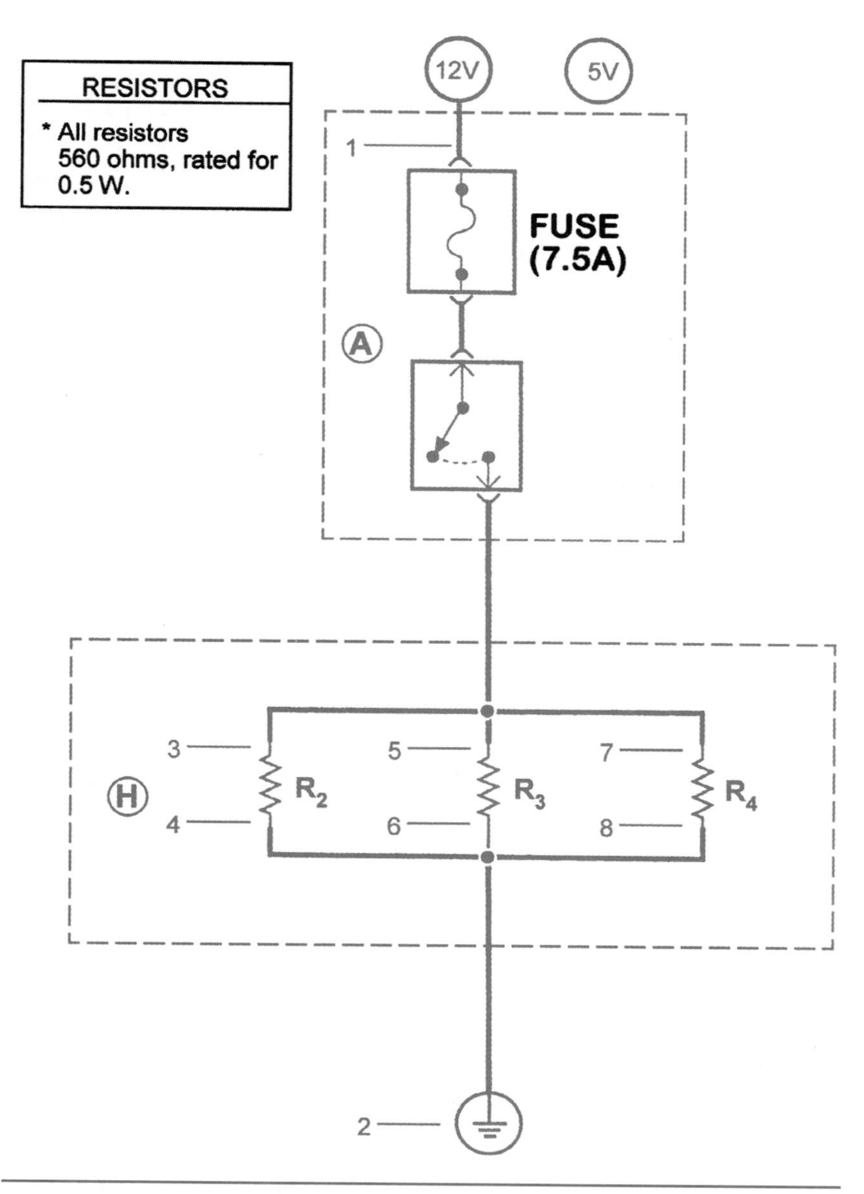

Figure 5-7. Parallel circuit voltage drop exercise. (GM Service and Parts Operations)

CALCULATING SERIES CIRCUIT TOTAL RESISTANCE

The way resistance behaves in a series circuit is easy to understand. Each resistance affects the entire circuit because there is only one current path. If a resistance is added or if an existing resistance increases, the current for the entire circuit decreases. To calculate the total resistance in a series circuit, add up the individual resistances. The total resistance of the series circuit in Figure 5-8 is 3 + 3 + 4 = 10 ohms. The total circuit resistance is usually called the equivalent resistance.

Open in Series Circuit

There is only one path for current flow in a series circuit. If that path is open, there is no current flow and the circuit loads cannot work. If there is an open in a series circuit, the voltage

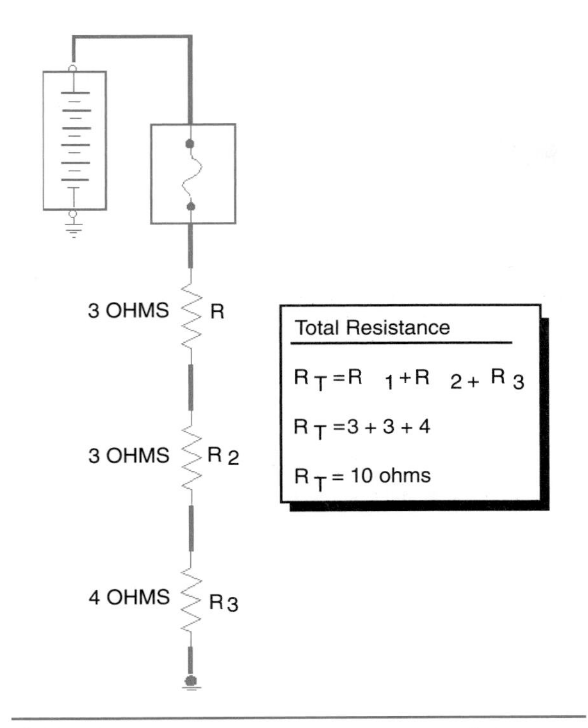

Figure 5-8. Series circuit total resistance. (GM Service and Parts Operations)

drop across the load in that circuit will be zero volts (Figure 5-9). At some point in that circuit, you will be able to measure applied voltage. If you measure the voltage across the open ends of the circuit (on either side of the break), you will measure the applied voltage. There will be no continuity (infinite resistance) between the source and ground. You will measure zero current flow at any point on the circuit. For more information about opens in a series circuit, see the "Circuit Faults" section in Chapter 5 of the *Shop Manual.*

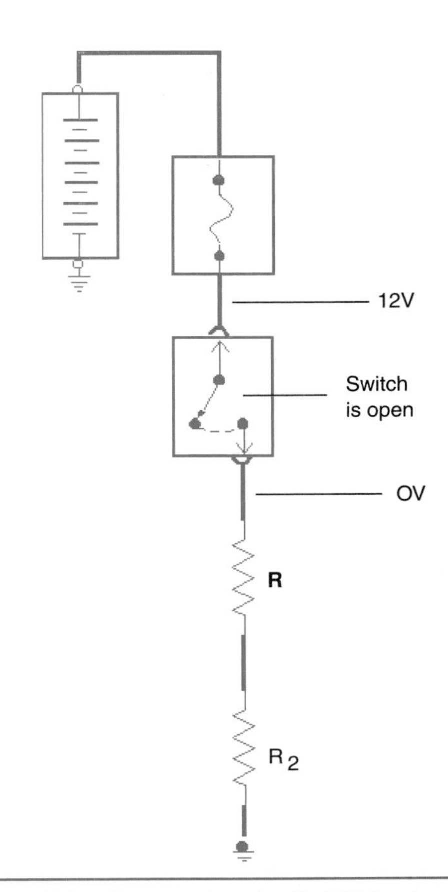

Figure 5-9. Open series circuit. (GM Service and Parts Operations)

Open in Series Circuit Exercise

Read each question carefully and fill in the blanks.

Turn to the "HVAC: Blower Controls, Manual C60" schematic in an OEM service manual and think about the following conditions.

Condition A

You suspect a problem in the blower resistor assembly and want to check the resistances across its various terminals.

1. Which DMM input terminal do you use?____
2. To which position do you turn the rotary switch?______
3. If you measure infinite resistance across terminals C and B of the blower resistor assembly, is it possible for the blower motor to work correctly in positions *LOW* and *M1*?
   a. Yes
   b. No

Condition B

The blower motor doesn't work with the blower switch in the *LOW* and *M1* positions. You know the blower switch is okay, and you suspect the problem is in the blower resistor assembly. You measure voltage between terminal 30 of the blower relay and ground while moving the blower switch through the various positions. The mode selector is turned to *MAX* and the engine is running.

4. Which DMM input terminal do you use?____
5. To which position do you turn the rotary switch?______

Current in a Series Circuit

You learned earlier that since there is a single current path in a series circuit, the current is the same in every part of the circuit. You can relate this back to the water pipe analogy. If water flows from a tank through a single pipe, the rate of flow will be the same in every part of the water circuit. It doesn't matter how many faucets or other parts are plumbed into the circuit, the flow of water has to be the same in every part. The same holds true for an electrical circuit.

CALCULATING PARALLEL CIRCUIT TOTAL RESISTANCE

Understanding the effect of resistance in a parallel circuit is more complicated than for a series circuit. The math for calculating the equivalent resistance for a parallel circuit is complex, and it's not likely you'll ever need to do such a calculation in your work. However, it is important for you to know that the total resistance of a parallel circuit is actually less than the resistance of its smallest resistor. For example, the circuit in Figure 5-10 contains 5-, 10-, and 30-ohm resistors. The smallest resistance is 5 ohms. The total resistance must be less than that. In fact, it's 3 ohms. If there are two loads in parallel of different resistance the following equation can be used to determine the total resistance:

$$R_t = \frac{R_1 \times R_2}{R_1 + R_2}$$

If you have more than two loads, because in a parallel circuit resistance is fractional, you use the following formula for total resistance:

$$R_t = \frac{1}{\frac{1}{R_1} + \frac{1}{R_2} + \frac{1}{R_3}} = \frac{1}{\frac{1}{6} + \frac{1}{3}} = \frac{1}{\frac{1}{6} + \frac{2}{6}}$$

$$= \frac{1}{\frac{3}{6}} = \frac{1}{\frac{1}{2}} = \frac{1}{0.5} = 2 \text{ ohms}$$

You can think of a parallel circuit in plumbing terms like water going through pipes. Consider a water tank with a pipe and two faucets (Figure 5-11). Assume the tank holds 10 gallons of water. Now assume that each faucet will allow water to flow out of the tank at about 1 gallon per minute: If you open one faucet, it flows 1 gallon per minute. It will take 10 minutes to empty 10 gallons from the tank. With both faucets open, the water flows at 1 gallon per minute through each faucet. With the faucets in parallel, a total

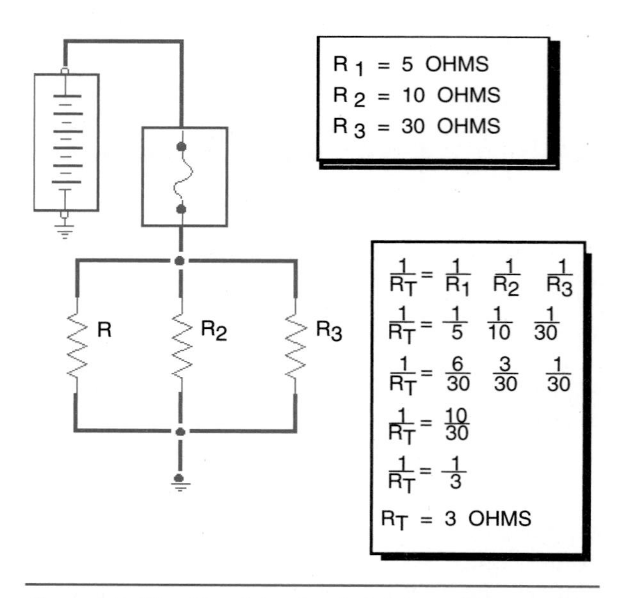

Figure 5-10. Parallel circuit total resistance. (GM Service and Parts Operations)

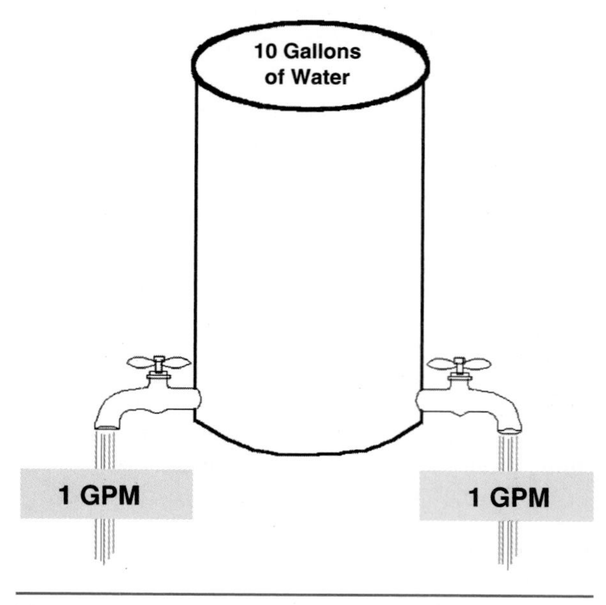

Figure 5-11. Water analogy for parallel circuits. (GM Service and Parts Operations)

of 2 gallons per minute flows out of the tank. It will take 5 minutes to empty 10 gallons from the tank.

Notice that the faucets are connected like resistors in a parallel circuit. Because each faucet offers a path for water to flow, two paths offer less resistance than one. In the same way, two parallel paths offer less resistance than a single path and allow more total current to flow.

Current in a Parallel Circuit

The current is not the same throughout a parallel or series-parallel circuit. It is true that the same voltage is applied to each branch. But, because the resistance in each branch can be different, the current for each branch can also be different. To find the total current in a parallel or series-parallel circuit, add up the currents in all of the circuit branches. For example, you will find the current to be $2\frac{1}{2}$ amps in the circuit shown in Figure 5-12. The current of the main line is always the same as the total current because it is the only path for that part of the circuit.

SERIES-PARALLEL CIRCUITS

A **series-parallel circuit** is a circuit that contains both series circuits and parallel circuits. This type of circuit is also known as a combination circuit as shown in a circuit in Figure 5-13. The simple circuit in Figure 5-13 has a 2-ohm resistor in series from the battery then splits into two parallel branches of first a 6-ohm resistor and then a 3-ohm resistor before recombining and returning to the battery. There is no specific law or formula that pertains to the whole series-parallel circuit for voltage, amperage, and resistance. Instead, it is a matter of determining

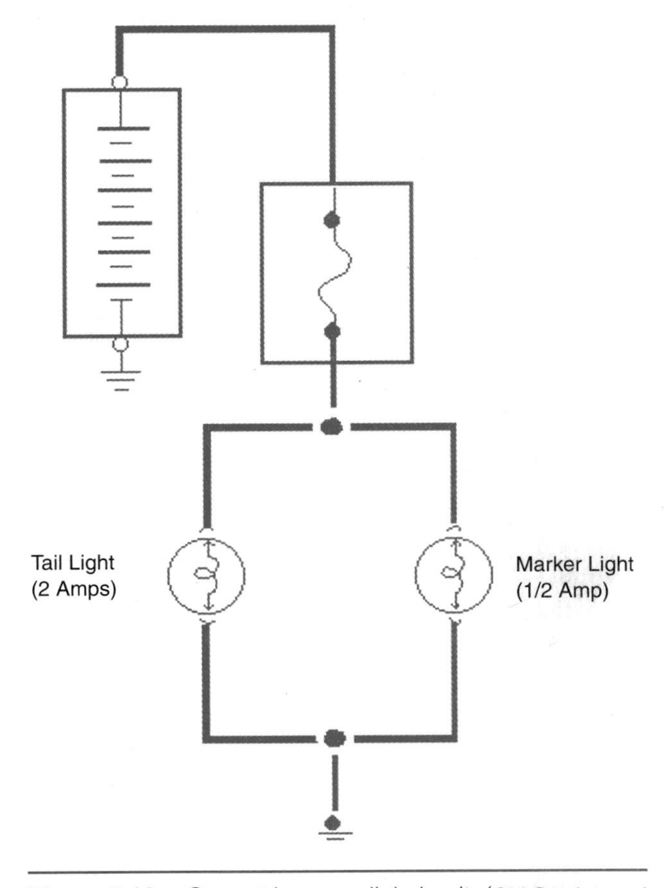

Figure 5-12. Current in a parallel circuit. (GM Service and Parts Operations)

which branch loads of the circuit are in series and which are in parallel, simplifying the circuit where possible, and using the circuit laws that apply to each of these branches to find the value totals. For more information about series-parallel circuits, see the "Circuit Faults" section in Chapter 5 of the *Shop Manual.*

Series-Parallel Circuits and Ohm's Law

Values in a series-parallel circuit are figured by reducing the parallel branches to equivalent values for single loads in series. Then the equivalent values and any actual series loads are combined. To calculate total resistance, first find the resistance of all loads wired in parallel. If the circuit is complex, it may be handy to group the parallel branches into pairs and treat each pair separately. Then add the values of all loads wired in series to the equivalent resistance of all the loads wired in parallel. In the circuit shown in Figure 5-13,

$$R_t = \frac{R_1 \times R_2}{R_1 + R_2} = \frac{6 \times 3}{6 + 3} + 2$$

$$= \frac{18}{9} + 2 = 4 \text{ ohms}$$

The equivalent resistance of the loads in parallel is

$$R_t = \frac{R_1 \times R_2}{R_1 + R_2} = \frac{6 \times 3}{6 + 3} = \frac{18}{9} = 2 \text{ ohms}$$

The total of the branch currents is $1 + 2 = 3$ amps, so the voltage drop is $E = IR = 3 \times 2 = 6$. The voltage drop across the load in series is $2 \times 3 = 6$ volts. Add these voltage drops to find the source voltage: $6 + 6 = 12$ volts.

To determine the source voltage in a series-parallel circuit, you must first find the equivalent resistance of the loads in parallel, and the total current through this equivalent resistance. Figure out the voltage drop across this equivalent resistance and add it to the voltage drops across all loads wired in series. To determine total current, find the currents in all parallel branches and add them together. This total is equal to the current at any point in the series circuit.

$$I = \frac{E}{R_1} + \frac{E}{R_2} = \frac{6}{6} + \frac{6}{3} = 1 + 2 = 3 \text{ amps}$$

In Figure 5-13, notice that there is only 6 volts across each of the branch circuits because another 6 volts has already been dropped across the 2-ohm resistor.

Figure 5-14 is a complete headlamp circuit with all bulbs and switches, which is an example of a series-parallel circuit.

Series-Parallel Circuit Exercise

Exercise Objective: Demonstrate that a series-parallel circuit has the characteristics of both a series circuit and a parallel circuit.

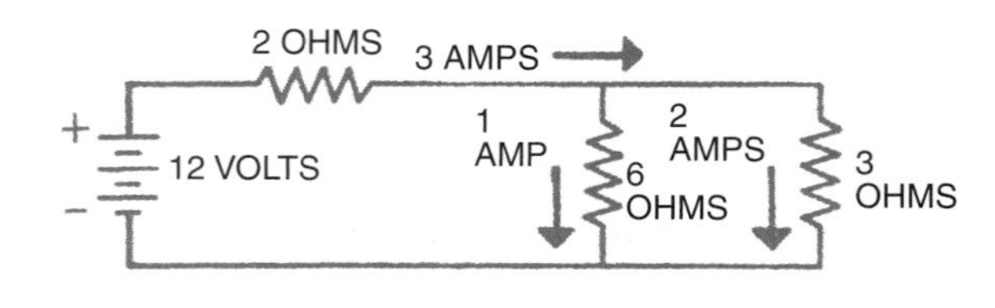

Figure 5-13. Series-parallel circuit.

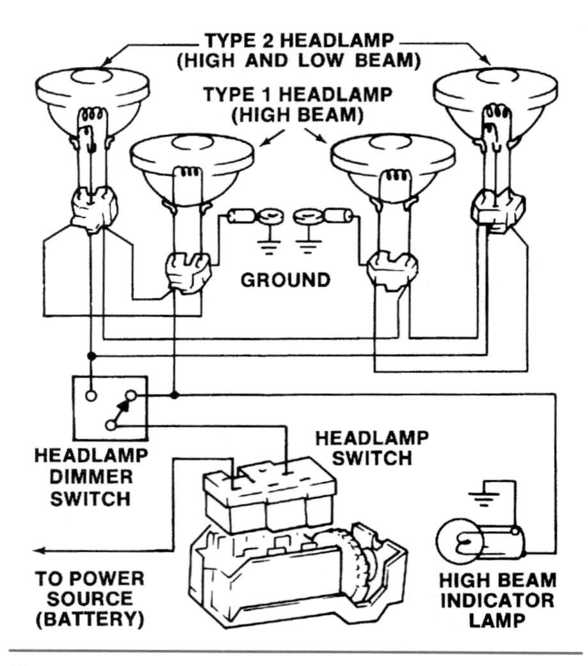

Figure 5-14. A complete headlamp circuit with all bulbs and switches, which is a series-parallel circuit.

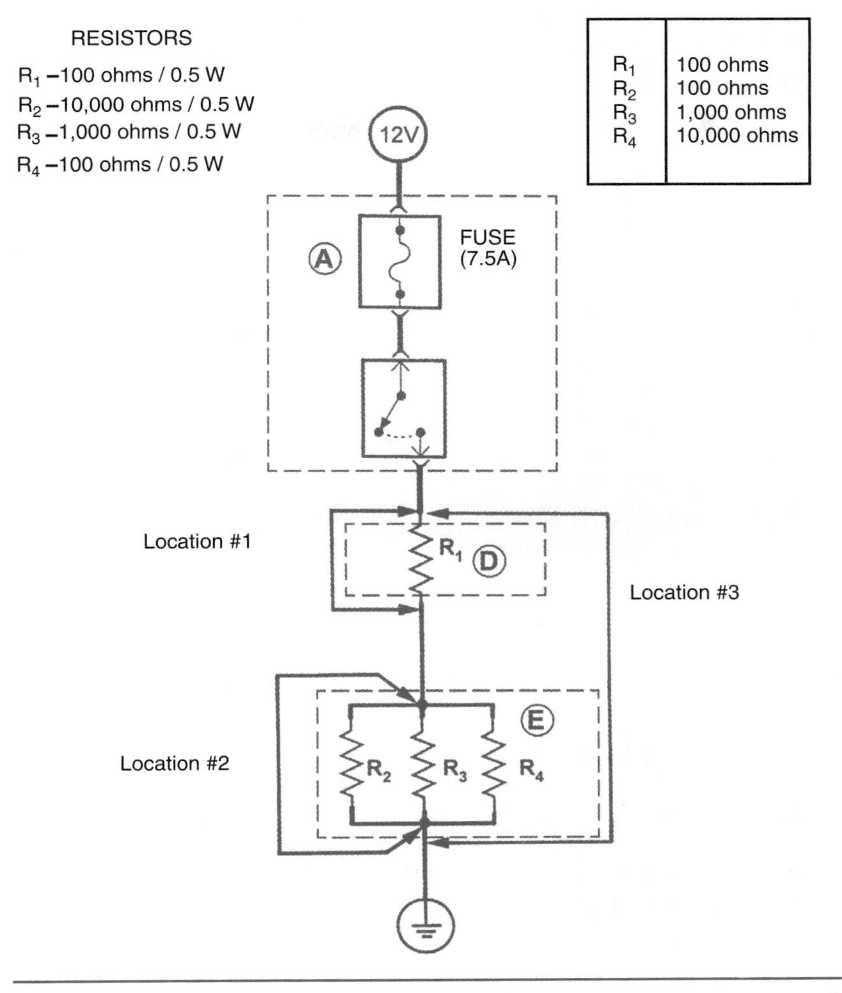

Figure 5-15. Series-parallel circuit exercise. (GM Service and Parts Operations)

Assemble the circuit shown in Figure 5-15 and answer the following questions.

1. Measure the voltage drop at location #1 and #2.

   ____

2. Measure the resistance at location #1.

   ____

3. Measure the resistance of each resistor in the parallel portion of the circuit.

   ____

4. Use Ohm's Law to figure out the total resistance for the parallel circuit.

   ____

5. Measure the resistance at location #2. Does it match your calculation?

   ____

6. Measure the resistance at location #3.

   ____

7. Does the resistance at location #1 and #2 add up to the resistance you measured at location #3?

   ____

8. Calculate the total circuit current using Ohm's Law.

   ____

9. Measure the total circuit current. Does it match your calculation?

   ____

10. Measure the current of each branch of the parallel portion of the circuit.

   ____

SERIES AND PARALLEL CIRCUIT FAULTS

Opens in a Parallel Circuit

The effect of an open on a parallel circuit (Figure 5-16) depends on where in the circuit the open is located and on the design of the circuit. If an open occurs in the main line, none of the loads on the circuit can work. In effect, all of the branches are open. If an open occurs on a branch below the main line, only the load on that branch is affected. All of the other branches still form closed circuits and still operate. For more information about opens in a parallel circuit, see the "Circuit Faults" section in Chapter 5 of the *Shop Manual.*

Opens in a Parallel Circuit Exercise

Using an OEM service manual, go to the "Exterior Lights" schematic to answer these questions. You should assume the following:

- The ignition is ON
- The turn signal switch is in LEFT
- The headlight switch is OFF

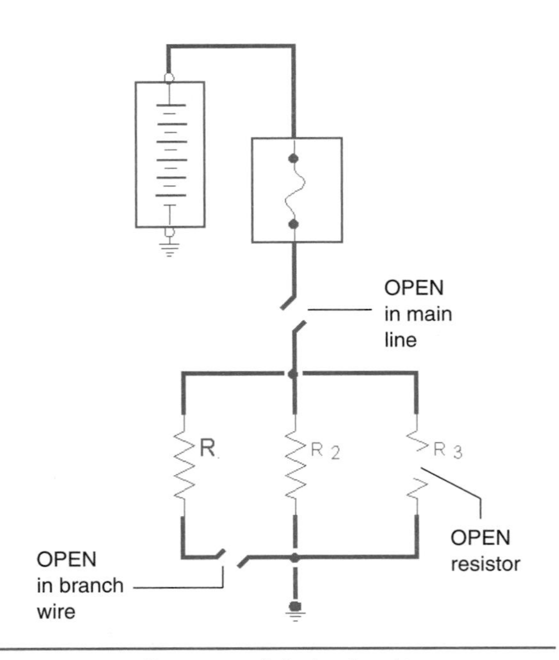

Figure 5-16. Open parallel circuits. (GM Service and Parts Operations)

1. Which loads would operate if the circuit was operating properly?

____

____

____

____

2. Which loads would operate if there was an open circuit between the turn flasher and the turn/hazard-headlight switch assembly in the circuit?

____

3. Which loads would operate if there was an open circuit between the turn/hazard-headlight switch assembly and the ground?

____

Which loads would not operate?

____

4. Which loads would operate if there was an open in the circuit between the turn/hazard-headlight switch assembly and the first connector?

____

Which loads would not operate?

____

Short to Voltage in a Parallel Circuit

A short to voltage happens when one circuit is shorted to the voltage of another circuit. Such a short can also occur between two separate branches of the same circuit. The cause is usually broken or damaged wire insulation. You can narrow down the location of a short to voltage by following the appropriate diagnostic steps, such as removing fuses and observing the results. We'll discuss these diagnostic steps in detail in Chapter 5 of the *Shop Manual.* You should also keep in mind that the OEM Service Manual often contains diagnostic procedures for specific symptoms.

The symptoms of a short to voltage depend on the location of the short in both circuits. One or both circuits may operate strangely. For example, in Figure 5-17A, the short is before the switches on both circuits. This means both switches control both loads. A different problem shows up if the

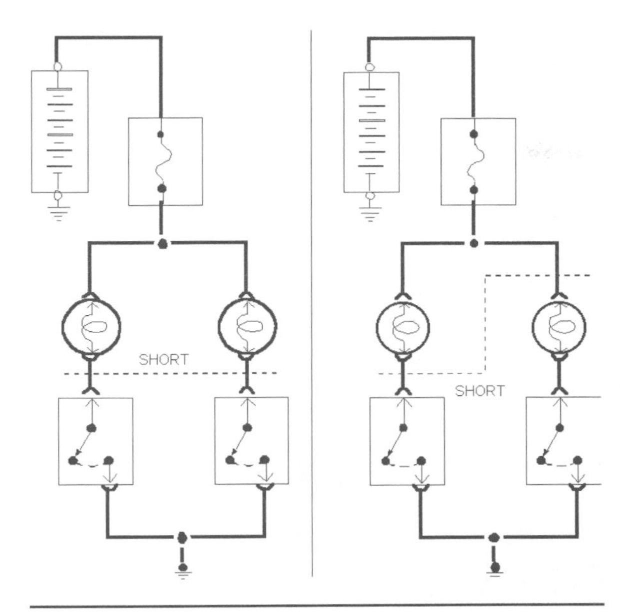

Figure 5-17. Short to voltage in a parallel circuit. (GM Service and Parts Operations)

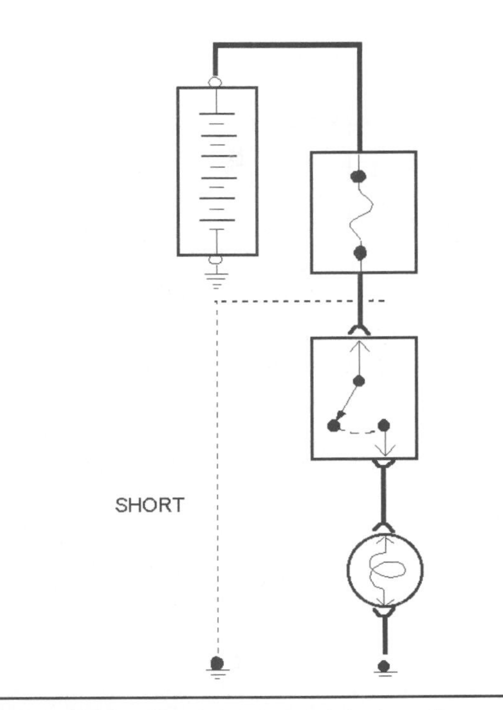

Figure 5-18. Short to ground before the switch. (GM Service and Parts Operations)

short is after the load on one branch and before the load on the second (Figure 5-17 B). The load in the second branch operates normally. The load in the first branch will not come on at all, or the current flow might be so high that the fuse blows. If there is no circuit protector, the wire could get so hot that it actually catches on fire.

Short to Ground

A short to ground (Figure 5-18) occurs when current flow is grounded before it was designed to be. This usually happens when wire insulation breaks and the wire touches a ground. The effect of a short to ground depends on the design of the circuit and on its location in relationship to the circuit control and load.

Figure 5-19 shows a short located between the switch and the load. The resistance is lower than it should be because the current is not passing through the loads. The fuse blows only after the switch is closed. Lower resistance means the current flow is higher than normal. The fuse or other circuit protector will open. An automatically resetting circuit breaker would repeatedly open and close. If there was no circuit protector at all, the wire might get hot enough to burn.

Figure 5-20 shows an example where a short to ground is after the load but before the control.

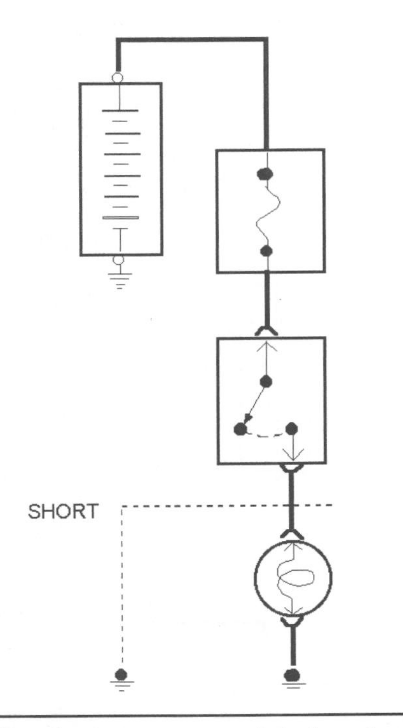

Figure 5-19. A short to ground before the load. (GM Service and Parts Operations)

This means the control switch is cut out of the circuit and the circuit is always closed. As a result, the bulb is lit all of the time. If a short to ground occurs close to the intended ground connection, you probably won't notice any effects.

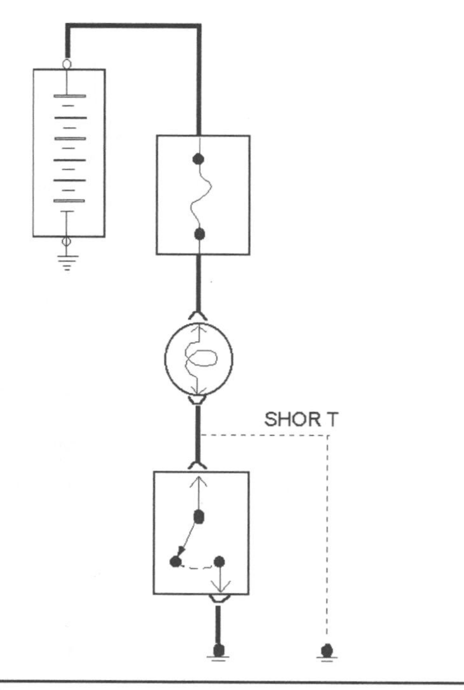

Figure 5-20. Short circuit before switch. (GM Service and Parts Operations)

SUMMARY OF SERIES CIRCUIT OPERATION

- The current flows through the circuit in only *one* path.
- The current flow is the same at any point in the circuit.
- The voltage drops in the circuit *always* add up to the source voltage.
- The total resistance is equal to the sum of the individual resistances.

SUMMARY OF PARALLEL CIRCUIT OPERATION

- The sum of the currents in each branch equals the total current in the circuit.
- The voltage drop will be the same across each branch in the circuit.
- The total resistance is always lower than the smallest branch resistance.

Review Questions

1. The total resistance is equal to the sum of all the resistance in:
   a. Series circuits
   b. Parallel circuits
   c. Series-parallel circuits
   d. Series and parallel circuits
2. What type of circuit does this figure illustrate?

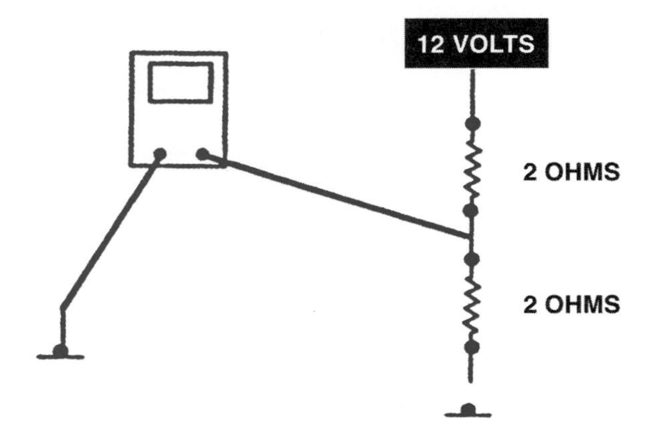

   a. Series
   b. Parallel
   c. Series-parallel
   d. Broken
3. The amperage in a series circuit conforms to which of these statements:
   a. It is the same anywhere in the circuit.
   b. It is always the same at certain points.
   c. It is the same under some conditions.
   d. It is never the same any where in the circuit.
4. Where current can follow more than one path to complete the circuit, the circuit is called:
   a. Branch
   b. Series
   c. Complete
   d. Parallel
5. If resistance in a parallel circuit is unknown, dividing the voltage by the branch ______ equals branch resistance.
   a. Amperage
   b. Conductance
   c. Voltage drops
   d. Wattage
6. In most modern automobiles, the chassis can act as a ground because it is connected to the ______
   a. Negative battery terminal
   b. Generator output bolt
   c. Generator ground
   d. Positive battery terminal
7. The following are all examples of loads:
   a. Switch, motor, bulb
   b. Bulb, fuse, resistor
   c. Bulb, motor, solenoid
   d. Fuse, wire, circuit breaker
8. A circuit has only one path for current.
   a. Parallel
   b. Series
   c. Series-parallel
   d. Ground
9. In a circuit with more than one resistor, the total resistance of a series circuit is ______ the resistance of any single resistor.
   a. Greater than
   b. Less than
   c. The same
   d. Equal to
10. In a circuit with more than one resistor, the total resistance of a parallel circuit is ______ the resistance of any single resistor.
    a. Greater than
    b. Less than
    c. The same
    d. Equal to
11. Technician A says an open in one of the branches of a parallel circuit will not affect the operation of the other branches. Technician B says an open in one of the branches of a parallel circuit will not affect the operation of the other branches. Who is right?
    a. A only
    b. B only
    c. Both A and B
    d. Neither A nor B
12. Which of the following describes characteristics of two resistances connected in series?
    a. They must have different resistances.
    b. They must have the same resistance.

c. The voltage drop will be equal across each.
d. There will be only one path for current to flow.

13. What happens when one resistance in a series circuit is open?
    a. The current in the other resistance is at maximum.
    b. The current is zero at all resistances.
    c. The voltage drop increases.
    d. The current stays the same.

14. Which of the following is true regarding a series circuit with unequal resistances?
    a. The highest resistance has the most current.
    b. The lowest resistance has the most current.
    c. The lowest resistance has the highest voltage drop.
    d. The highest resistance has the highest voltage drop.

15. Which of the following is true regarding a parallel circuit with unequal branch resistance?
    a. The current is equal in all branches.
    b. The current is higher in the highest resistance branch.
    c. The voltage is higher in the lowest resistance branch.
    d. The current is higher in the lowest resistance branch.

16. The total resistance of a series circuit is:
    a. Equal to the current
    b. The sum of the individual resistances
    c. Always a high resistance
    d. Each resistance multiplied together

17. In a circuit with three parallel branches, if one branch opens, the total current will:
    a. Increase
    b. Decrease
    c. Stay the same
    d. Blow the fuse

18. Which of the following is true regarding series-parallel circuits?
    a. Voltages are always equal across each load.
    b. Current is equal throughout the circuit.
    c. Only one current path is possible.
    d. Voltage applied to the parallel branches is the source voltage minus any voltage drop across loads wired in series.

19. The amperage in a series circuit is:
    a. The same throughout the entire circuit
    b. Different, depending on the number of loads
    c. Sometimes the same, depending on the number of loads
    d. Never the same anywhere in the circuit

20. What does a short circuit to ground before the load cause?
    a. An increase in circuit resistance
    b. Voltage to increase
    c. Current flow to Increase
    d. Current flow to decrease

21. Three lamps are connected in parallel. What would happen if one lamp burns out?
    a. The other two lamps would go out.
    b. Current flow would increase through the "good" lamps.
    c. Total circuit resistance would go up.
    d. Voltage at the other two lamps would increase.

22. Total resistance in a series circuit is equal to the:
    a. Sum of the individual resistances
    b. Voltage drop across the resistor with the highest value
    c. Current in the circuit divided by the source voltage
    d. Percent of error in the voltmeter itself

23. Parallel circuits are being discussed. Technician A says that adding more branches to a parallel circuit reduces total circuit resistance. Technician B says that adding more branches to a parallel circuit increases the total current flowing in the circuit. Who is right?
    a. Technician A
    b. Technician B
    c. Both A and B
    d. Neither A nor B

24. The sum of all voltage drops in a series circuit equals the:
    a. Voltage across the largest load
    b. Voltage across the smallest load
    c. Source or applied voltage

25. What is the name for a circuit that allows two or more paths for current flow?
    a. Series circuit
    b. Parallel circuit
    c. Both A and B
    d. Neither A nor B

26. What is the name for a circuit that allows only one path for current to flow?
    a. Series circuit
    b. Parallel circuit
    c. Series-parallel circuit
    d. Integrated circuit

6

Electrical Diagrams and Wiring

LEARNING OBJECTIVES

Upon completion and review of this chapter, you should be able to:

- Identify the wire types and materials used in automotive wiring.
- Explain how wire size is determined by both the American Wire Gauge (AWG) system and the metric system.
- Explain the use of a wiring harness and define the different types of connectors and terminal ends.
- Define the ground, parallel data, serial data, and multiplexing paths.
- Identify common electrical parts and explain their operation.
- Explain the color-coding of automotive wiring.
- Explain the terms used in the language of automotive wiring diagrams.
- Identify the component symbols used in automotive wiring schematics.
- Explain the purpose of a wiring diagram or schematic.

KEY TERMS

Circuit Number
Color Coding
Component Symbols
Connectors
Ground Cable
Installation Diagram
Metric Wire Sizes
Multiplexing
Primary Wiring
Schematic Diagram
Solenoid
Switches
Weatherproof Connectors
Wire Gauge Diagram
Wire Gauge Number
Wiring Harness

INTRODUCTION

Now that we have discussed current flow, voltage, sources, electrical loads, and series and parallel circuits, in this chapter we start to build some automotive circuits. To build a complete circuit, we must have conductors to carry the current from the voltage source to the electrical loads. The conductors are the thousands of feet of wire and cable used in the complete electrical system. The vehicle chassis is also a conductor for the ground side of the circuits, as we will see later. We will begin our study by looking at the wiring harnesses, connectors, and terminals of the system.

The preceding chapters used symbols to show some of the components in an automotive electrical system. After studying the basic parts of the system (voltage source, conductors, and loads), it is time to put them together into complete circuits.

In real-world cases, diagrams of much greater complexity are used. Technicians must be able to identify each component by its symbol and determine how current travels from the power source to ground. Technicians use electrical circuit diagrams to locate and identify components on the vehicle and trace the wiring in order to make an accurate diagnosis of any malfunctions in the system.

WIRING AND HARNESSES

An automobile may contain as much as half a mile of wiring, in as many as 50 harnesses, with more than 500 individual connections (Figure 6-1). This wiring must perform under very poor working conditions. Engine heat, vibration, water, road dirt, and oil can damage the wiring and its connections. If the wiring or connections break down, the circuits will fail.

To protect the many wires from damage and to keep them from becoming a confusing tangle, the automotive electrical system is organized into bundles of wire known as **wiring harnesses** that serve various areas of the automobile. The wires are generally wrapped with tape or plastic covering, or they may be enclosed in insulated tubing. Simple harnesses are designed to connect two components; complex harnesses are collections of simple harnesses bound together (Figure 6-2).

Main wiring harnesses are located behind the instrument panel (Figure 6-3), in the engine compartment (Figure 6-4 and Figure 6-5), and along the body floor. Branch harnesses are routed from the main harness to other parts of the system. Items 1, 2, and 3 in Figure 6-4 are ground connections. The colored insulation used on individual wires makes it easier to trace them through

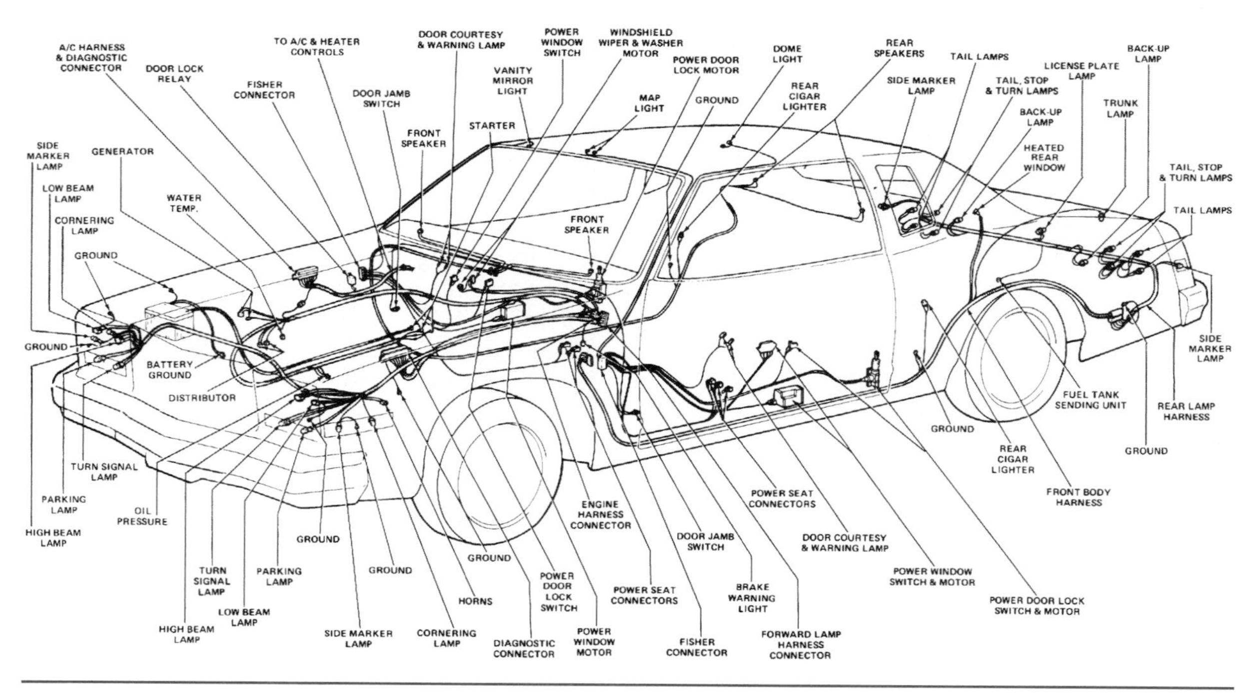

Figure 6-1. The wiring harness in this vehicle is typical of those in most late-model cars. (GM Service and Parts Operations)

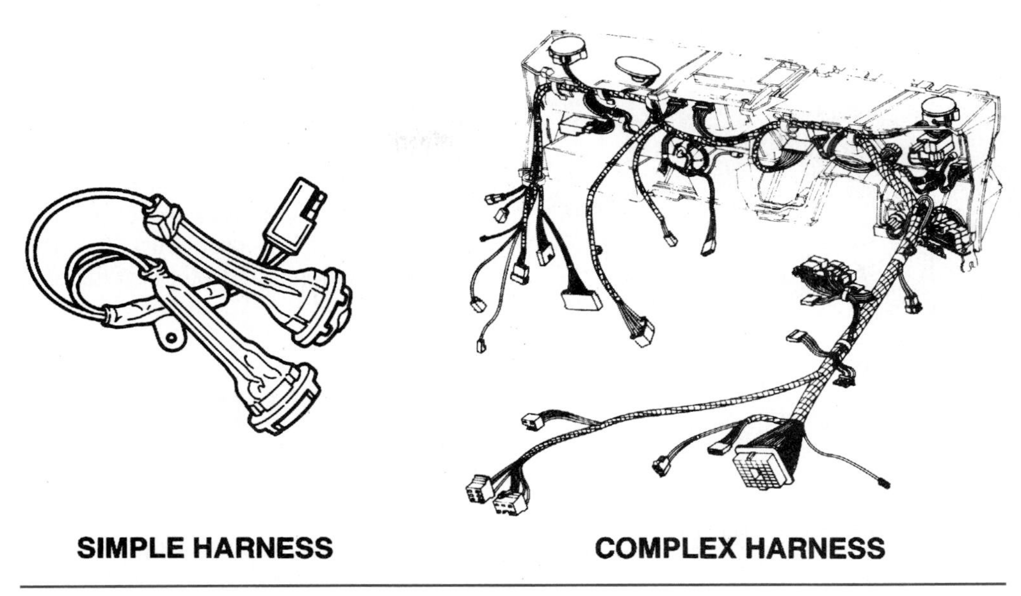

Figure 6-2. Wiring harnesses range from the simple to the complex. (DaimlerChrysler Corporation)

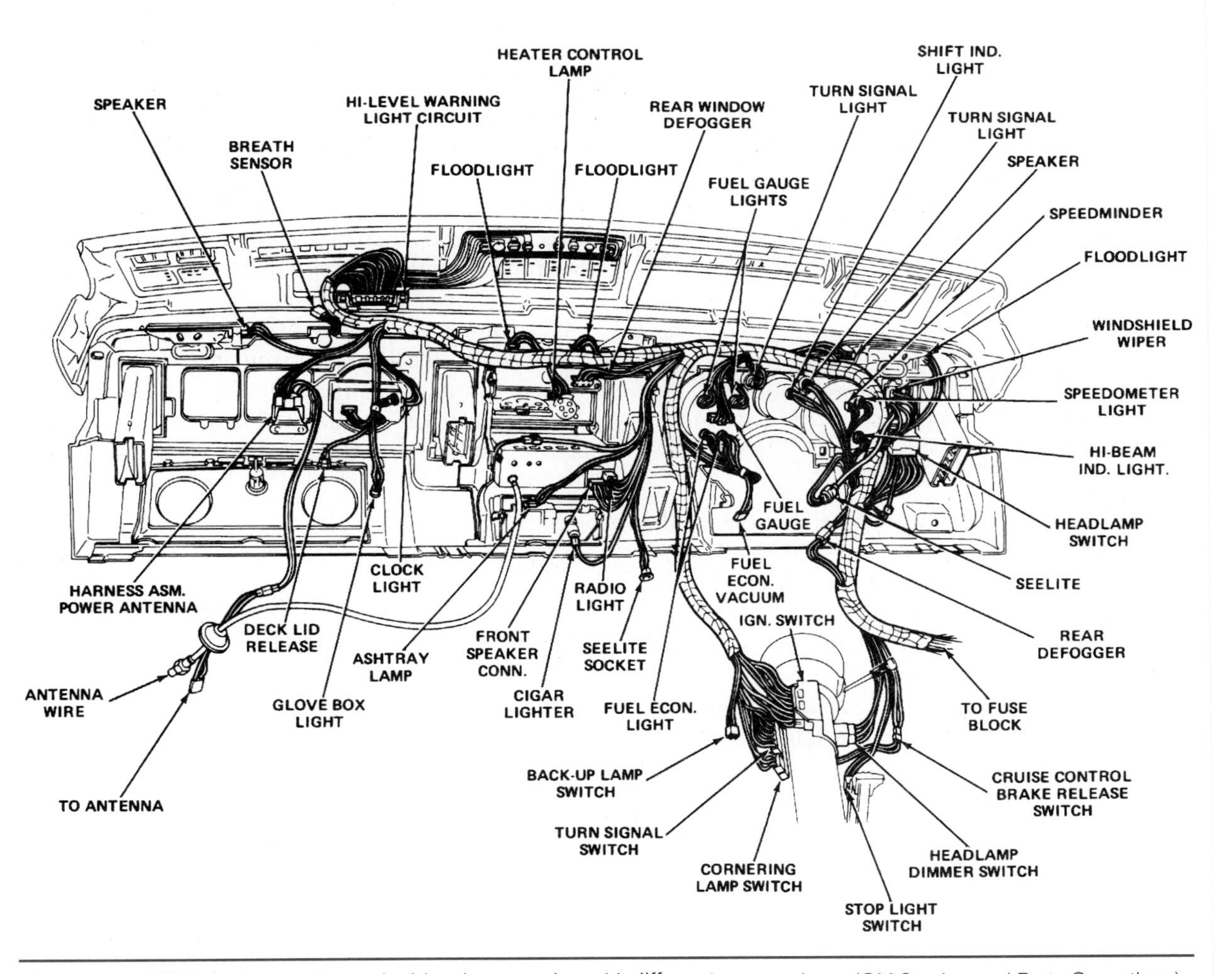

Figure 6-3. This instrument panel wiring harness has 41 different connectors. (GM Service and Parts Operations)

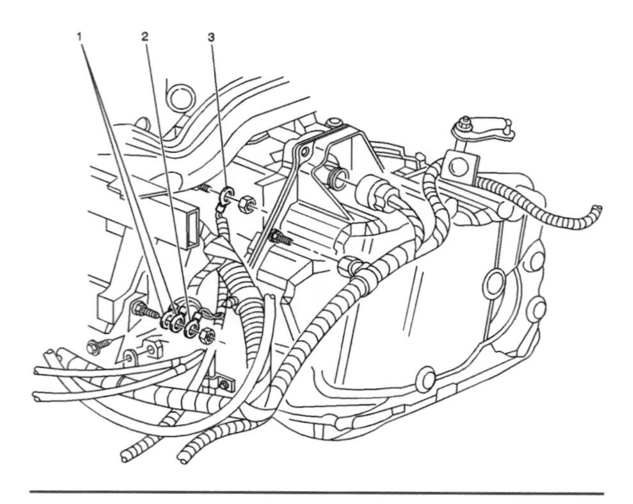

Figure 6-4. The engine compartment wiring harnesses. (GM Service and Parts Operations)

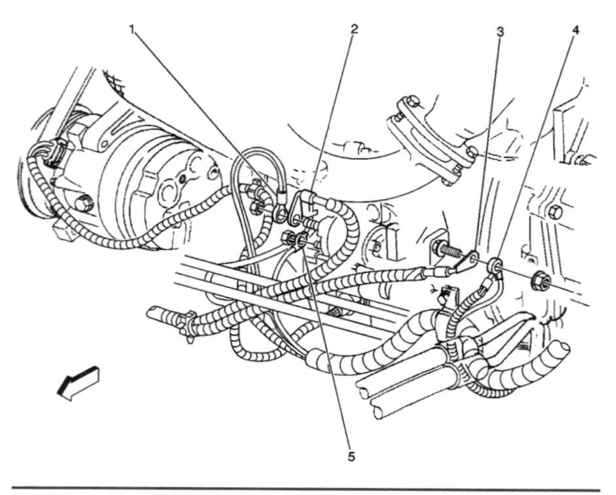

Figure 6-5. The engine wiring harnesses connects to the individual engine components to the engine compartment wiring harness. (GM Service and Parts Operations)

these harnesses, especially where sections of the wire are hidden from view.

A loose or corroded connection, or a replacement wire that is too small for the circuit, will add extra resistance and an additional voltage drop to the circuit. For example, a 10-percent extra drop in voltage to the headlamps will cause a 30-percent voltage loss in candlepower. The same 10-percent voltage loss at the power windows or windshield wiper motor can reduce, or even stop, motor operation. All automotive electrical circuits, except the secondary circuit of the ignition system (from the coil to the spark plugs), operate on 12 to 14 volts and are called low-voltage systems. (Six-volt systems on older cars and 24-volt systems on trucks also are considered low-voltage systems.) The low-voltage wiring of a vehicle, with the exception of the battery cables, is called the **primary wiring.** This usually includes all lighting, accessory, and power distribution circuits. By 2003, we will see 42-volt systems in some hybrid and mybrid applications. For more information about diagnosing wiring problems, see the "Tracing Circuits" section in Chapter 6 of the *Shop Manual.*

WIRE TYPES AND MATERIALS

Most automotive wiring consists of a conductor covered with an insulator. Copper is the most common conductor used. It has excellent conductivity, is flexible enough to be bent easily, solders readily, and is relatively inexpensive. A conductor must be surrounded with some form of protective covering to prevent it from contacting other conductors. This covering is called insulation. High-resistance plastic compounds have replaced the cloth or paper insulation used on older wiring installations.

Stainless steel is used in some heavy wiring, such as battery cables and some ignition cables. Some General Motors cars use aluminum wiring in the main body harness. Although less expensive, aluminum is also less conductive and less flexible. For these reasons, aluminum wires must be larger than comparable copper wires and they generally are used in the lower forward part of the vehicle where flexing is not a problem. Brown plastic wrapping indicates aluminum wiring in GM cars; copper wiring harnesses in the cars have a black wrapping.

Wire Types

Automotive wiring or circuit conductors are used in one of three forms, as follows:

- Solid wires (single-strand)
- Stranded wires (multistrand)
- Printed circuitry

Solid or single-strand wire is used where current is low and flexibility is not required. In automotive electrical systems, it is used inside components such as alternators, motors, relays, and other devices with only a thin coat of enamel or shellac for insulation. Stranded or multistrand wire is made by braiding or twisting a number of solid wires together into a single conductor insulated with a covering of colored plastic, as shown in Figure 6-6. Most automotive electrical system wiring uses stranded wire, either as single conductors or grouped together in harnesses or looms. For more information about wire types, see the section on "Copper Wiring Repair" in Chapter 6 of the *Shop Manual*.

Printed circuitry is a thin film of copper or other conductor that has been etched or embedded on a flat insulating plate (Figure 6-7). A complete printed circuit consists of conductors, insulating material, and connectors for lamps and other components, and is called a printed circuit (PC) board. It is used in places where space for individual wires or harnesses is limited, such as behind instrument panels.

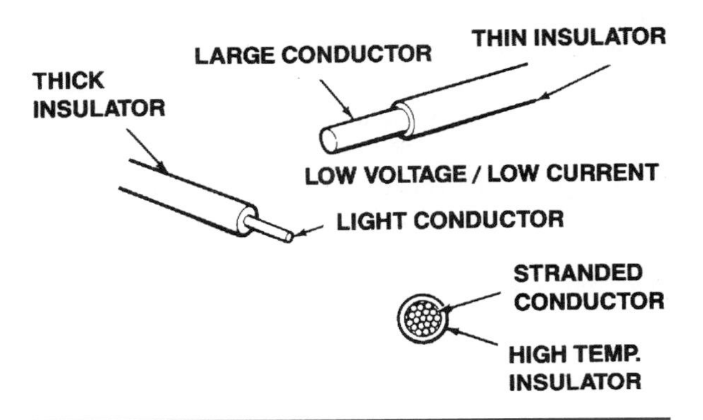

Figure 6-6. Automotive wiring may be solid-wire conductors or multistrand-wire conductors. (DaimlerChrysler Corporation)

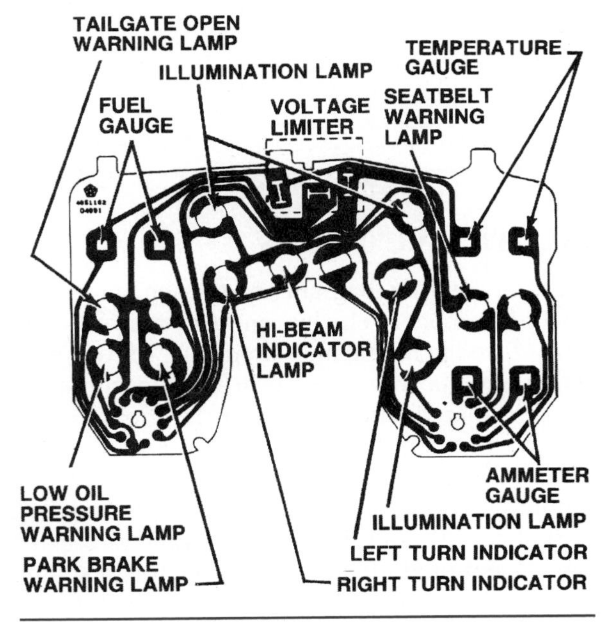

Figure 6-7. Printed circuit boards are used in automotive instrument panels and elsewhere. (DaimlerChrysler Corporation)

WIRE SIZE

Automotive electrical systems are very sensitive to changes in resistance. This makes the selection of properly sized wires critical whenever systems are designed or circuits repaired. There are two important factors to consider: wire gauge number and wire length.

Wire Gauge Number

A **wire gauge number** is an expression of the cross-sectional area of the conductor. The most common system for expressing wire size is the American Wire Gauge (AWG) system. Figure 6-8 is a table of AWG wire sizes commonly used in automotive systems. Wire cross-sectional area is measured in circular mils; a mil is one-thousandth of an inch (0.001), and a circular mil is the area of a circle 1 mil (0.001) in diameter. A circular mil measurement is obtained by squaring the diameter of a conductor measured in mils. For example, a conductor 1/4 inch in diameter is 0.250 inch, or 250 mils, in diameter. The circular mil cross-sectional area of the wire is 250 squared, or 62,500 circular mils.

American Wire Gauge Sizes

Gauge size	Conductor diameter (inches)	Cross-section area (circular mils)
20	.032	1,020
18	.040	1,620
16	.051	2,580
14	.064	4,110
12	.081	6,530
10	.102	10,400
8	.128	16,500
6	.162	26,300
4	.204	41,700
2	.258	66,400
1	.289	83,700
0	.325	106,000
2/0	.365	133,000
4/0	.460	211,600

Figure 6-8. This table lists the most common wire gauge sizes used in automotive electrical systems. (DaimlerChrysler Corporation)

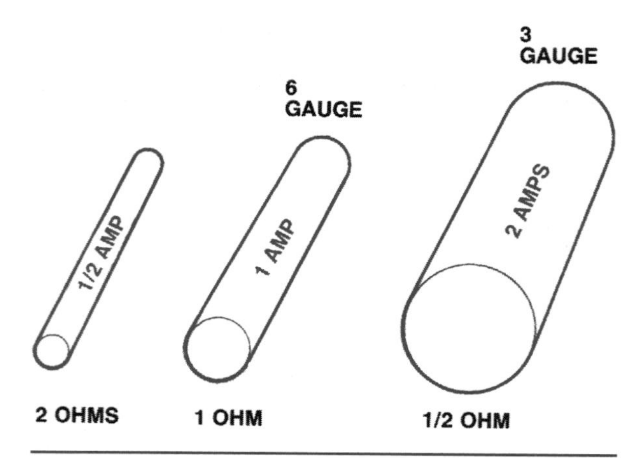

Figure 6-9. This figure shows the relationship between current capacity and resistance as the cross-section of a conductor changes.

Gauge numbers are assigned to conductors of various cross-sectional areas. As gauge number increases, area decreases and the conductor becomes smaller (Figure 6-9). A 6-gauge conductor is smaller than a 3-gauge conductor, and a 12-gauge conductor is smaller than a 6-gauge conductor. You learned in Chapter 1 that as the cross-sectional area of a conductor decreases, its resistance increases. As resistance increases, so does the gauge number. Also, because the current-carrying ability of a conductor decreases as the resistance increases, a conductor with a higher gauge number will carry less current than a conductor with a lower gauge number.

Remember that the wire gauge number refers to the size of the conductor, not the size of the complete wire (conductor plus insulation). For example, it is possible to have two 16-gauge wires of different outside diameters because one has a thicker insulation than the other. Twelve-volt automotive zelectrical systems generally use 14-, 16-, and 18-gauge wire. Main power distribution circuits between the battery and alternator, ignition switch, fuse box, headlamp switch, and larger accessories use 10- and 12-gauge wire. Low-current electronic circuits may use 20-gauge wire. Lighting other than the headlamps, as well as the cigarette lighter, radio, and smaller accessories, use 14-, 16-, and 18-gauge wire. Battery cables, however, generally are listed as 2-, 4-, or 6-AWG wire size.

The gauge sizes used for various circuits in an automobile are generally based on the use of copper wire. A larger gauge size is required when aluminum wiring is used, because aluminum is not as good a conductor as copper. Similarly, 6-volt electrical systems require larger-gauge wires than 12-volt systems for the same current loads. This is because the lower source voltage requires lower resistance in the conductors to deliver the same current. Generally, 6-volt systems use wires two sizes larger than 12-volt systems for equivalent current loads. Future 42-volt systems will not require as large a wire diameter as the current 12-volt system. Generally, a 42-volt system will use two sizes smaller than 12-volt systems for equivalent current loads.

■ Wire Size Matters

The following drawing shows how a large wire easily conducts a high-amperage current, such as you would find going to a starter motor. The heaviest wires are often called cables, but their purpose is the same. On the other hand, a comparatively light wire tends to restrict current flow, which may generate excess heat if the wire is too small for the job. Too much current running though a light wire may cause the insulation to melt, leading to a short circuit or even a fire.

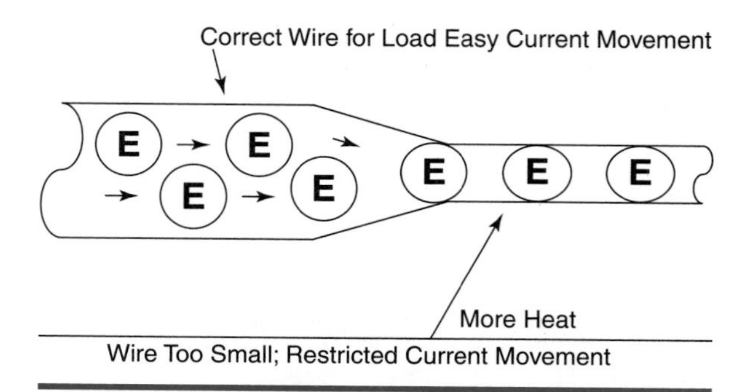

Wire Too Small; Restricted Current Movement

Metric Wire Sizes

Look at a wiring diagram or a service manual for most late model vehicles, and you may see wire sizes listed in metric measurements. **Metric wire sizes** have become the norm in domestic automotive manufacturing due to the global economy. For example, if you look at a wiring diagram for an import or late-model domestic vehicle, you will see wire sizes listed as 0.5, 1.0, 1.5, 4.0, and 6.0. These numbers are the cross-sectional area of the conductor in square millimeters (mm^2). Metric measurements are not the same as circular-mil measurements; they are determined by calculating the cross-sectional area of the conductor with the following formula: Area = Radius^2 × 3.14. A wire with a 1-mm cross-sectional area actually has a 1.128-mm diameter. The following table lists AWG sizes and equivalent metric wire sizes.

AWG Size Metric Size Table

AWG Size (Gauge)	Metric Size (mm^2)
20	0.5
18	0.8
14	2.0
12	3.0
10	5.0
8	8.0
6	13.0
4	19.0

Wire Length

Wire length also must be considered when designing electrical systems or repairing circuits. As conductor length increases, so does resistance. An 18-gauge wire can carry a 10-ampere load for 10 feet without an excessive voltage drop. However, to carry the same 10-ampere load for 15 feet, a 16-gauge wire will be required. Figure 6-10 is a table showing the gauge sizes required for wires of different lengths to carry various current loads. Wire lengths are based on circuits that are grounded to the vehicle chassis.

Total approximate circuit amperes (12 volts)

Wire Gauge Table

Circuit length (feet)

	3	5	7	10	15	20	25	30	40	50	75	100
1	18	18	18	18	18	18	18	18	18	18	18	18
1.5	18	18	18	18	18	18	18	18	18	18	18	18
2	18	18	18	18	18	18	18	18	18	18	16	16
3	18	18	18	18	18	18	18	18	18	18	14	14
4	18	18	18	18	18	18	18	18	16	16	12	12
5	18	18	18	18	18	18	18	18	16	14	12	12
6	18	18	18	18	18	18	16	16	16	14	12	10
7	18	18	18	18	18	18	16	16	14	14	10	10
8	18	18	18	18	18	16	16	16	14	12	10	10
10	18	18	18	18	16	16	16	14	12	12	10	10
11	18	18	18	18	16	16	14	14	12	12	10	8
12	18	18	18	18	16	16	14	14	12	12	10	8
15	18	18	18	18	14	14	12	12	12	10	8	8
18	18	18	16	16	14	14	12	12	10	10	8	8
20	18	18	16	16	14	12	10	10	10	10	8	6
22	18	18	16	16	12	12	10	10	10	8	6	6
24	18	18	16	16	12	12	10	10	10	8	6	6
30	18	16	16	14	10	10	10	10	10	6	4	4
40	18	16	14	12	10	10	8	8	6	6	4	2
50	16	14	12	12	10	10	8	8	6	6	2	2
100	12	12	10	10	6	6	4	4	4	2	1	1/0
150	10	10	8	8	4	4	2	2	2	1	2/0	2/0
200	10	8	8	6	4	4	2	2	1	1/0	4/0	4/0

Figure 6-10. Wire gauge table: As wire length increases, larger-gauge wire must be used to carry the same amount of current.

Special Wiring

Although most of the electrical system is made up of low-voltage primary wiring, special wiring is required for the battery and the spark plugs. Since these wires are larger in size than primary wiring, they are often called cables. Battery cables are low-resistance, low-voltage conductors. Ignition cables are high-resistance, high-voltage conductors.

Battery Cables

The battery is connected to the rest of the electrical system by very large cables. Large cables are necessary to carry the high current required by the starter motor. Figure 6-11 shows several kinds of battery cables. Twelve-volt systems generally use number 4 or number 6 AWG wire cables; 6-volt systems and some 12-volt diesel systems require number 0 or number 1 AWG wire cables. Cables designed for a 6-volt system can be used on a 12-volt system, but the smaller cable intended for a 12-volt system cannot be used on a 6-volt system without causing too much voltage drop.

Battery installations may have an insulated ground cable or one made of braided, uninsulated wire. The braided cables or straps are flat instead of round; however, they have the same resistance and other electrical properties of a round cable of equivalent gauge. Most battery cables are fitted at one end with a lead terminal clamp to connect to the battery, although many import cars use a spring-clamp terminal. The lead terminal is used to reduce corrosion when attached to the lead battery post. A tinned copper terminal is attached to the other end of the cable to connect to the starter motor or ground, as required.

Ignition Cables

The ignition cables, or spark plug cables, are often called high-tension cables. They carry current at 10,000 to 40,000 volts from the coil to the distributor cap, and then to the spark plugs. Because of the high voltage, these cables must be very well insulated.

Years ago, all ignition cables were made with copper or steel wire conductors. During the past 30

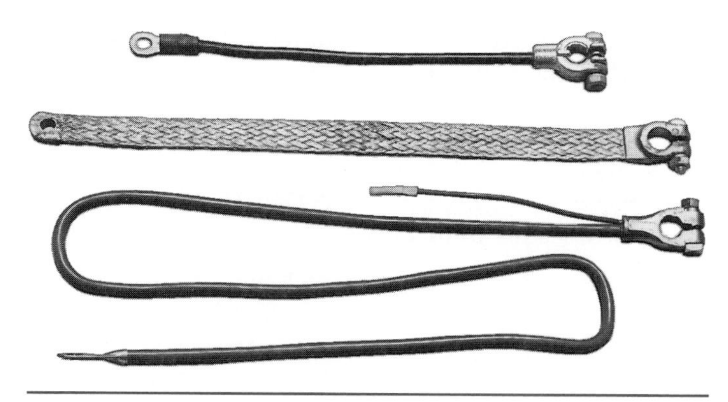

Figure 6-11. Assorted battery cables.

years, however, high-resistance, non-metallic cables have replaced metallic conductor cables as original equipment on cars and light trucks. Although metallic-conductor ignition cables are still made, they are sold for special high-performance or industrial applications and are not recommended for highway use. The conductors used in high-resistance, non-metallic ignition cables are made of carbon, or of linen or fiberglass impregnated with carbon. These cables evolved for the following reasons:

- High-voltage ignition pulses emit high-frequency electrical impulses or radio frequency interference (RFI) that interfere with radio and television transmission, as described in Chapter 2. The principal method used to limit this interference is the use of high-resistance ignition cables, often referred to as suppression cables.
- The extra resistance in the cable decreases the current flow and thus reduces the burning of spark plug electrodes. The higher resistance also helps take advantage of the high-voltage capabilities of the ignition system, as shown in Part Five of this manual.

The high-voltage current carried by ignition cables requires that they have much thicker insulation than low-voltage primary wires. Ignition cables are 7 or 8 millimeters in diameter, but the conductor in the center of the cable is only a small core. The rest of the cable diameter is the heavy insulation used to contain the high voltage and protect the core from oil, dirt, heat, and moisture.

One type of cable insulation material is known by its trade name, *Hypalon,* but the type most commonly used today is silicone rubber. Silicone is generally thought to provide greater high-voltage insulation while resisting heat and moisture better than other materials. However, silicone insulation is softer and more pliable than other materials and thus more likely to be torn or damaged by rough handling. Cables often have several layers of insulation over the conductor to provide the best insulating qualities with strength and flexibility.

CONNECTORS AND TERMINALS

Electrical circuits can be broken by the smallest gap between conductors. The gaps can be caused by corrosion, weathering, or mechanical breaks. One of the most common wear points in an automobile electrical system is where two conductors have been joined. Their insulation coats have been opened and the conductive material exposed. Special **connectors** are used to provide strong, permanent connections and to protect these points from wear.

These simple connectors are usually called wiring terminals. They are metal pieces that can be crimped or soldered onto the end of a wire. Terminals are made in many shapes and sizes for the many different types of connections required. They can be wrapped with plastic electrical tape or covered with special pieces of insulation. The simplest wire terminals join a single wire to a device, to another single wire, or to a few other wires (Figure 6-12). Terminals for connecting to a device often have a lug ring, a spade, or a hook, which can be bolted onto the device. Male and female spade terminals or bullet connectors are often used to connect two individual wires

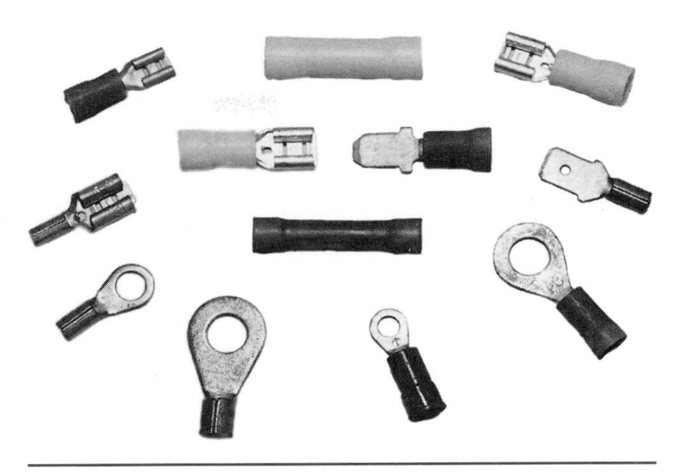

Figure 6-12. Some common single-wire terminals (connectors).

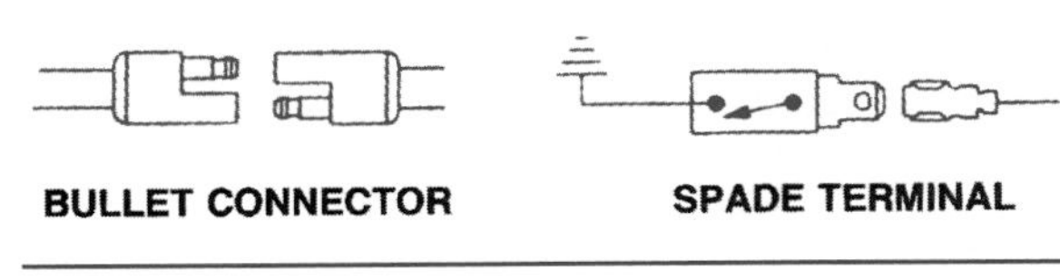

Figure 6-13. Male and female bullet connectors and spade terminals are common automotive connectors. (DaimlerChrysler Corporation)

(Figure 6-13). For more information about the use of different types of connectors, see the "Connector Repair" section in Chapter 6 of the *Shop Manual.*

Multiple Wire Connectors

Although the simple wiring terminals just described are really wire connectors, the term *connector* is normally used to describe multiple-wire connector plugs. This type of plug is used to connect wiring to switches, as shown in Figure 6-14, or to other components. It also is used to join wiring harnesses.

Multiple-wire connectors are sometimes called junction blocks. On older vehicles, a junction block was a stationary plastic connector with terminals set into it, in which individual wires were plugged or screwed in place. Because of the time required to connect this type of junction block on the assembly line, it has been replaced by a modem version that accepts several plugs from different harnesses (Figure 6-15).

Some multiple-connector plugs have as many as 40 separate connections in a single plug. They provide a compact, efficient way to connect wires for individual circuits while still grouping them together in harnesses. Wiring connections can be made quickly and accurately with multiple connectors, an important consideration in assembly-line manufacturing.

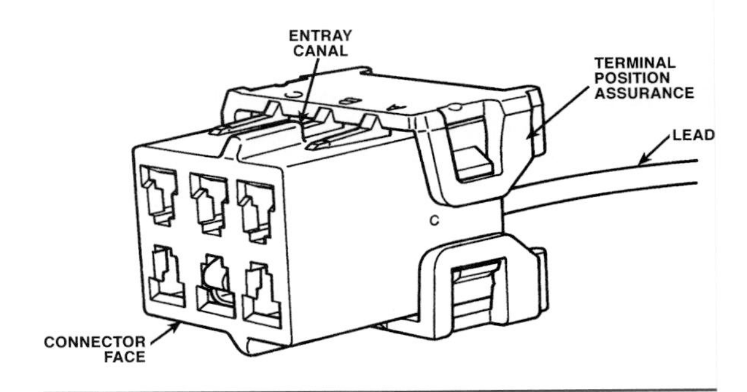

Figure 6-14. Multiple connectors are used to make complex switch connections. (DaimlerChrysler Corporation)

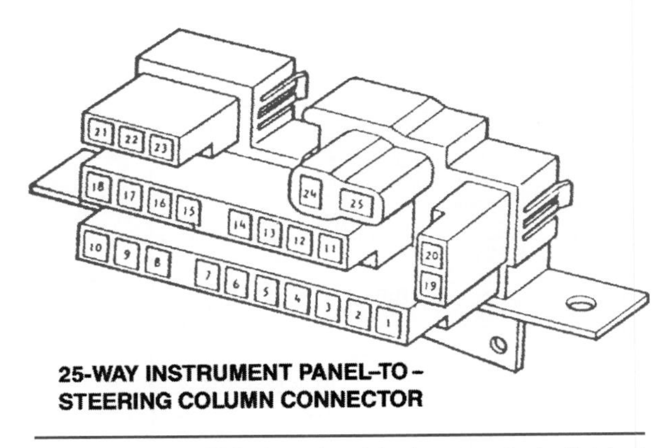

Figure 6-15. This junction block accepts individual wires on one side and connectors on the other. (DaimlerChrysler Corporation)

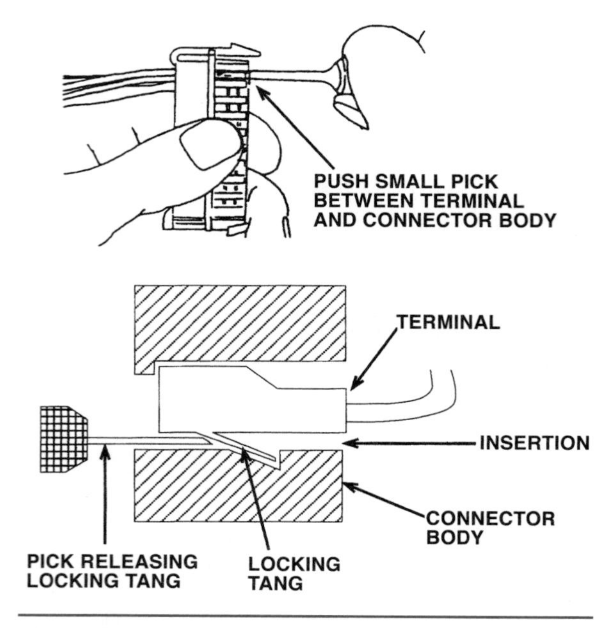

Figure 6-16. Connectors have some form of lock to prevent accidental separation. Individual terminals and wires can be removed from some connectors; other connectors are replaced as an entire assembly. (GM Service and Parts Operations)

Such connector plugs generally have hard plastic shells, with one half of the connector containing the male terminals or pins, and the other half containing the female terminals or sockets. Probing the rear of the individual connections without separating the connector can test circuit operation. A locking tab of some type is used to prevent the connector halves from separating. Separation or removal of the plug may require the locking tab to be lifted or depressed (Figure 6-16).

Although many hard-shell connector designs allow removal of the individual wires or their terminals for repair, as shown in Figure 6-17, manufacturers are now using plugs that are serviced as an assembly. If a wire or terminal is defective, the entire plug is cut from the harness. The replacement plug is furnished with 2 or 3 inches of wires extending from the rear of the plug. These plugs are designed to be replaced by matching and soldering their wire leads to the harness.

Bulkhead Connectors

A special multiple connector, called a bulkhead connector or bulkhead disconnect, is used where a

CAVITY	DESCRIPTION
1	WINDSHIELD WIPER
2	WINDSHIELD WIPER
3	BRAKE WARNING LAMP
4	VACANT
5	WINDSHIELD WIPER
6	WINDSHIELD WIPER
7	BACK-UP LAMP
8	BACK-UP LAMP
9	WINDSHIELD WIPER
10	HAZARD FLASHERS
11	RIGHT TURN SIGNAL
12	HORN
13	LEFT TURN SIGNAL
14	HIGH BEAM
15	LOW BEAM
16	TACHOMETER
17	IGNITION RUN
18	VACANT
19	IGNITION SWITCH
20	AMMETER
21	AMMETER
22	VACANT
23	IGNITION SWITCH
24	VACANT
25	HEADLAMP SWITCH
26	IGNITION SWITCH
27	HEADLAMP SWITCH
28	A/C HIGH BLOWER
29	OIL PRESSURE
30	VACANT
31	A/C CLUTCH
32	TEMPERATURE

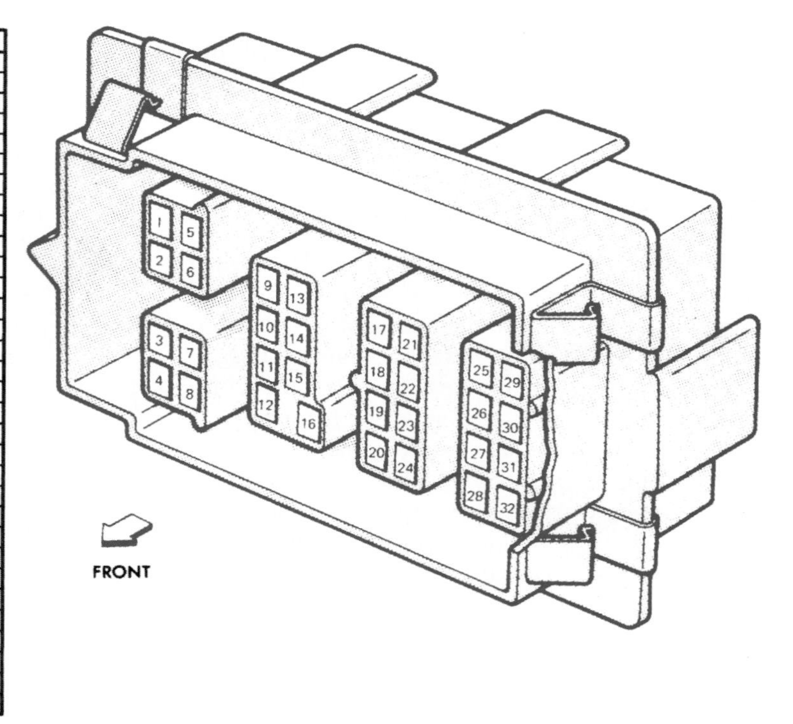

Figure 6-17. A bulkhead connector, or disconnect, is mounted on many firewalls. Multiple-wire connectors plug into both sides. (DaimlerChrysler Corporation)

number of wiring circuits must pass through a barrier such as the firewall (Figure 6-17). The bulkhead connector is installed in the firewall and multiple connectors are plugged into each side of it to connect wires from the engine and front accessories to wires in the rest of the car.

Weatherproof Connectors

Special **weatherproof connectors** are used in the engine compartment and body harnesses of late-model GM cars. This type of connector has a rubber seal on the wire ends of the terminals, with secondary sealing covers on the rear of each connector half. Such connectors are particularly useful in electronic systems where moisture or corrosion in the connector can cause a voltage drop. Some Japanese carmakers use a similar design (Figure 6-18).

GROUND PATHS

We have spoken as if wiring carried all of the current in an automotive electrical system. In fact, wiring is only about half of each circuit. The other half is the automobile engine, frame, and body, which provide a path for current flow. This side of the circuit is called the *ground* (Figure 6-19). Automotive electrical systems are called single-wire or ground-return systems.

The cable from one battery post or terminal is bolted to the car engine or frame. This is called the **ground cable.** The cable from the other battery terminal provides current for all the car's electrical loads. This is called the insulated, or hot, cable. The insulated side of every circuit in the vehicle is the wiring running from the battery to the devices in the circuit. The ground side of every circuit is the vehicle chassis (Figure 6-19).

The hot battery cable is always the insulated type of cable described earlier. The ground cable may be an insulated type of cable, or it may be a braided strap. On many vehicles additional grounding straps or cables are connected between the engine block and the vehicle body or frame. The battery ground cable may be connected to either the engine or the chassis, and the additional ground cable ensures a good, low-resistance ground path between the engine and the chassis. This is necessary for proper operation of the circuits on the engine and elsewhere in the vehicle. Late-model vehicles, which rely heavily on computerized components, often use additional ground straps whose sole purpose is to minimize or eliminate electromagnetic interference (EMI), as shown in Chapter 4.

The resistances in the insulated sides of all the circuits in the vehicle will vary depending on the number and kinds of loads and the length of the wiring. The resistance on the ground side of all circuits, that is, between each load and its ground connection, must be virtually zero. For more information about ground paths, see the "Copper Wiring Repair" section in Chapter 6 of the *Shop Manual*.

Early Wiring Problems

Early automobiles had many problems with their electrical systems, usually the result of poor electrical insulation. For example, high-tension cable

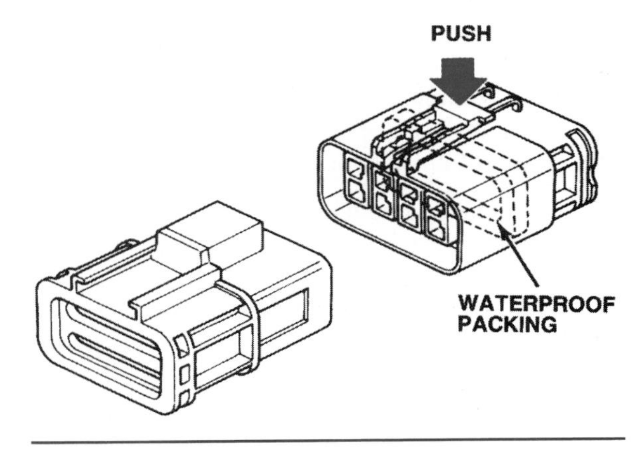

Figure 6-18. Nissan uses this type of waterproof connector. (Courtesy of Nissan North America, Inc.)

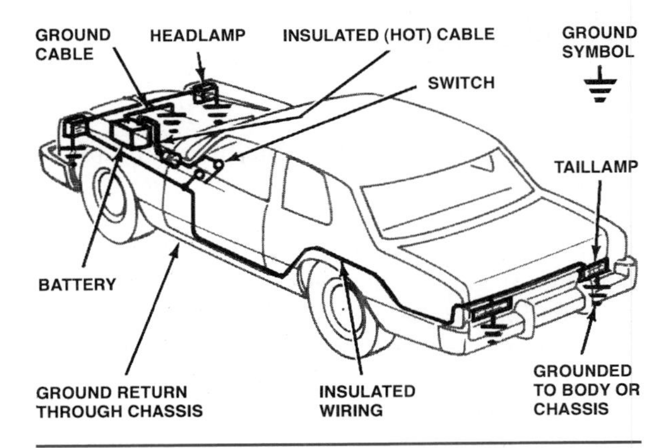

Figure 6-19. Half of the automotive electrical system is the ground path through the vehicle chassis.

insulation, made by wrapping cotton or silk around wire and then coating it with rubber, was easily hardened by heat. The insulation often broke off, leaving bare wire exposed.

A common problem in cars that used dry-cell batteries was moisture penetration through the battery's paper insulation. Current design would flow to ground and the batteries would become discharged.

Even washing a car sometimes caused trouble. Water got into the distributor terminals and made the engine hard to start. Some technicians poured melted wax into the space between the plug wires and the distributor cap terminals. For protection from heat, moisture, oil, and grease, wiring was often run through a metal conduit. Armored cable-insulated wire enclosed in a permanent, flexible metal wrapping was also used, especially in a circuit where any voltage drop was critical.

This is an important point to remember. It may be helpful at this time to review the explanations in Chapters 3 and 5 of voltage drops and current flow in various circuits from the source, through all the loads, and back to the source. Every electrical load is attached to the chassis so that current can pass through the ground and back to the grounded battery terminal. Grounding connections must be secure for the circuit to be complete. In older cars where plastics were rarely used, most loads had a direct connection to a metal ground. With the increased use of various plastics, designers have had to add a ground wire from some loads to the nearer metal ground. The ground wires in most circuits are black for easy recognition.

MULTIPLEX CIRCUITS

The use of **multiplexing,** or multiplex circuits, is becoming a necessity in late-model automobiles because of the increasing number of conventional electrical circuits required by electronic control systems. Wiring harnesses used on such vehicles have ballooned in size to 60 or more wires in a single harness, with the use of several harnesses in a vehicle not uncommon. Simply put, there are too many wires and too limited space in which to run them for convenient service. With so many wires in close proximity, they are subject to electromagnetic interference (EMI), which you learned about in Chapter 4. To meet the almost endless need for electrical circuitry in the growing and complex design of automotive control systems, engineers are gradually reducing the size and number of wire and wiring harnesses by using a multiplex wiring system.

The term *multiplexing* means different things to different people, but generally it is defined as a means of sending two or more messages simultaneously over the same channel. Different forms of multiplexing are used in automotive circuits. For example, windshield wiper circuits often use multiplex circuits. The wiper and washer functions in such circuit work though a single input circuit by means of different voltage levels. In this type of application, data is sent in parallel form. However, the most common form of multiplexing in automotive applications is serial data transmission, also known as time-division multiplex. In the time-division type of circuit, information is transmitted between computers through a series of digital pulses in a program sequence that can be read and understood by each computer in the system. The three major approaches to a multiplex wiring system presently in use are as follows:

- Parallel data transmission
- Serial data transmission
- Optical data links

We will look at each of these types of system, and then we will discuss the advantages of multiplexing over older systems of wiring.

Parallel Data Transmission

The most common parallel data multiplexing circuits use differentiated voltage levels as a means of controlling components. The multiplex wiring circuit used with a Type C General Motors pulse wiper-washer unit is shown in Figure 6-20. The circuit diagram shows several major advantages over other types of pulse wiper circuits, as follows:

- Eliminating one terminal at the washer pump reduces the wiring required between the wiper and control switch.
- Using a simple grounding-type control switch eliminates a separate 12-volt circuit to the fuse block.
- Eliminating a repeat park cycle when the wash cycle starts with the control switch in the OFF position—in standard circuits, the

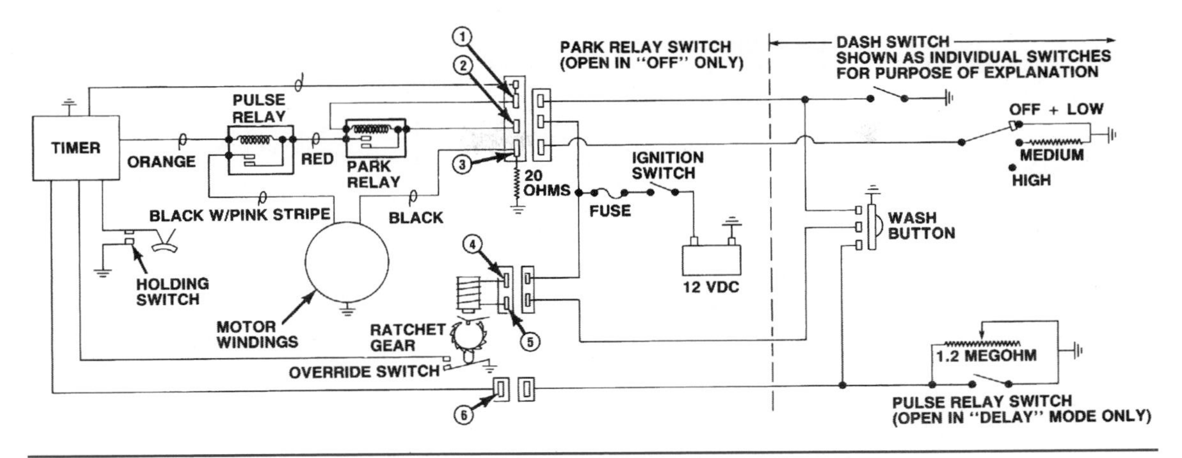

Figure 6-20. Parallel data transmission through differentiated voltage levels reduces the amount of wiring in this multiplex wiper-washer circuit. (DaimlerChrysler Corporation)

blades begin a wash cycle from the park position and return to park before continuing the cycle—simplifies operation.

An electronic timer controls the park and pulse relays. The timer consists of a capacitor, a variable resistor in the control switch, and electronic switching circuitry. The variable resistor controls the length of time required to charge the capacitor. Once the capacitor reaches a certain level of charge, it energizes the electronic switching circuit, completing the ground circuit to the pulse relay. This energizes the 12-volt circuit to the motor windings and the motor operates. When the driver presses the wash button, it grounds the washer pump ratchet relay coil circuits, starting a wash cycle. The electronic timer circuitry uses a high-voltage signal for wiper operation and a low-voltage signal for the wash cycle.

A multiplex circuit that functions with parallel data transmission is a good tool for simple circuit control. However, transmitting data in parallel form is slower and more cumbersome than transmitting in serial form. This is important when the signal is to be used by several different components or circuits at the same time.

Serial Data Transmission

Serial data transmission has become the most frequently used type of multiplex circuit in automotive applications. It is more versatile than parallel transmission but also more complex. A single circuit used to transmit data in both directions also is called a bus data link.

Sequencing voltage inputs transmitted in serial form can operate several different components, or elements within a single component. This allows each component or element to receive input for a specified length of time before the input is transmitted to another component or element. A four-element light-emitting diode (LED) display in the instrument cluster is a typical example. By rotating the applied voltage from left to right rapidly enough, each segment of the display is illuminated 25 percent of the time, but the human eye cannot detect that fact. To the eye, the entire display appears to be uniformly illuminated 100 percent of the time.

To prevent interference between the various signals transmitted, a multiplex system using bus data links must have a central transmitter (microprocessor) containing a special encoder. The system also requires a receiver with a corresponding decoder at each electrical load to be controlled. The transmitter and each receiver are connected to battery power and communicate through a two-way data link called a peripheral serial bus. Operational switches for each circuit to be controlled have an individual digital code or signal and are connected to the transmitter. When the transmitter receives a control code, it determines which switch is calling and sends the control signal to the appropriate receiver. The receiver then carries out the command. If a driver operates the headlamp switch, the transmitter signals the proper receiver to turn the headlights on or off, according to the switch position.

On the Chrysler application shown in Figure 6-21, each module has its own microprocessor

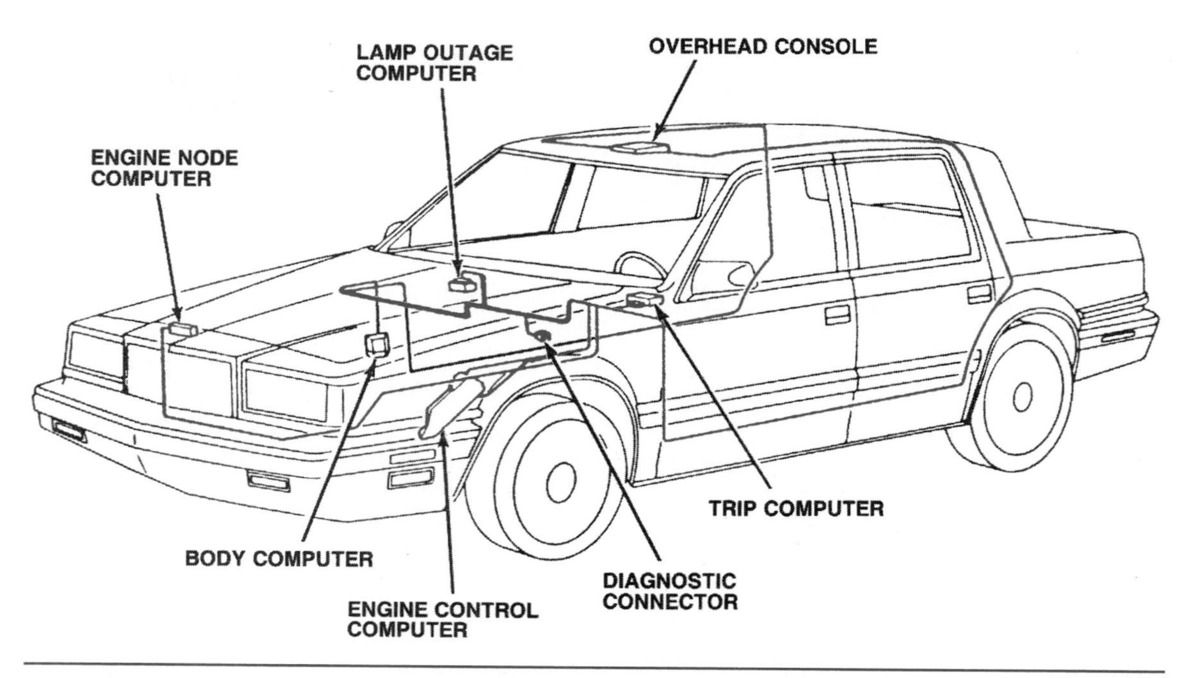

Figure 6-21. The DaimlerChrysler EVIC system is an example of a vehicle data communications network that allows separate computers to share data and communicate with each other through serial data transmission. (DaimlerChrysler Corporation)

connected to the data bus through the Chrysler Collision Detection (CCD) integrated circuit, which sends and receives data. The CCD circuit acts like a traffic control officer at a four-way intersection. If the data bus is not in use, it allows unrestricted transmission from a module. However, if one module is transmitting, it blocks the transmission of data from another module until the bus (intersection) is clear. If two or more modules start to transmit at the same time, or almost at the same time, the CCD circuit assigns a priority to the messages according to the identification code at the beginning of the transmission. If the CCD circuit blocks a message, the module that originally sent it retransmits the signal until it is successful.

Receivers work in one of two ways: they operate the electrical load directly, or they control a relay in the circuit to operate the load indirectly. They are not capable of making decisions on their own, but only carry out commands from the transmitter. However, they can send a feedback signal informing the transmitter that something is wrong with the system.

Optical Data Links

A variation of the serial data transmission approach to multiplexing substitutes optical data links or fiber-optic cables for the peripheral serial bus. The concept is the same, but light signals are substituted for voltage signals. An optical data link system operates with the transmitter and receivers described earlier, but a light-emitting diode (LED) in the transmitter sends light signals through the fiber-optic cables to a photo diode in the receiver. The light signals are decoded by the receiver, which then performs the required control function. Primarily Toyota and other foreign manufacturers have used this form of multiplexing. Because it uses light instead of voltage to transmit signals, system operation is not affected by EMI, nor does the system create interference that might have an adverse influence on other electrical systems in the vehicle.

Multiplex Advantages

Regardless of the type of multiplex system used, such a circuit offers several advantages over conventional wiring circuits used in the past, as follows:

- The size and number of wires required for a given circuit can be greatly reduced. As a result, the complexity and size of wiring harnesses also are reduced.
- The low-current-capacity switches used in a multiplex circuit allow the integration of

various touch-type switches into the overall vehicle design.
- The master computer or transmitter can be programmed with timing functions for convenience features, such as locking doors above a given speed or unlocking them when the ignition is shut off.

ELECTRICAL SYSTEM POLARITY

We discussed positive (+) and negative (−) electrical charges in Chapter 3. We learned that like charges repel each other and unlike charges attract each other. We also noted that the terminals of a voltage source are identified as positive and negative. In Chapter 2, we defined magnetic polarity in terms of the north and south poles of a magnet and observed that unlike poles of a magnet attract each other, just as unlike charges do. Similarly, like poles repel each other.

The polarity of an electrical system refers to the connections of the positive and negative terminals of the voltage source, the battery, to the insulated and ground sides of the system. All domestic cars and trucks manufactured since 1956 have the negative battery terminal connected to ground and the positive terminal connected to the insulated side of the system. These are called negative-ground systems and are said to have positive polarity.

Before 1956, 6-volt Ford and Chrysler vehicles had the positive battery terminal connected to ground and the negative terminal connected to the insulated side of the system. These are called positive-ground systems and are said to have negative polarity. Foreign manufacturers used positive-ground systems as late as 1969. In both kinds of systems, we say that current leaves the hot side of the battery and returns through the ground path to the grounded battery terminal.

In your service work, it is very important to recognize system polarity negative or positive ground before working on the electrical system. Some electrical components and test equipment are sensitive to the system polarity and must be installed with their connections matching those of the battery. Reversing polarity can damage alternators, cause motors to run backwards, ruin electronic modules, and cause relays or solenoids to malfunction.

COMMON ELECTRICAL PARTS

Many common electrical parts are used in various circuits in an electrical system. All circuits have switches of some kind to control current flow. Most circuits have some form of protective device, such as a fuse or circuit breaker, to protect against too much current flow. Various kinds of solenoids, relays, and motors are used in many circuits, and whatever their purpose, they operate in similar ways wherever they are used.

Before we look at complete circuits and system diagrams later in this next chapter, we should learn about some of the common devices used in many circuits.

Switches

Switches are used in automobile electrical systems to start, stop, or redirect current flow. They can be operated manually by the driver or remotely through mechanical linkage. Manual switches, such as the ignition switch and the headlamp switch, allow the driver to control the operation of the engine and accessories. Examples are shown in Figure 6-22; the driver or the passengers control a remotely operated switch indirectly. For example, a mechanical switch called a neutral safety switch on automatic transmission gear selectors will not let the engine start if the automobile is in gear. Switches operated by opening and closing the doors control the interior lights. For more information about switches, see the "Copper Wiring Repair" section in Chapter 6 of the *Shop Manual*.

Toggle Push-Pull Push Button

Switches exist in many forms but have common characteristics. They all depend upon physical movement for operation. A simple switch contains one or more sets of contact points, with half of the points stationary and the other half movable. When the switch is operated, the movable points change position.

Switches can be designed so that the points are normally open and switch operation closes them to allow current flow. Normally closed switches allow the operator to open the points and stop current flow. For example, in an automobile with a seatbelt warning buzzer, the switch points are opened when

Figure 6-22. Many different types of switches are used in the complete electrical system of a modern automobile.

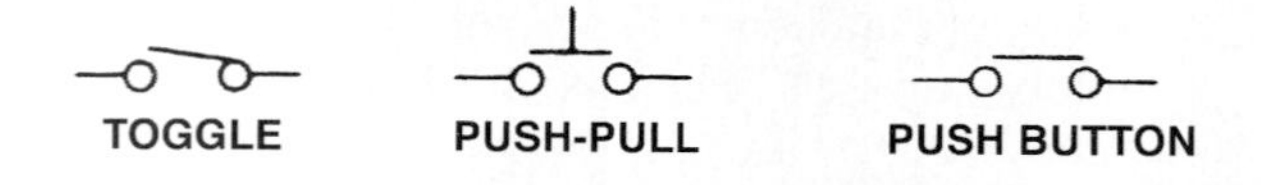

Figure 6-23. These symbols for normally open switches are used on electrical system diagrams.

the seatbelt is buckled; this stops current flow to the buzzer. Figure 6-23 shows the electrical symbols for some simple normally open switches.

A switch may lock in the desired position, or it may be spring-loaded so that a constant pressure is required to keep the points out of their normal position. Switches with more than one set of contact points can control more than one circuit. For example, a windshield wiper switch might control a low, medium, and high wiper speed, as well as a windshield washer device (Figure 6-24).

Switches are shown in simplified form on electrical diagrams so that current flow through them can easily be traced (Figure 6-25). Triangular contact points generally indicate a spring-loaded return, with circular contacts indicating a locking-position switch. A dashed line between the movable parts of a switch means that they are mechanically connected and operate in unison, as shown in Figure 6-26.

In addition to manual switches, automotive electrical circuits use a variety of other switch designs. Switches may be operated by temperature or pressure. Switches designed to sense engine coolant temperature contain a bimetal arm that flexes as it heats and cools, opening or closing the switch contacts (Figure 6-26). Oil pressure and vacuum switches respond to changes in pressure.

Mercury and inertia switches are motion-detector switches, that is, they open and close circuits automatically when their position is disturbed. A mercury switch uses a capsule containing two electrical contacts at one end. The other end is partially filled with mercury, which is a good conductor (Figure 6-27).

When the capsule moves a specified amount in a given direction, the mercury flows to the opposite end of the capsule and makes a circuit between the contacts. This type of switch often is used to turn on engine compartment or trunk lamps. It can also be used as a rollover switch to open an electric fuel pump or other circuit in an accident.

An inertia switch is generally a normally closed switch with a calibrated amount of spring pressure or friction holding the contacts together. Any sharp physical movement (a sudden change in inertia) sufficient to overcome the spring pressure or friction will open the contacts and break the circuit. This type of switch is used to open the fuel pump circuit in an impact collision. After the switch has opened, it must be reset manually to its normally closed position.

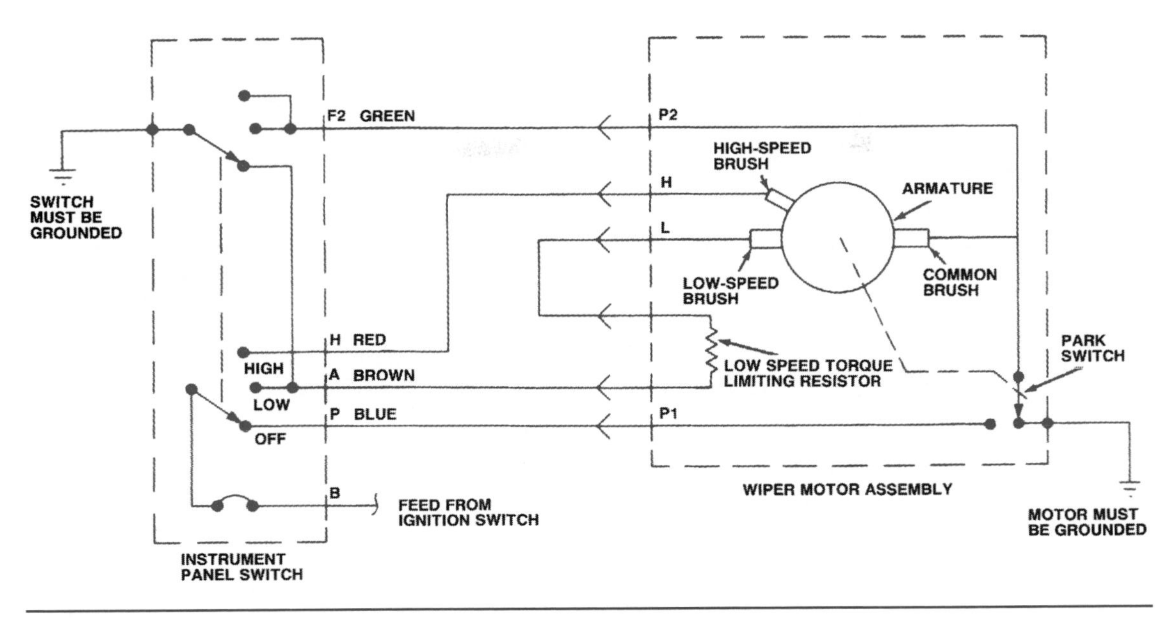

Figure 6-24. The instrument panel switch in this two-speed windshield wiper circuit has two sets of contacts linked together, as shown by the broken line. The Park switch is operated by mechanical linkage from the wiper motor armature. (DaimlerChrysler Corporation)

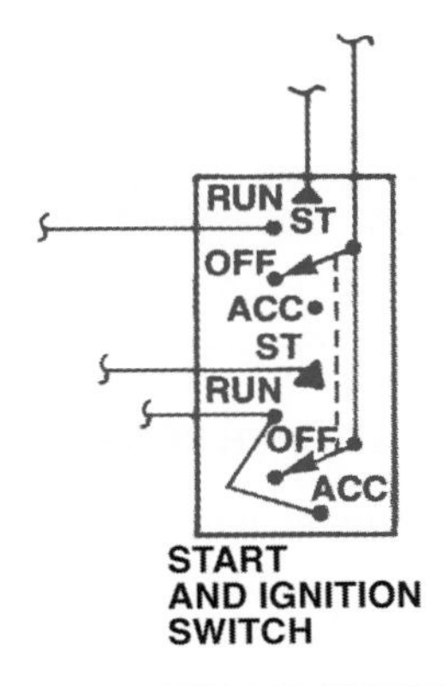

Figure 6-25. This starting and ignition switch has two sets of contacts linked together by the dashed line. Triangular terminals in the start (ST) position indicate that this position is spring-loaded and that the switch will return to RUN when the key is released. (DaimlerChrysler Corporation)

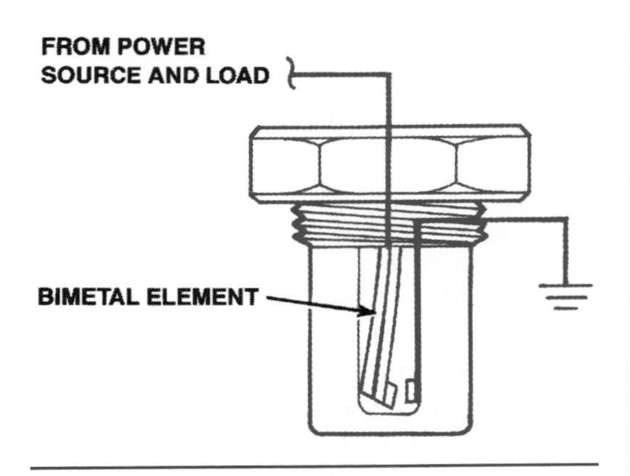

Figure 6-26. A coolant temperature switch in its normally open position.

Relays

A *relay* is a switch that uses electromagnetism to physically move the contacts. It allows a small current to control a much larger one. As you remember from our introduction to relays in Chapter 2, a small amount of current flow through the relay coil moves an armature to open or close a set of contact points. This is called the control circuit because the points control the flow of a much larger amount of current through

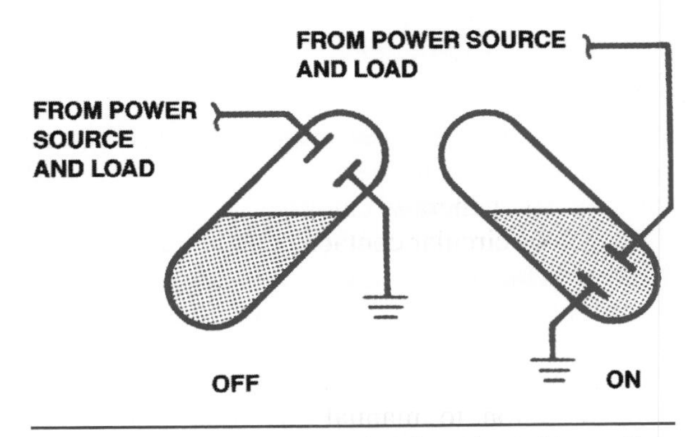

Figure 6-27. A mercury switch is activated by motion.

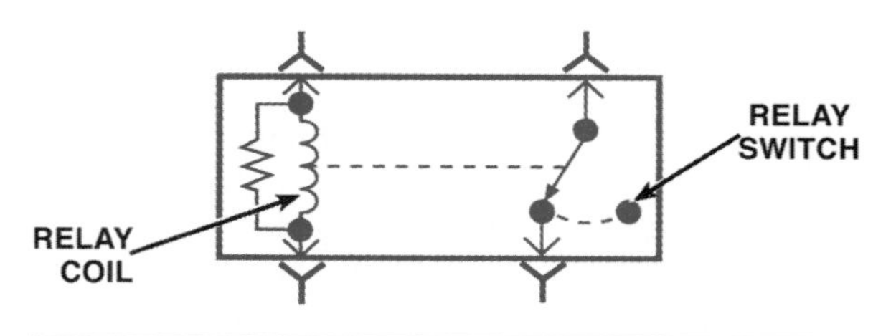

Figure 6-28. A relay contains a control circuit and a power circuit.

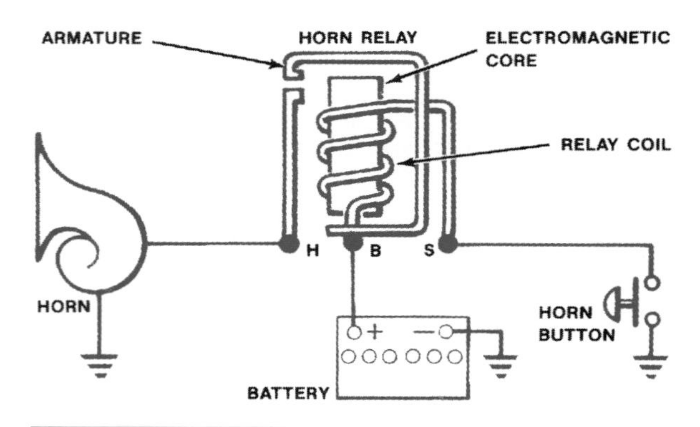

Figure 6-29. When the horn button is pressed, low current through the relay coil magnetizes the core. This pulls the armature down and closes the contacts to complete the high-current circuit from the battery to the horn.

a separate circuit, called the power circuit (Figure 6-28).

A relay with a single control winding is generally used for a short duration, as in a horn circuit (Figure 6-29). Relays designed for longer periods or continuous use require two control windings. A heavy winding creates the magnetic field necessary to move the armature; a lighter second winding breaks the circuit on the heavy winding and maintains the magnetic field to hold the armature in place with less current drain.

Solenoids

A **solenoid** is similar to a relay in the way it operates. The major difference is that the solenoid core moves instead of the armature, as in a relay. This allows the solenoid to change current flow into mechanical movement.

Solenoids consist of a coil winding around a spring-loaded metal plunger (Figure 6-30). When the switch is closed and current flows through the windings, the magnetic field of the coil attracts the movable plunger, pulling it

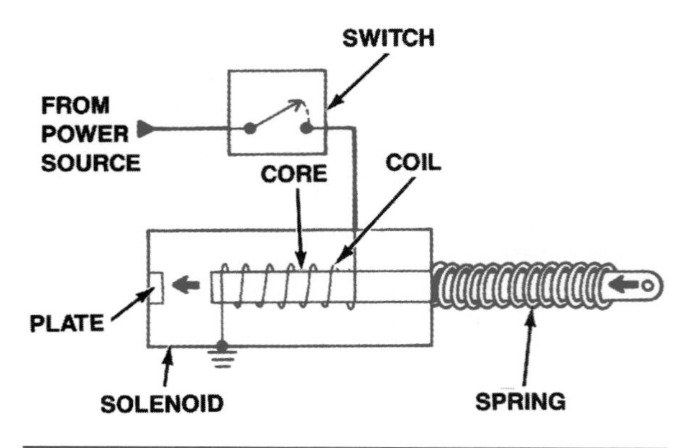

Figure 6-30. Energizing a solenoid moves its core, converting current flow into mechanical movement. (GM Service and Parts Operations)

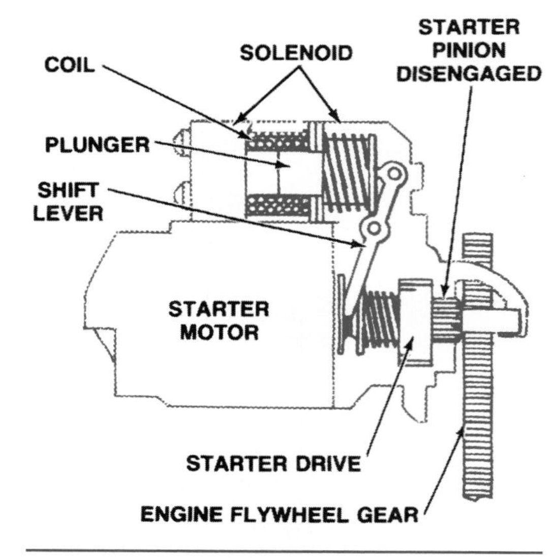

Figure 6-31. A starter solenoid mounted on the starter motor. Solenoid movement engages the starter drive with the engine flywheel gear.

against spring pressure into the center of the coil toward the plate. Once current flow stops, the magnetic field collapses and spring pressure moves the plunger out of the coil. This type of solenoid is used to operate remote door locks and to control vacuum valves in emission control and air conditioning systems.

The most common automotive use of a solenoid is in the starter motor circuit. In many systems, the starter solenoid is designed to do two jobs. The movement of the plunger engages the starter motor drive gear with the engine flywheel ring gear so that the motor can crank the engine (Figure 6-31). The starter motor requires high current, so the solenoid also acts as a relay. When the plunger moves into the coil, a large contact point on the plunger meets a large stationary contact point (Figure 6-32). Current flow across these contact points completes the battery-to-starter motor circuit. The plunger must remain inside the coil for as long as the starter motor needs to run.

A large amount of current is required to draw the plunger into the coil, and the starter motor also requires a large amount of current. To conserve battery energy, starting circuit solenoids have two coil windings, the primary or pull-in winding and the secondary or hold-in winding (Figure 6-33). The pull-in winding is made of very large diameter wire, which creates a magnetic field strong enough to pull the plunger into the coil. The hold-in winding is made of much smaller diameter wire. Once the plunger is inside the coil, it is close enough to the hold-in winding that a weak magnetic field will hold it there. The large current flow through the pull-in winding is stopped when the plunger is completely inside the coil, and only the smaller hold-in winding draws current from the battery. The pull-in winding on a starter solenoid may draw from 25 to 45 amperes. The hold-in winding may draw only 7 to 15 amperes. Some starter motors do not need the solenoid movement to engage gears; circuits for these motors use a solenoid primarily as a current switch. The physical movement of the plunger brings it into contact with the battery and starter terminals of the motor (Figure 6-34).

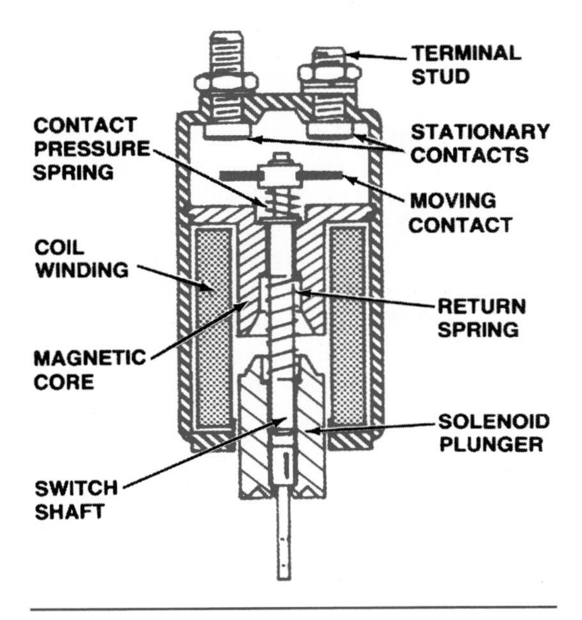

Figure 6-32. A starter solenoid also acts as a relay.

Buzzers and Chimes

Buzzers are used in some automotive circuits as warning devices. Seatbelt buzzers and door-ajar buzzers are good examples. A buzzer is similar in construction to a relay but its internal connections differ. Current flow through a coil magnetizes a core to move an armature and a set of contact points. However, in a buzzer, the coil is in series with the armature and the contact points are normally closed.

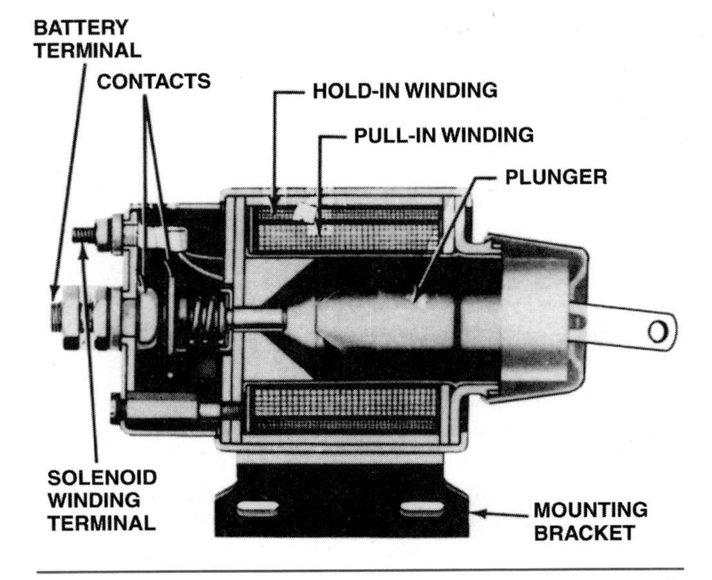

Figure 6-33. A starter solenoid, showing the pull-in and hold-in windings. (Delphi Corporation)

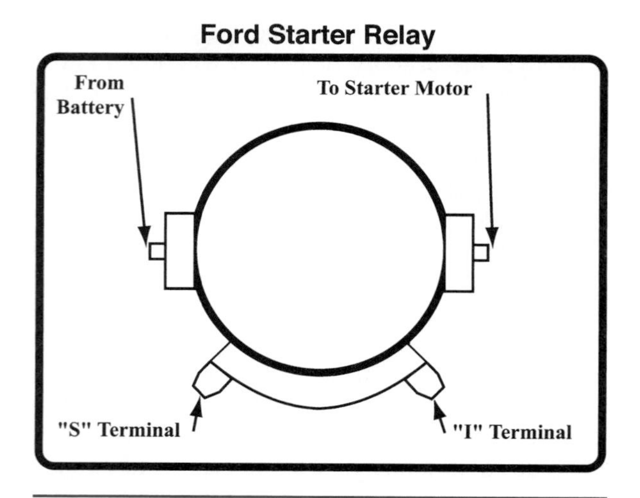

Figure 6-34. When the Ford starter relay is energized, the plunger contact disk moves against the battery and starter terminals to complete the circuit.

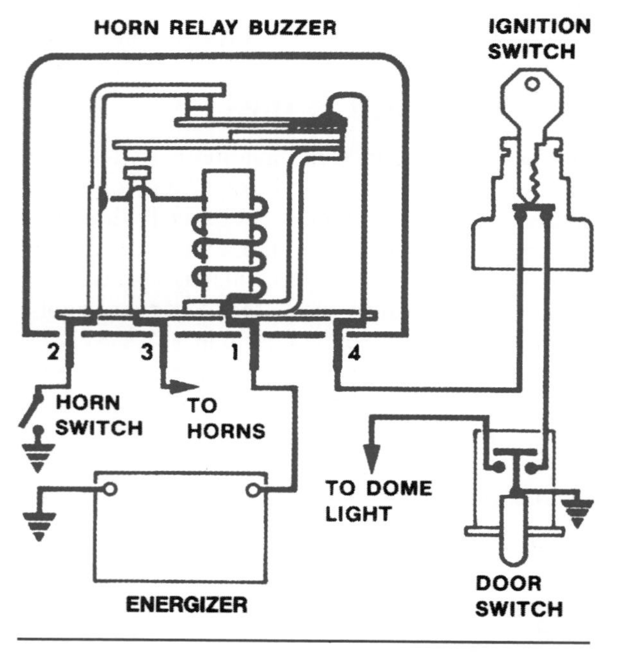

Figure 6-35. Typical horn relay and buzzer circuits. (Delphi Corporation)

When the switch is closed, current flow through the buzzer coil reaches ground through the normally closed contacts. However, current flow also magnetizes the buzzer core to move the armature and open the contacts. This breaks the circuit, and current flow stops. Armature spring tension then closes the contacts, making the circuit again (Figure 6-35). This action is repeated several hundred times a second, and the vibrating armature creates a buzzing sound.

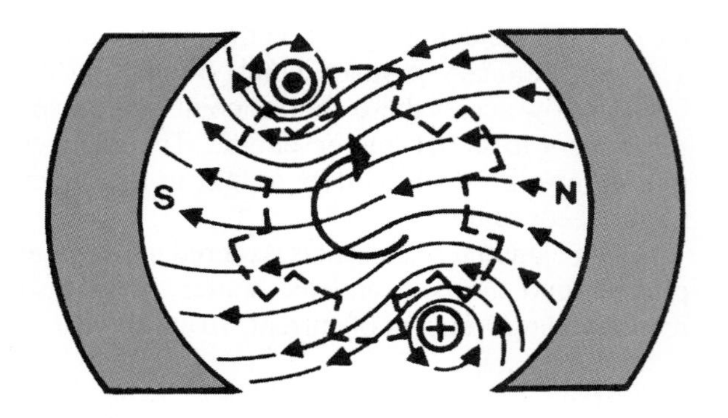

Figure 6-36. The motor principle.

Most simple automotive buzzers are sealed units and simply plug into their circuits. Some buzzers are combined in a single assembly with a relay for another circuit (Figure 6-35), such as a horn relay. This application is used on some General Motors cars. While mechanical buzzers are still in use, they are comparatively heavy and draw a relatively high current compared to the lighter solid-state chimes and buzzers provided by electronic technology and tone generators.

Motors

The typical automotive electrical system includes a number of motors that perform various jobs. The most common is the starter motor (also called a cranking motor), which rotates the automobile's crankshaft until the engine starts and can run by itself. Smaller motors run windshield wipers, power windows, and other accessories. Whatever job they do, all electric motors operate on the same principles of electromagnetism.

We explained the motor principle in terms of magnetic field interaction in Chapter 4. When a current-carrying conductor is placed in an external magnetic field, it tends to move out of a strong field area and into a weak field area (Figure 6-36). This motion can be used to rotate an armature. Now we will see how automotive electrical motors are constructed and used.

A simple picture of electric motor operation (Figure 6-37) looks much like the operation of a simple generator. Instead of rotating the looped conductor to induce a voltage, however, we are applying a current to force the conductor to rotate. As soon as the conductor has made a half-revolution, the field interaction would tend to force it back in the opposite direction. To keep the conductor rotating in one direction, the current flow through the conductor must be reversed.

This is done with a split-ring commutator, which rotates with the conductor as shown in Figure 6-37. Current is carried to the conductor through carbon brushes. At the point where current direction must be reversed, the commutator has rotated so that the opposite half of the split ring is in contact with the current-feeding brush. Current flow is reversed in the conductor and rotation continues in the original direction. In actual motors, many more conductor loops are mounted on an armature (Figure 6-38).

Electric motors can be manufactured with several brushes and varying combinations of series and parallel connections for armature windings and electromagnetic field windings. The design depends upon the use to which the motor will be put. Electric motors generally use electromagnetic field poles because they can produce a strong field in a limited space. Field strength in such a motor is determined by the current flow through the field windings. The starter motor is the most common automotive application of this design.

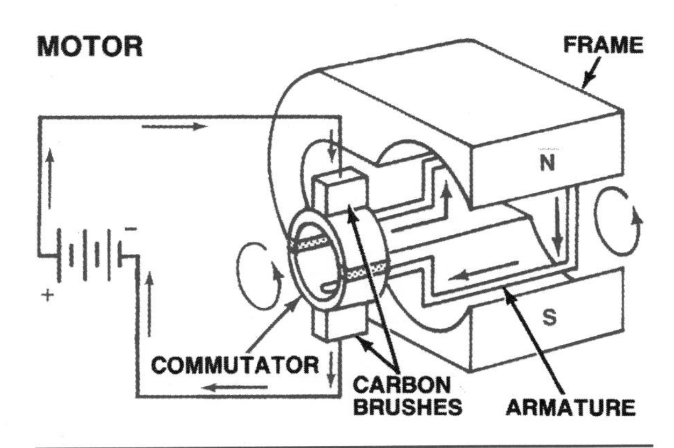

Figure 6-37. A simple motor.

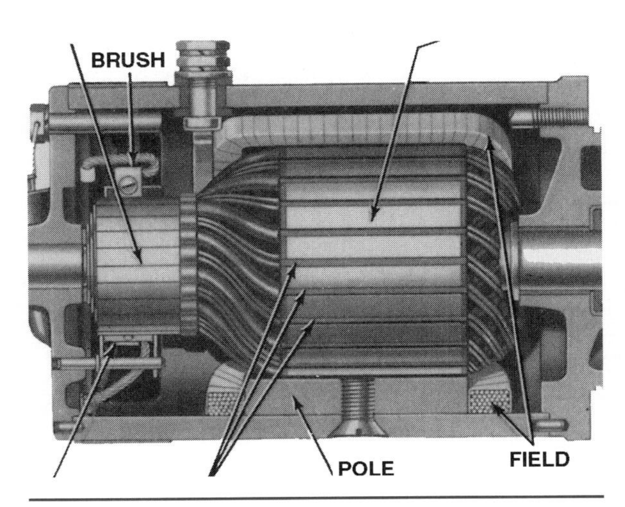

Figure 6-38. An electric motor. (Delphi Corporation)

Most small motors used in automotive applications, however, are built with permanent magnet fields. These motors are inexpensive, lightweight, can reverse direction of operation if necessary, and can be equipped with up to three operating speeds. They are ideal for constant light loads, such as a small electric fan.

Regardless of how they are built, all motors work on these principles. Understanding the internal connections of a motor is essential for testing and repair. Figure 6-39 shows the circuit symbol for a motor.

Figure 6-39. The electrical symbol for a motor.

WIRE COLOR CODING

Figure 6-40 shows current flows through a simple circuit consisting of a 12-volt battery for power, a fuse for protection, a switch for control, and a lamp as the load. In this example, each component is labeled and the direction of current is marked. Manufacturers use **color coding** to help technicians follow wires in a circuit. We have explained how most automotive wires are covered with a colored polyvinyl chloride (PVC), or plastic, insulation. The color of the insulation helps identify a particular wire in the system. Some drawings of a circuit have letters and numbers printed near each wire (Figure 6-41). The code table accompanying the drawing

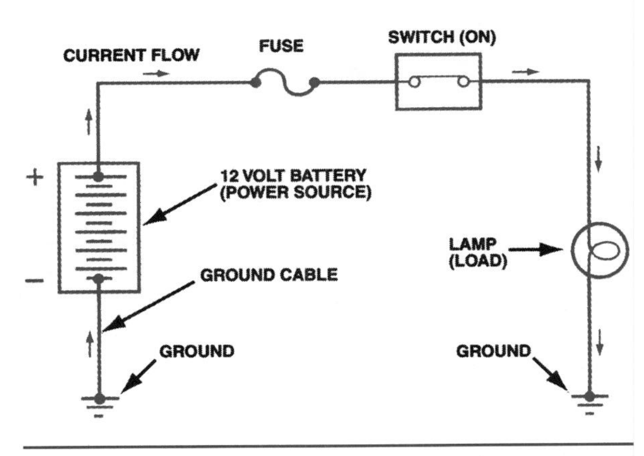

Figure 6-40. Diagram of a simple circuit.

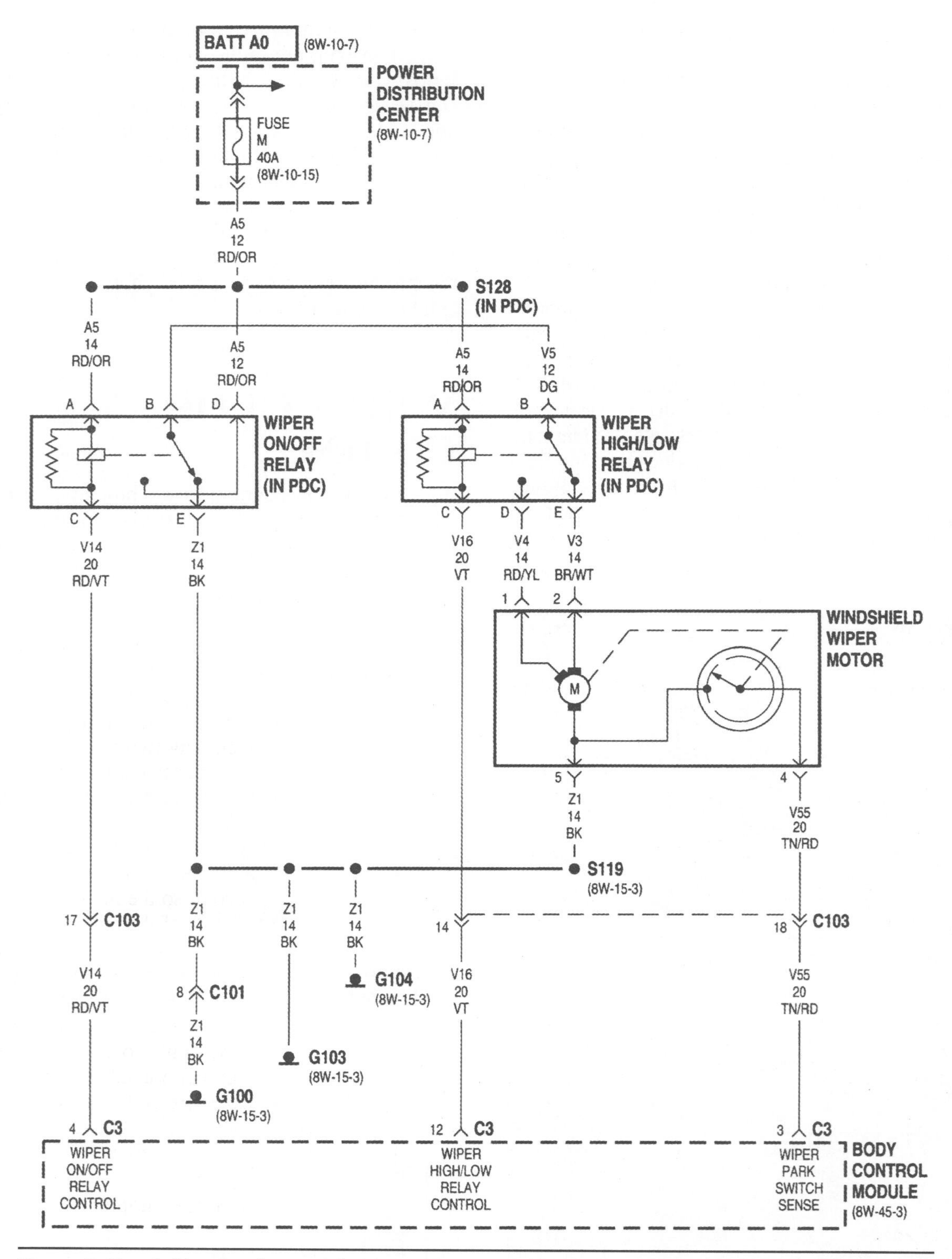

Figure 6-41. A Chrysler diagram showing circuits identified by number and wire color. (DaimlerChrysler Corporation)

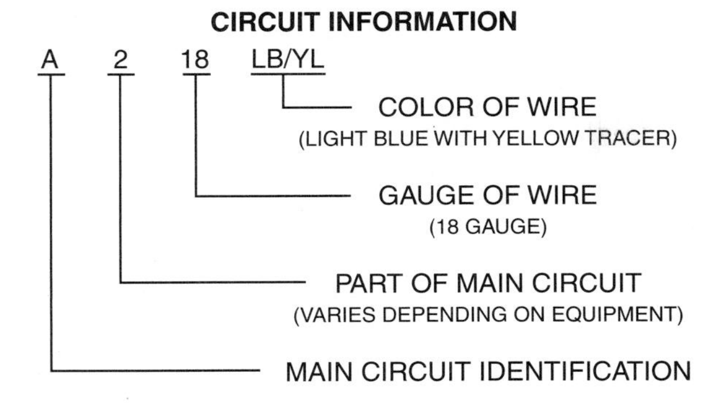

WIRE COLOR CODE CHART

COLOR CODE	COLOR	STANDARD TRACER COLOR
BL	BLUE	WT
BK	BLACK	WT
BR	BROWN	WT
DB	DARK BLUE	WT
DG	DARK GREEN	WT
GY	GRAY	BK
LB	LIGHT BLUE	BK
LG	LIGHT GREEN	BK
OR	ORANGE	BK
PK	PINK	BK or WT
RD	RED	WT
TN	TAN	WT
VT	VIOLET	WT
WT	WHITE	BK
YL	YELLOW	BK
*	WITH TRACER	

Figure 6-42. Chrysler circuit identification and wire color codes. (DaimlerChrysler Corporation)

WIRE IDENTIFICATION CHART

COLOR	SYMBOL	COLOR	SYMBOL
ALUMINUM	AL	NATURAL	NAT
BLACK	BLK	ORANGE	ORN
BLUE-LIGHT	BLU LT	PINK	PINK
BLUE-DARK	BLU DK	PURPLE	PPL
BROWN	BRN	RED	RED
GLAZED	GLZ	TAN	TAN
GREEN-LIGHT	GRN LT	VIOLET	VLT
GREEN-DARK	GRN DK	WHITE	WHT
GRAY	GRA	YELLOW	YEL
MAROON	MAR		

Figure 6-43. GM diagrams printed in color in the service manual include this table of color abbreviations. (GM Service and Parts Operations)

explains what the letters and numbers stand for. This Chrysler diagram contains code information on wire gauge, circuit numbers, and wire color. Circuit numbers are discussed later in this chapter. Figures 6-41, 6-42, 6-43, and 6-44 show how Chrysler, GM, and Ford may present color-code information. Note that the Toyota diagram in Figure 6-45 simply has the color name printed on the wires; wire gauge is not identified in this drawing.

For more information about wire color coding, see the "Copper Wiring Repair" in Chapter 6 of the *Shop Manual*.

THE LANGUAGE OF ELECTRICAL DIAGRAMS

In this chapter, illustrations from GM, Chrysler, Ford, Toyota, and Nissan show how different manufacturers present electrical information. Note that many component symbols and circuit identification do not look exactly the same among different vehicle manufacturers. Once you become familiar with the diagrams, the differences become less confusing.

Circuit Numbers

If the wire is labeled with a **circuit number,** as in Figures 6-41 and 6-44, those circuits are identified in an accompanying table. The top half of Figure 6-42 shows the Chrysler method of identifying circuits with a letter and number. Any two wires with the same circuit number are connected within the same circuit. Some General Motors service manuals contain current-flow diagrams developed by SPX Valley Forge Technical Information Systems; However, GM no longer uses these diagrams. Electrical circuit diagrams are printed in color so the lines match the color of the wires. The name of the color is printed beside the wire (Figure 6-46). The metric wire gauge may also be printed immediately before the color name. Other GM drawings contain a statement that all wires are of a certain gauge, unless otherwise identified. If this is the case, only some wires in the drawing have a gauge number printed on them.

The Ford circuit and table in Figure 6-44 are for a heater and air conditioner electrical circuit. The wire numbers are indicated by code numbers, which are also circuit numbers. Again, no wire gauges are identified in this example.

Wire Sizes

Another piece of information found in some electrical diagrams is the wire size. In the past, vehicles built in the United States used wire sizes specified by gauge. Gauge sizes typically vary from 2 for a

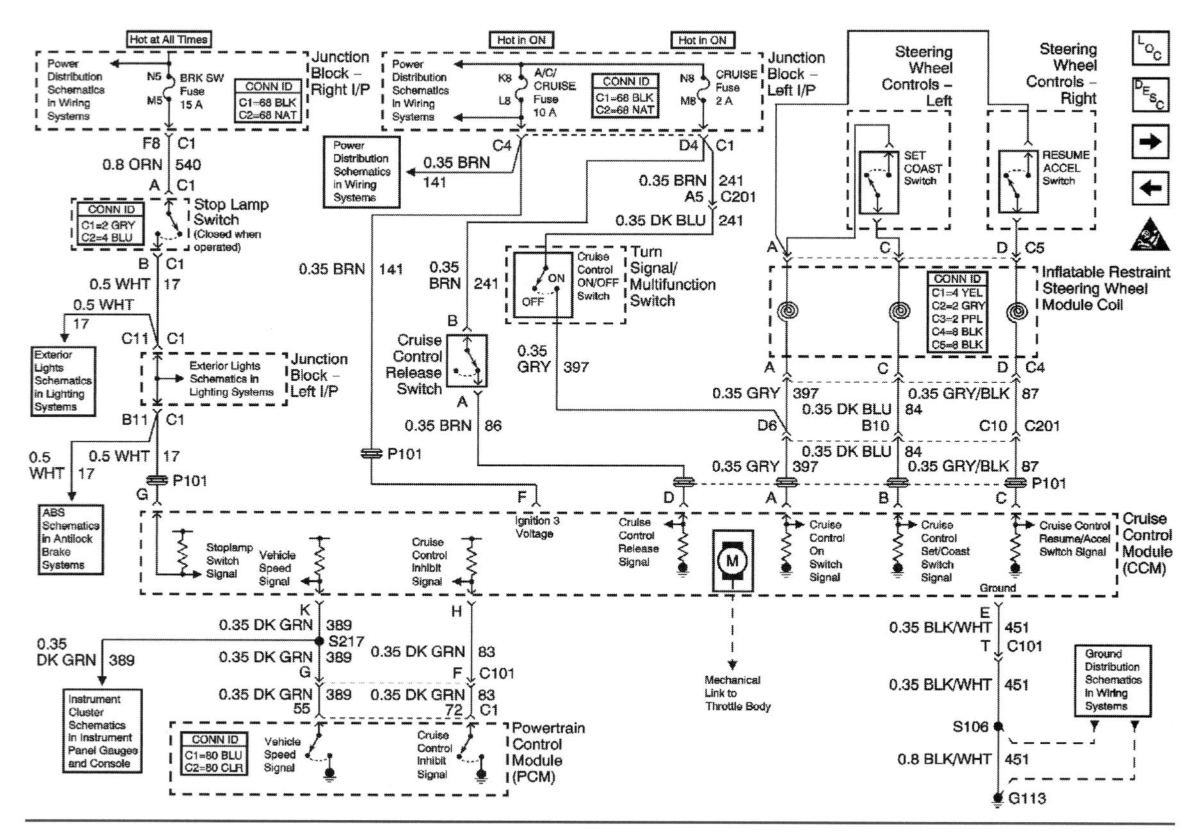

Figure 6-44. The GM accessory circuit is color coded by circuit number. (GM Service and Parts Operations)

starter cable to 20 for a license plate lamp. Note that gauge-size numbers are the reverse of physical wire sizes: a lower gauge number for heavy wires and a higher one for light wires. Figure 6-47 shows a typical circuit using 20-gauge wire.

Most vehicles built in recent years specify wire sizes by their diameter in millimeters (mm). In this case, a starter cable might be 32 mm while a typical circuit might be 1 mm or 0.8 mm. The wire size appears next to the color and on the opposite side of the wire from the circuit number, as shown in Figure 6-46. Note that the "mm" abbreviation does not appear in the diagram. An advantage to using the metric system is that wire size corresponds directly to thickness.

Component Symbols

It is time to add new symbols to the basic **component symbols** list (Figure 6-48). Figures 6-49A and 6-49B show additional symbols for many of the electrical devices on GM vehicles. Figure 6-50 illustrates symbols used by DaimlerChrysler Corporation. Nissan, like other manufacturers, often includes the symbols with its components, connector identification, and switch continuity positions (Figure 6-51). Switch continuity diagrams are discussed later in this chapter. For more information about component symbols, see the "Copper Wiring Repair" section in Chapter 6 of the *Shop Manual*.

Figure 6-52 is a basic diagram of a Toyota Celica sunroof control relay, which controls the sunroof motor operation. Figure 6-53 shows how the circuit is activated to tilt the sunroof open. The current travels to the motor through relay number one and transistor one when the "up" side of the tilt switch is pressed.

DIAGRAMS

The color codes, circuit numbers, and symbols just illustrated are combined to create a variety of electrical diagrams. Most people tend to refer to any electrical diagram as a "wiring diagram," but

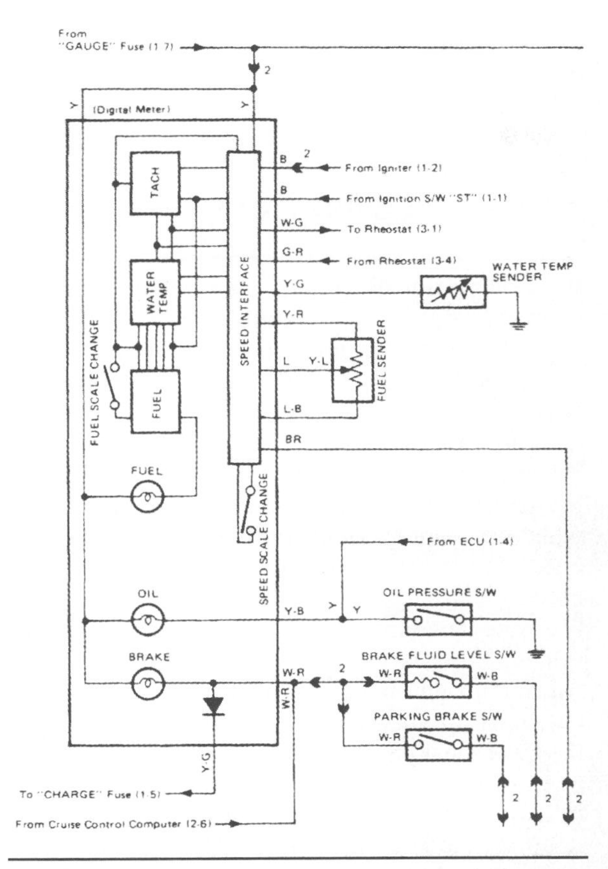

Figure 6-45. This Toyota diagram has no color code table; wire color abbreviations are printed directly on the drawing. (Reprinted by permission of Toyota Motor Corporation)

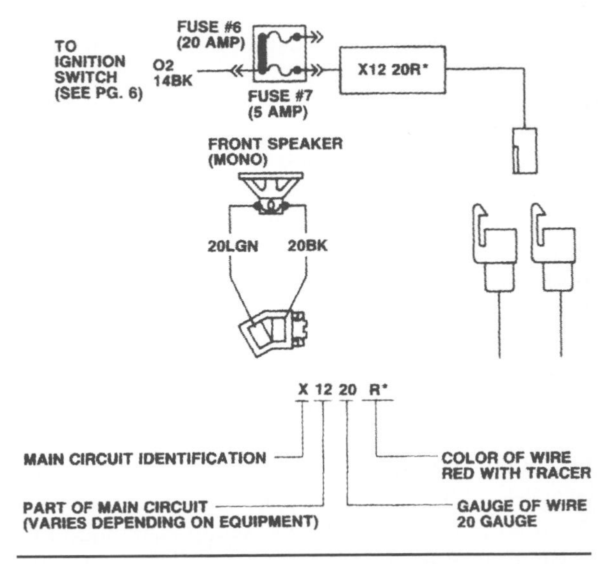

Figure 6-47. In this example from Chrysler, the "X12" in the wire code stands for the #12 part of the main circuit. (DaimlerChrysler Corporation)

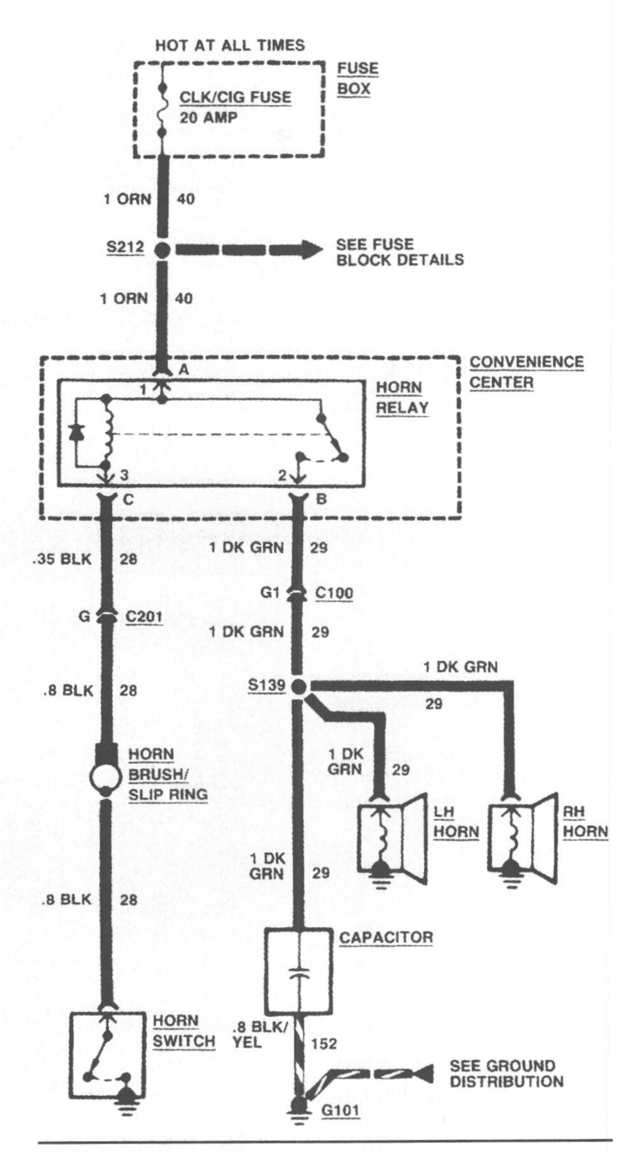

Figure 6-46. General Motors Valley Forge schematics are provided in color with the name of the color printed beside the wire. (GM Service and Parts Operations)

there are at least three distinct types with which you should be familiar:

- System diagrams (also called "wiring" diagrams)
- Schematic diagrams (also called "circuit" diagrams)
- Installation diagrams (also called "pictorial" diagrams)

System Diagrams

A system diagram is a drawing of the entire automobile electrical system. This may also properly

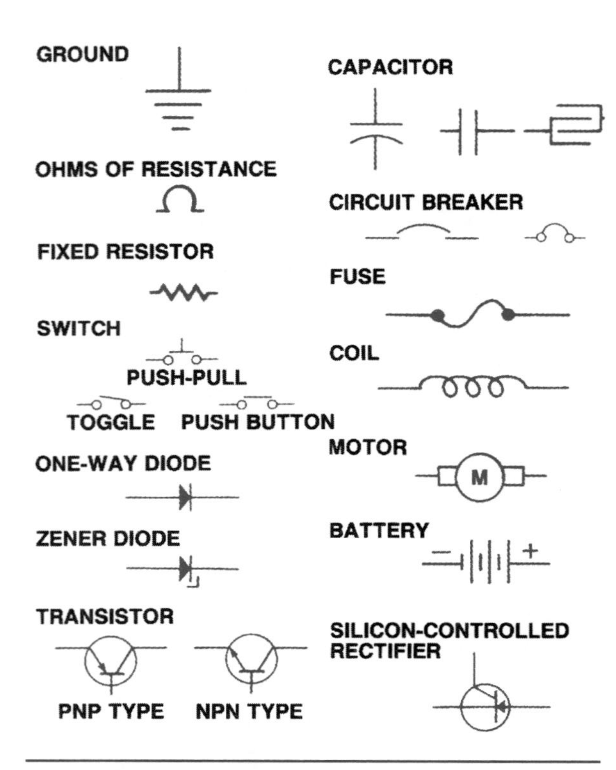

Figure 6-48. These electrical symbols are discussed in the *Classroom Manual.*

be called a "wiring diagram." System diagrams show the wires, connections to loads, switches, and the type of connectors used, but not how the loads or switches work. Installation diagrams express where and how the loads and wires are installed. This is covered later in this chapter. Figure 6-54 shows the same warning lamp circuit as Figure 6-55, but in a different format. System diagrams may cover many pages of a system and grounds are identified for all circuits. The diagram is also organized by individual subsystems at the top. This variation on the grid theme is another tool to quickly locate the desired part of the diagram. A Chrysler Corporation shop manual may not supply an entire system diagram for a vehicle, but may instead illustrate all circuits work. Figure 6-54 shows the same warning lamp circuit as Figure 6-55 but in a different format. System diagrams may cover many pages of a manual as ground points are identified for all circuits (Figure 6-56). The diagram is also organized by individual subsystems at the top. This variation on the grid theme is another tool to quickly locate the desired part of the diagram. A DaimlerChrysler Corporation shop manual may not supply an entire system diagram for a vehicle, but may instead illustrate all circuits separately, as shown in Figure 6-57.

Schematic Diagrams

A **schematic diagram,** also called a "circuit diagram" describes the operation of an individual circuit. Schematics tell you how a circuit works and how the individual components connect to each other (Figure 6-55). Engineers commonly use this type of diagram.

Some schematics are *Valley Forge diagrams,* which present current moving vertically. The power source is at the top and the ground at the bottom of the page (Figure 6-58). Figure 6-57 illustrates the circuit for a DaimlerChrysler radio system. Some of the wires are fully identified with two circuit numbers, wire gauge, and wire color. Other wires, such as the two wires connected to the front speaker, are identified only by wire gauge and color. The "20LGN" indicates a 20-gauge, light green wire. Figure 6-59 is the fuel economy lamp circuit in a GM vehicle. Here, neither wire gauge nor wire color is indicated. The "green" and "amber" refer to the color of the lamp bulbs.

Figure 6-60 shows a Ford side marker lamp circuit. Again, wire size and color are not identified. The numbers on the wires are circuit numbers. Note that the ground wires on the front and rear lamps may not be present depending upon the type of lamp socket used on the automobile.

Switches

Some manufacturers, such as Nissan, extend the system diagram to include major switches, as in the headlight circuit shown in Figure 6-53. This illustration shows the current traveling from the fuse block, through the switch, and to the headlights. If a switch does not work properly, it causes a malfunction in the electrical system. Switch diagrams may take extra time to understand, but they are indispensable in testing and diagnosis.

Each connection is shown as two circles joined by a line. The grid diagram shows which individual circuits have power at each switch position. A drawing of the headlight switch is included to explain the meaning of OFF, 1ST, 2ND, A, B, and C. Normally, a drawing of the switch action does not accompany the system diagram. If the switch

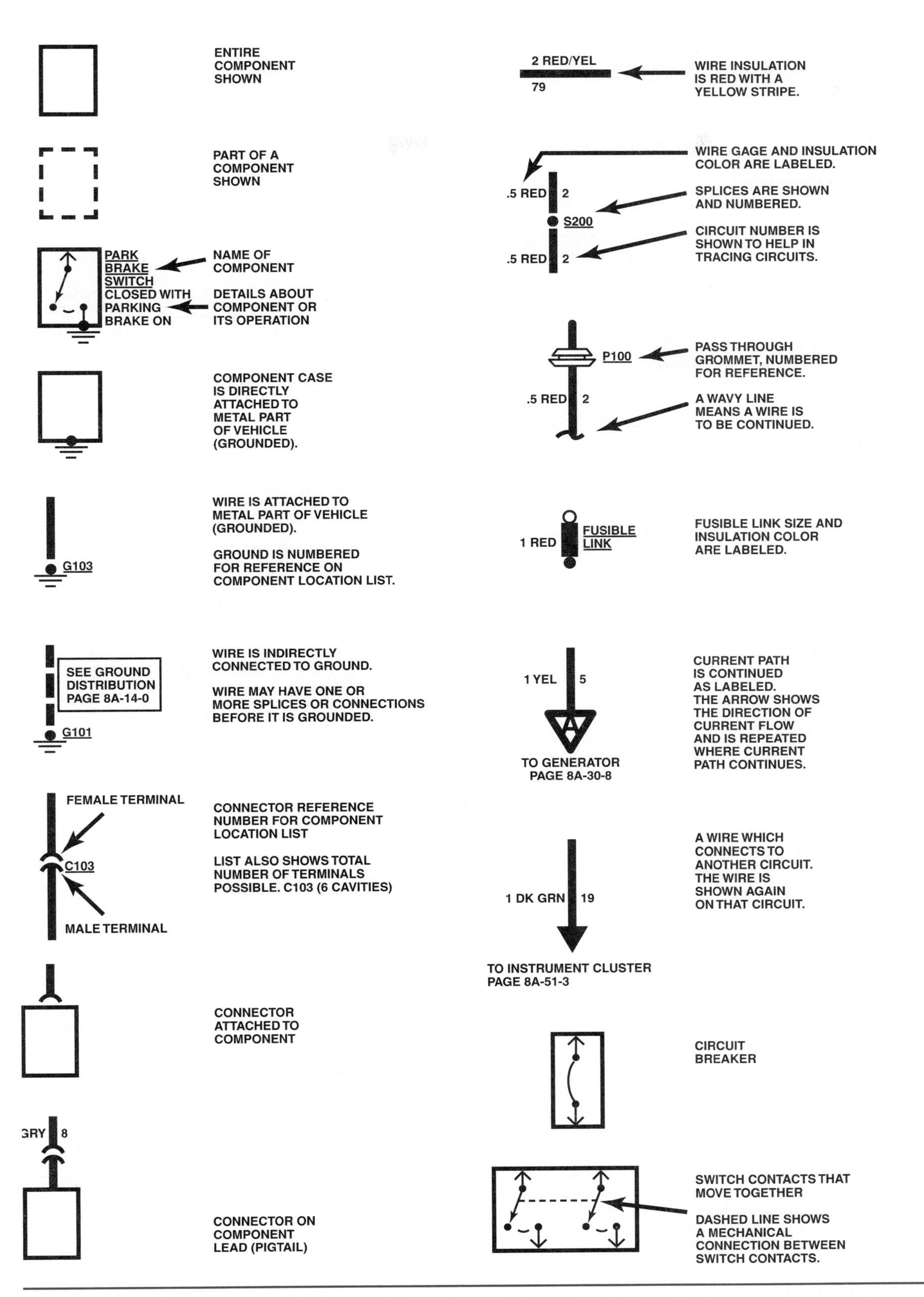

Figure 6-49A. Component symbols used by GM. (GM Service and Parts Operations)

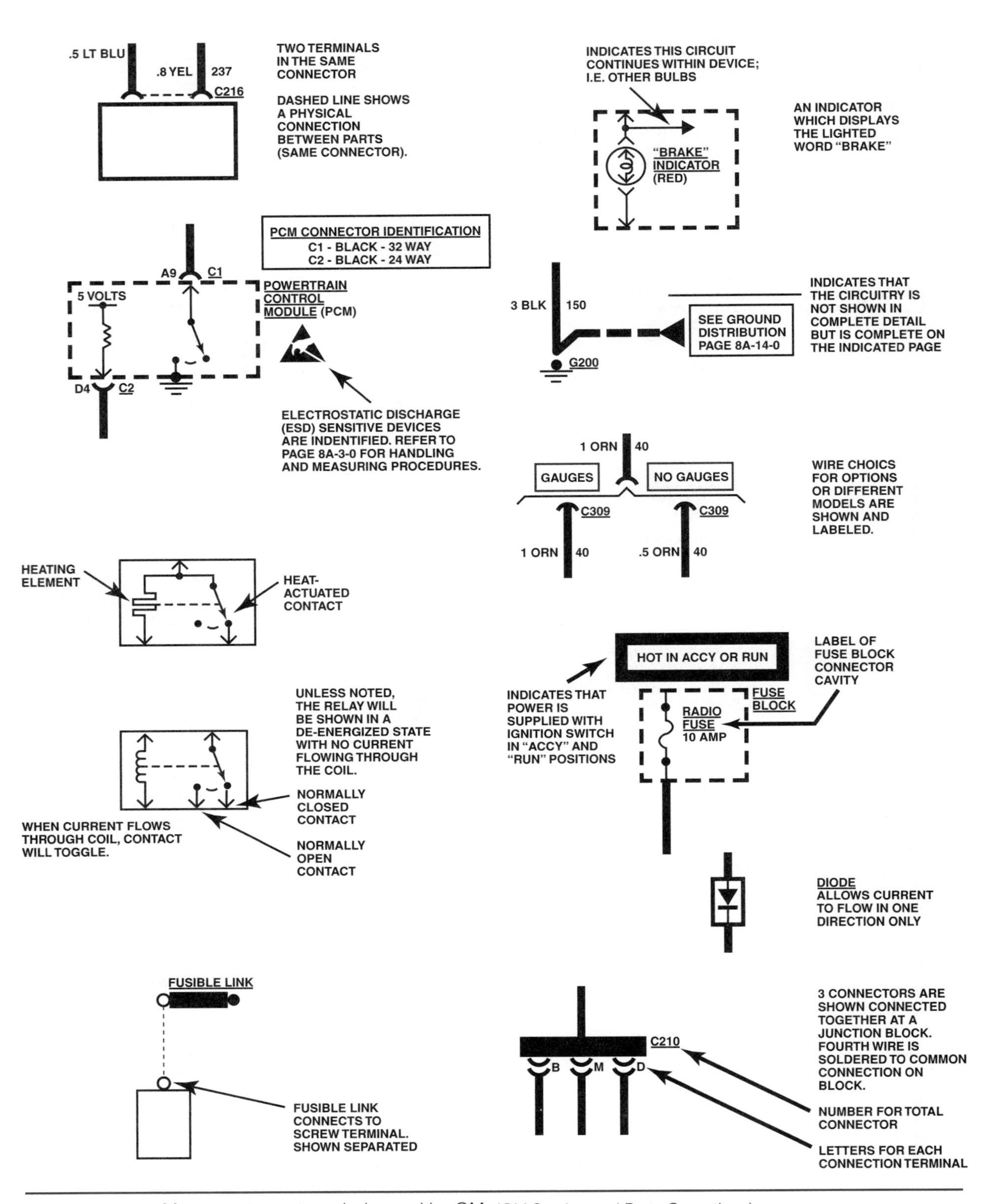

Figure 6-49B. More component symbols used by GM. (GM Service and Parts Operations)

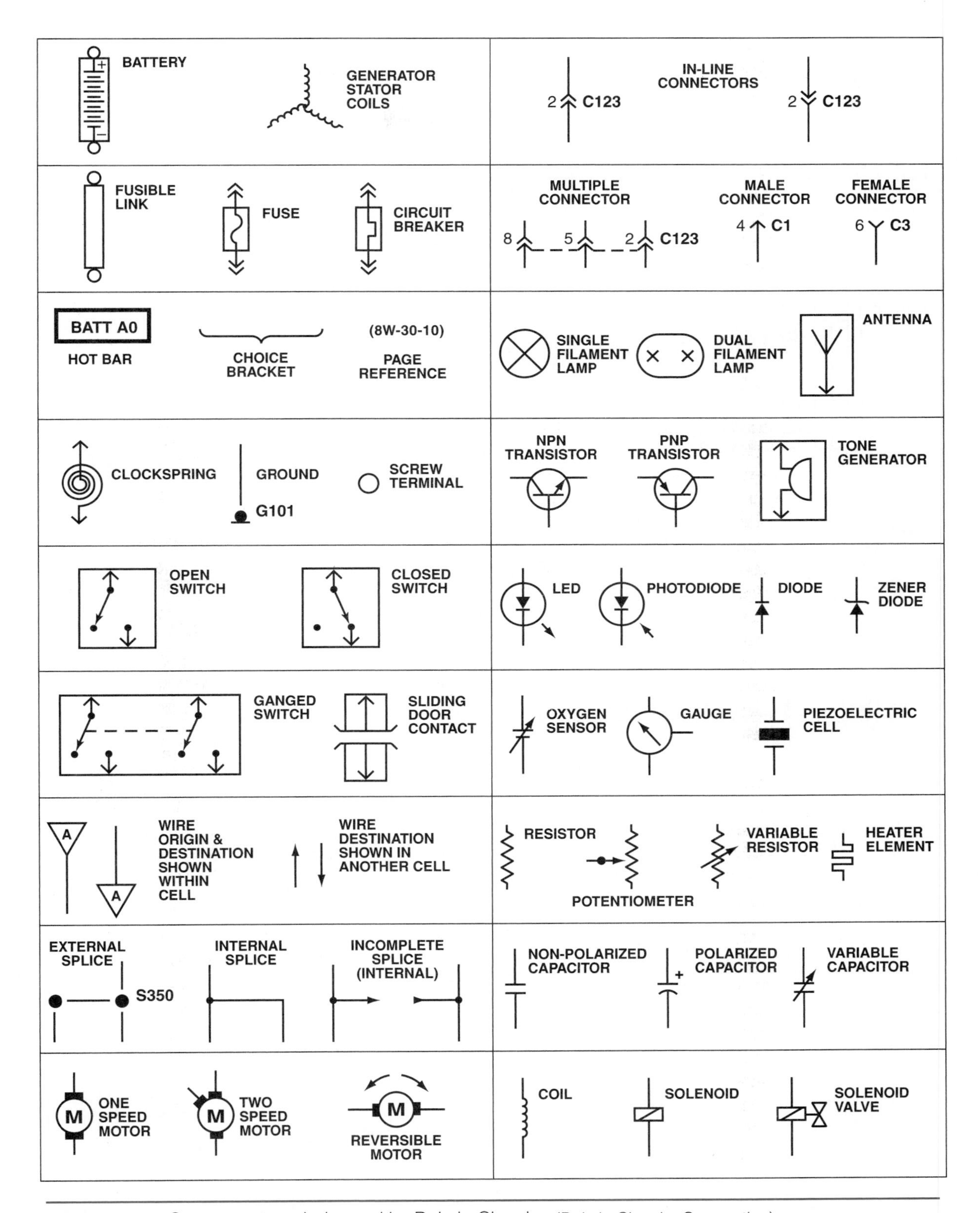

Figure 6-50. Component symbols used by DaimlerChrysler. (DaimlerChrysler Corporation)

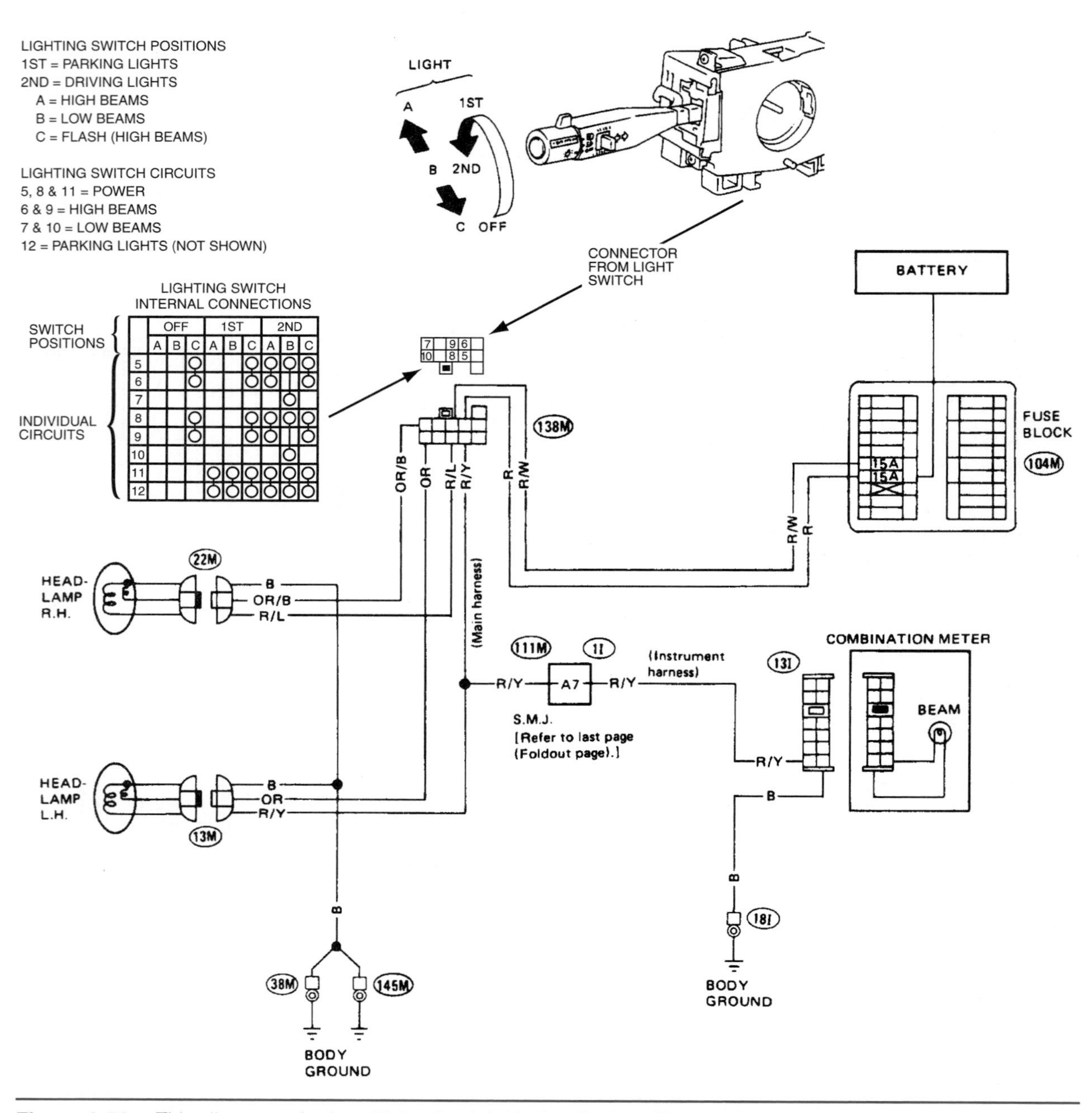

Figure 6-51. This diagram of a headlight circuit includes the headlight switch internal connections. (Courtesy of Nissan North America, Inc.)

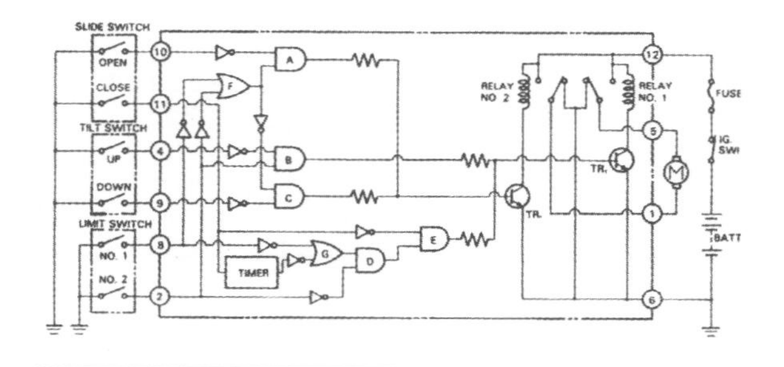

Figure 6-52. The advance computer technology makes logic symbols like these a typical part of an automotive wiring diagram.

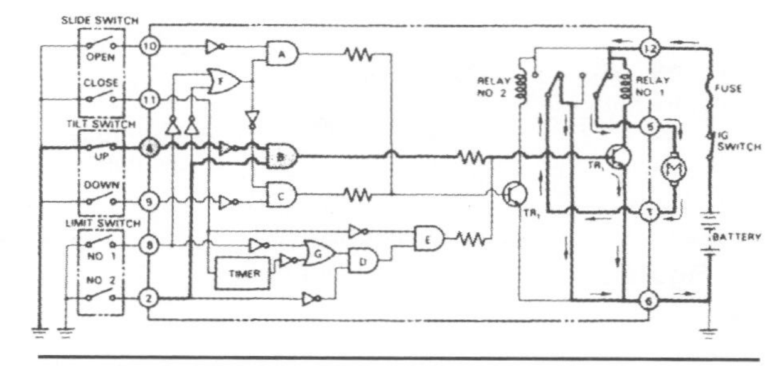

Figure 6-53. This circuit uses logic symbols to show how the sunroof motor operates to tilt the mechanism open.

Warning Lamps / Wiring Diagram

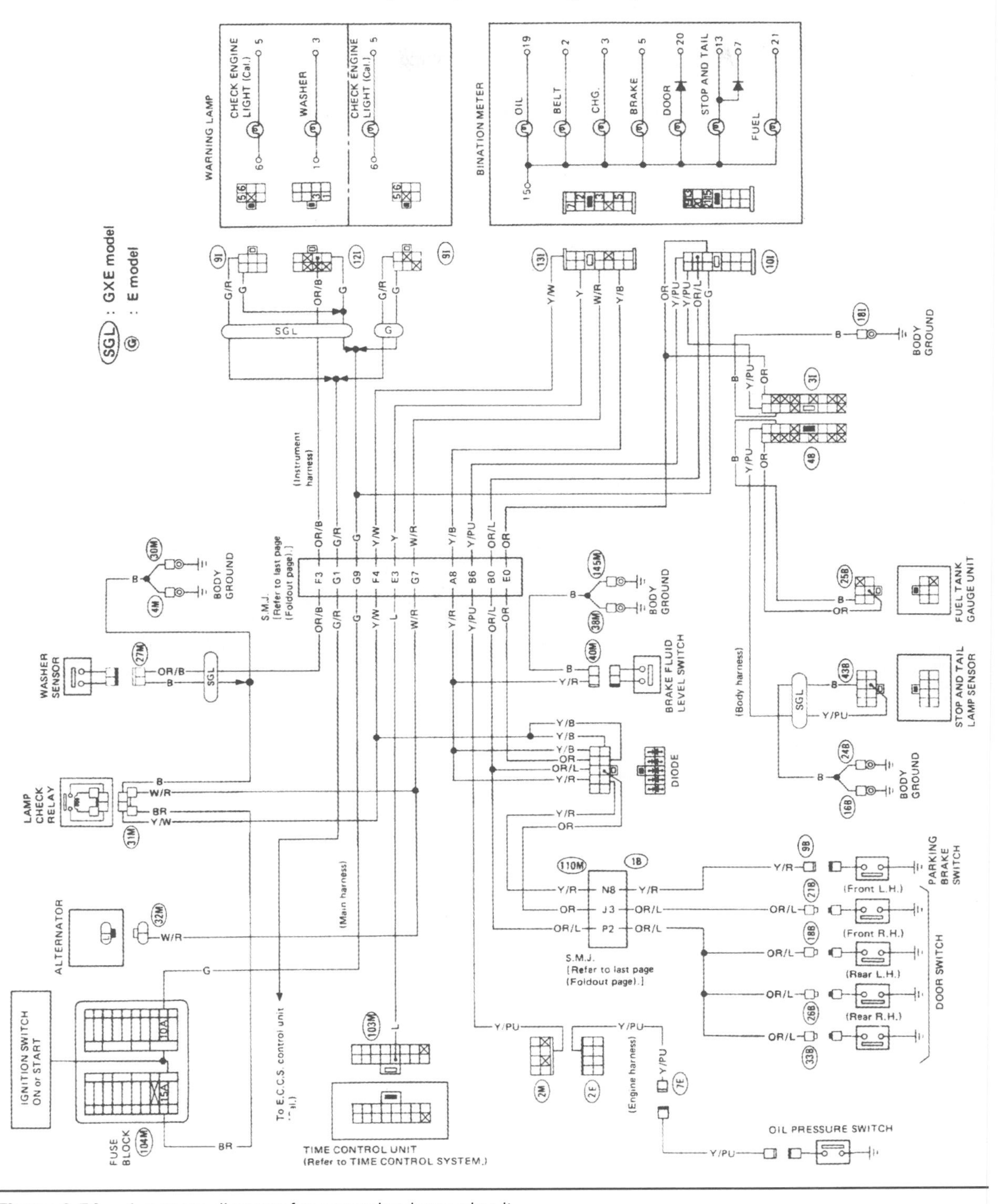

Figure 6-54. A system diagram for a warning lamp circuit.

Warning Lamps / Schematic

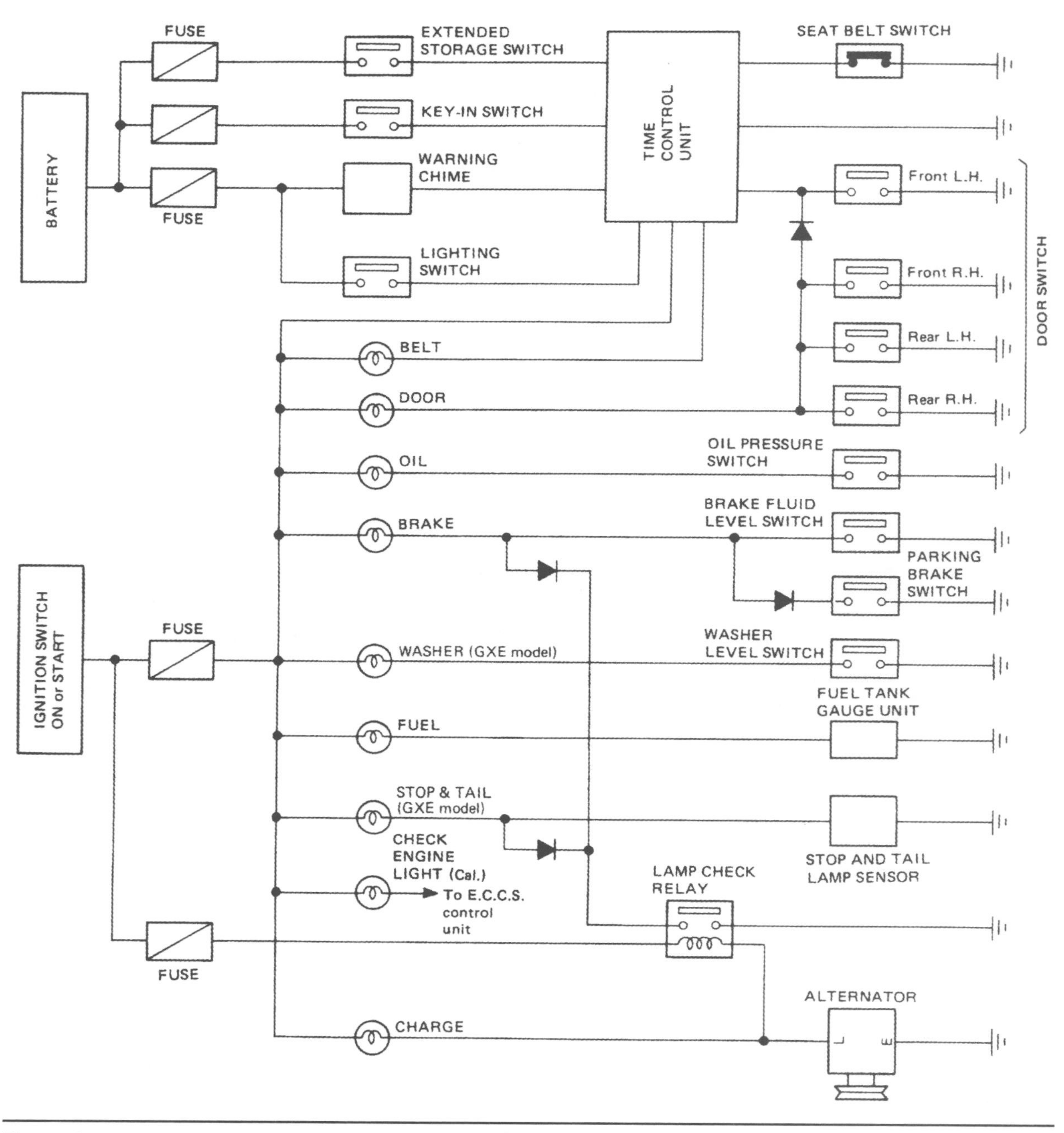

Figure 6-55. A schematic diagram for a warning lamp circuit.

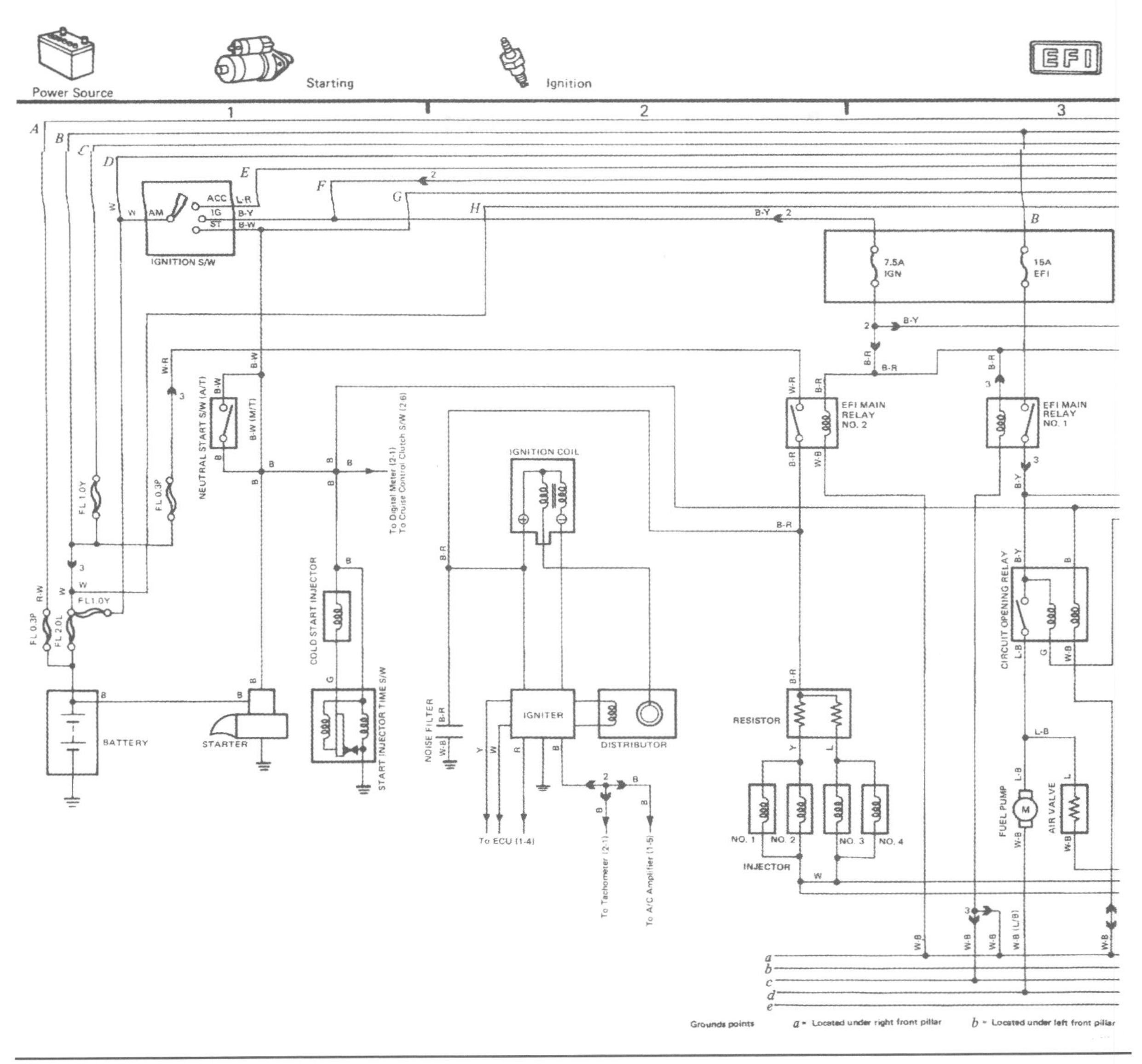

Figure 6-56. Toyota system diagrams are organized by individual systems and include ground points. (Reprinted by permission of Toyota Motor Corporation)

positions are not clear, find that information elsewhere in the electrical section of the manufacturer's shop manual.

Installation Diagrams

None of the diagrams shown so far have indicated *where* or *how* the wires and loads are installed in the automobile. Many manufacturers provide **installation diagrams,** or pictorial diagrams that show these locations. Some original equipment manufacturers (OEM) call these diagrams product description manuals (PDMs). Figures 6-61 and 6-62 show different styles of installation diagrams that help locate the general harness or circuit.

The DaimlerChrysler installation diagram in Figure 6-61 includes the wiring for radio speakers. The circuit diagram for the speakers was shown in Figure 6-57. Compare these two diagrams and notice that the installation diagram highlights the location of circuits while the circuit diagram gives more of a detailed picture of the circuit.

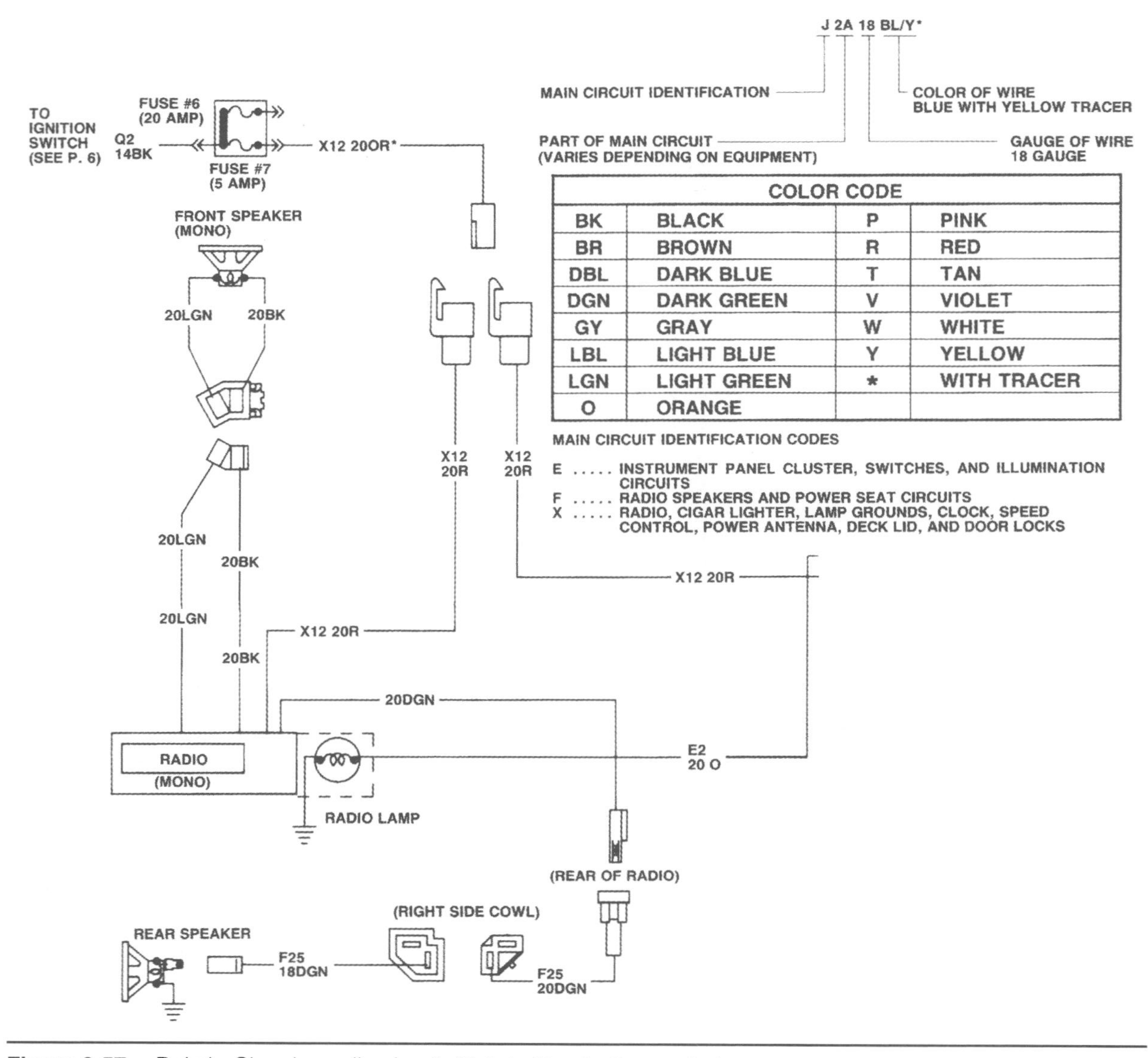

Figure 6-57. DaimlerChrysler radio circuit. (DaimlerChrysler Corporation)

Figure 6-62 is a GM installation diagram for the fuel-economy indicator switch. The circuit diagram for this accessory was given in Figure 6-59. Compare these two diagrams and note that the installation diagram focuses on the harness and circuit location while the schematic diagram shows a specific circuit current reading top to bottom.

Troubleshooting with Schematic Diagrams

It is quicker and easier to diagnose and isolate an electrical problem using a schematic diagram than by working with a system diagram, because schematic diagrams do not distract or confuse with wiring that is not part of the circuit being tested. A schematic diagram shows the paths that electrical current takes in a properly functioning circuit. It is important to understand how the circuit is supposed to work before determining why it is malfunctioning. For more information about troubleshooting with schematic diagrams, see the "Copper Wiring Repair" section in Chapter 6 of the *Shop Manual.*

General Motors incorporates a special troubleshooting section in each shop manual for the entire electrical system, broken down by individual circuits (Figures 6-48 and 6-60). Each schematic contains all the basic information necessary to trace the circuit it covers: wire size

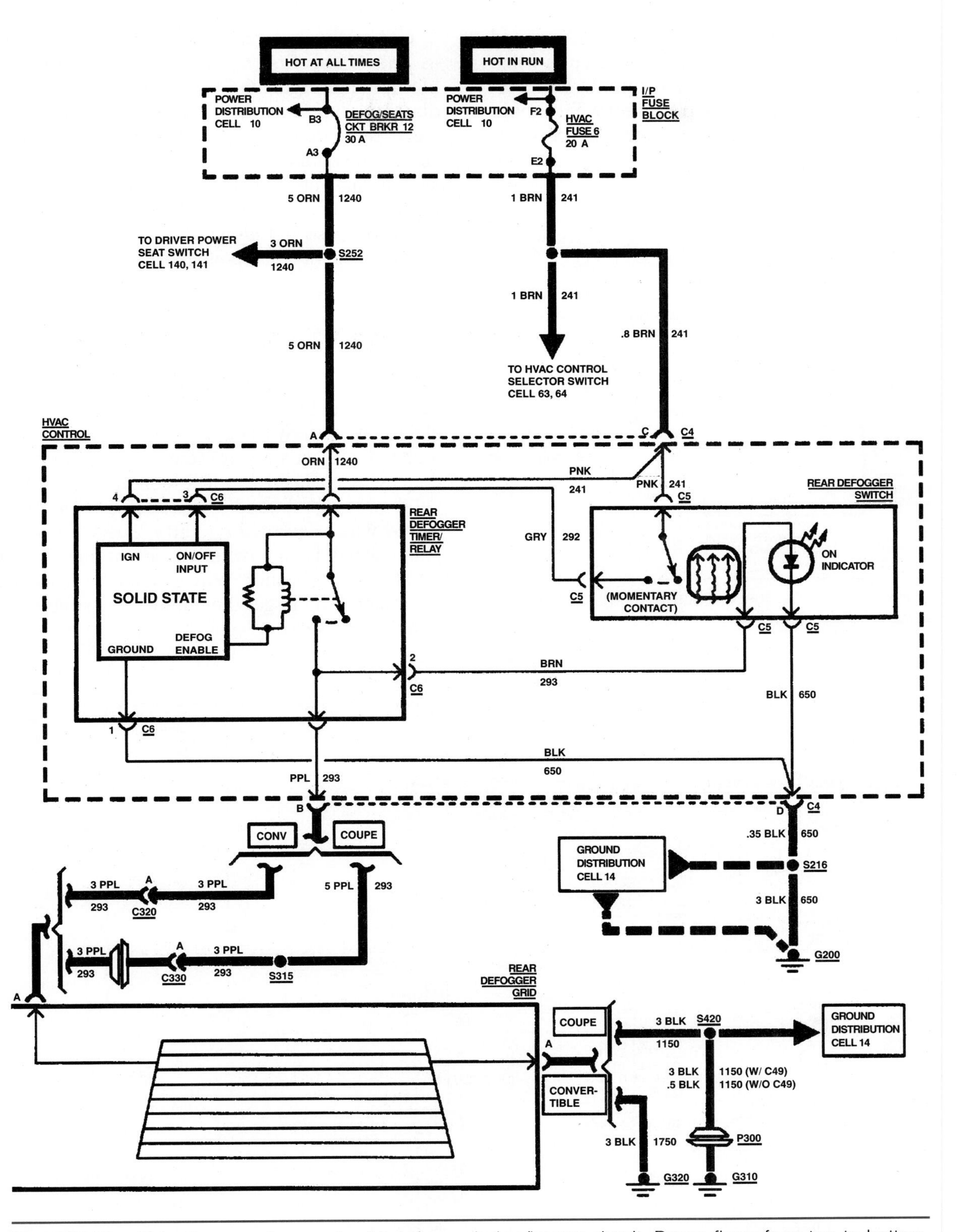

Figure 6-58. This diagram shows a backlight/rear window/heater circuit. Power flows from top to bottom, which is typical of a SPX-Valley Forge current flow diagram. Only components or information that belong to this specific circuit are shown. (GM Service and Parts Operations)

and color, components, connector and ground references, and references to other circuits when necessary. In addition, a quick summary of system operation is provided to explain what should happen when the system is working properly.

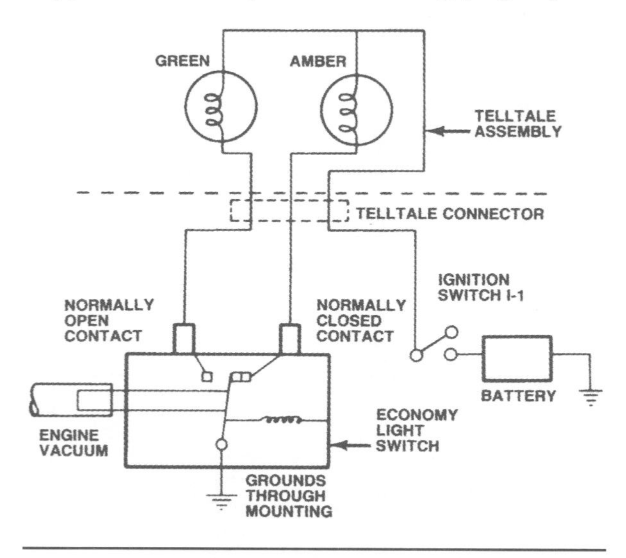

Figure 6-59. GM fuel-economy lamp circuit. (GM Service and Parts Operations)

GM and includes a power distribution diagram as one of the first overall schematics for troubleshooting (Figure 6-64). The power distribution diagram represents the "front end" of the overall electrical system. As such, it includes the battery, starter solenoid or relay, alternator, ignition switch, and fuse panel. This diagram is useful in locating short circuits that blow fusible links or fuses, because it follows the power distribution wiring to the first component in each major circuit.

SUMMARY

Many of the conductors in an automobile are grouped together into harnesses to simplify the electrical system. The conductors are usually made of copper, stainless steel, or aluminum covered with an insulator. The conductor can be a solid or single-strand wire, multiple or multistrand wire or printed circuitry. The wire size or gauge depends on how much current must be carried for what distance. Wire gauge is expressed as a number—the larger the number, the smaller the wire's cross-section.

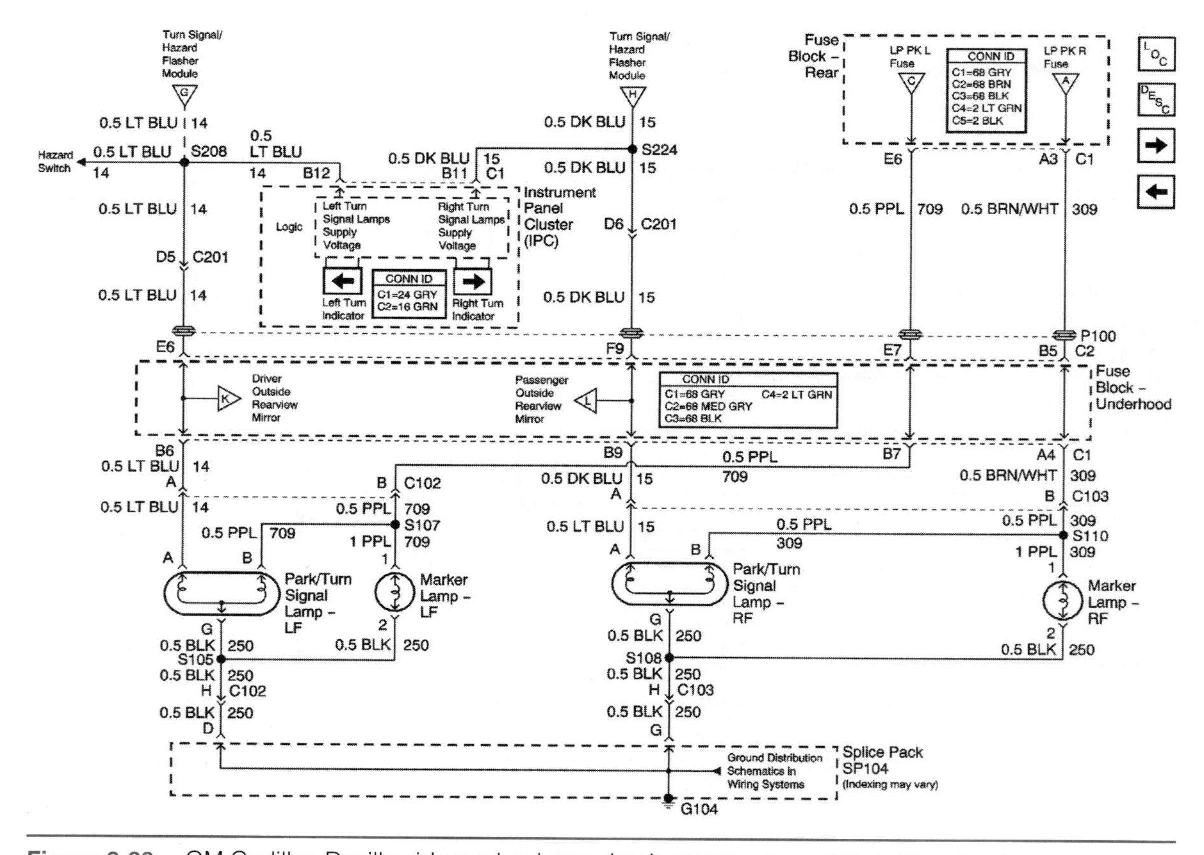

Figure 6-60. GM Cadillac Deville side marker lamp circuit. (GM Service and Parts Operations)

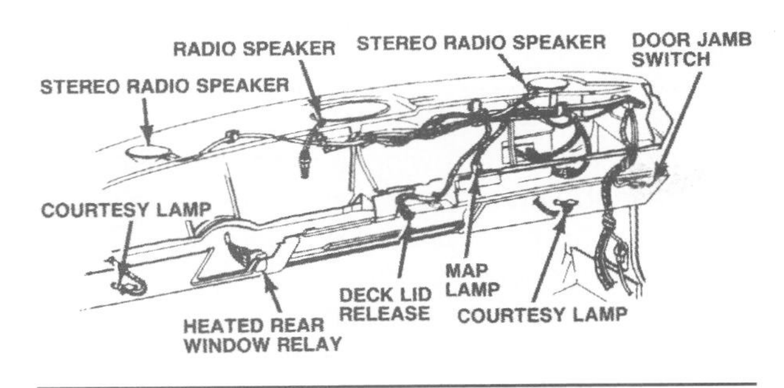

Figure 6-61. DaimlerChrysler installation diagram. (DaimlerChrysler Corporation)

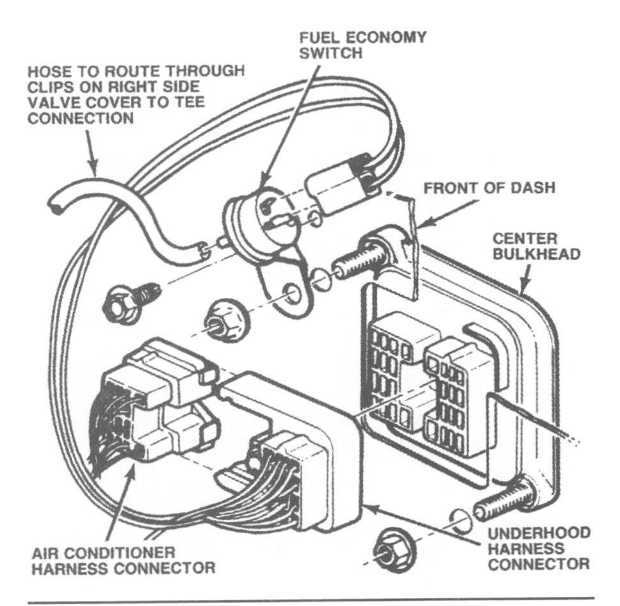

Figure 6-62. GM installation diagram. (GM Service and Parts Operations)

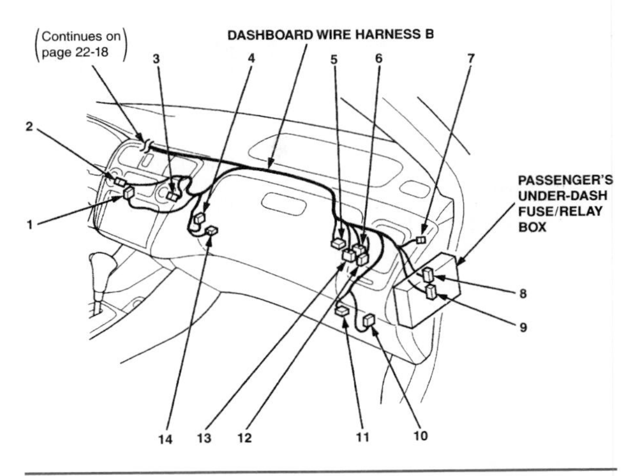

Figure 6-63. Acura installation diagram. (Courtesy of American Honda Motor Co., Inc.)

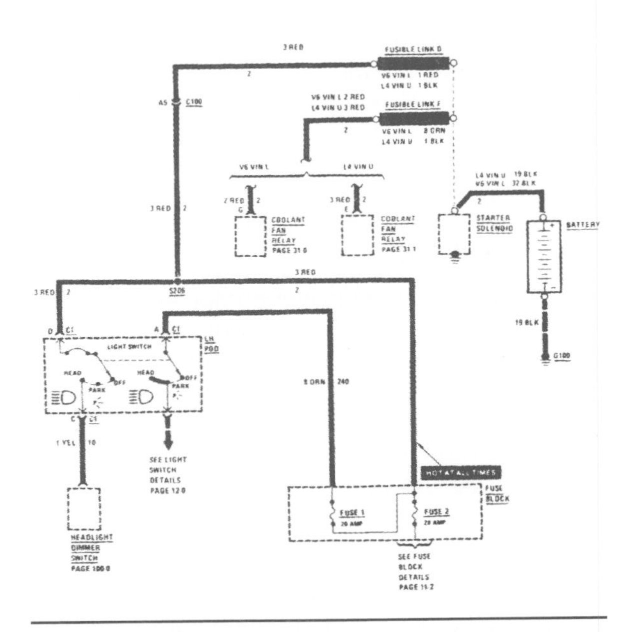

Figure 6-64. A typical GM power distribution schematic. (GM Service and Parts Operations)

Cars use some special types of wire, especially in battery cables and ignition cables. Terminals and connectors join different wires in the car; these can join single wires or 40 or more wires. Part of every automotive circuit is the ground path through the car's frame and body. The battery terminal that is connected to ground determines the electrical system's polarity. Most modem automobiles have a negative-ground system.

Multiplexing simplifies wiring by sending two or more electric signals over a single channel. Along with conductors, connectors and the ground path, each automotive circuit has controlling or working parts. These include switches, relays, solenoids, buzzers and motors.

There are three types of electrical circuit diagrams: system, schematic, and installation. System diagrams typically present an overall view, while schematic diagrams isolate a single circuit and are more useful for troubleshooting individual problems. Installation diagrams show locations and harness routing. To better understand these diagrams, a variety of electrical symbols are used to represent electrical components. Other tools to aid in successfully reading these diagrams are the color coding and the circuit numbering of wires, which identify the wire and its function. Manufacturers publish diagrams of each vehicle's electrical systems, often using these color codes and circuit numbers. A technician cannot service a circuit without knowing how to read and use these diagrams.

Review Questions

1. Which of the following is *not* considered part of the primary wiring system of an automobile?
   a. Spark plug cables
   b. Lighting circuits
   c. Accessory wiring circuits
   d. Power distribution circuits
2. Automotive wiring, or circuit conductors, exist as all of the following, *except*:
   a. Single-strand wire
   b. Multistrand wire
   c. Printed circuitry
   d. Enameled chips
3. Which of the following wires are known as suppression cables?
   a. Turn signal wiring
   b. Cables from the battery to the starter motor
   c. Cables from the distributor cap to the spark plugs
   d. Wiring harnesses from the fuse panel to the accessories
4. High-resistance ignition cables are used to do all of the following, except:
   a. Reduce radio frequency interference
   b. Provide extra resistance to reduce current flow to the spark plugs
   c. Provide more current to the distributor
   d. Boost the voltage being delivered to the spark plugs
5. One of the most common wear points in an automobile electrical system is:
   a. At the ground connecting side
   b. The point where a wire has been bent
   c. Where two connectors have been joined
   d. At a Maxifuse connection
6. The symbol below indicates which of these:

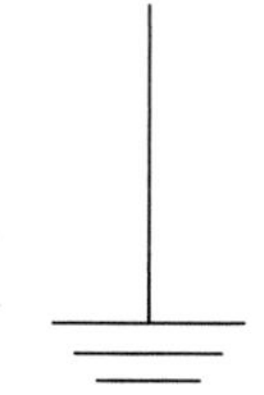

   a. Battery
   b. Capacitor
   c. Diode
   d. Ground
7. Which of the following is *not* a term used to describe an automobile wiring system?
   a. Hot-return system
   b. Single-wire system
   c. Ground-return system
   d. Negative-ground system
8. For easy identification, ground wires on most automotive electrical systems are color-coded:
   a. Red
   b. White
   c. Black
   d. Brown
9. Which of the following is *not* used to switch current flow?
   a. Relay
   b. Solenoid
   c. Transistor
   d. Coil Windings
10. Two separate windings are used in starter solenoids to:
    a. Increase resistance in the circuit
    b. Decrease resistance in the circuit
    c. Increase current being drawn from the battery
    d. Decrease current being drawn from the battery
11. Which of the following reverses the flow of current through the conductor of a motor?
    a. The armature
    b. The terminals
    c. The field coils
    d. The commutator
12. The symbol shown below is for which of these:

    a. Fuse
    b. Relay
    c. Motor
    d. Resistor
13. Which of the following is NOT used to protect a circuit from too much current flow?
    a. Fuse
    b. Buss bar

c. Fusible link
d. Circuit breaker

14. What usually causes a fuse to "blow"?
 a. Too much voltage
 b. Too much current
 c. Too little voltage
 d. Too little resistance

15. Fuses are rated by _________ capacity.
 a. Current
 b. Voltage
 c. Resistance
 d. Power

16. Which of the following is true of circuit breakers?
 a. Made of a single metal strip
 b. Must be replaced after excess current flow
 c. Less expensive than fuses are
 d. Used for frequent temporary overloads

17. This symbol represents which of these items:

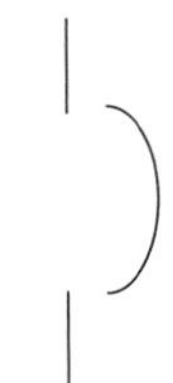

 a. Circuit breaker
 b. Variable resistor
 c. Capacitor
 d. Solenoid

18. The following two symbols represent which two devices?

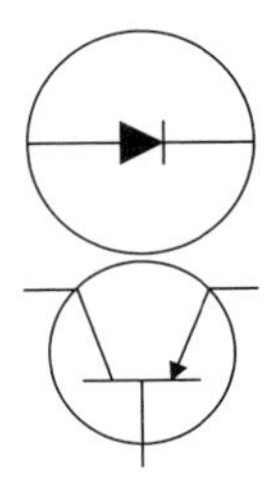

 a. Zener diode and a PNP transistor
 b. Zener diode and a NPN transistor
 c. One-way diode and a PNP transistor
 d. One-way diode and a NPN transistor

19. The electrical diagram that shows where the wires and loads are physically located on the vehicle is the:
 a. Schematic diagram
 b. Electrical system diagram
 c. Installation diagram
 d. Alternator circuit diagram

20. Automobile manufacturers color-code the wires in the electrical system to:
 a. Help trace a circuit
 b. Identify wire gauge
 c. Speed the manufacturing process
 d. Identify replacement parts

21. A wire in a Valley Forge (system) diagram is identified as .8 PUR/623 which of the following:
 a. 8-gauge wire, power plant, circuit number 623
 b. 8-gauge wire, purple, vehicle model 623
 c. 0.8-mm wire, purple, circuit number 623
 d. 0.8 points per length, vehicle model 623

22. Which of the following electrical symbols indicates a lamp?

 a.

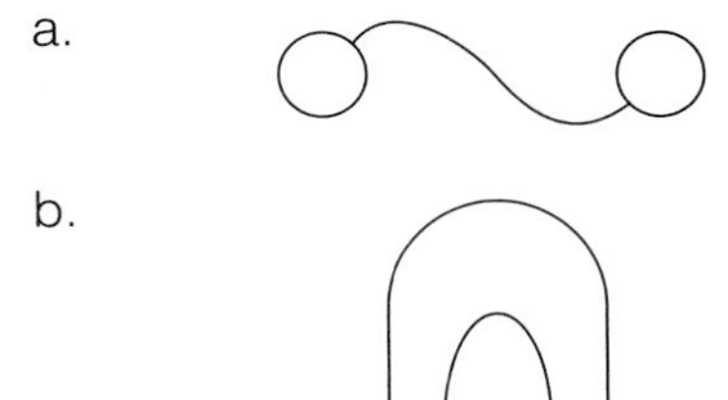

 b.

 c.

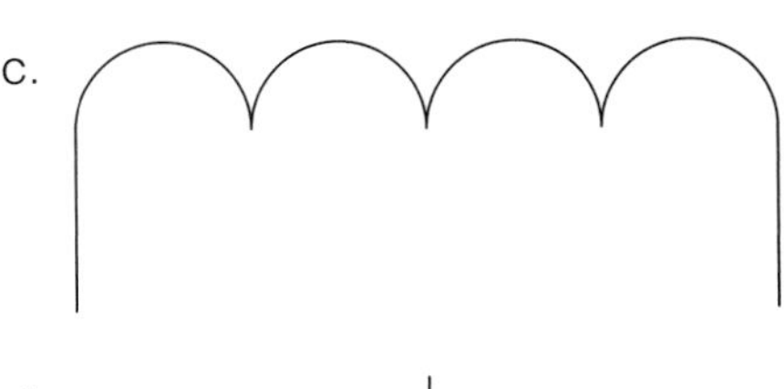

 d.

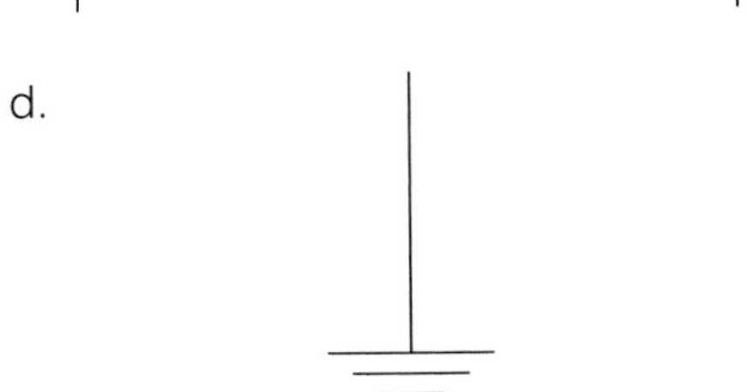

23. Two technician are discussing the meaning of different types of wiring diagrams. Technician A says schematic diagrams tell you how a circuit works and how the individual components connect to each other and they are commonly used by engineers. Technician B says a system diagram is drawing of a circuit or any part of

a circuit that shows circuit numbers, wire size, and color-coding. Who is right?
a. A only
b. B only
c. Both A and B
d. Neither A nor B

24. Technician A says the battery cable provides a connection from the vehicle chassis to the battery. Technician B says that automotive electrical systems are called double-wire or positive return systems. Who is right?
a. A only
b. B only
c. Both A and B
d. Neither A nor B

25. Technicians are discussing wire gauge sizes. Technician A says wire size numbers are based on the cross-sectional area of the conductor and larger wires have lower gauge numbers. Technician B says wire cross-section is measured in circular rods. Who is right?
a. A only
b. B only
c. Both A and B
d. Neither A nor B

7

Automotive Battery Operation

LEARNING OBJECTIVES

Upon completion and review of this chapter, you should be able to:

- Identify the purpose of the battery.
- Describe battery operation.
- Explain battery capacity.
- Identify battery safety procedures.
- Explain battery ratings.

KEY TERMS

Battery
Cell
Cold Cranking Amperes (CCA)
Cycling
Electrolyte
Element
Plates
Primary Battery
Secondary Battery
Sealed Lead-Acid (SLA)
Specific Gravity
State-of-charge Indicator

INTRODUCTION

The automotive **battery** does not actually store electricity, as is often believed. The battery is a quick-change artist. It changes electrical current generated by the vehicle's charging system into chemical energy. Chemicals inside the battery store the electrical energy until it is needed to perform work. It is then changed back into electrical energy and sent through a circuit to the system where it is needed. We will begin our study of batteries by listing their functions and looking at the chemical action and construction of a battery.

Just as you are made up of a bunch of cells, so is the battery in a car. Each battery contains a number of **cells** made up of alternating positive and negative **plates.** Between each plate is a separator that keeps the plates from touching, yet lets electrolyte pass back and forth between it. The separators are made of polyvinyl chloride (PVC). An automotive battery does the following:

- Operates the starter motor
- Provides current for the ignition system during cranking

- Supplies power for the lighting systems and electrical accessories when the engine is not operating
- Acts as a voltage stabilizer for the entire electrical system
- Provides current when the electrical demand of the vehicle exceeds the output of the charging system

ELECTROCHEMICAL ACTION

All automotive wet-cell batteries operate because of the chemical action of two dissimilar metals in the presence of a conductive and reactive solution called an **electrolyte.** Because this chemical action produces electricity, it is called electrochemical action. The chemical action of the electrolyte causes electrons to be removed from one metal and added to the other. This loss and gain of electrons causes the metals to be oppositely charged, and a potential difference, or voltage, exists between them.

The metal plate that has lost electrons is positively charged and is called the positive plate. The plate that has gained electrons is negatively charged and is called the negative plate. If conductors and a load are connected between the two plates, current will flow through the conductor and the load (Figure 7-1). For simplicity, battery current flow is assumed to be conventional current flow (+ to −) through the external circuit connected to the battery.

Primary and Secondary Batteries

There are two general types of batteries: primary and secondary. The action within a **primary battery** causes one of the metals to be totally destroyed after a period of time. When the battery has delivered all of its voltage to an outside circuit, it is useless and must be replaced. Many small dry-cell batteries, such as those for flashlights and radios, are primary batteries.

In a **secondary battery,** both the electrolyte and the metals change their atomic structure as the battery supplies current to an outside circuit. This is called discharging. The action can be reversed, however, by applying an outside current to the battery terminals and forcing current through the battery in the opposite direction. This current causes a chemical action that restores the battery materials to their original condition, and the battery can again supply current. This is called charging the battery. The condition of the battery materials is called the battery's state of charge.

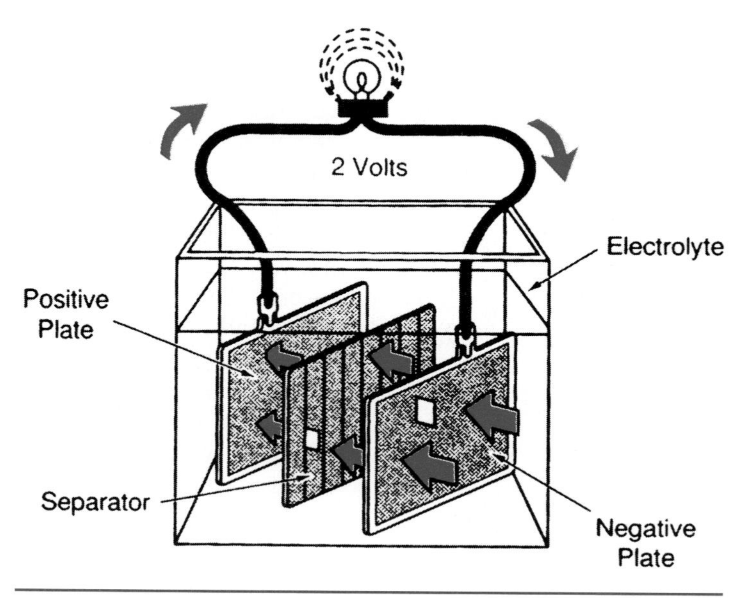

Figure 7-1. The potential difference between the two plates of a battery can cause current to flow in an outside circuit. (GM Service and Parts Operations)

Electrochemical Action in Automotive Batteries

A fully charged automotive battery contains a series of negative plates of chemically active sponge lead (Pb), positive plates of lead dioxide (PbO_2) and an electrolyte of sulfuric acid (H_2SO_4) and water (H_2O_2) (Figure 7-2). As the battery discharges, the chemical action taking place reduces the acid content in the electrolyte and increases the water content. At the same time, both the negative and the positive plates gradually change to lead sulphate ($PbSO_4$).

A discharged battery (Figure 7-2) has a very weak acid solution because most of the electrolyte has changed to water. Both series of plates are mostly lead sulfate. The battery now stops functioning because the plates are basically two similar metals in the presence of water, rather than two dissimilar metals in the presence of an electrolyte.

During charging (Figure 7-2) the chemical action is reversed. The lead sulfate on the plates gradually decomposes, changing the negative plates back to sponge lead and the positive plates to lead dioxide. The sulfate is re-deposited in the water, which increases the sulfuric acid content and returns the electrolyte to full strength. Now, the battery is again able to supply current.

This electrochemical action and battery operation from fully charged to discharged and back to fully charged is called **cycling.**

WARNING: Hydrogen and oxygen gases are formed during battery charging. Hydrogen gas is explosive. Never strike a spark or bring a flame near a battery, particularly during or after charging. *This could cause the battery to explode.*

Battery Construction

There are four types of automotive batteries currently in use, as follows:

- Vent-cap (requires maintenance)
- Low-maintenance (requires limited maintenance)
- Maintenance-free (requires no maintenance)
- Recombinant (requires no maintenance)

The basic physical construction of all types of automotive batteries is similar, but the materials used are not. We will look at traditional vent-cap construction first and then explain how the other battery types differ.

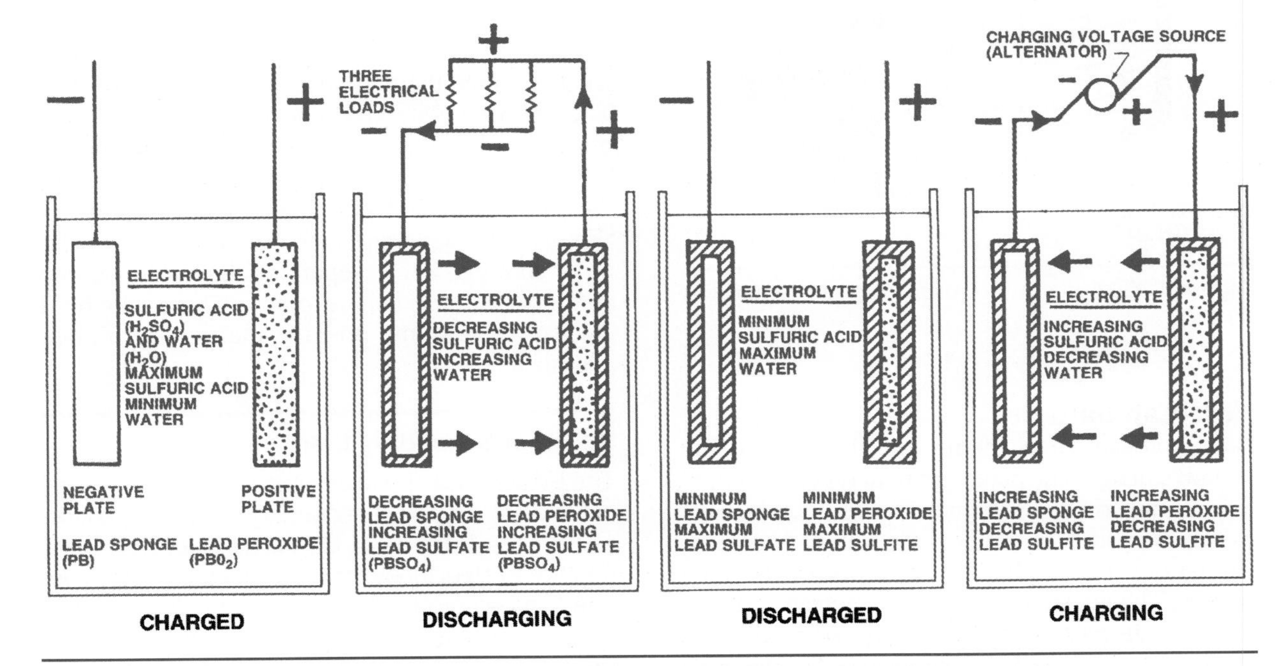

Figure 7-2. Battery electrochemical action from charged to discharged and back to charged.

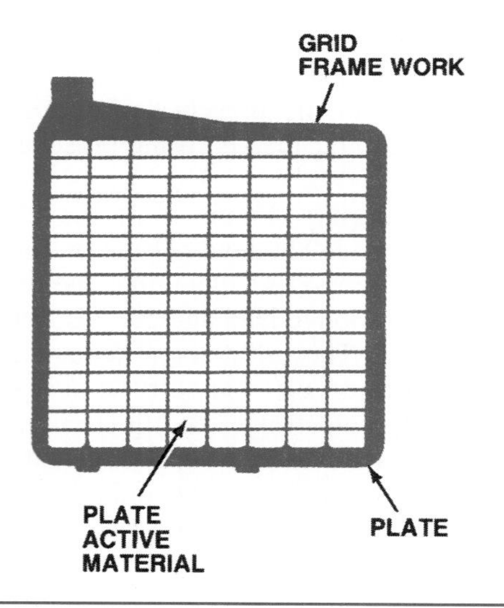

Figure 7-3. The grid provides a support for the plate active material.

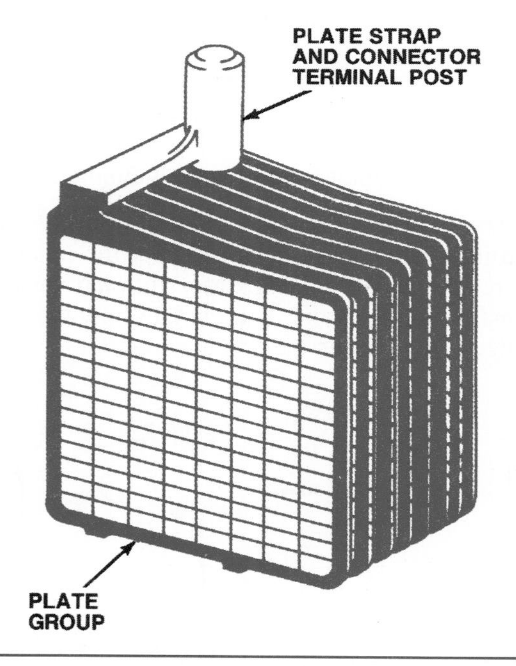

Figure 7-4. A number of plates are connected into a group.

Vent-Cap Batteries

Battery construction begins with the positive and negative plates. The plates are built on grids of conductive materials, as shown in Figure 7-3, which act as a framework for the dissimilar metals. These dissimilar metals are called the active materials of the battery. The active materials, sponge lead and lead dioxide, are pasted onto the grids. When dry, the active materials are very porous, so that the electrolyte can easily penetrate and react with them.

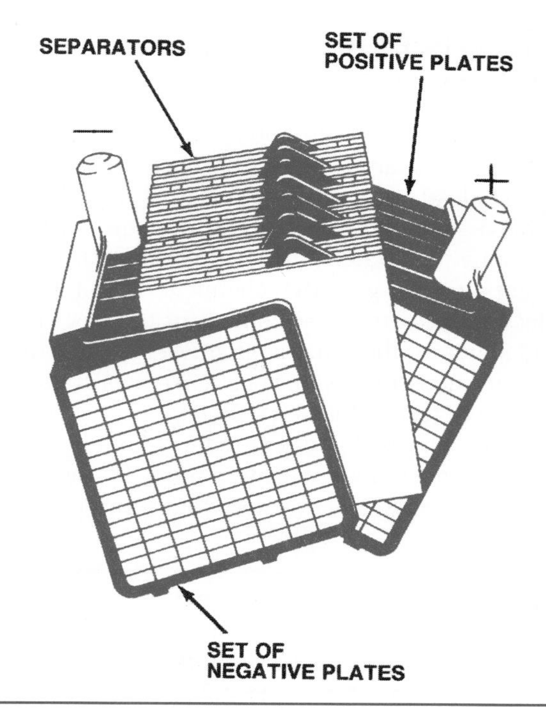

Figure 7-5. Two groups are interlaced to form a battery element.

A number of similar plates, all positive or all negative, are connected together into a plate group (Figure 7-4). The plates are joined to each other by welding them to a plate strap through a process called lead burning. The plate strap has a connector or a terminal post for attaching plate groups to each other.

A positive and a negative plate group are interlaced so that their plates alternate, Figure 7-5. The negative plate group normally has one more plate than the positive group. To reduce the possibility of a short between plates of the two groups, they are separated by chemically inert separators (Figure 7-5). Separators are usually made of plastic or fiberglass. The separators have ribs on one side next to the positive plates. These ribs hold electrolyte near the positive plates for efficient chemical action.

Other Secondary Cells

The Edison (nickel-iron alkali) cell and the silver cell are two other types of secondary cells. The positive plate of the Edison cell is made of pencil-shaped, perforated steel tubes that contain nickel hydroxide. These tubes are held in a steel grid. The negative plate has pockets that hold iron oxide. The electrolyte used in this cell is a solution of potassium hydroxide and a small amount of lithium hydroxide.

An Edison cell weighs about one-half as much as a lead-acid cell of the same ampere-hour capacity. This cell has a long life and is not damaged by short circuits or overloads. It is however, more costly than a lead-acid cell. The silver cell has a positive plate of silver oxide and a negative plate of zinc. The electrolyte is a solution of potassium hydroxide or sodium. For its weight, this cell has a high ampere-hour capacity. It can withstand large overloads and short circuits. It, too, is more expensive than a lead-acid cell.

A complete assembly of positive plates, negative plates, and separators is called an **element.** It is placed in a cell of a battery case. Because each cell provides approximately 2.1 volts, a 12-volt battery has six cells and actually produces approximately 12.6 volts when fully charged. The elements are separated from each other by cell partitions, and rest on bridges at the bottom of the case that form chambers where sediment can collect. These bridges prevent accumulated sediment from shorting across the bottoms of the plates. Once installed in the case, the elements (cells) are connected to each other by connecting straps that pass over or through the cell partitions (Figure 7-6). The cells are connected alternately in series (positive to negative to positive to negative, and so on), and the battery top is bonded onto the case to form a watertight container.

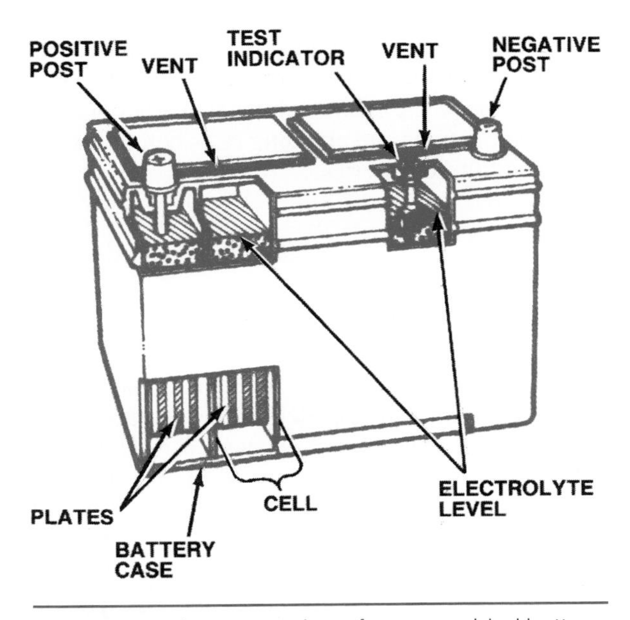

Figure 7-6. A cutaway view of an assembled battery. (DaimlerChrysler Corporation)

Vent caps in the battery top provide an opening for adding electrolyte and for the escape of gases that form during charging and discharging. The battery is connected to the car's electrical system by two external terminals. These terminals are either tapered posts on top of the case or internally threaded connectors on the side. The terminals, which are connected to the ends of the series of elements inside the case, are marked positive (+) or negative (−), according to which end of the series each terminal represents.

Sealed Lead-Acid (SLA) Batteries

Most new batteries today are either partially sealed, *low-maintenance* or completely sealed, *maintenance-free* batteries. Low-maintenance batteries provide some method of adding water to the cells, such as the following:

- Individual slotted vent caps installed flush with the top of the case
- Two vent panel covers, each of which exposes three cells when removed
- A flush-mounted strip cover that is peeled off to reveal the cell openings

Sealed lead-acid (SLA) batteries have only small gas vents that prevent pressure buildup in the case. A low-maintenance battery requires that water be added much less often than with a traditional vent-cap battery, while a SLA battery will never need to have water added during its lifetime.

These batteries differ from vent-cap batteries primarily in the materials used for the plate grids. For decades, automotive batteries used antimony as the strengthening ingredient of the grid alloy. In low-maintenance batteries, the amount of antimony is reduced to about three percent. In maintenance-free batteries, the antimony is eliminated and replaced by calcium or strontium.

Reducing the amount of antimony or replacing it with calcium or strontium alloy results in lowering the battery's internal heat and reduces the amount of gassing that occurs during charging. Since heat and gassing are the principal reasons for battery water loss, these changes reduce or eliminate the need to periodically add water. Reduced water loss also minimizes terminal corrosion, since the major cause of this corrosion is condensation from normal battery gassing.

In addition, non-antimony lead alloys have better conductivity, so a maintenance-free battery has about a 20 percent higher cranking performance rating than a traditional vent-cap battery of comparable size.

Recombinant Batteries

More recently, completely sealed maintenance-free batteries were introduced. These new batteries do not require—and do not have—the small gas vent used on previous maintenance-free batteries. Although these batteries are basically the same kind of lead-acid voltage cells used in automobiles for decades, a slight change in plate and electrolyte chemistry reduces hydrogen generation to almost nothing.

During charging, a vent-cap or maintenance-free battery releases hydrogen at the negative plates and oxygen at the positive plates. Most of the hydrogen is released through electrolysis of the water in the electrolyte near the negative plates as the battery reaches full charge. In the sealed maintenance-free design, the negative plates never reach a fully charged condition and therefore cause little or no release of hydrogen. Oxygen is released at the positive plates, but it passes through the separators and recombines with the negative plates. The overall effect is virtually no gassing from the battery. Because the oxygen released by the electrolyte recombines with the negative plates, some manufacturers call these batteries "recombination" or recombinant electrolyte batteries.

Recombination electrolyte technology and improved grid materials allow some sealed, maintenance-free batteries to develop fully charged, open-circuit voltage of approximately 2.1 volts per cell, or a total of 12.6 volts for a six-cell 12-volt battery. Microporous fiberglass separators reduce internal resistance and contribute to higher voltage and current ratings.

In addition, the electrolyte in these new batteries is contained within plastic envelope-type separators around the plates (Figure 7-7). The entire case is not flooded with electrolyte. This eliminates the possibility of damage due to sloshing or acid leaks from a cracked battery. This design feature reduces battery damage during handling and installation, and allows a more compact case design. Because the battery is not vented, terminal corrosion from battery gassing and electrolyte spills or spray is also eliminated.

The envelope design also catches active material as it flakes off the positive plates during discharge. By holding the material closer to the plates, envelope construction ensures that it will be more completely redeposited during charging. Although recombinant batteries are examples of advanced technology, test and service requirements are basically the same as for other maintenance-free, lead-acid batteries. Some manufacturers caution, however, that fast charging at high current rates may overheat the battery and can cause damage. Always check the manufacturer's instructions for test specifications and charging rates before servicing one of these batteries.

Figure 7-7. Many maintenance-free batteries have envelope separators that hold active material near the plates.

BATTERY ELECTROLYTE

For the battery to become chemically active, it must be filled with an electrolyte solution. The electrolyte in an automotive battery is a solution of sulfuric acid and water. In a fully charged battery, the solution is approximately 35 to 39 percent acid by weight (25 percent by volume) and 61 to 65 percent water by weight. The state of charge of a battery can be measured by checking the specific gravity of the electrolyte.

WARNING: When lifting a battery, excessive pressure on the end walls could cause acid spill through the vent caps, resulting in personal injury. Lift with a battery carrier or with your hands at opposite corners. For more information, see the "Battery Safety" section in Chapter 7 of the *Shop Manual*.

Specific gravity is the weight of a given volume of liquid divided by the weight of an equal volume of water. Since the acid is heavier than water, and water has a specific gravity of 1.000, the specific gravity of a fully charged battery is greater than 1.000 (approximately 1.260 when weighed in a hydrometer). As the battery discharges, the specific gravity of the electrolyte decreases because the acid is changed into water. The specific gravity of the electrolyte can tell you approximately how discharged the battery has become:

- 1.265 specific gravity: 100% charged
- 1.225 specific gravity: 75% charged
- 1.190 specific gravity: 50% charged
- 1.155 specific gravity: 25% charged
- 1.120 specific gravity or lower: discharged

These values may vary slightly, according to the design factors of a particular battery. Specific gravity measurements are based on a standard temperature of 80°F (26.7°C). At higher temperatures, specific gravity is lower. At lower temperatures, specific gravity is higher. For every change of 10°F, specific gravity changes by four points (0.004). That is, you should compensate for temperature differences as follows:

- For every 10°F above 80°F, add 0.004 to the specific gravity reading.
- For every 10°F below 80°F, subtract 0.004 from the specific gravity reading

When you study battery service in Chapter 7 of the *Shop Manual*, you will learn to measure specific gravity of a vent-cap battery with a hydrometer. See the section on "Inspection, Cleaning and Replacement" for more information.

STATE-OF-CHARGE INDICATORS

Many low-maintenance and maintenance-free batteries have a visual **state-of-charge indicator** or built-in hydrometer installed in the battery top. The indicator shows whether the electrolyte has fallen below a minimum level, and it also functions as a *go/no-go* hydrometer.

The indicator shown in Figure 7-8 is a plastic rod inserted in the top of the battery and extending into the electrolyte. In the design used by Delco (now Delphi), a green plastic ball is suspended in a cage from the bottom of the rod. Depending upon the specific gravity of the electrolyte, the ball will float or sink in the cage, changing the appearance of the indicator "eye" from green to dark. When the eye is dark, the battery should be recharged.

Other manufacturers use either the "Delco Eye" under license, or one of several variations of the design. One variation contains a red ball and a blue ball side by side in the cage. When the specific gravity is high, only the blue ball can be seen in the "eye". As the specific gravity falls, the blue ball sinks in the cage, allowing the red ball to take its place. When the battery is recharged, the increasing specific gravity causes the blue ball to move upward, forcing the red ball back into the side of the cage.

Another variation is the use of a small red ball on top of a larger blue ball. When the specific gravity is high, the small ball is seen as a red spot surrounded by blue. As the specific gravity falls, the blue ball sinks, leaving the small ball to be

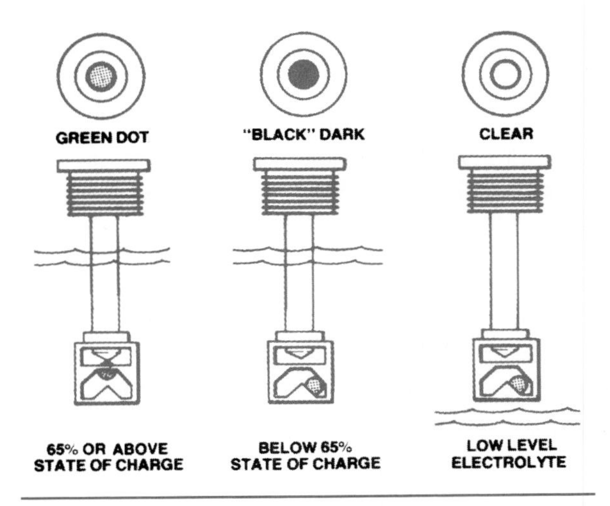

Figure 7-8. Delco (Delphi) "Freedom" batteries have this integral hydrometer built into their tops. (Delphi Corporation)

seen as a red spot surrounded by a clear area. The battery then should be recharged.

If the electrolyte drops below the level of the cage in batteries using a state-of-charge indicator, the "eye" will appear clear or light yellow. This means that the battery must be replaced because it has lost too much electrolyte. For more information, see the "Battery Testing" section in Chapter 7 of the *Shop Manual.*

WET-CHARGED AND DRY-CHARGED BATTERIES

Batteries may be manufactured and sold as either wet-charged or dry-charged batteries. Before maintenance-free batteries became widely used, dry-charged batteries were very common. A wet-charged battery is completely filled with an electrolyte when it is built. A dry-charged battery is shipped from the factory without electrolyte. During manufacture, the positive and negative plates are charged and then completely washed and dried. The battery is then assembled and sealed to keep out moisture. It will remain charged as long as it is sealed, and it can be stored for a long time in any reasonable environment. A dry-charged battery is put into service by adding electrolyte, checking the battery state of charge, and charging if needed.

Even when a wet-charged battery is not in use, a slow reaction occurs between the plates and the electrolyte. This is a self-discharging reaction, and will eventually discharge the battery almost completely. Because this reaction occurs faster at higher temperatures, wet-charged batteries should be stored in as cool a place as possible when not in use. A fully charged battery stored at a room temperature of 100°F (38°C) will almost completely discharge after 90 days. If the battery is stored at a temperature of 60°F (16°C), very little discharge will take place.

BATTERY CHARGING VOLTAGE

Forcing current through it in the direction opposite from its discharge current charges a battery. In an automobile, the generator or alternator supplies this charging current. The battery offers some resistance to this charging current, because of the battery's chemical voltage and the resistance of the battery's internal parts. The battery's chemical voltage is another form of counterelectromotive force (CEMF) that you studied in Chapter 4.

When a battery is fully charged, its CEMF is very high. Very little charging current can flow through it. When the battery is discharged, its CEMF is very low, and charging current flows freely. For charging current to enter the battery, the charging voltage must be higher than the battery's CEMF plus the voltage drop caused by the battery's internal resistance.

Understanding this relationship of CEMF to the battery state of charge is helpful. When the battery is nearly discharged, it needs, and will accept, a lot of charging current. When the battery is fully charged, the high CEMF will resist charging current. Any additional charging current could overheat and damage the battery materials. Charging procedures are explained in Chapter 3 of your *Shop Manual.* See also the section on "Battery Testing" in Chapter 7 of the *Shop Manual.*

The temperature of the battery affects the charging voltage because temperature affects the resistance of the electrolyte. Cold electrolyte has higher resistance than warm electrolyte, so a colder battery is harder to charge. The effects of temperature must be considered when servicing automotive charging systems and batteries, as we shall see later in this chapter.

BATTERY SELECTION AND RATING METHODS

Automotive batteries are available in a variety of sizes, shapes, and current ratings. They are called "starting batteries" and are designed to deliver a large current output for a brief time to start an engine. After starting, the charging system takes over to supply most of the current required to operate the car. The battery acts as a system stabilizer and provides current whenever the electrical loads exceed the charging current output. An automotive battery must provide good cranking power for the car's engine and adequate reserve power for the electrical system in which it is used.

Manufacturers also make 12-volt automotive-type batteries that are not designed for automotive use. These are called "cycling batteries" and are designed to provide a power source for a vehicle or accessory without continual recharging. Cycling batteries provide a constant low current for a long period of time. They are designed for industrial, marine, and recreational vehicle (RV) or motor home use. Most of their current capacity is exhausted in each cycle before recharging.

The brief high current flow required of a starting battery is produced by using relatively thin plates, compared to those used in a cycling battery. The thicker plates of the cycling battery will provide a constant current drain for several hours. Using a starting battery in an application calling for a cycling battery will shorten its life considerably, as we shall see later in the chapter. The use of a cycling battery to start and operate a car will cause excessive internal heat from the brief but high current draw, resulting in a shorter service life.

Test standards and rating methods devised by the Battery Council International (BCI) and the Society of Automotive Engineers (SAE) are designed to measure a battery's ability to meet the requirements of its intended service.

The BCI publishes application charts that list the correct battery for any car. Optional heavy-duty batteries are normally used in cars with air conditioning or several major electrical accessories or in cars operated in cold climates. To ensure adequate cranking power and to meet all other electrical needs, a replacement battery may have a higher rating, but never a lower rating, than the original unit. The battery must also be the correct size for the car, and have the correct type of terminals. BCI standards include a coding system called the group number. BCI battery rating methods are explained in the following paragraphs.

Cold Cranking Amperes (CCA)

The primary duty of the battery is to start the engine. It cranks, or rotates, the crankshaft while it maintains sufficient voltage to activate the ignition system until the engine starts. This requires a high discharge over a very short time span. Cold engines require more power to turn over, but batteries have difficulty delivering power when it is cold. **Cold cranking amperes (CCA)** are an important measurement of battery capacity because they measure the discharge load, in amps, that a battery can supply for 30 seconds at 0°F while maintaining a voltage of 1.2 volts per cell (7.2 volts per battery) or higher. The CCA rating generally falls between 300 and 970 for most passenger cars; it is identified as 300 CCA, 400 CCA, 500 CCA, and so on. The rating is typically higher for commercial vehicles. For more information about cold cranking amps, see the section on "Battery Testing" in Chapter 7 of the *Shop Manual.* Some batteries are rated as high as 1,100 CCA.

Cranking Amps (CA)

Cranking amps (CA) represent the discharge load (in amps) that a fully charged battery can supply for 30 seconds at 32°F while maintaining a voltage of 1.2 volts per cell (7.2 volts per 12 volt battery) or higher.

NOTE: *CA* (Cranking Amps) is nearly the same as *CCA* (Cold Cranking Amps), but the two ratings should not be confused. *CCA* is rated at 0°F. *CA*, on the other hand is measured at a temperature of 32°F. The difference in temperature will produce a considerable amount of additional current when measured as a CA rating.

Reserve Capacity Rating

Reserve capacity is the time required (in minutes) for a fully charged battery at 80°F under a constant 25-amp draw to reach a voltage of 10.5 volts. This rating helps determine the battery's ability to sustain a minimum vehicle electrical load in the event of a charging system failure. The minimum electrical load under the worst possible conditions (winter driving at night) would likely require current for the ignition, low-beam headlights, windshield wipers, and the defroster at low speed. Reserve capacity is useful for measuring the battery's ability to power a vehicle that has small, long-term parasitic electrical loads but enough reserve to crank the engine.

Battery reserve capacity ratings range from 30 to 175 minutes, and correspond approximately to the length of time a vehicle can be driven after the charging system has failed. For more information, see the section on "Battery Testing" in Chapter 7 of the *Shop Manual.*

■ Historical Ampere-Hour Rating

The oldest battery rating method, no longer used to rate batteries, was the ampere-hour rating. This rating method was the industry standard for decades. It was replaced, however, years ago by the cranking performance and reserve capacity ratings, which provide better indications of a battery's performance. The ampere-hour method was also called the 20-hour discharge rating method. This rating represented the steady current flow that a battery delivered at a temperature of 80°F (27°C) without cell voltage falling below 1.75 volts (a total of 10.5 volts for a 12-volt battery). For example, a battery that continuously delivered 3 amperes for 20 hours was rated as a 60 ampere-hour battery (3 amperes × 20 hours = 60 ampere-hours).

■ Historical Watt-Hour Rating

The starter motor converts the electrical power supplied by the battery into mechanical power, so some battery manufacturers rate their batteries using watt-hour rating. The watt-hour rating of the battery is determined at 0°F (−17.7°C) because the battery's capability to deliver wattage varies with temperature. Watt-hour rating is determined by calculating the ampere-hour rating of the battery times the battery voltage.

Battery Size Selection

Some of the aspects that determine the battery rating required for a vehicle include engine size, engine type, climatic conditions, vehicle options, and so on. The requirement for electrical energy to crank the engine increases as the temperature decreases. Battery power drops drastically as temperatures drop below freezing. The engine also becomes harder to crank due to the tendency of oils to thicken when cold, which results in increased friction. As a general rule, it takes 1 ampere of cold cranking power per cubic inch of engine displacement. Therefore, a 200 cubic inch displacement (CID) engine should be fitted with a battery of at least 200 CCA. To convert this into metric, it takes 1 amp of cold cranking power for every 16 cc of engine displacement: A 1.6-liter engine requires a battery rated at 100 CCA. This rule may not apply to vehicles that have several electrical accessories. The best method of determining the correct battery is to refer to the manufacturer's specifications.

The battery that is selected should fit the battery holding fixture and the hold-down must be able to be installed. It is also important that the height of the battery not allow the terminals to short across the vehicle hood when it is shut. BCI group numbers are used to indicate the physical size and other features of the battery. This group number does not indicate the current capacity of the battery.

Group Number

Manufacturers provide a designated amount of space, usually in the engine compartment, to accommodate the battery. Since battery companies build batteries of various current-capacity ratings in a variety of sizes and shapes, it is useful to have a guide when replacing a battery, because it must fit into the space provided. The BCI size group number identifies a battery in terms of its length, width, height, terminal design, and other physical features.

BATTERY INSTALLATIONS

Most automobiles use one 6-cell, 12-volt battery installed in the engine compartment. Certain factors influence battery location as follows:

- The distance between the battery and the alternator or starter motor determines the length of the cables used. Cable length is important because of electrical system resistance. The longer the cables, the greater the resistance.
- The battery should be located away from hot engine components in a position where it can be cooled by airflow.
- The battery should be in a location where it can be securely mounted as protection against internal damage from vibration.
- The battery should be positioned where it can be easily serviced.

The decrease in size of late-model vehicles has resulted in lighter, smaller batteries with greater capacity. The use of new plastics and improved grid and plate materials has contributed to the new battery designs.

Some older cars and a few new imported and domestic models have the battery located in the trunk. For example, the battery used with the Ford Escort diesel is mounted in the trunk beneath a trim cover and encased in a protective bag (Figure 7-9). The bag will retain battery acid in case of an accident that might damage it. A tube and seal assembly connected to the battery vents allows gassing to the atmosphere.

This venting device should be inspected periodically and replaced, if necessary, because proper venting is essential for safety. Such locations require the use of long cables of heavy-gauge wire. The size of such cables offsets their greater length in keeping resistance manageable, but increases cost and weight while reducing convenience.

Late-model GM diesel cars and Ford light trucks use two 12-volt batteries connected in parallel (Figure 7-10). Both battery positive

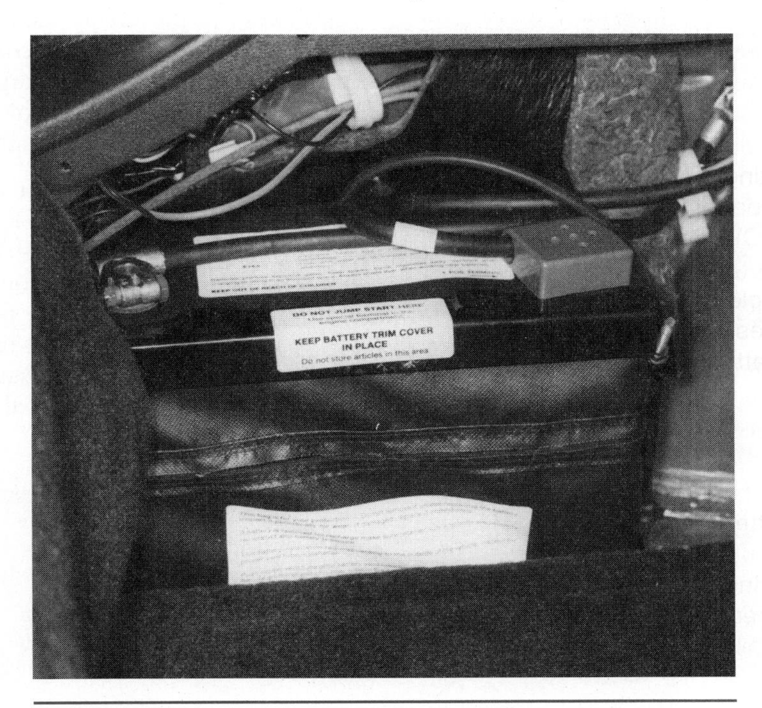

Figure 7-9. This Ford Escort diesel battery is encased in a protective bag and housed in the trunk.

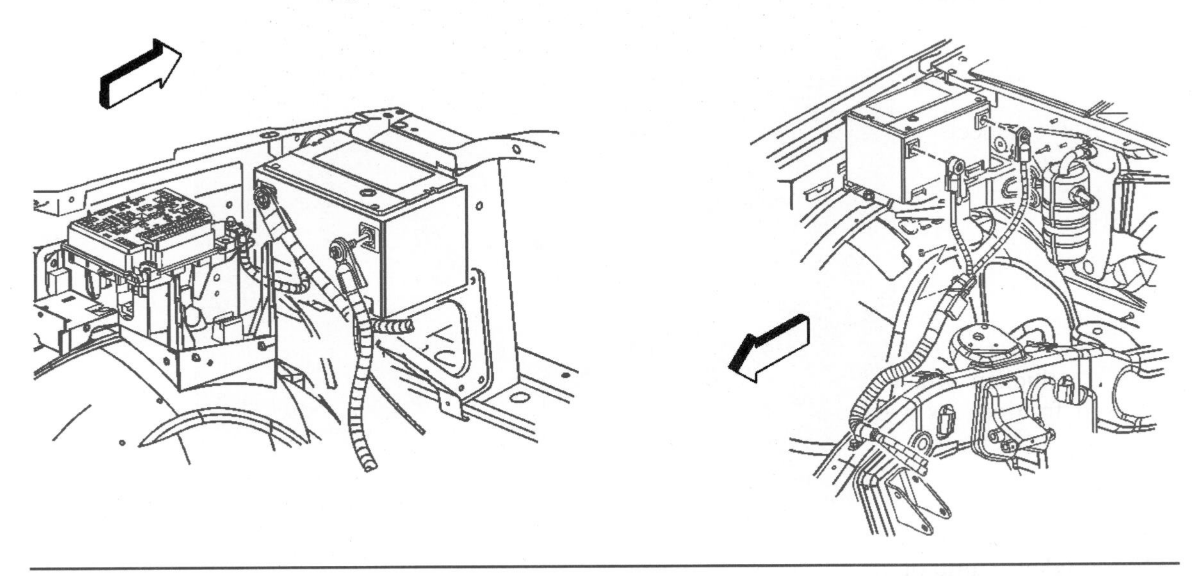

Figure 7-10. Diesel vehicles generally have two 12-volt batteries for better cranking with a 12-volt starter. (GM Service and Parts Operations)

terminals are connected to each other and to the positive battery cable attached to the starter motor (GM) or to the relay (Ford). The battery terminals are connected to each other in a similar manner, and to the ground cable. The use of a parallel installation doubles the current available for starting the high-compression diesel without increasing system voltage. If the batteries were connected in series, the voltage would double. Both batteries are charged simultaneously by the alternator. For more information, see the section on "Battery Changing" in Chapter 7 of the *Shop Manual.*

BATTERY INSTALLATION COMPONENTS

Selecting and maintaining properly designed battery installation components is necessary for good battery operation and service life.

Connectors, Carriers, and Holddowns

Battery cables are very large-diameter multistrand wire, usually 0 to 6 gauge. Diesel engine vehicles generally use the larger 0, while gasoline engine vehicles use 6. A new battery cable should always be the same gauge as the one being replaced.

Battery terminals may be tapered posts on the top or internally threaded terminals on the side of the battery. To prevent accidental reversal of battery polarity (incorrectly connecting the cables), the positive terminal is slightly larger than the negative terminal. Three basic styles of connectors are used to attach the battery cables to the battery terminals:

- A bolt-type clamp is used on top-terminal batteries, Figure 7-11. The bolt passes through the two halves of the cable end into a nut. When tightened, it squeezes the cable end against the battery post.
- A bolt-through clamp is used on side-terminal batteries. The bolt threads through the cable end and directly into the battery terminal, Figure 7-12.

Figure 7-11. The most common type of top terminal battery clamp.

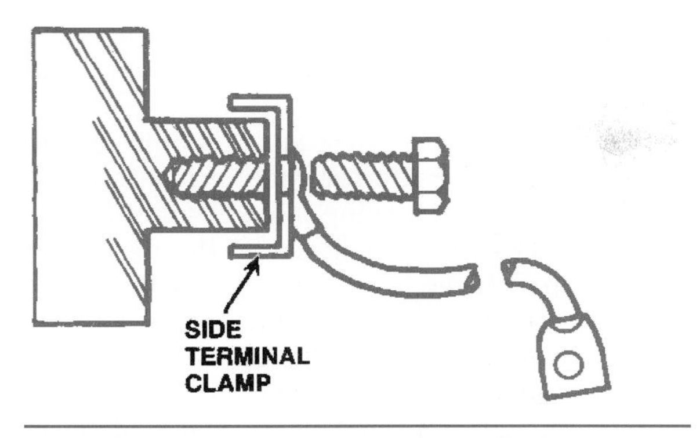

Figure 7-12. The side terminal clamp is attached with a bolt. (DaimlerChrysler Corporation)

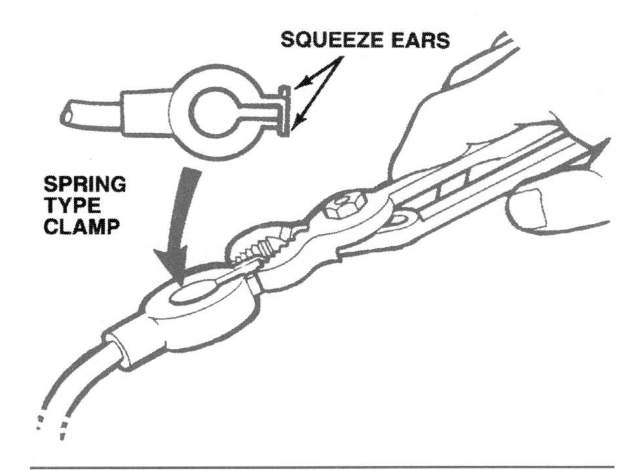

Figure 7-13. The spring-type clamp generally is found on non-domestic cars.

- A spring-type clamp is used on some top-terminal batteries. A built-in spring holds the cable end on the battery post (Figure 7-13).

Batteries are usually mounted on a shelf or tray in the engine compartment, although some manufacturers place the battery in the trunk, under the seat, or else where in the vehicle. The shelf or tray that holds the battery is called the carrier (Figure 7-14). The battery is mounted on the carrier with brackets called holddowns (Figures 7-14 and 7-15). These keep the battery from tipping over and spilling acid. A battery must be held securely in its carrier to protect it from vibration that can damage the plates and internal plate connectors. For more information, see the section on "Battery Cable Service" in Chapter 7 of the *Shop Manual.*

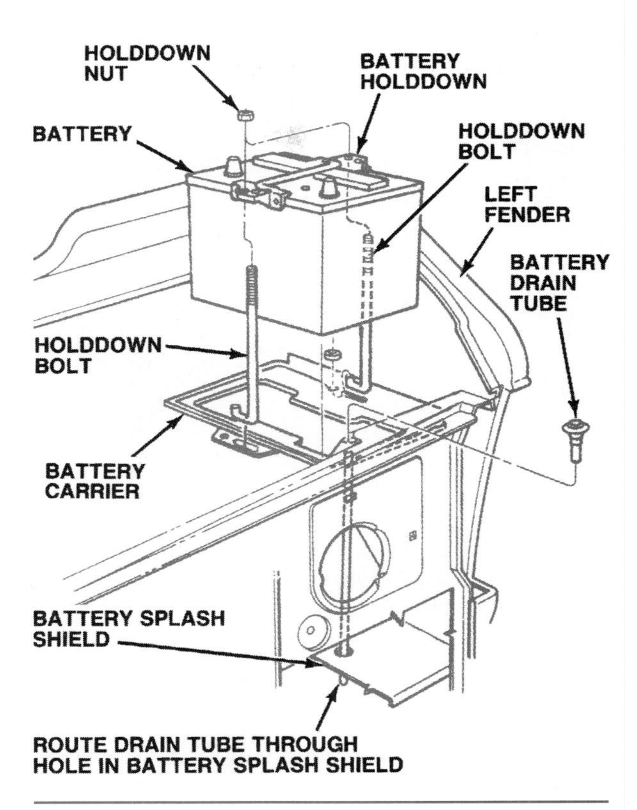

Figure 7-14. A common type of battery carrier and holddown.

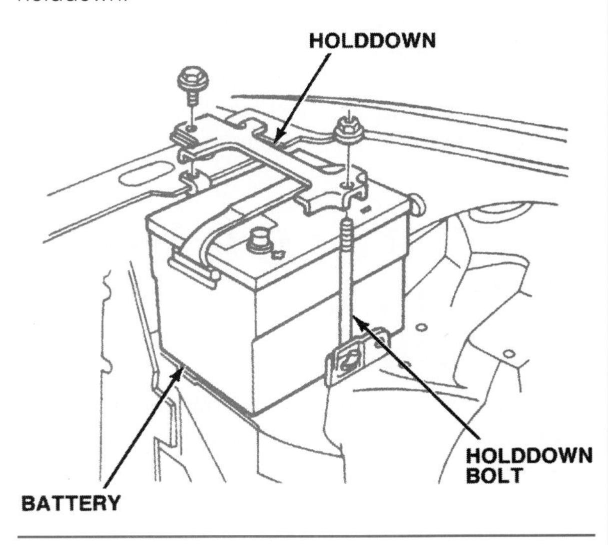

Figure 7-15. Another common battery holddown.

WARNING: Don't Pull The Plugs

Do you make a practice of removing the vent plugs from a battery before charging it? Prestolite says you shouldn't, at least with many late-model batteries. "A great number of batteries manufactured today will have safety vents," says Prestolite. "If these are removed, the batteries are open to external sources of explosion ignition." Prestolite recommends that, on batteries with safety vents, the vent plugs should be left in place when charging.

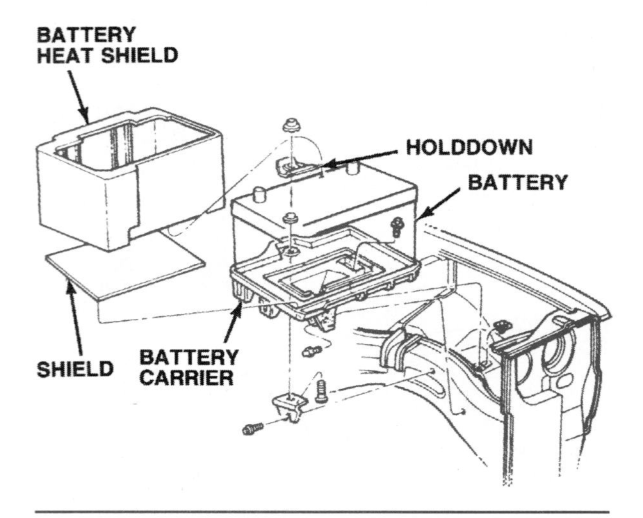

Figure 7-16. A molded heat shield that fits over the battery is used by Chrysler and some other manufacturers. (DaimlerChrysler Corporation)

Battery Heat Shields

Many late-model cars use battery heat shields (Figure 7-16) to protect batteries from high underhood temperatures. Most heat shields are made of plastic, and some are integral with the battery holddown. Integral shields are usually large plastic plates that sit alongside the battery. Heat shields do not require removal for routine battery inspection and testing, but must be removed for battery replacement.

BATTERY LIFE AND PERFORMANCE FACTORS

All batteries have a limited life, but certain conditions can shorten that life. The important factors that affect battery life are discussed in the following paragraphs.

Electrolyte Level

As we have seen, the design of maintenance-free batteries has minimized the loss of water from electrolyte so that battery cases can be sealed. Given normal use, the addition of water to such batteries is not required during their service life. However, even maintenance-free batteries will lose some of their water to high temperature, overcharging, deep cycling, and recharging all factors in battery gassing and resultant water loss.

With vent-cap batteries, and to some extent, low- maintenance batteries, water is lost from the electrolyte during charging in the form of hydrogen and oxygen gases. This causes the electrolyte level to drop. If the level drops below the top of the plates, active material will be exposed to the air. The material will harden and resist electrochemical reaction. Also, the remaining electrolyte will have a high concentration of acid, which can cause the plates to deteriorate quickly. Even the addition of water will not restore such hardened plates to a fully active condition.

Parasitic Losses

Parasitic losses are small current drains required to operate electrical systems, such as the clock, that continue to work when the car is parked and the ignition is off. The current demand of a clock is small and not likely to cause a problem.

The advent of computer controls, however, has made parasitic losses more serious. Many late-model cars have computers to control such diverse items as engine operation, radio tuning, suspension leveling, climate control, and more. Each of these microprocessors contains random access memory (RAM) that stores information relevant to its job. To "remember," RAM requires a constant voltage supply, and therefore puts a continuous drain on the car's electrical system.

The combined drain of several computer memories can discharge a battery to the point where there is insufficient cranking power after only a few weeks. Vehicles with these systems that are driven infrequently, put into storage, or awaiting parts for repair will require battery charging more often than older cars with lower parasitic voltage losses.

Because of the higher parasitic current drains on late-model cars, the old test of removing a battery cable connection and tapping it against the terminal while looking for a spark is both dangerous and no longer a valid check for excessive current drain. Furthermore, every time the power source to the computer is interrupted, electronically stored information, such as radio station presets, is lost and will have to be reprogrammed when the battery is reconnected.

On engine control systems with learning capability, like GM's Computer Command Control, driveability may also be affected until the computer relearns the engine calibration modifications that were erased from its memory when the battery was disconnected. For more information, see the section on "Battery Testing" in Chapter 7 of the *Shop Manual.*

Corrosion

Battery corrosion is caused by spilled electrolyte and by electrolyte condensation from gassing. The sulfuric acid attacks and can destroy not only connectors and terminals, but metal holddowns and carriers as well. Corroded connectors increase resistance at the battery connections. This reduces the applied voltage for the car's electrical system. Corrosion also can cause mechanical failure of the holddowns and carrier, which can damage the battery. Spilled electrolyte and corrosion on the battery top also can create a current leakage path, which can allow the battery to discharge.

Overcharging

Batteries can be overcharged either by the automotive charging system or by a separate battery charger. In either case, there is a violent chemical reaction in the battery. The water in the electrolyte is rapidly broken down into hydrogen and oxygen gases. These gas bubbles can wash active material off the plates, as well as lower the level of the electrolyte. Overcharging can also cause excessive heat, which can oxidize the positive grid material and even buckle the plates. For more information, see the section on "Battery Changing" in Chapter 7 of the *Shop Manual.*

Undercharging and Sulfation

If an automobile is not charging its battery, either because of stop-and-start driving or a fault in the charging system, the battery will be constantly discharged. As we saw in the explanation of electrochemical action, a discharged plate is covered with lead sulfate. The amount of lead sulfate on the plate will vary according to the state of charge. As the lead sulfate builds up in a constantly undercharged battery, it can crystallize and not recombine with the electrolyte. This is called battery sulfation. The crystals are difficult to break down by normal recharging and the battery becomes useless. Despite the chemical additives sold as "miracle cures" for sulfation, a completely sulfated battery cannot be effectively recharged.

Cycling

As we learned at the beginning of this chapter, the operation of a battery from charged to discharged and back to charged is called cycling. Automotive batteries are not designed for continuous deep-cycle use (although special marine and RV batteries are). If an automotive battery is repeatedly cycled from a fully charged condition to an almost discharged condition, the active material on the positive plates may shed and fall into the bottom of the case. If this happens, the material cannot be restored to the plates. Cycling thus reduces the capacity of the battery and shortens its useful service life.

Temperature

Temperature extremes affect battery service life and performance in a number of ways. High temperature, caused by overcharging or excessive engine heat, increases electrolyte loss and shortens battery life.

Low temperatures in winter can also harm a battery. If the electrolyte freezes, it can expand and break the case, ruining the battery. The freezing point of electrolyte depends upon its specific gravity and thus, on the battery's state of charge. A fully charged battery with a specific gravity of 1.265 to 1.280 will not freeze until its temperature drops below −60°F (−51°C). A discharged battery with electrolyte that is mostly water can freeze at −18°F (−28°C).

As we saw earlier, cold temperatures make it harder to keep the battery fully charged, yet this is when a full charge is most important. Figure 7-17 compares the energy levels available from a fully charged battery at various temperatures. As you can see, the colder a battery is, the less energy it can supply. Yet the colder an engine gets, the more energy it requires for cranking. This is why battery care is especially important in cold weather.

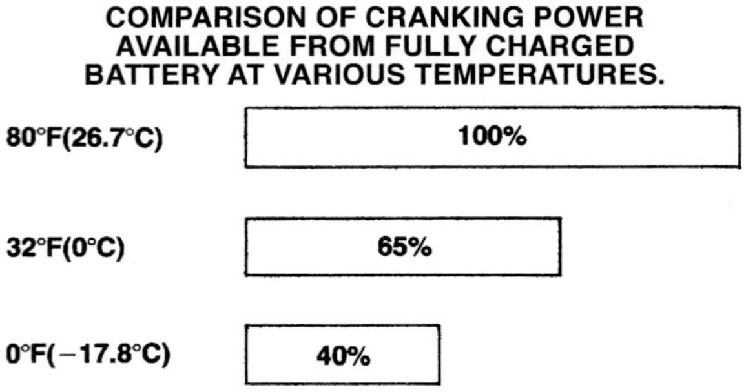

COMPARISON OF CRANKING POWER AVAILABLE FROM FULLY CHARGED BATTERY AT VARIOUS TEMPERATURES.

Temperature	Cranking power
80°F(26.7°C)	100%
32°F(0°C)	65%
0°F(−17.8°C)	40%

Figure 7-17. Battery power decreases as temperature decreases.

Vibration

As mentioned earlier, a battery must be securely mounted in its carrier to protect it from vibration. Vibration can shake the active materials off the plates and severely shorten a battery's life. Vibration can also loosen the plate connections to the plate strap and damage other internal connections. Some manufacturers now build batteries with plate straps and connectors in the center of the plates to reduce the effects of vibration. Severe vibration can even crack a battery case and loosen cable connections.

See the section on "Jump Starting" in Chapter 7 of the *Shop Manual*.

SUMMARY

Automotive batteries are lead-acid secondary batteries containing a number of electrochemical cells that can be recharged after discharging. Batteries not only store power, they generate voltage and current through the electrochemical action between dissimilar plates in the presence of an electrolyte. Each lead-acid cell generates about 2.1 volts regardless of the number of positive and negative plates. Cells are connected in series, allowing six cells to produce about 12.6 volts in a fully charged 12-volt battery. Current output of a cell depends upon the total surface area of all the plates. Batteries with higher current or capacity ratings have larger plate areas.

The battery state of charge is determined by electrolyte specific gravity. In a fully charged battery, electrolyte should have a specific gravity of 1.260 to 1.265. Maintenance-free batteries contain calcium-alloy grids to reduce battery heat and water loss. Since such batteries are sealed, their electrolyte cannot be checked and water cannot be added to their cells.

Automotive batteries are designed for starting the engine, not for continual cycling from fully charged to discharged and back to fully charged. Batteries have cranking performance and reserve capacity ratings, and BCI group numbers indicate their size and physical characteristics. Battery service life is affected by electrolyte level, corrosion, overcharging or undercharging, cycling, vibration, and temperature variations.

■ How the Battery Got Its Name

The word "battery" means a group of like things used together. An automobile battery is a group of electrochemical cells connected and working together. Battery voltage is determined by the number of cells connected in series in the battery.

Early automobile batteries could be taken apart for service. Cases were made of wood, and the tops were sealed with tar or a similar material. The top could be opened and the plate element could be removed from a single cell and replaced with a new one.

Deep-Cycle Service

Some batteries, like those in golf carts and electric vehicles, are used for deep-cycle service. This means that as they provide electrical current, they go from a fully charged state to an almost fully discharged state, and are then recharged and used again.

Maintenance-free batteries should never be used in deep-cycle service. Deep-cycle service promotes shedding of the active materials from the battery plates. This action drastically reduces the service life of a maintenance-free battery.

Review Questions

1. Which of the following occurs within an automobile battery?
   a. The positive plate gains electrons and is positively charged.
   b. The negative plate loses electrons and is negatively charged.
   c. The positive plate loses electrons and the negative plate gains electrons.
   d. The positive plate gains electrons and the negative plate loses electrons.

2. Battery electrolyte is a mixture of water and:
   a. Lead peroxide
   b. Sulfuric acid
   c. Lead sulfate
   d. Sulfur crystals

3. The plates of a *discharged* battery are:
   a. Two similar metals in the presence of an electrolyte
   b. Two similar metals in the presence of water
   c. Two dissimilar metals in the presence of an electrolyte
   d. Two dissimilar metals in the presence of water

4. Which of the following is true about a "secondary" battery?
   a. It can be recharged.
   b. Neither the electrolyte nor the metals change their atomic structure.
   c. One of the metals is totally destroyed by the action of the battery.
   d. The action of the battery cannot be reversed.

5. Which of the following does *not* occur during battery recharging?
   a. The lead sulfate on the plates gradually decomposes.
   b. The sulfate is redeposited in the water.
   c. The electrolyte is returned to full strength.
   d. The negative plates change back to lead sulfate.

6. Each cell of an automobile battery can produce about ________ volts.
   a 1.2
   b 2.1
   c 4.2
   d 6

7. Which of the following is true of a 6-volt automobile battery?
   a. It has six cells connected in series.
   b. It has three cells connected in series.
   c. It has six cells connected in parallel.
   d. It has three cells connected in parallel.

8. The correct ratio of water to sulfuric acid in battery electrolyte is approximately:
   a. 80 percent water to 20 percent sulfuric acid
   b. 60 percent water to 40 percent sulfuric acid
   c. 40 percent water to 60 percent sulfuric acid
   d. 20 percent water to 80 percent sulfuric acid

9. At 80°F, the correct specific gravity of electrolyte in a fully charged battery is:
   a. 1.200 to 1.225
   b. 1.225 to 1.265
   c. 1.265 to 1.280
   d. 1.280 to 1.300

10. A specific gravity of 1.170 to 1.190 at 80°F indicates that a battery's state of charge is about:
    a. 75 percent
    b. 50 percent
    c. 25 percent
    d. 10 percent

11. Which of the following materials is *not* used for battery separators?
    a. Lead
    b. Wood
    c. Paper
    d. Plastic

12. Batteries are rated in terms of:
    a. Amperes at 65°F
    b. Resistance at 32°F
    c. Voltage level at 80°F
    d. Cranking performance at 0°F

13. Maintenance-free batteries:
    a. Have individual cell caps
    b. Require water infrequently
    c. Have three pressure vents
    d. Use non-antimony lead alloys

14. Which of the following statements is *not* true of a replacement battery?
    a. It may have the same rating as the original battery.
    b. It may have a higher rating than the original battery.
    c. It may have a lower rating than the original battery.
    d. It should be selected according to an application chart.
15. An automobile battery with a cranking performance rating of 380 can deliver 380 amps for:
    a. 30 seconds at 0°F
    b. 60 seconds at 0°F
    c. 90 seconds at 32°F
    d. 90 seconds at 0°F
16. The principal cause of battery water loss is:
    a. Spillage from the vent caps
    b. Leakage through the battery case
    c. Conversion of water to sulfuric acid
    d. Evaporation due to heat of the charging current
17. Which of the following is *not* true of a maintenance-free battery?
    a. It will resist overcharging better than a vent-cap battery.
    b. It will lose water slower than a vent-cap battery.
    c. It will produce a greater voltage than a vent-cap battery.
    d. It has a greater electrolyte capacity than a vent-cap battery.
18. The electrolyte in a fully charged battery will generally not freeze until the temperature drops to:
    a. 32°F
    b. 0°F
    c. −20°F
    d. −50°F
19. The grid material used in a maintenance-free battery is alloyed with:
    a. Silicon
    b. Antimony
    c. Calcium
    d. Germanium
20. Low-maintenance batteries:
    a. Have no cell caps
    b. Have a higher proportion of sulfuric acid
    c. Have no gas-pressure vents
    d. Require infrequent water addition
21. Recombinant batteries are:
    a. Rebuilt units
    b. Completely sealed
    c. Vented to release gassing
    d. Able to produce a higher cell voltage
22. Which of these parts of a battery hold the electrical charge?
    a. Side-post types
    b. Positive and negative plates
    c. Top-post type
    d. Bottom plates
23. Technician A says that the battery term ampere-hour refers to stored charge capacity of a battery. Technician B says that a 75-ampere-hour charge applied to a 200-ampere-hour battery should turn the charge indicator green. Who is right?
    a. A only
    b. B only
    c. Both A and B
    d. Neither A nor B
24. On a vehicle with the two battery 12-volt system, the battery's connection is which one of the following?
    a. Series circuit
    b. Parallel circuit
    c. DC circuit
    d. AC circuit
25. During normal operation, the battery(s) perform all of the following functions, *except:*
    a. Provides electrical energy for the accessories when the engine is not running
    b. Acts as voltage storage for the truck electrical system
    c. Serves as the voltage source for starting
    d. Provides voltage for the injection solenoid when running
26. The subject of battery ratings is being discussed. Technician A says reserve capacity is a rating that represents the time in minutes that a battery can operate a truck at night with minimum electrical load. Technician B says that cranking amps (CA) is basically the same as cold cranking amps (CCA) but at a temperature of 32°F. Who is right?
    a. A only
    b. B only
    c. Both A and B
    d. Neither A nor B

8

Charging System Operation

LEARNING OBJECTIVES

Upon completion and review of this chapter, you should be able to:

- Identify charging system development and principles.
- Explain the operation of a DC generator.
- Identify AC charging system components and explain charging voltage.
- Explain diode rectification.
- Identify AC generator components and explain their function.
- Explain current production in an AC generator (alternator) operation.
- Identify different OEM AC generators and explain those differences.
- Explain how voltage is regulated in an AC generator.
- Identify the different types of voltage regulators and explain how they operate.

KEY TERMS

AC Generators
Delta-Type Stator
Diode
Field Circuit
Full-Wave Rectification
Half-Wave Rectification
Output Circuit
Rotor
Sine Wave Voltage
Single-Phase Current
Single-Phase Voltage
Sliprings and Brushes
Stator
Three-phase Current
Voltage Regulator
Y-Type Stator

INTRODUCTION

The charging system converts the engine's mechanical energy into electrical energy. This electrical energy is used to maintain the battery's state of charge and to operate the loads of the automotive electrical system. In this chapter we will use the conventional theory of current: Electrons move from positive to negative (+ to −).

During cranking, all electrical energy for the vehicle is supplied by the battery. After the engine is running, the charging system must produce enough electrical energy to recharge the battery and to supply the demands of other loads in the electrical system. If the starting system is in poor condition and draws too much current, or if the charging system cannot recharge the battery and supply the additional loads, more energy must be drawn from the battery for short periods of time.

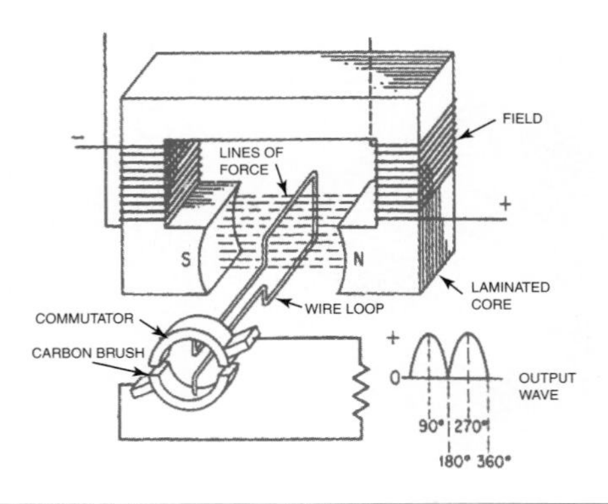

Figure 8-1. DC generator.

CHARGING SYSTEM DEVELOPMENT

For many years, automotive charging systems used only direct current (DC) generators to provide electrical energy. Internally, generators produce an alternating current voltage, which is mechanically rectified by the commutator into direct current voltage. Systems using DC generators are called DC charging systems. Vehicles with DC generators are very rare today.

AC generators or alternators also produce alternating current (AC), but there was no simple way to rectify the current until semiconductor technology finally provided the answer in the form of diodes, or one-way electrical valves. Since the mid-1960s, virtually all new automobiles have diode-rectified AC generators (alternators) in their charging systems.

AC generators (alternators) replaced DC generators back in the late fifties, except for Volkswagen, which used DC generators until 1975. In the automotive charging system, they have the following advantages:

- Weigh less per ampere of output
- Can be operated at much higher speed
- Pass less current through the brushes with only a few amperes of field current, reducing brush wear
- Govern their own maximum current output, requiring no external current regulation
- Can produce current when rotated in either direction, although their cooling fans usually are designed for one-way operation.

DC GENERATOR

The principles of electromagnetic induction are employed in generators for producing DC current. The basic components of a DC generator are shown in Figure 8-1. A framework composed of laminated iron sheets or other ferromagnetic metal has a coil wound on it to form an electromagnet. When current flows through this coil, magnetic fields are created between the pole pieces, as shown. Permanent magnets could also be employed instead of the electromagnet.

To simplify the initial explanation, a single wire loop is shown between the north and south pole pieces. When this wire loop is turned within the magnetic fields it cuts the lines of force and a voltage is induced. If there is a complete circuit from the wire loop, current will flow. The wire loop is connected to a split ring known as a commutator, and carbon brushes pick off the electric energy as the commutator rotates. Connecting wires from the carbon brushes transfer the energy to the load circuit.

When the wire loop makes a half-turn, the energy generated rises to a maximum level and drops to zero, as shown in Figure 8-1. If the wire loop completed a full rotation, the induced voltage would reverse itself and the current would flow in the opposite direction (AC current) after the initial half-turn. To provide for an output having a single polarity (DC current), a split-ring commutator is used. Thus, for the second half-turn, the carbon brushes engage commutator segments opposite to those over which they slid for the first half-turn, keeping the current in the same direction. The output waveform is not a steady-level DC, but rises and falls to form a pattern referred to as *pulsating DC*. Thus, for a complete 360-degree turn of the wire loop, two waveforms are produced, as shown in Figure 8-1.

CHARGING VOLTAGE

Although the automotive electrical system is called a 12-volt system, the AC generator (alternator) must produce more than 12 volts. A fully charged battery produces about 2.1 volts per cell; this means the open-circuit voltage of a fully charged 12-volt battery, which has six cells, is approximately 12.6 volts. If the AC generator cannot produce more than 12.6 volts, it cannot charge the battery until the system voltage drops under 12.6 volts. This would leave nothing extra to serve the other electrical demands put on the system by lights, air conditioning, and power accessories.

Alternating-current charging systems are generally regulated to produce a maximum output of 14.5 volts. Output of more than 16 volts will overheat the battery electrolyte and shorten its life. High voltage also damages components that rely heavily on solid-state electronics, such as fuel injection and engine control systems. On the other hand, low voltage output causes the battery to become sulfated. The charging system must be maintained within the voltage limits specified by the manufacturer if the vehicle is to perform properly.

AC Charging System Components

The automotive charging system. (Figure 8-2) contains the following:

- A battery, which provides current to initiate the magnetic field required to operate the AC generator (alternator) and, in turn, is charged and maintained by the AC generator.
- An AC generator (alternator), which is belt-driven by the engine and converts mechanical motion into charging voltage and current.

The simple AC generator (alternator) shown in Figure 8-3 consists of a magnet rotating inside a fixed-loop stator, or conductor. The alternating current produced in the conductor is rectified by diodes for use by the electrical system.

- A **voltage regulator,** which limits the field current and thus the AC generator (alternator) output voltage according to the electrical system demand. A regulator can be either an electromechanical or a solid-state device. Some late-model, solid-state regulators are part of the vehicle's onboard computer.
- An ammeter, a voltmeter, or indicator warning lamp mounted on the instrument panel to give a visual indication of charging system operation.

Charging System Circuits

The charging system consists of the following major circuits (Figure 8-4):

- The **field circuit,** which delivers current to the AC generator (alternator) field
- The **output circuit,** which sends voltage and current to the battery and other electrical components

Single-Phase Current

AC Generators (alternators) induce voltage by rotating a magnetic field inside a fixed conductor. The greatest current output is produced when the rotor is parallel to the stator with its magnetic field at right angles to the stator, as in Figure 8-3. When the rotor makes one-quarter of a revolution and is at right angles to the stator with its magnetic field parallel to the stator, as in Figure 8-5, there is no

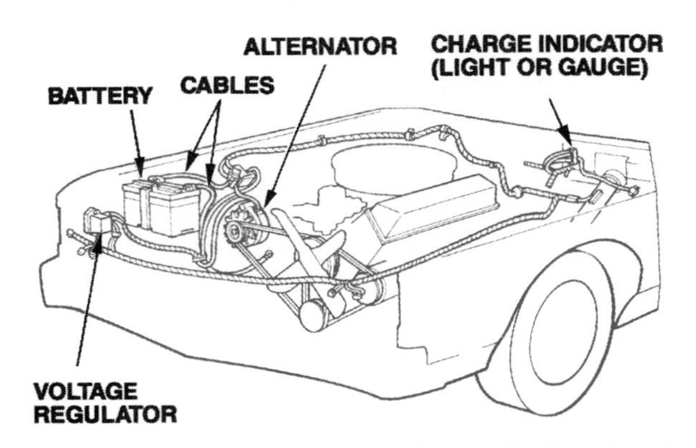

Figure 8-2. Major components of an automotive charging system. (DaimlerChrysler Corporation)

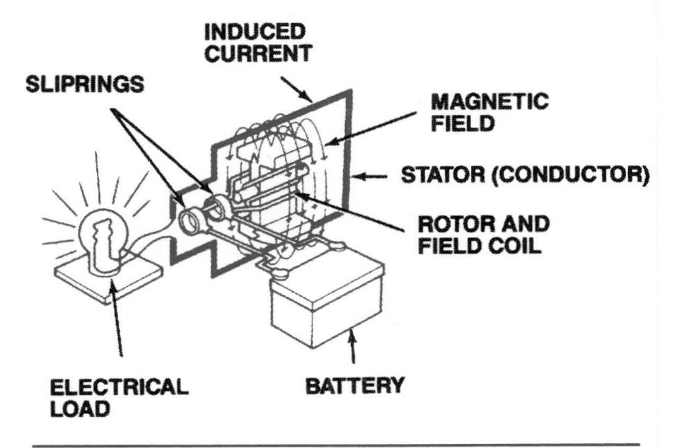

Figure 8-3. An AC generator (alternator) is based on the rotation of a magnet inside a fixed-loop conductor. (DaimlerChrysler Corporation)

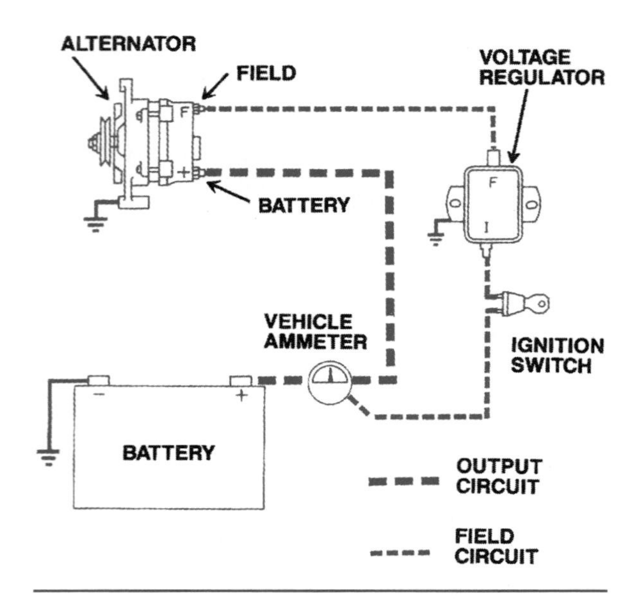

Figure 8-4. The output circuit and the field circuit make up the automotive charging system. (DaimlerChrysler Corporation)

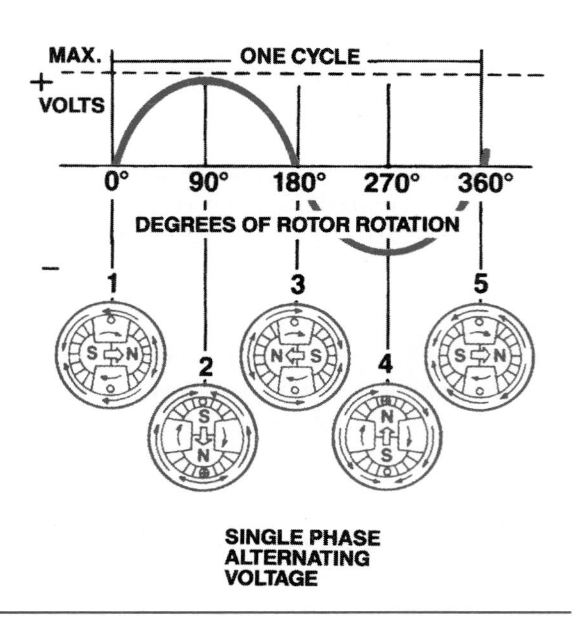

Figure 8-6. These are the voltage levels induced across the upper half of the conductor during one rotor revolution. (DaimlerChrysler Corporation)

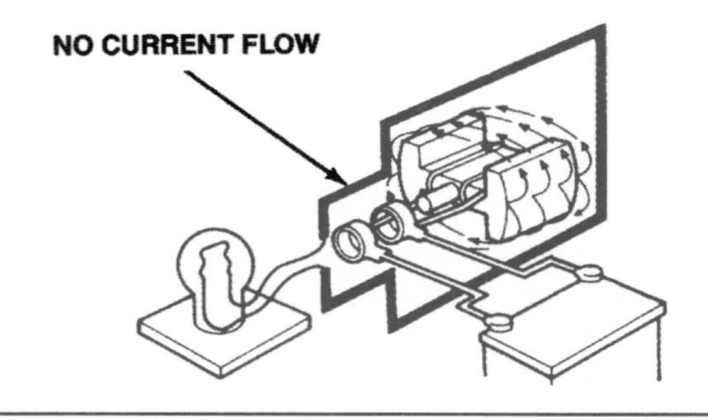

Figure 8-5. No current flows when the rotor's magnetic field is parallel to the stator. (DaimlerChrysler Corporation)

current output. Figure 8-6 shows the voltage levels induced across the upper half of the looped conductor during one revolution of the rotor.

The constant change of voltage, first to a positive peak and then to a negative peak, produces a **sine wave voltage.** This name comes from the trigonometric sine function. The wave shape is controlled by the angle between the magnet and the conductor. The sine wave voltage induced across one conductor by one rotor revolution is called a **single-phase voltage.** Positions 1 through 5 of Figure 8-6 show complete sine wave single-phase voltage.

This single-phase voltage causes alternating current to flow in a complete circuit because the voltage switches from positive to negative as the rotor turns. The alternating current caused by a single-phase voltage is called **single-phase current.**

DIODE RECTIFICATION

If the single-phase voltage shown in Figure 8-6 made current travel through a simple circuit, the current would flow first in one direction and then in the opposite direction. As long as the rotor turned, the current would reverse its flow with every half revolution. The battery cannot be recharged with alternating current. Alternating current must be rectified to direct current to recharge the battery. This is done with diodes.

A **diode** acts as a one-way electrical valve. If a diode is inserted into a simple circuit, as shown in Figure 8-7, one-half of the AC voltage is blocked. That is, the diode allows current to flow from *X* to *Y*, as shown in position A. In position B, the current cannot flow from *Y* to *X* because it is blocked by the diode. The graph in Figure 8-7 shows the total current.

The first half of the current, from *X* to *Y*, was allowed to pass through the diode. It is shown on the graph as curve *XY*. The second half of the current, from *Y* to *X*, was not allowed to pass through the diode. It does not appear on the graph because it never traveled through the circuit. When the voltage reverses at the start of the next rotor revolution, the current is again allowed through the diode from *X* to *Y*.

An AC generator with only one conductor and one diode would show this current output pattern. However, this output would not be very useful because half of the time there is no current available. This is called **half-wave rectification,** since only half of the AC voltage produced by the AC generator is allowed to flow as DC voltage.

Adding more diodes to the circuit, as shown in Figure 8-8, allows more of the AC voltage to be rectified to DC. In position A, current moves from *X* to *Y*. It travels from *X*, through diode 2, through the load, through diode 3, and back to *Y*. In position B, current moves from *Y* to *X*. It travels from *Y*, through diode 4, through the load, through diode I, and back to *X*.

Notice that in both cases current traveled through the load in the same direction. This is because the AC has been rectified to DC. The graph in Figure 8-8 shows the current output of an AC generator with one conductor and four diodes.

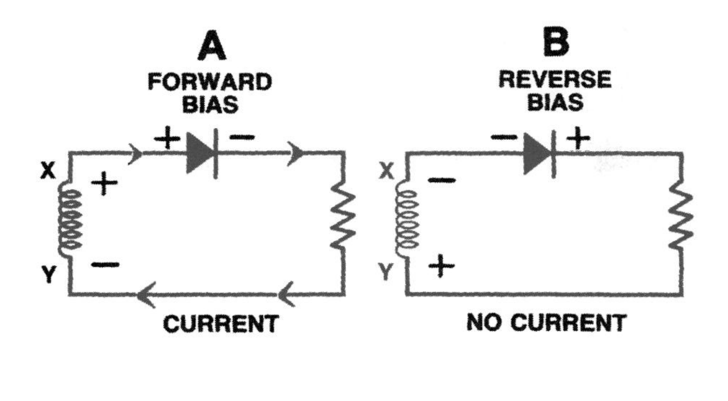

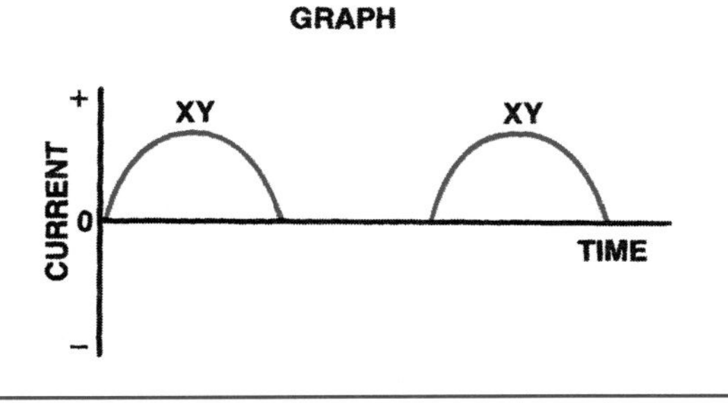

Figure 8-7. A single diode in the circuit results in half-wave rectification. (Delphi Automotive Systems)

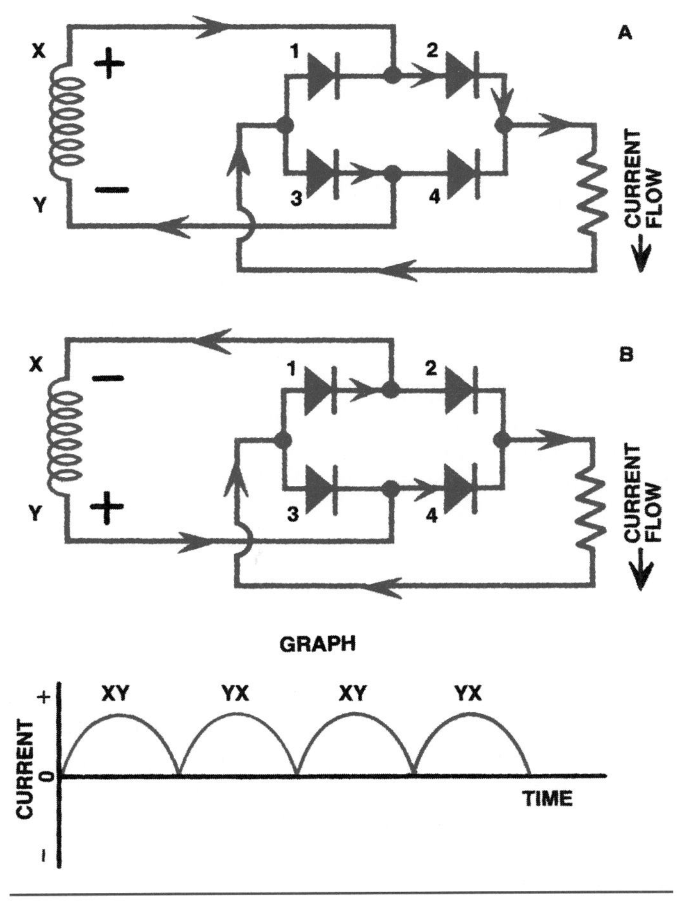

Figure 8-8. More diodes are needed in the circuit for full-wave rectification. (Delphi Automotive Systems)

There is more current available because all of the voltage has been rectified. This is called **full-wave rectification.** However, there are still moments when current is at zero. Most automotive AC generators use three conductors and six diodes to produce overlapping current waves so that current output is *never* at zero.

Heat Sinks

The term *heat sink* is commonly used to describe the block of aluminum or other material in which the AC generator diodes are mounted. The job of the heat sink is to absorb and carry away the heat in the diodes caused by electrical current through them. This action keeps the diodes cool and prevents damage. An internal combustion engine is also a heat sink. The engine is designed so that the combustion and friction heat are carried away and dissipated to the atmosphere. Although they are not thought of as heat sinks, many individual parts of an automobile—such as the brake drums—are also designed to do this important job.

AC GENERATOR (ALTERNATOR) COMPONENTS

The previous illustrations have shown the principles of AC generator operation. To provide enough direct current for an automobile, AC generators must have a more complex design. But no matter how the design varies, the principles of operation remain the same.

The design of the AC generator limits its maximum output. To change this maximum value for different applications, manufacturers change the design of the stator, rotor, and other components. The following paragraphs describe the major parts of an automotive AC generator.

Rotor

The **rotor** carries the magnetic field. Unlike a DC generator, which usually has only two magnetic poles, the AC generator rotor has several north (N) and south (S) poles. This increases the number of flux lines within the AC generator and increases the voltage output. A typical automotive rotor (Figure 8-9) has 12 poles: 6 N and 6 S. The rotor consists of two steel rotor halves, or pole pieces,

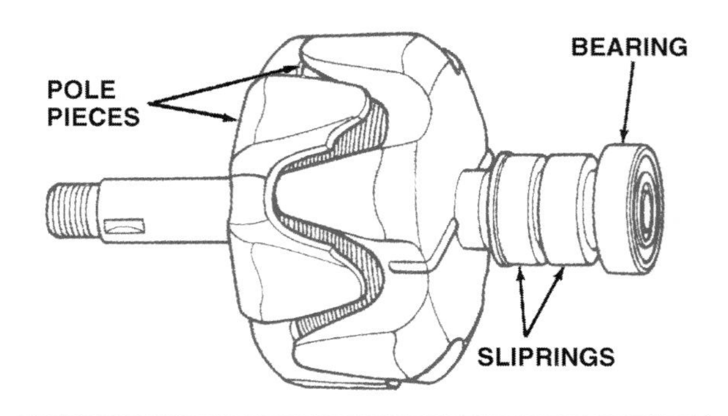

Figure 8-9. Rotor. (DaimlerChrysler Corporation)

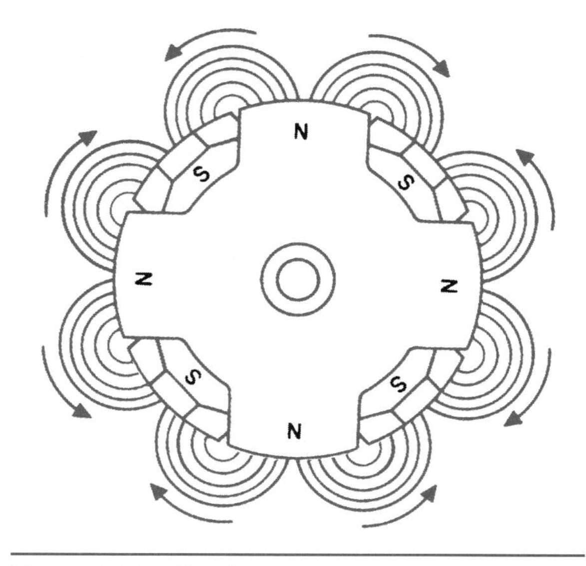

Figure 8-10. The flux lines surrounding an 8-pole rotor. (DaimlerChrysler Corporation)

with fingers that interlace. These fingers are the poles. Each pole piece has either all N or all S poles. The magnetic flux lines travel between adjacent N and S poles (Figure 8-10). Keep in mind that an alternator is an AC generator; in some European manufacturers' service manuals, the AC generator (alternator) is referred to as a generator.

Along the outside of the rotor, note that the flux lines point first in one direction and then in the other. This means as the rotor spins inside the AC generator, the fixed conductors are being cut by flux lines, which point in alternating directions. The induced voltage alternates, just as in the example of a simple AC generator with only two poles. Automotive AC generators may have any number of poles, as long as they are placed N-S-N-S. Common designs use eight to fourteen poles.

The rotor poles may retain some magnetism when the AC generator is not in operation, but this

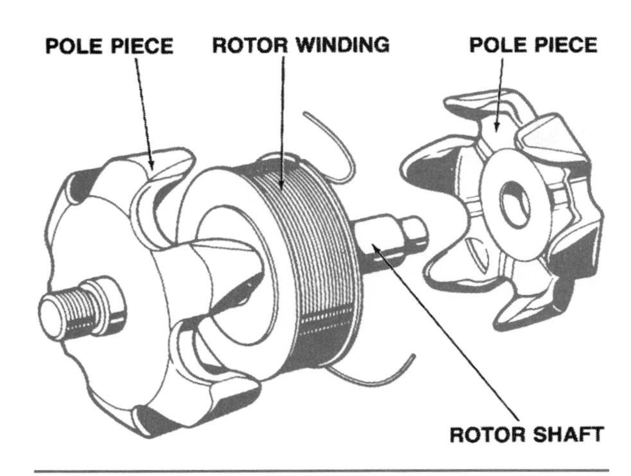

Figure 8-11. The magnetic field of the rotor is created by current through the rotor winding. (Reprinted by permission of Robert Bosch GmbH)

residual magnetism is not strong enough to induce any voltage across the conductors. Current produces the magnetic field of the rotor through the rotor winding, which is a coil of wire between the two pole pieces (Figure 8-11). This is also called the excitation winding, or the field winding. Varying the amount of field current through the rotor winding varies the strength of the magnetic field, which affects the voltage output of the AC generator. A soft iron core is mounted inside the rotor-winding (Figure 8-12). One pole piece is attached to either end of the core; when field current travels through the winding, the iron core is magnetized, and the pole pieces take on the magnetic polarity of the end of the core to which they are attached. Current is supplied to the winding through **sliprings and brushes.**

The combination of a soft iron rotor core and steel rotor halves provides better localization and permeability of the magnetic field. The rotor pole pieces, winding, core, and sliprings are pressed onto a shaft. The ends of this shaft are supported by bearings in the AC generator housing. Outside the housing, a drive pulley is attached to the shaft, as shown in Figure 8-13. A belt, driven by the engine crankshaft-pulley, passes around the drive pulley to turn the AC generator shaft and rotor assembly.

Stator

The three AC generator conductors are wound onto a cylindrical, laminated core. The lamination prevents unwanted eddy currents from forming in the core. The assembled piece is called a

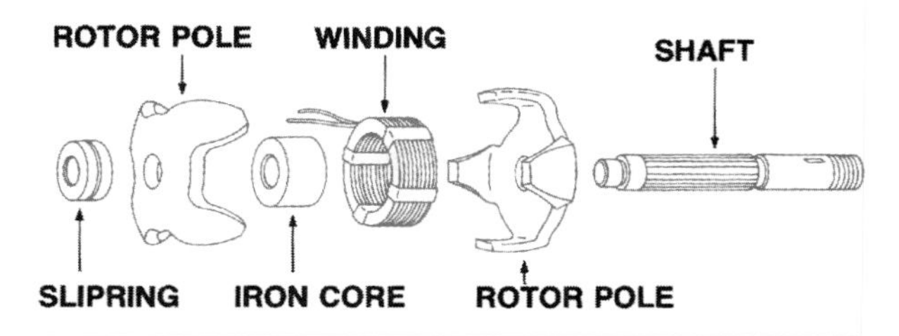

Figure 8-12. Exploded view of the parts of the complete rotor assembly. (DaimlerChrysler Corporation)

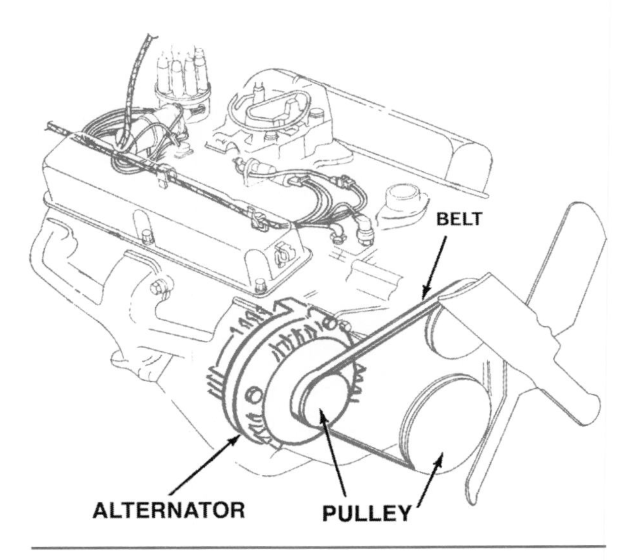

Figure 8-13. AC generator (alternator) and drive pulley. (DaimlerChrysler Corporation)

stator (Figure 8-14). The word stator comes from the word "stationary" because it does not rotate, as does the commutator conductor of a DC generator. Each conductor, called a stator winding, is formed into a number of coils spaced evenly around the core. There are as many coil conductors as there are pairs of N-S rotor poles. Figure 8-15 shows an incomplete stator with only one of its conductors installed: There are seven coils in the conductor, so the matching rotor would have seven pairs of N-S poles, for a total of fourteen poles. There are two ways to connect the three-stator windings.

Housing

The AC generator housing, or frame, is made of two pieces of cast aluminum (Figure 8-16). Aluminum is lightweight and non-magnetic and conducts heat well. One housing piece holds a bearing for the end of the rotor shaft where the

Figure 8-14. An AC generator (alternator) stator.

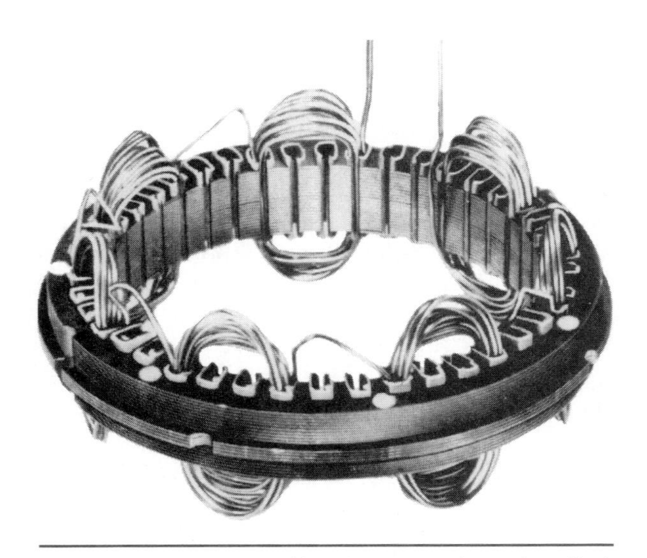

Figure 8-15. A stator with only one conductor installed. (Delphi Automotive Systems)

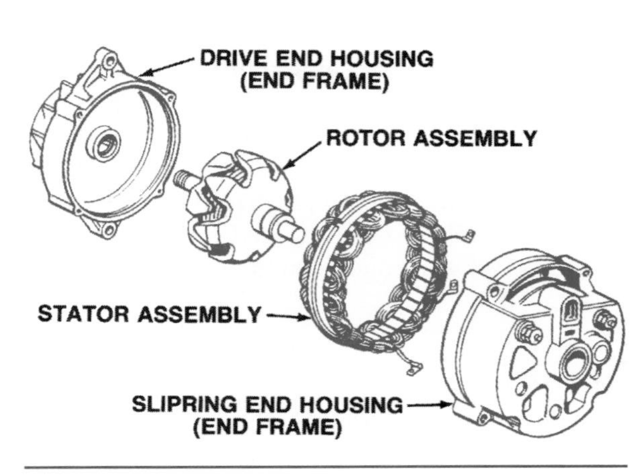

Figure 8-16. AC generator (alternator) housing encloses the rotor and stator.

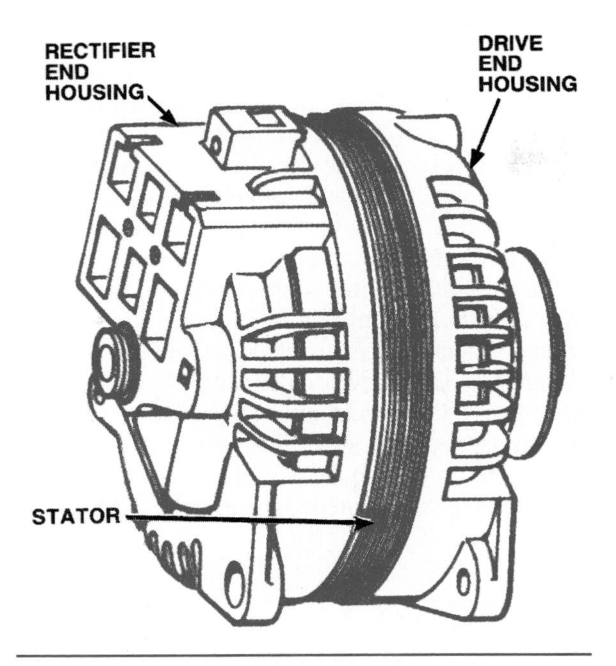

Figure 8-17. AC generator (alternator) with exposed stator core.

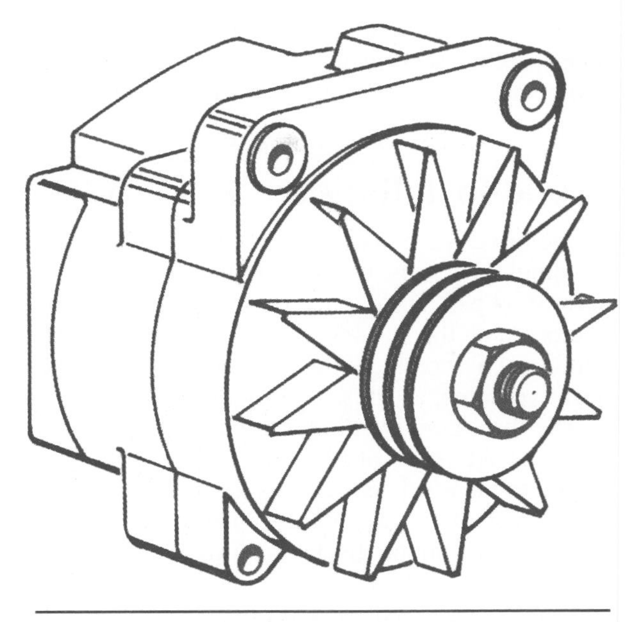

Figure 8-18. AC generator (alternator) with the stator core enclosed.

drive pulley is mounted. This is often called the drive-end housing, or front housing, of the AC generator. The other end holds the diodes, the brushes, and the electrical terminal connections. It also holds a bearing for the slipring end of the rotor shaft. This is often called the slipring-end housing, or rear housing. Together, the two pieces completely enclose the rotor and the stator windings.

The end housings are bolted together. Some stator cores have an extended rim that is held between the two housings (Figure 8-17). Other stator cores provide holes for the housing bolts, but do not extend to the outside of the housings (Figure 8-18). In both designs, the stator is rigidly bolted in place inside the AC generator housing. The housing is part of the electrical ground path because it is bolted directly to the engine. Anything connected to the housing that is not insulated from the housing is grounded.

Sliprings and Brushes

The sliprings and brushes conduct current to the rotor winding. Most automotive AC generators have two sliprings mounted on the rotor shaft. The sliprings are insulated from the shaft and from each other. One end of the rotor winding is connected to each slipring (Figure 8-19). One brush rides on each ring to carry current to and

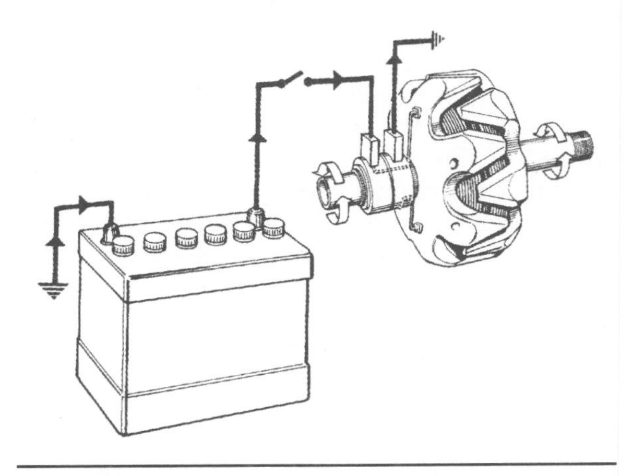

Figure 8-19. The sliprings and brushes carry current to the rotor windings.

from the winding. A brush holder supports each brush and a spring applies force to keep the brush in constant contact with the rotating slipring. The brushes are connected parallel with the AC generator output circuit. They draw some of the AC generator current output and route it through the rotor winding. Current through the winding must be DC. Field current in an AC generator is usually about 1.5 to 3.0 amperes. Because the brushes carry so little current, they do not require as much maintenance as DC generator brushes, which must conduct all of the current output.

For more information about generator maintenance, see the section on "Disassembly, Cleaning, and Inspection" in Chapter 8 of the *Shop Manual*.

Diode Installation

Automotive AC generators that have three stator windings generally use six diodes to rectify the current output. The connections between the conductors and the diodes vary slightly, but each conductor is connected to one positive and one negative diode, as shown in Figure 8-20.

The three positive diodes are always insulated from the AC generator housing. They are connected to the insulated terminal of the battery and to the rest of the automotive electrical system. The battery cannot discharge through this connection because the bias of the diodes blocks any current from the battery. The positive diodes only conduct current traveling from the conductors toward the battery.

The positive diodes are mounted together on a conductor called a heat sink (Figure 8-21). The heat sink carries heat away from the diodes, just as the radiator carries heat away from the engine. Too much heat damages the diodes.

In the past, the three negative diodes were pressed or threaded into the AC generator rear housing. On high-output AC generators, they may be mounted in a heat sink for added protection. In either case, the connection to the AC generator housing is a ground path; the negative diodes conduct only the current traveling from ground into the conductors. Each group of three or more negative or positive diodes can be called a diode bridge, a diode trio, or a diode plate. Some manufacturers use complete rectifier assemblies containing all the diodes and connections on a printed circuit board. This assembly is replaced as a unit if any of the individual components fail.

Each stator winding connects to its proper negative diode through a circuit in the rectifier. A capacitor generally is installed between the output terminal at the positive diode heat sink to ground at the negative diode heat sink. This capacitor is used to eliminate voltage-switching transients at the stator, to smooth out the AC voltage fluctuations, and to reduce electromagnetic interference (EMI).

CURRENT PRODUCTION IN AN AC GENERATOR

After studying the principles of AC generator operation and its components, the total picture of how an automotive AC generator produces current becomes clear.

Three-Phase Current

The AC generator stator has three windings. Each is formed into a number of coils, which are spaced evenly around the stator core. The voltages induced across each winding by one rotor revolution are shown in the graphs of Figure 8-22. The

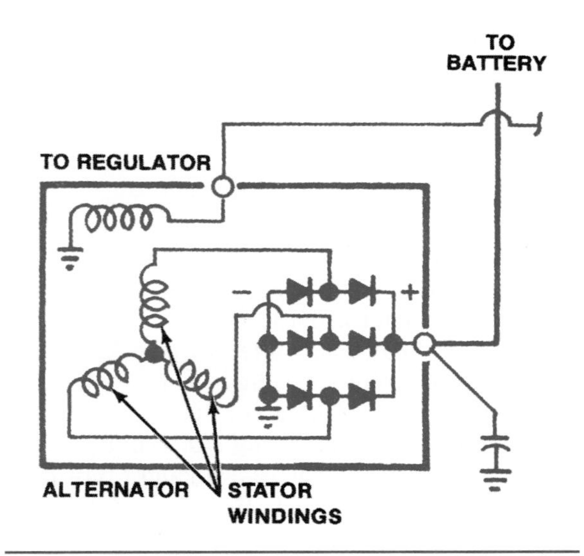

Figure 8-20. Each conductor is attached to one positive and one negative diode.

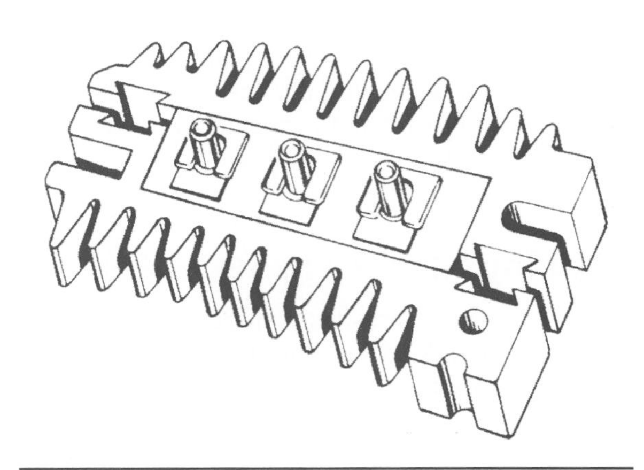

Figure 8-21. Positive and negative diodes may be mounted in a heat sink for protection.

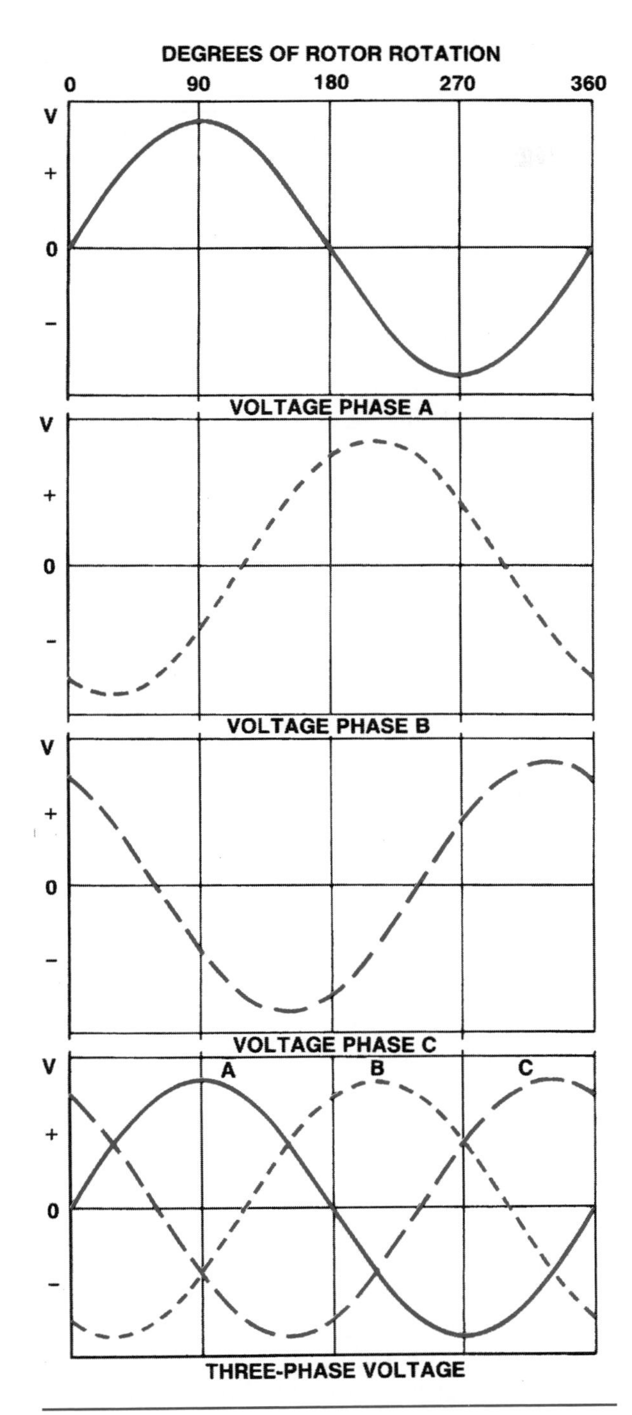

Figure 8-22. The single-phase voltages of three conductors create a three-phase voltage output. (Reprinted by permission of Robert Bosch GmbH)

total voltage output of the AC generator is three overlapping, evenly spaced, single-phase voltage waves, as shown in the bottom graph of the illustration. If the stator windings are connected into a complete circuit, the three-phase voltages cause an AC output called **three-phase current.**

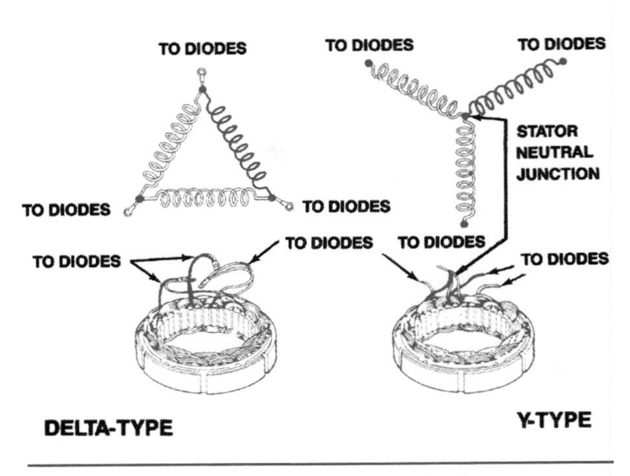

Figure 8-23. Two types of stator windings. (DaimlerChrysler Corporation)

Stator Types

When the three conductors are completely wound on the stator core, six loose ends remain. The way in which these ends are connected to the diode rectifier circuitry determines if the stator is a Y-type or a Delta-type (Figure 8-23). Both kinds of stators produce three-phase current and the rectification produces DC output. However, the voltage and current levels within the stators differ.

Y-Type Stator Design

In the **Y-type stator** or Y-connected stator, one end of each of the three windings is connected at a neutral junction (Figure 8-23). The circuit diagram of the Y-type stator (Figure 8-24) looks like the letter *Y*. This is also sometimes called a wye or a star connection. The free end of each conductor is connected to a positive and a negative diode.

In a Y-type stator (Figure 8-24), two windings always form a series circuit between a positive and a negative diode. At any given instant, the position of the rotor determines the direction of current through these two windings. Current flows from the negative voltage to the positive voltage. A complete circuit from ground, through a negative diode, through two of the windings, and through a positive diode to the AC generator output terminal, exists throughout the 360-degree rotation of the rotor. The induced voltages across the two windings added together produce the total voltage at the output terminal. The majority of AC generators in use today have Y-type stators because of the need for high voltage output at low speeds.

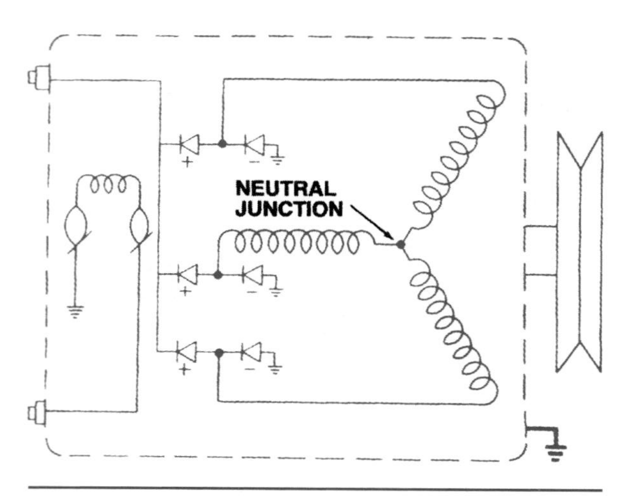

Figure 8-24. Y-type stator circuit diagram. (DaimlerChrysler Corporation)

■ Unrectified AC Generators

Although the battery cannot be recharged with AC, other automotive accessories can be designed to run on unrectified alternator output. Motorola has made AC generators with separate terminals for AC output. Ford has offered a front-and-rear-window defroster that heats the windows with three-phase, 120-volt AC. An additional AC generator supplies the high-voltage current. This second AC generator is mounted above the standard 12-volt AC generator and driven by the same belt.

The Ford high-voltage AC generator has a Y-type stator. Field current draw is more than 4 amps, and there is no regulator in the field circuit. Output is 2,200 watts at high engine speed. All of the wiring between the AC generator and the defrosters is special, shielded wiring with warning tags at all connectors. Ford test procedures use only an ohmmeter, because trying to test such high output could be dangerous.

Some AC generators include a center tap lead from the neutral junction to an insulated terminal on the housing (Figure 8-25). The center tap may control the field current, to activate an indicator lamp, to control the electric choke on a carburetor, or for other functions.

Delta-Type Stator Design

The **delta-type stator** or delta-connected stator has the three windings connected end-to-end (Figure 8-26). The circuit diagram of a delta-type stator (Figure 8-[illegible]oks like the Greek letter delta

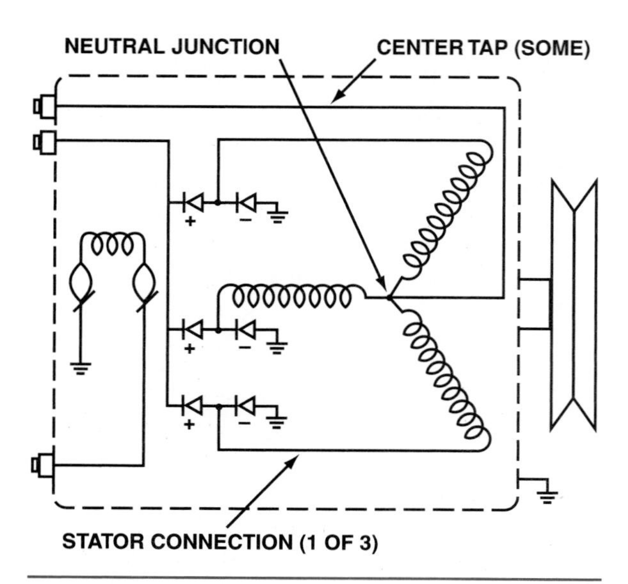

Figure 8-25. A Y-type stator circuit diagram. The center tap connects to the neutral junction of the Y-type stator of some AC generators. (DaimlerChrysler Corporation)

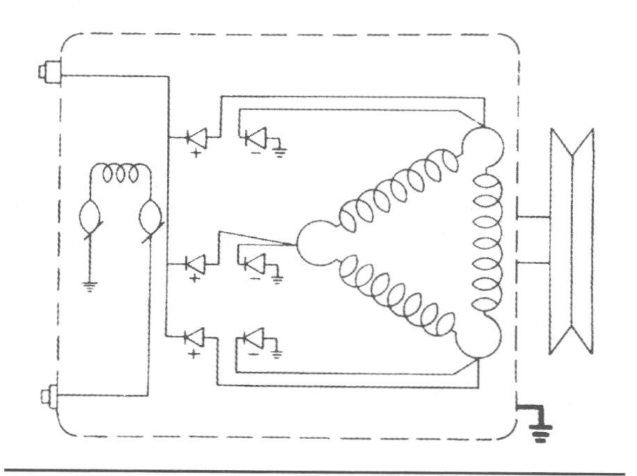

Figure 8-26. Circuit diagram of a delta-type stator.

(Δ), a triangle. There is no neutral junction in a delta-type stator. The windings always form two parallel circuit paths between one negative and two positive diodes. Current travels through two different circuit paths between the diodes (Figure 8-27). The current-carrying capacity of the stator is double because there are two parallel circuit paths. Delta-type stators are used when a high current output is needed.

Phase Rectification

The current pattern during rectification is similar in any automotive AC generator. The only differences are specific current paths through Y-type and delta-type stators as shown in Figures 8-27 and 8-28.

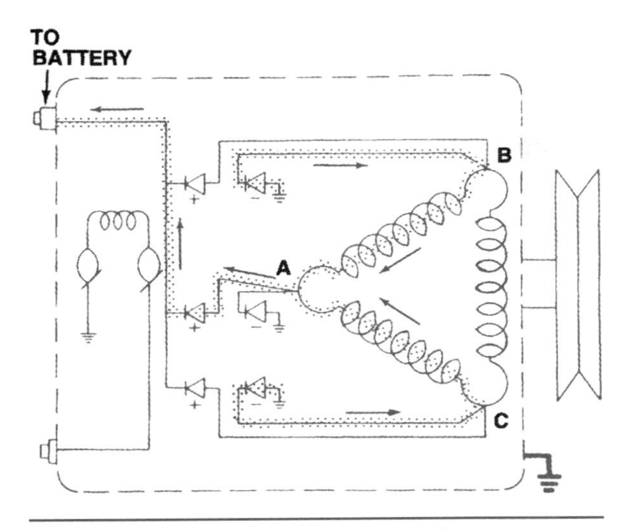

Figure 8-27. A typical current path during rectification in a delta-type stator. (DaimlerChrysler Corporation)

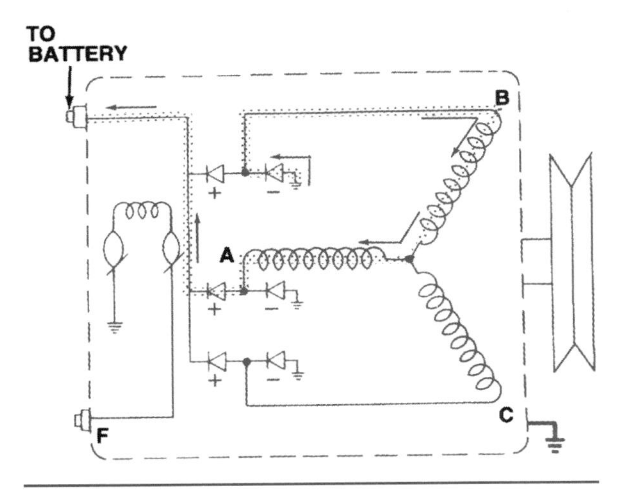

Figure 8-28. A typical current path during rectification in a Y-type stator. (DaimlerChrysler Corporation)

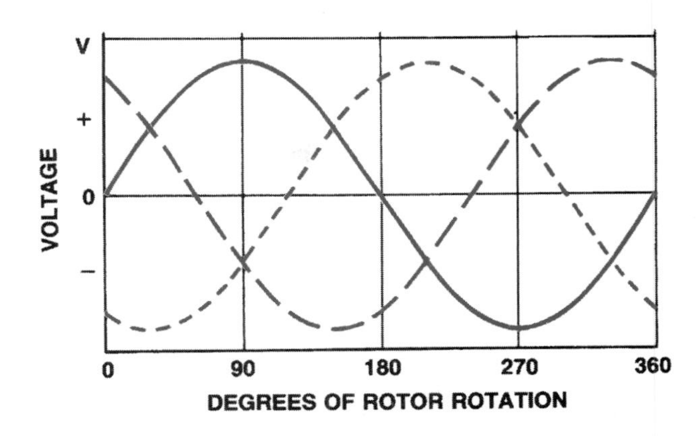

Figure 8-29. The three-phase voltage from one revolution of the rotor. (Reprinted by permission of Robert Bosch GmbH)

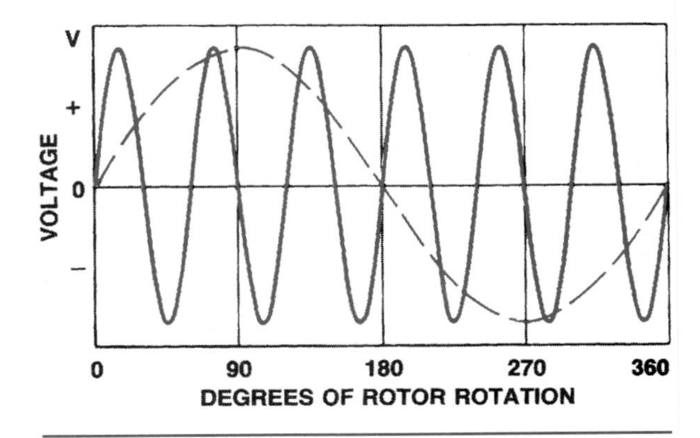

Figure 8-30. An AC generator (alternator) with six pairs of N-S poles would produce the solid line voltage trace. The dashed line represents an alternator with one pair. (Reprinted by permission of Robert Bosch GmbH)

Rectification with Multiple-Pole Rotors

The three-phase voltage output used in the examples (Figure 8-29) is the voltage, which would result if the rotor had only one N and one S pole. Actual AC generator rotors have many N and S poles. Each of these N-S pairs produces one complete voltage sine wave per rotor revolution, across each of the three windings. One complete sine wave begins at zero volts, climbs to a positive, peak, drops past zero to a negative peak, then returns to zero, or baseline voltage. In Figure 8-30, the sine wave voltage caused by a single pole is shown as a dashed line. The actual voltage trace from one winding of a 12-pole AC generator is shown as a solid line. The entire stator output is three of these waves, evenly spaced and overlapping (Figure 8-31). The maximum voltage value from these waves pushes current through the diodes (Figure 8-32).

After rectification, AC generator (alternator) output is DC voltage, which is slightly lower than the maximum voltage peaks of the stator output. The positive portion of the AC sine wave greater than the DC output voltage is viewable on an oscilloscope in what is called an AC generator (alternator) ripple pattern (Figure 8-33).

Excitation Field Circuit

Field current through the rotor windings creates the magnetic field of the rotor. Field current is drawn from the AC generator output circuit once

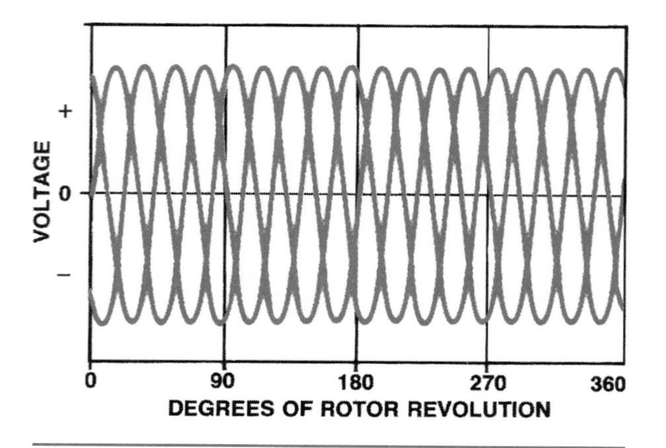

Figure 8-31. This is the total output of a three-winding, multiple-pole AC generator.

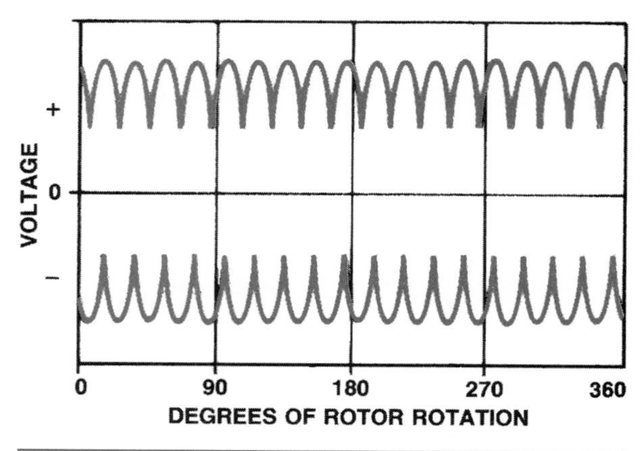

Figure 8-32. The diodes receive the maximum voltage values.

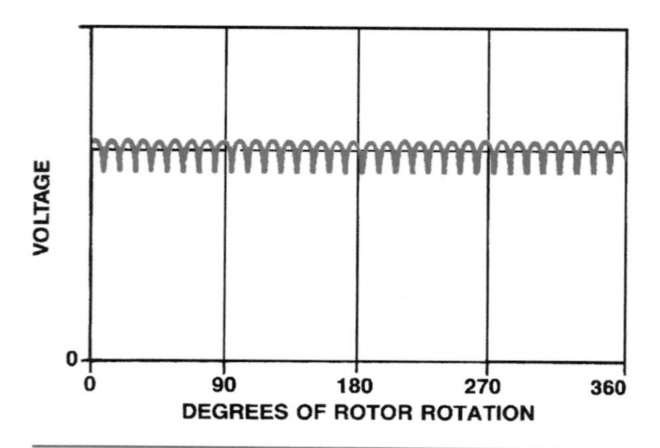

Figure 8-33. AC generator (alternator) ripple is the AC voltage exceeding DC output voltage.

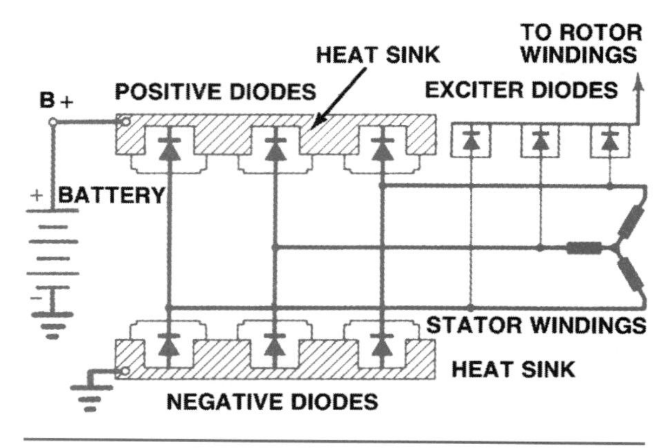

Figure 8-34. Some AC generators (alternators) have additional diodes to rectify field current. (Reprinted by permission of Robert Bosch GmbH)

the AC generator has begun to produce current. However, there is not enough residual magnetism in the rotor poles to induce voltage during start up. An AC generator cannot start operation independently. Field current must be drawn from another source in order to magnetize the rotor and begin AC generator output.

The other source is the vehicle battery connected to the rotor winding through the excitation, or field, circuit. Battery voltage "excites" the rotor magnetic field and begins output. When the engine is off, the battery must be disconnected from the excitation circuit. If not, it could discharge through the rotor windings to ground. Some AC generators use a relay to control this circuit. Other systems use the voltage regulator or a part of the ignition switch to control the excitation circuit.

Once the AC generator has started to produce current, field current is drawn from the AC generator output. The current may be drawn after it has been rectified by the output diodes. Some AC generators draw field current from unrectified AC output, which is then rectified by three additional diodes to provide DC to the rotor winding (Figure 8-34). These additional diodes are called the exciter diodes or the field diodes.

Circuit Types

AC generators (alternators) are designed with different types of field circuits. The two most common types are A-circuits and B-circuits. Circuit types are determined by where the voltage regulator is connected and from where the field current is drawn.

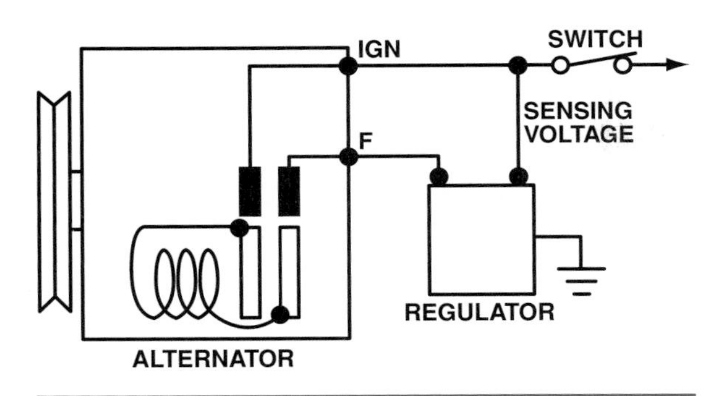

Figure 8-35. An A-circuit. (Regulator is after the field.)

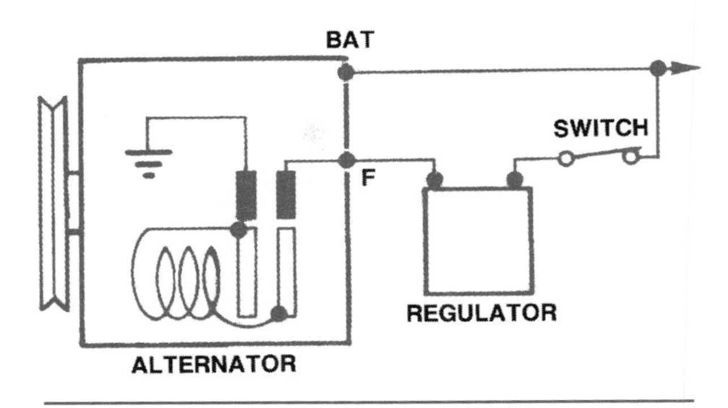

Figure 8-36. A B-circuit. (Regulator is before the field.)

A-Circuit

The A-circuit AC generator (Figure 8-35) is also called an externally grounded field AC generator. Both brushes are insulated from the AC generator housing. One brush connects to the voltage regulator, where it is grounded. The second brush connects to the output circuit within the AC generator, where it draws current for the rotor winding. The regulator connects between the rotor field winding and ground. This type of circuit is often used with solid-state regulators, which are small enough to be mounted on the AC generator housing.

B-Circuit

The B-circuit AC generator (Figure 8-36) is also called an internally grounded field AC generator. One brush is grounded within the AC generator housing. The other brush is insulated from the housing and connected through the insulated voltage regulator to the AC generator output circuit. The rotor field winding is between the regulator and ground. This type of circuit is most often used with electromagnetic voltage regulators, which are mounted away from the AC generator housing.

Self-Regulation of Current

The maximum current output of an AC generator is limited by its design. As induced voltage causes current to travel in a conductor, a counter-voltage is also induced in the same conductor. The counter-voltage is caused by the expanding magnetic field of the original induced current. The counter-voltage tends to oppose any change in the original current.

The more current the AC generator puts out, the greater this counter-voltage becomes. At a certain point, the counter-voltage is great enough to completely stop any further increase in the AC generator's current output. At this point, the AC generator has reached its maximum current output. Therefore, because the two voltages continue to increase as AC generator speed increases, a method of regulating AC generator voltage is required.

For more information about testing AC generators, see the following sections in Chapter 8 of the *Shop Manual:* "Testing Specific Models," "Unit Removal," "Disassembly, Cleaning, and Inspection," and "Bench Tests."

VOLTAGE REGULATION

AC generator regulators limit voltage output by controlling field current. The location of the regulator in the field circuit determines whether the AC generator is an A-circuit or a B-circuit. Voltage regulators basically add resistance to field current in series.

AC generator output voltage is directly related to field strength and rotor speed. An increase in either factor increases voltage output. Similarly, a decrease in either factor decreases voltage output. Rotor speed is controlled by engine speed and cannot be changed simply to control the AC generator. Controlling the field current in the rotor winding can change field strength; this is how AC generator voltage regulators work. Figure 8-37 shows how the field current (dashed line) is lowered to keep AC generator output (solid line) at a constant maximum, even when the rotor speed increases.

At low rotor speeds, the field current is at full strength for relatively long periods of time, and is reduced only for short periods (Figure 8-38A). This causes a high average field current. At high

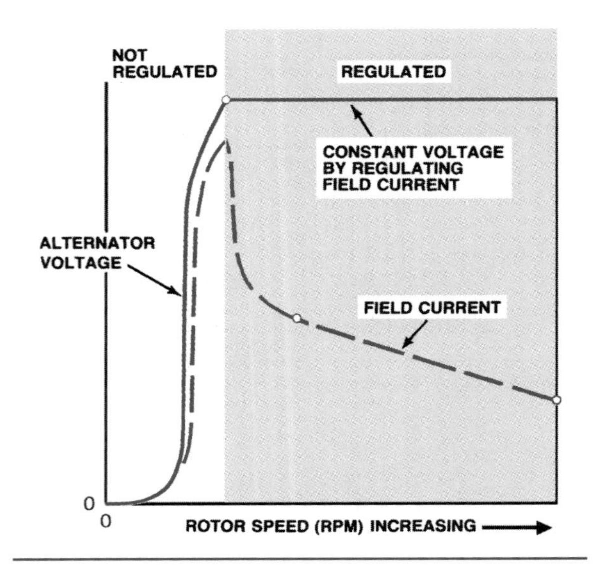

Figure 8-37. Field current is decreased as rotor speed increases, to keep AC generator (alternator) output voltage at a constant level.

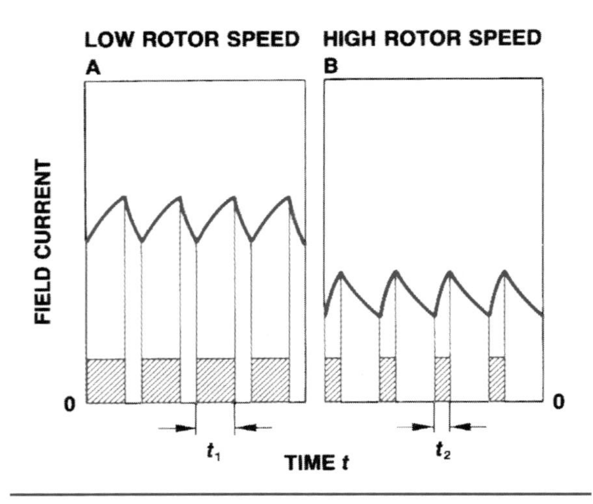

Figure 8-38. The field current flows for longer periods of time at low speeds (t_1) than at high speeds (t_2). (Reprinted by permission of Robert Bosch GmbH)

rotor speeds, the field current is reduced for long periods of time and is at full strength only for short periods (Figure 8-38B). This causes a low average field current.

On older vehicles, an electromagnetic regulator controlled the field circuit. However, in the 1970s, semiconductor technology made solid-state voltage regulators possible. Because they are smaller and have no moving parts, solid-state regulators replaced the older electromagnetic types in AC charging systems. On newer vehicles, the solid-state regulator may be a separate component built

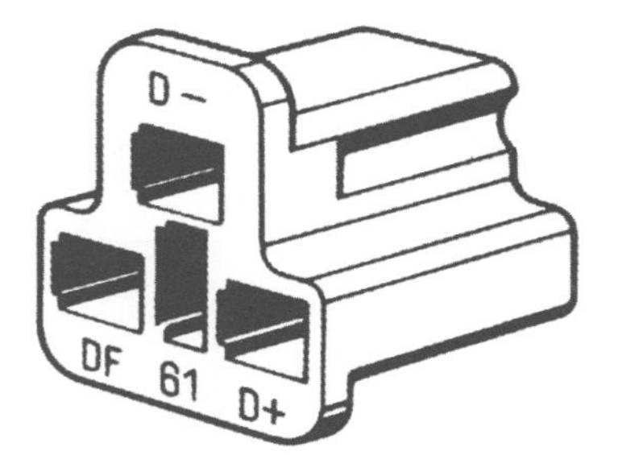

Figure 8-39. Most regulators use a multiple-plug connector to ensure connections are properly made. (Reprinted by permission of Robert Bosch GmbH)

into the AC generator or incorporated into the powertrain control module (PCM).

Some solid-state regulators are mounted on the inside or outside of the AC generator housing. Remotely mounted voltage regulators often use a multiple-plug connector (Figure 8-40) to ensure all connections are properly made. This eliminates exposed wiring and connections, which are prone to damage.

ELECTROMAGNETIC REGULATORS

Electromagnetic voltage regulators, sometimes called electromechanical regulators, operate the same whether used with old DC generators or more common AC generators. The electromagnetic coil of the voltage regulator is connected from the ignition switch to ground. This forms a parallel branch receiving system voltage, either from the AC generator output circuit or from the battery. The magnetic field of the coil acts upon an armature to open and close contact points controlling current to the field.

Double-Contact Voltage Regulator

At high rotor speeds, the AC generator may be able to force too much field current through a single-contact regulator and exceed the desired output. This is called voltage creep, or voltage drift. Single-contact regulators are used only with low-current-output alternators. Almost all electromagnetic voltage regulators used with automotive

AC generators are double-contact units (Figure 8-40). When the first set of contacts opens at lower rotor speeds, current passes through a resistor wired in series with the field circuit. These contacts are called the series contacts. The value of the regulating resistor is kept very low to permit high field current when needed. At higher rotor speeds, the coil further attracts the armature and a second set of contacts is closed. This grounds the field circuit, stopping the field current. These contacts are called the shorting contacts because they short-circuit the field to ground. The double-contact design offers consistent regulation over a broad range of AC generator speeds.

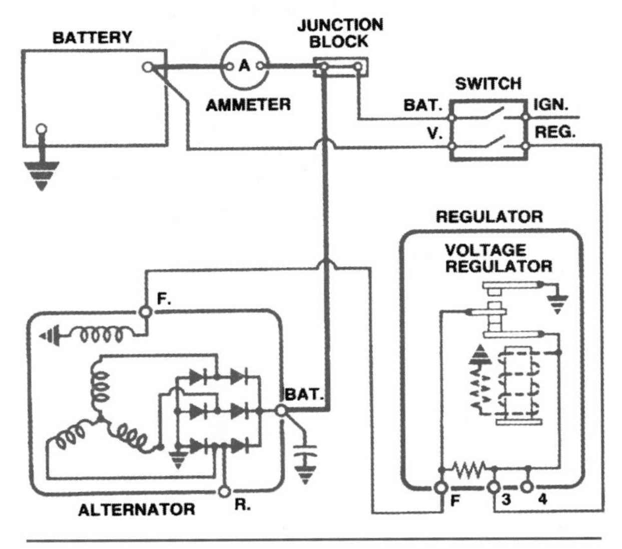

Figure 8-40. The circuit diagram of a double-contact regulator. (Delphi Automotive Systems)

SOLID-STATE REGULATORS

Solid-state regulators completely replaced the older electromagnetic design on late-model vehicles. They are compact, have no moving parts, and are not seriously affected by temperature changes. The early solid-state designs combine transistors with the electromagnetic field relay. The latest and most compact is the integrated circuit (IC) regulator (Figure 8-41). This combines all control circuitry and components on a single

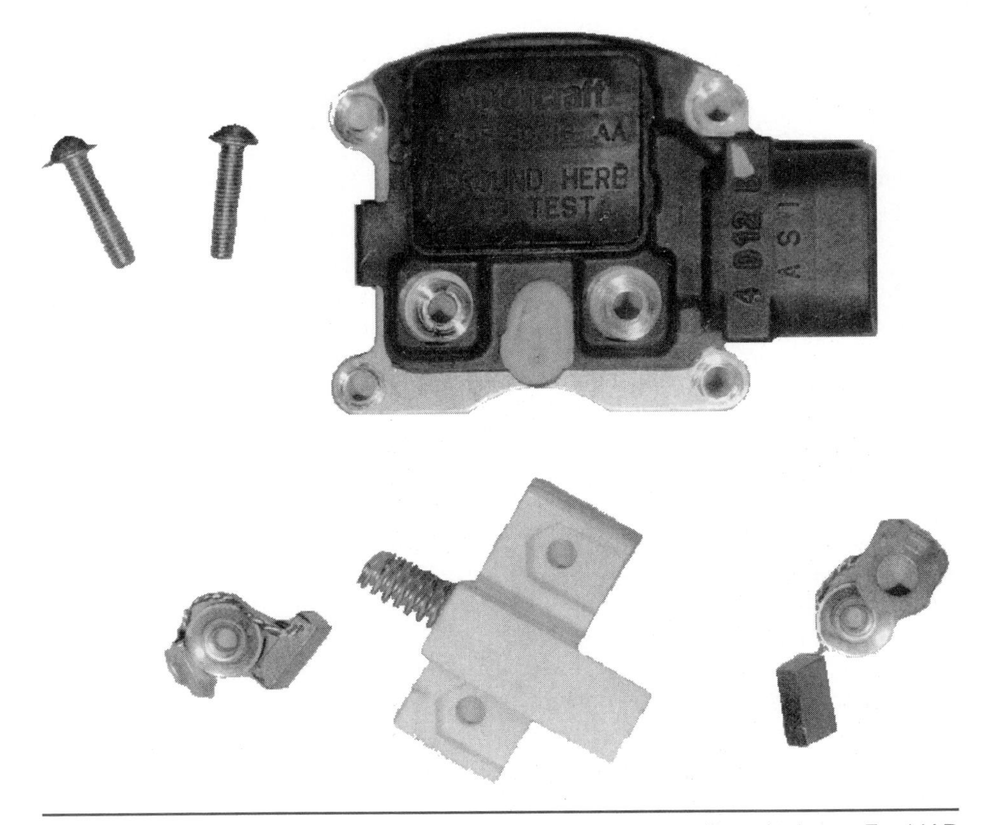

Figure 8-41. An example of the latest integrated circuit regulator design, a Ford IAR regulator and brush holder.

silicon chip. Attaching terminals are added and the chip is sealed in a small plastic module that mounts inside, or on the back of, the AC generator. Because of their construction, however, all solid-state regulators are non-serviceable and must be replaced if defective. No adjustments are possible.

Here is a review of most components of a solid-state regulator:

- Diodes
- Transistors
- Zener diodes
- Thermistors
- Capacitors

Diodes are one-way electrical check valves. Transistors act as relays. A Zener diode is specially doped to act as a one-way, electrical check valve until a specific reverse voltage level is reached. At that point, the Zener diode allows reverse current to pass through it.

The electrical resistance of a thermistor, or thermal resistor, changes as temperature changes. Most resistors used in automotive applications are called negative temperature coefficient (NTC) resistors because their resistance decreases as temperature increases. The thermistor in a solid-state regulator reacts to temperature to ensure proper battery charging voltage. Some manufacturers, in order to smooth out any abrupt voltage surges and protect the regulator from damage, use a capacitor. Diodes may also be used as circuit protection.

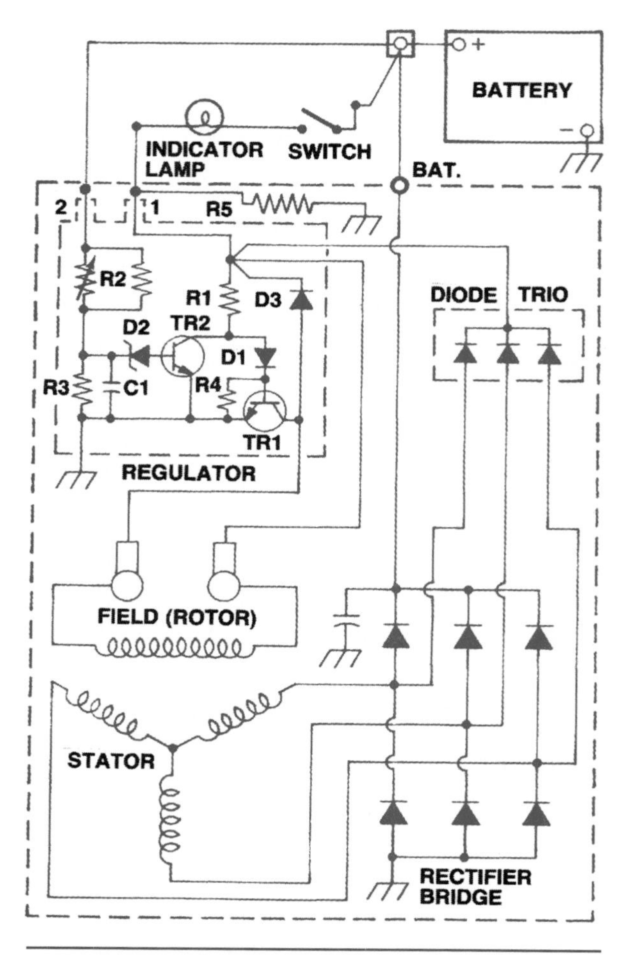

Figure 8-42. Circuit diagram of a typical solid-state voltage regulator. (Delphi Automotive Systems)

General Regulator Operation

Figure 8-42 is a simplified circuit diagram of a solid-state regulator. This A-circuit regulator is contained within the housing. Terminal 2 on the AC generator is always connected to the battery, but battery discharge is limited by the high resistance of R2 and R3. The circuit allows the regulator to sense battery voltage.

When the ignition switch is closed in the circuit shown in Figure 8-43, current travels from the battery to ground through the base of the TR1 transistor. This causes the transistor to conduct current through its emitter-collector circuit from the battery to the low-resistance rotor winding, which energizes the AC generator field and turns on the warning lamp. When the AC generator begins to produce current (Figure 8-44), field current is drawn from unrectified AC generator output and rectified by the diode trio, which is charging voltage. The warning lamp is turned off by equal voltage on both sides of the lamp.

When the AC generator has charged the battery to a maximum safe voltage level (Figure 8-45), the battery voltage between R2 and R3 is high enough to cause Zener diode D2 to conduct in reverse bias. This turns on TR2, which shorts the base circuit of TR1 to ground. When TR1 is turned off, the field circuit is turned off at the ground control of TR1.

With TR1 off, the field current decreases and system voltage drops. When voltage drops low enough, the Zener diode switches off and current is no longer applied to TR2. This opens the field circuit ground and energizes TR1. TR1 turns back on. The field current and system voltage increase. This cycle repeats many times per second to limit the AC generator voltage to a predetermined value.

The other components within the regulator perform various functions. Capacitor C1 provides

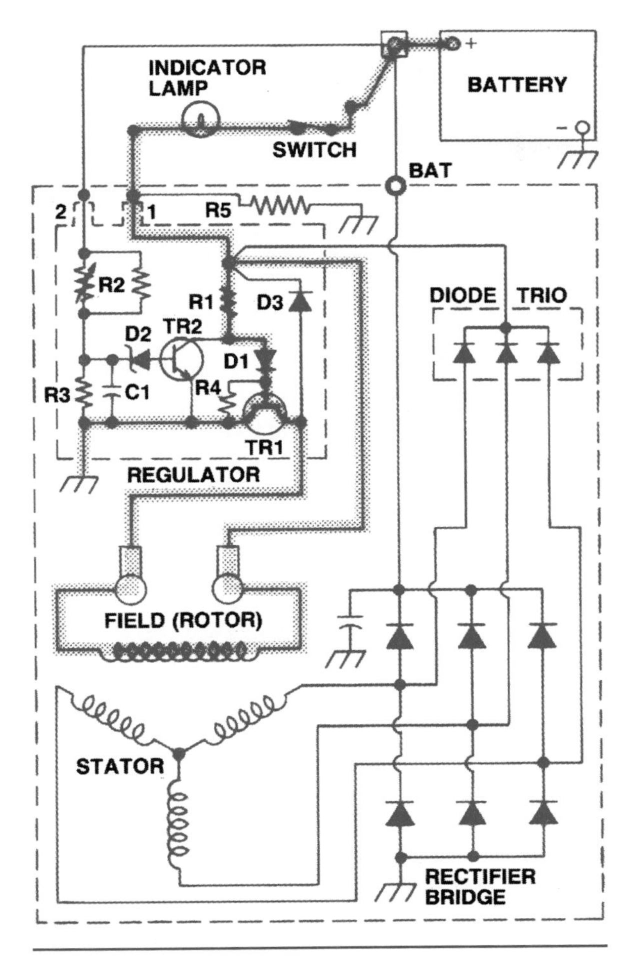

Figure 8-43. Field current in a typical solid-state regulator during starting. (Delphi Automotive Systems)

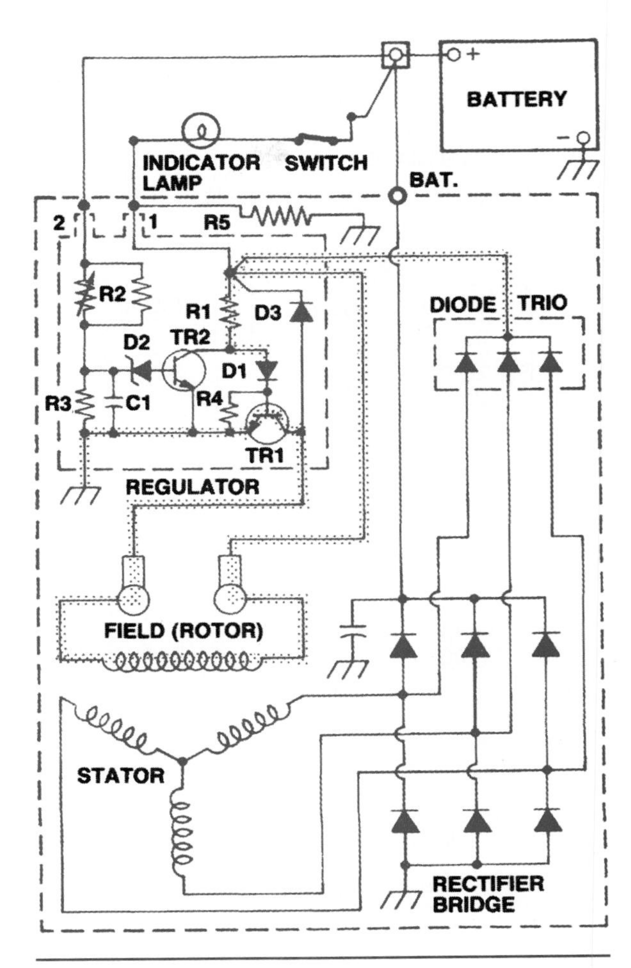

Figure 8-44. Field current drawn from AC generator (alternator) output. (Delphi Automotive Systems)

stable voltage across resistor R3. Resistor R4 prevents excessive current through TR1 at high temperatures. To prevent circuit damage, diode D3 bypasses high voltages induced in the field windings when TR1 turns off. Resistor R2 is a thermistor, which causes the regulated voltage to vary with temperature. R5 allows the indicator lamp to turn off if the field circuit is open.

For more information about voltage regulators, see the "Voltage Regulator Service Section" in Chapter 8 of the *Shop Manual.*

Specific Solid-State Regulator Designs

GM Delco-Remy (Delphi)

The Delco-Remy solid-state automotive regulator is used in Figures 8-43 to 8-46 to explain the basic operation of these units. This is the integral regulator unit of the SI series AC generators. The 1 and 2 terminals on the housing connect directly to the regulator. The 1 terminal conducts field current from the battery or the AC Generator (alternator) and controls the indicator lamp. The 2 terminal receives battery voltage and allows the Zener diode to react to it.

Unlike other voltage regulators, the multifunction IC regulator used with Delco-Remy CS-series AC generators (Figure 8-46) switches the field current on and off at a fixed frequency of 400 Hz (400 times per second). The regulator varies the duty cycle, or percentage of on time to total cycle time, to control the average field current and to regulate voltage. At high speeds, the on time might be 10 percent with the off time 90 percent. At low speeds with high electrical loads, this ratio could be reversed: 90 percent on time and 10 percent off time. Unlike the SI series, CS AC generators have

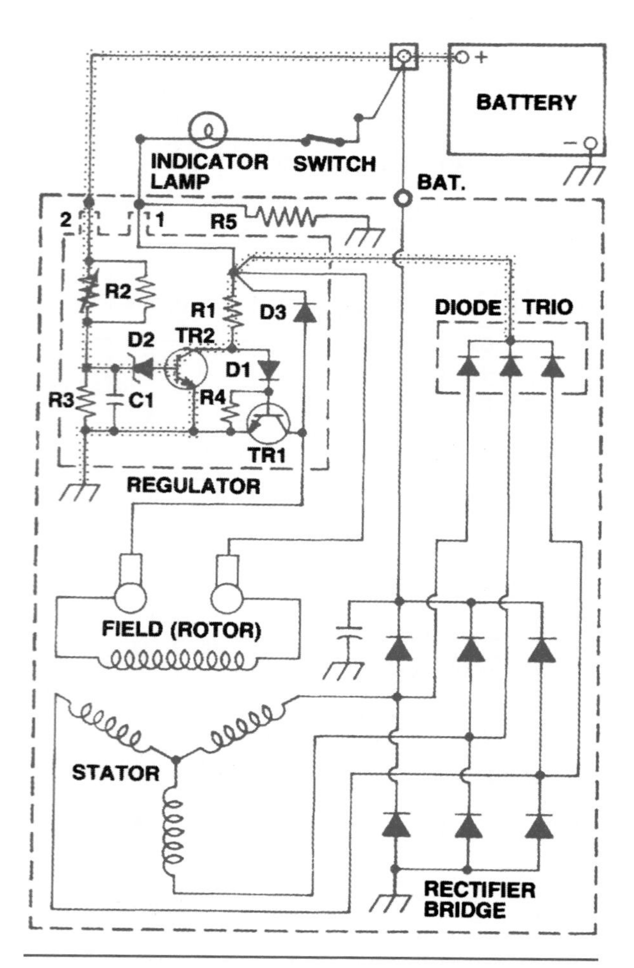

Figure 8-45. When AC generator (alternator) output voltage reaches a maximum safe level, no current is allowed in the rotor winding. (Delphi Automotive Systems)

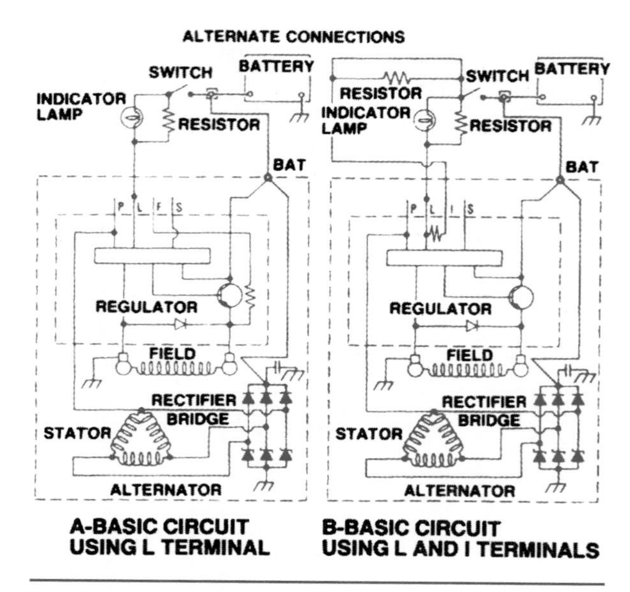

Figure 8-46. GM CS AC generator (alternator) regulator circuit. (Delphi Automotive Systems)

no test hole to ground the regulator for full-field testing. The regulator cannot be tested with an ohmmeter; a special tester is required.

Motorcraft

At one time, Motorcraft AC generators used both remote-mounted and integral solid-state regulators. Motorcraft regulator terminals are designated as follows:

- A+ or A− connects the battery to the field relay contacts.
- S− connects the AC generator output to the field relay coil.
- F− connects the field coil to the regulator transistors.
- I− connects the ignition switch to the field relay and regulator contacts (only on vehicles with a warning lamp).

Ford began using a remote-mounted, fully solid-state regulator on its intermediate and large models in 1978. The functions of the I, A+, 8, and F terminals are identical to those of the transistorized regulator. On systems with an ammeter, the regulator is color coded blue or gray, and the T terminal is not used. On warning lamp systems, the regulator is black, and all terminals can be used. The two models are not interchangeable, and cannot be substituted for the earlier solid-state unit with a relay or for an electromagnetic regulator. However, Ford does provide red or clear service replacement solid-state regulators, which can be used with both systems.

The Motorcraft integral alternator/regulator (JAR) introduced in 1985 uses an IC regulator, which is also mounted on the outside of the rear housing. This regulator differs from others because it contains a circuit indicating when the battery is being overcharged. It turns on the charge indicator lamp if terminal A voltage is too high or too low, or if the terminal 8 voltage signal is abnormal.

DaimlerChrysler

The DaimlerChrysler solid-state regulator depends on a remotely mounted field relay to open and close the isolated field or the A-circuit AC generator field. The relay closes the circuit only when the ignition switch is turned on. The voltage regulator (Figure 8-47) contains two transistors that are turned on and off by a Zener diode. The Zener diode reacts to system voltage to start and stop field

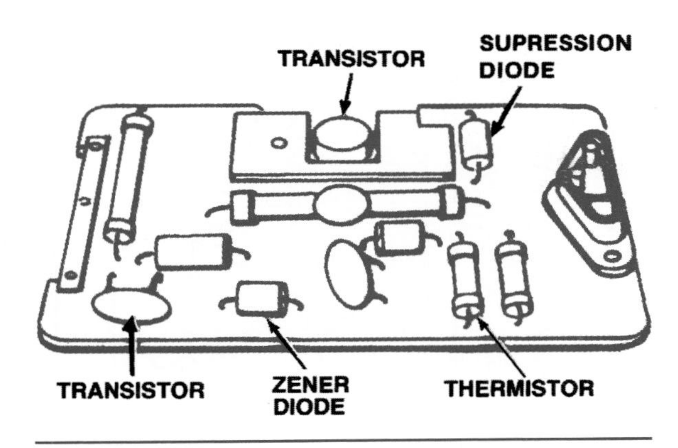

Figure 8-47. DaimlerChrysler solid-state voltage regulator. (DaimlerChrysler Corporation)

current. The field current travels through what DaimlerChrysler calls a field-suppression diode, which limits current to control AC generator output. The regulator also contains a thermistor to control battery charging voltage at various temperatures. The regulator has two terminals: One is connected to the ignition system; the other is connected to the alternator field.

Computer-Controlled Regulation

DaimlerChrysler Corporation eliminated the separate regulator by moving its function to the powertrain control module (PCM) in 1985. When the ignition is turned on, the PCM logic module or logic circuit checks battery temperature to determine the control voltage (Figure 8-48). A pre-driver transistor in the logic module or logic circuit then signals the power module or power circuit driver transistor to turn the AC generator current on (Figure 8-49). The logic module or logic circuit continually monitors battery temperature and system voltage. At the same time, it transmits output signals to the power module or power circuit driver to adjust the field current as required to maintain output between 13.6 and 14.8 volts. Figure 8-50 shows the complete circuitry involved.

General Motors has taken a different approach to regulating CS charging system voltage electronically. Turning the ignition switch on supplies voltage to activate a solid-state digital regulator, which uses pulse-width modulation (PWM) to supply rotor current and thus control output voltage. The rotor current is proportional to the PWM pulses from the digital regulator.

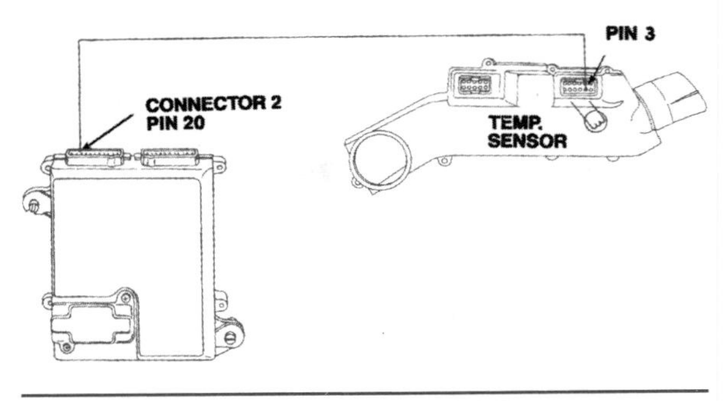

Figure 8-48. DaimlerChrysler solid-state voltage regulator. (DaimlerChrysler Corporation)

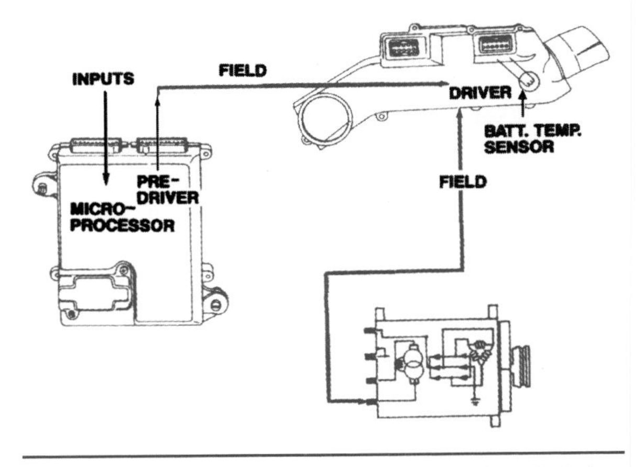

Figure 8-49. Battery temperature circuit of the DaimlerChrysler computer regulated charging system. (DaimlerChrysler Corporation)

With the ignition on, narrow width pulses are sent to the rotor, creating a weak magnetic field. As the engine starts, the regulator senses AC generator rotation through AC voltage detected on an internal wire. Once the engine is running, the regulator switches the field current on and off at a fixed frequency of about 400 cycles per second (400 Hz). By changing the pulse width, or on-off time, of each cycle, the regulator provides a correct average field current for proper system voltage control.

A lamp driver in the digital regulator controls the indicator warning lamp, turning on the bulb when it detects an under- or over-voltage condition. The warning lamp also illuminates if the AC generator is not rotating. The PCM does not directly control charging system voltage, as in the DaimlerChrysler application. However, it does monitor battery and system voltage through an ignition switch circuit. If the PCM reads a voltage above 17 volts, or less than 9 volts for longer than

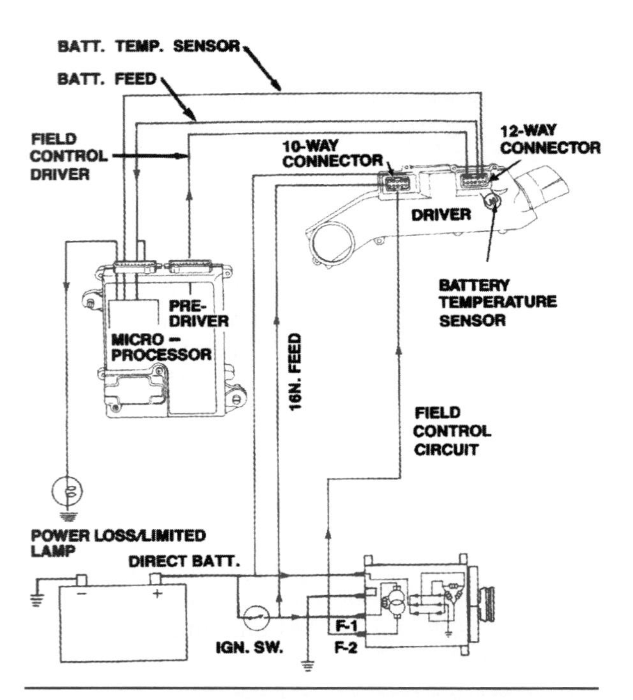

Figure 8-50. DaimlerChrysler computer-regulated charging system internal field control. (DaimlerChrysler Corporation)

10 seconds, it sets a code 16 in memory and turns on the malfunction indicator lamp (MIL).

Diagnostic Trouble Codes (DTC)

On late-model DaimlerChrysler vehicles, the onboard diagnostic capability of the engine control system detects charging system problems and records up to five diagnostic trouble codes (DTC) in the system memory. Some of the codes light a MIL on the instrument panel; others do not. Problems in the General Motors CS charging system cause the PCM to turn on the indicator lamp and set a single code in memory.

CHARGE/VOLTAGE/ CURRENT INDICATORS

A charging system failure cripples an automobile. Therefore, most manufacturers provide some way for the driver to monitor the system operation. The indicator may be an ammeter, a voltmeter, or an indicator lamp.

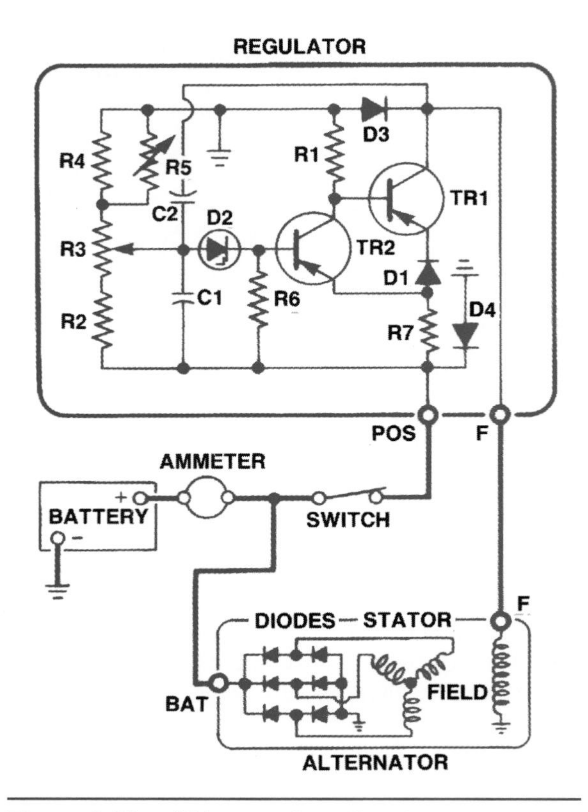

Figure 8-51. GM ammeter. (GM Service and Parts Operations)

Ammeter

An instrument panel ammeter measures charging system current into and out of the battery and the rest of the electrical system (Figure 8-51). The ammeter reads the voltage drop of the circuit. When current is traveling from the AC generator into the battery, the ammeter moves in a positive or charge direction. When the battery takes over the electrical system's load, current travels in the opposite direction and the needle moves into the negative, or discharge, zone. The ammeter simply indicates which is doing the most work in the electrical system, the battery or the AC generator (alternator). Some ammeters are graduated to indicate the approximate current in amperes, such as 5, 10, or 20. Others simply show an approximate rate of charge or discharge, such as high, medium, or low. Some ammeters have a resistor parallel so the meter does not carry all of the current, these are called shunt ammeters. While the ammeter tells the driver whether the charging system is functioning normally, it does not give a good picture of the battery condition. Even when the ammeter indicates a charge, the current output may not be high enough to fully charge the battery while supplying other electrical loads.

Voltmeter

The instrument panels of many late-model vehicles contain a voltmeter instead of an ammeter (Figure 8-52). A voltmeter measures electrical pressure and indicates regulated generator voltage output or battery voltage, whichever is greater. System voltage is applied to the meter through the ignition switch contacts. Figure 8-53 shows a typical voltmeter circuit.

The voltmeter tells a driver more about the condition of the electrical system of a vehicle than an ammeter. When a voltmeter begins to indicate lower-than-normal voltage, it is time to check the battery and the voltage regulator.

Indicator Lamps

Most charging systems use an instrument panel indicator, or warning lamp, to show general charging system operation. Although the lamp usually does not warn the driver of an overcharged battery or high charging voltage, it lights to show an undercharged battery or low voltage from the AC generator.

The lamp also lights when the battery supplies field current before the engine starts. The lamp is often connected parallel to a resistor; therefore, field current travels even if the bulb fails. The lamp is wired so it lights when battery current travels through it to the AC generator field. When the alternator begins to produce voltage, this voltage is applied to the side of the lamp away from the battery. When the two voltages are equal, no voltage drops are present across the lamp and it goes out. When indicator lamps are used, the regulator must be able to monitor when the AC generator is charging. One method is to use "stator" or neutral voltage. This signal is present only when the AC generator is charging, and is one-half of charging voltage. When stator voltage is about three volts, it energizes a relay to open the indicator lamp ground circuit (Figure 8-53).

Figure 8-54 shows a typical warning lamp circuit installation. In figure 9-14, a 500-ohm resistor is used for warning lamp systems and a 420-ohm resistor for electronic display clusters. In Figure 8-54, a 40-ohm resistor (R5) is installed near the integral regulator. In each case, the grounded path ensures the warning lamp lights if an open occurs in the field circuitry.

As previously discussed, indicator lamps can also be controlled by the field relay. The indicator lamp for a Delco-Remy CS system works differently than most others. It lights if charging voltage is either too low or too high. Any problem in the charging system causes the lamp to light at full brilliance.

Figure 8-52. Automotive voltmeter in instrument panel.

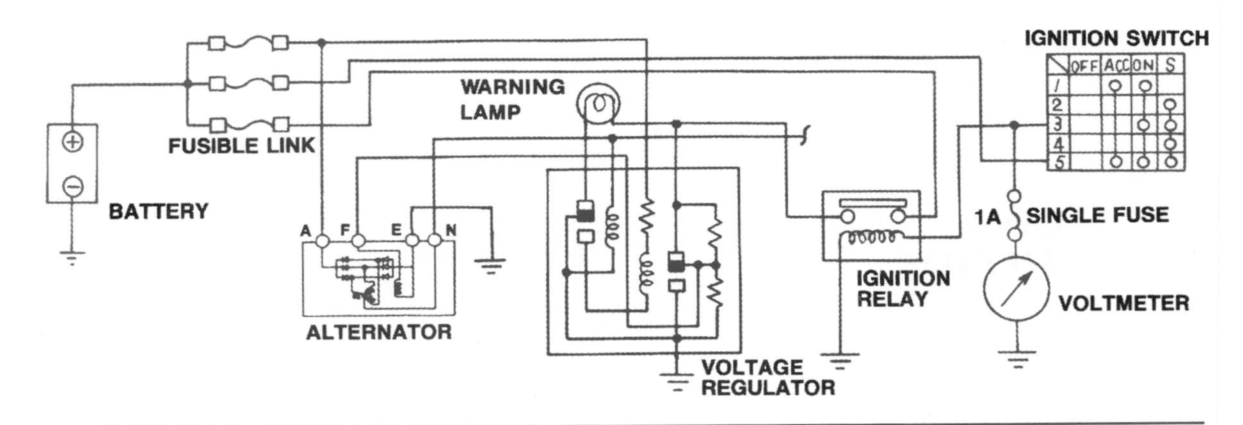

Figure 8-53. A typical voltmeter circuit.

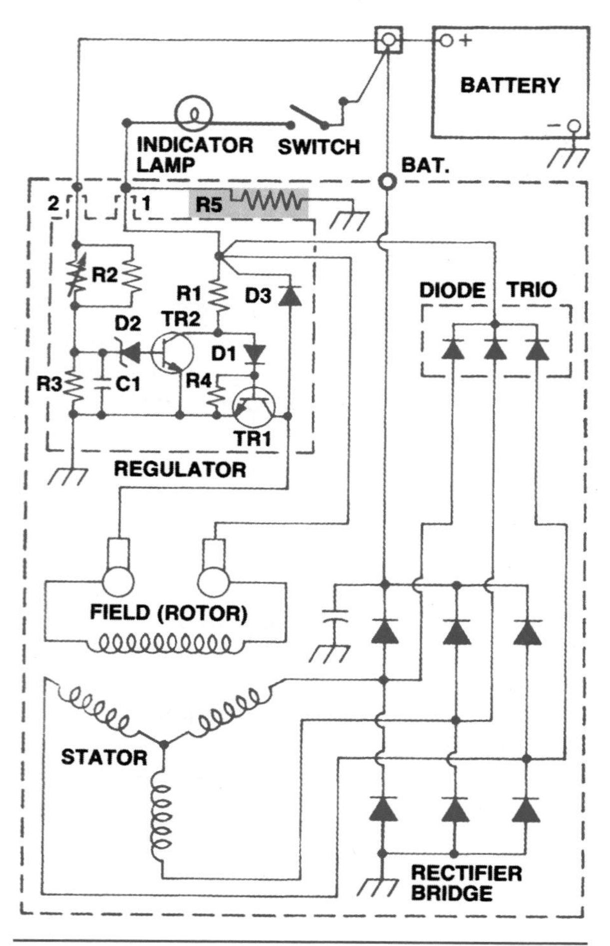

Figure 8-54. GM warning lamp circuit. (GM Service and Parts Operations)

CHARGING SYSTEM PROTECTION

If a charging system component fails or malfunctions, excessive current or heat, voltage surges, and other uncontrolled factors could damage wiring and other units in the system. To protect the system from high current, fusible links are often wired in series at various places in the circuitry. Figure 8-55 shows some typical fusible link locations.

COMPLETE AC GENERATOR OPERATION

When the ignition is first switched from off to on, before cranking the engine, a charge lamp comes on. This indicates, of course, that the AC generator is not generating a voltage. At the same time, battery voltage is applied to the rotor coil so that when the rotor begins to spin, the magnetic fields cut across the stator windings and produce current (Figure 8-56).

After the engine starts, the rotor is spinning fast enough to induce current from the stator. The current travels through the diodes and out to the battery and electrical system (Figure 8-57). Once the IC regulator senses system voltage is greater than

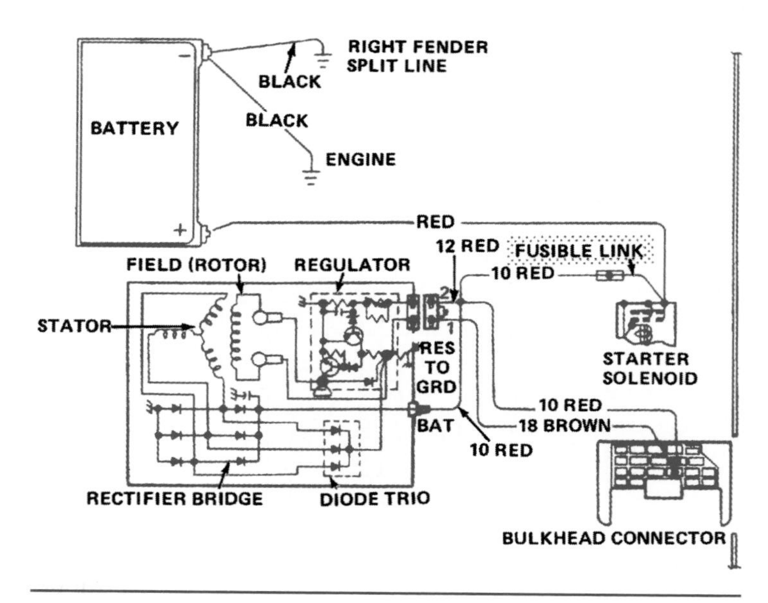

Figure 8-55. GM fusible links.

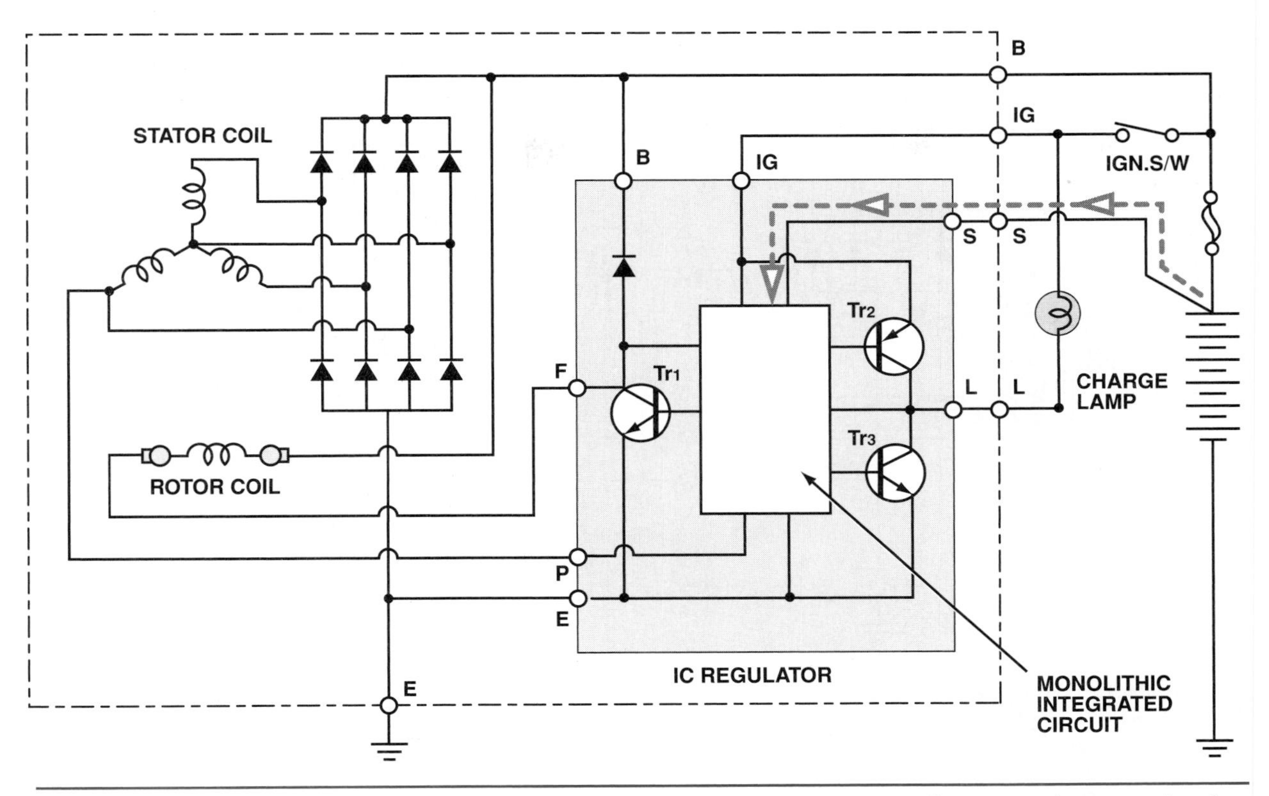

Figure 8-56. When AC generator voltage is too high, the regulator momentarily cuts off current to the rotor coils, eliminating the magnetic field. The rotor continues to spin, but no voltage is generated. (Reprinted by permission of Toyota Motor Corporation)

battery voltage, it redirects current to switch off the charge lamp.

During normal operation, AC generator voltage exceeds the typically specified 14.5 volts at times. To protect the battery and delicate components in the electrical system, the IC regulator shuts off current to the rotor, cutting AC generator output to zero (Figure 8-58). Note that even though the AC generator is momentarily "turned off," the charge lamp does not come on. Within a split second, the IC regulator re-energizes the rotor again once output falls below the minimum. The IC regulator switches battery voltage on and off this way to control output and maintain system voltage at an ideal level.

AC GENERATOR (ALTERNATOR) DESIGN DIFFERENCES

Original equipment manufacturers (OEM) use various AC generator designs for specific applications. It has been noted that such factors as maximum current output and field circuit types affect AC generator construction. The following paragraphs describe some commonly used automotive AC generators.

Delphi (Delco-Remy) General Motors Applications

Delphi, formerly the Delco-Remy division of General Motors Corporation, is now a separate corporation that supplies most of the electrical devices used on GM vehicles, as well as those of some other manufacturers. The trademark name for Delco-Remy alternators was Delcotron generators. The alternator model number and current output can be found on a plate attached to, or stamped into, the housing.

DN-Series

The 10-DN series AC generator or alternator uses an external electromagnetic voltage regulator. Six individual diodes are mounted in the rear housing (Figure 8-59) with a capacitor for protection. A 14-pole rotor and Y-type stator

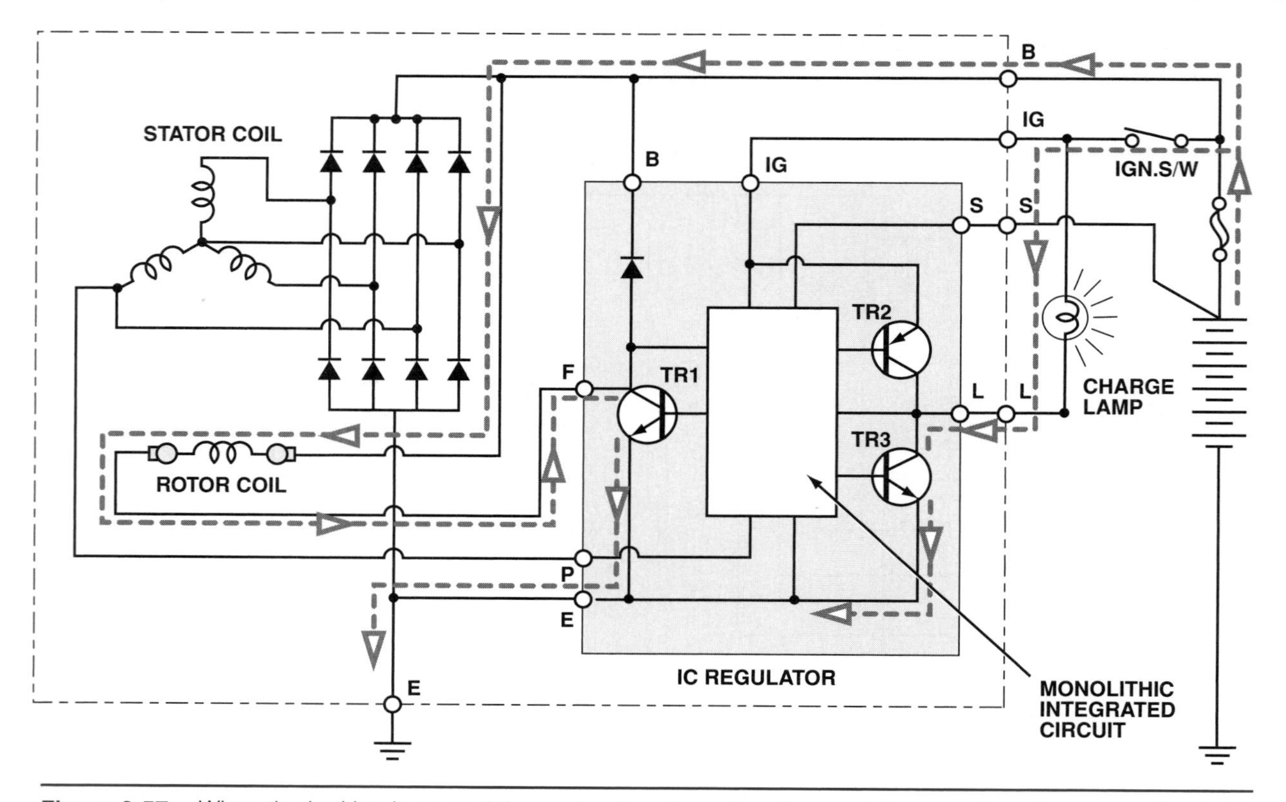

Figure 8-57. When the ignition is on and the engine is not running, the regulator energizes the rotor coil to build a magnetic field in the stator. The regulator turns on the charge light, indicating that the AC generator is not generating a voltage. (Reprinted by permission of Toyota Motor Corporation)

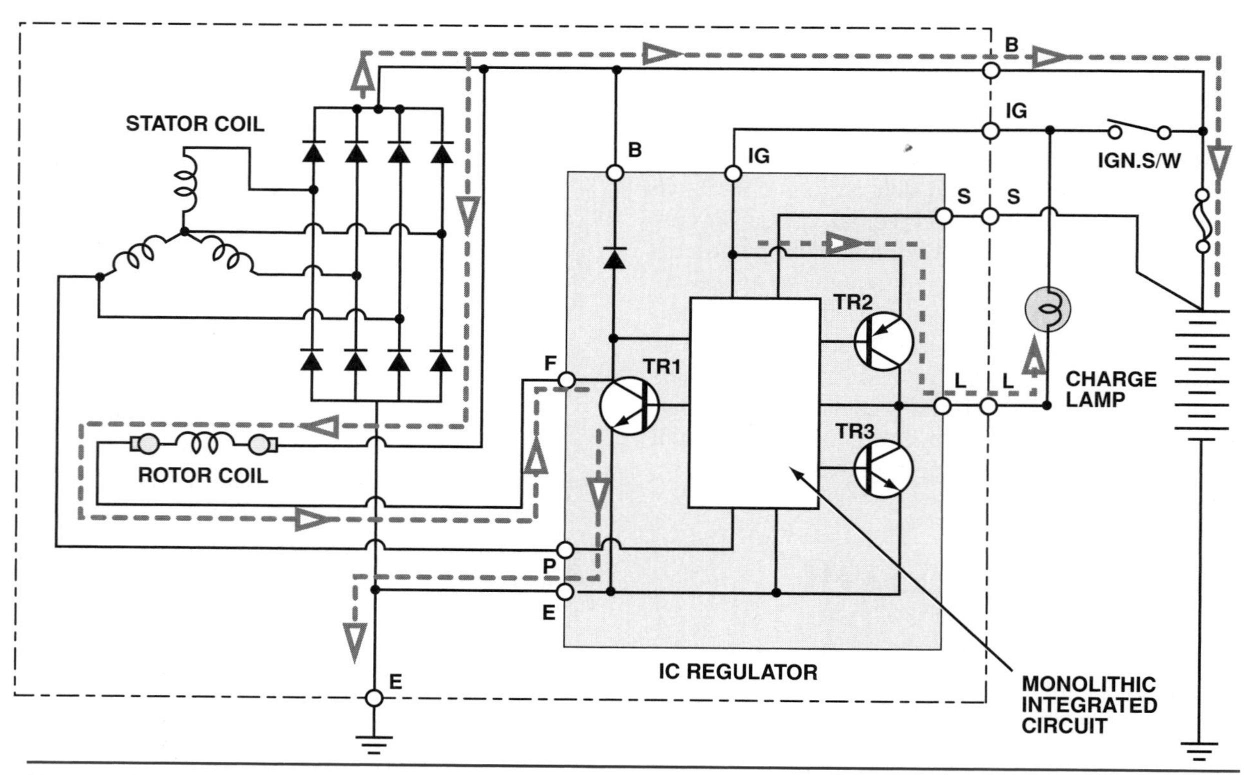

Figure 8-58. After the engine is running, the AC generator generates a voltage greater than the battery. The charge light goes out, indicating normal operation. (Reprinted by permission of Toyota Motor Corporation)

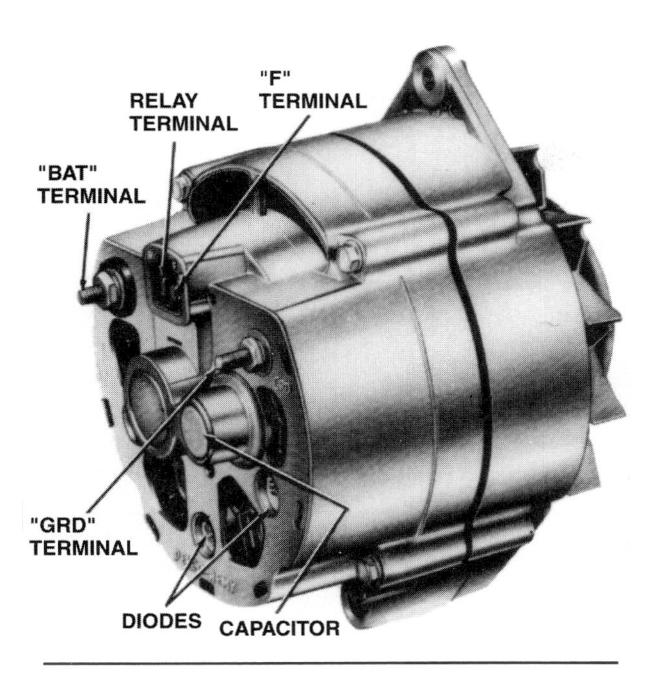

Figure 8-59. A 10-SI series AC generator (alternator). (Delphi Automotive Systems)

Figure 8-60. A 10-SI series AC generator [Delcotron]. (Delphi Automotive Systems)

provide current output. Field current is drawn from rectified output and travels through a B-circuit. The terminals on a 10-DN are labeled BAT, GRD, R, and F. If the AC generator is used with an external electromagnetic regulator, the following applies:

- BAT connects AC generator output to the insulated terminal of the battery.
- GRD, if used, is an additional ground path.
- R, if used, is connected to a separate field relay controlling the indicator lamp.
- F connects the rotor winding to the voltage regulator.

Some 10-DN alternators are used with a remotely mounted solid-state regulator. The voltage control level of this unit is usually adjustable. The terminal connections are the same for electromagnetic and solid-state regulators.

SI Series

The 10-SI series AC generator uses an internally mounted voltage regulator and came into use in the early 1970s. The most common early model Delcotron alternators are part of the SI series and include models 10, 12, 15, and 27 as shown in Figure 8-60. A 14-pole rotor is used in most models. The 10-SI and 12-SI models have Y-type stators, and the 15-SI and 27-SI models have delta-type stators. Two general SI designs have been used, with major differences appearing in the rear housing diode installation, regulator appearance, field circuitry, and ground path.

Most SI models have a rectifier bridge that contains all six rectifying diodes (Figure 8-61). The regulator is a fully enclosed unit attached by screws to the housing. Field current is drawn from unrectified AC generator output and rectified by an additional diode trio. All SI models have A-circuits, their terminals are labeled BAT, No. 1, and No. 2.

- The BAT terminal connects AC generator output to the insulated terminal of the battery.
- The No. 1 terminal conducts battery current to the rotor winding for the excitation circuit and is connected to the indicator lamp.
- The No. 2 terminal receives battery voltage so the voltage regulator can react to system operating conditions.

All SI models have a capacitor installed in the rear housing to protect the diodes from sudden voltage surges and to filter out voltage ripples that could produce EMI. The 27-SI, which is intended principally for commercial vehicles, has an adjustable voltage regulator (Figure 8-62). The voltage is adjusted by removing the adjustment cap, rotating it until the desired setting of low,

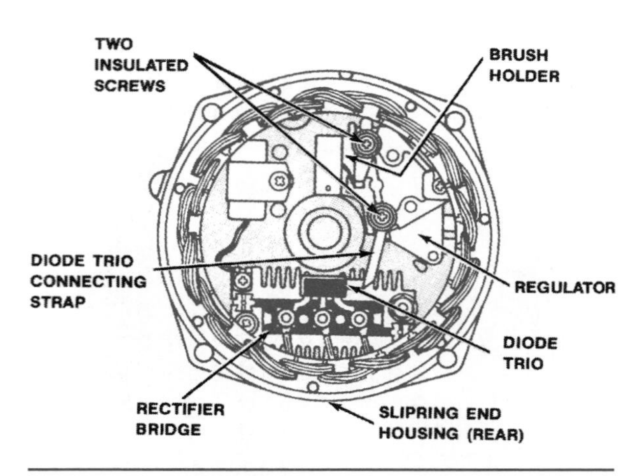

Figure 8-61. The rear end housing of the later-model 10-SI. (Delphi Automotive Systems)

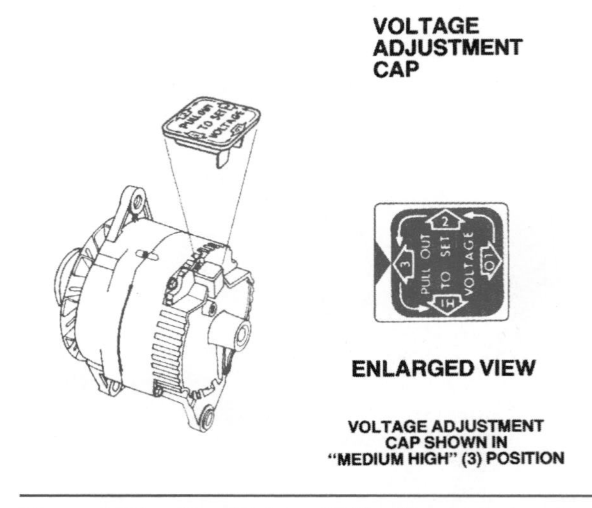

Figure 8-62. The Delco 27-SI alternator has an adjustable voltage regulator. (Delphi Automotive Systems)

medium, medium-high, or high is opposite the arrow on the housing, and reinstalling it in the new position. Repair or replace a 27-SI AC generator only if it fails to pass an output test after the regulator has been adjusted.

CS Series

The smaller Delco-Remy CS series AC generators introduced on some 1986 GM cars (Figure 8-63) maintain current output similar to larger AC generators. This series includes models CS-121, CS-130, and CS-144. The number following the CS designation denotes the outer diameter of the stator lamination in millimeters. All models use a delta-type stator. Field current is taken directly from the stator, eliminating the field diode trio. An integral cooling fan is used on the CS-121 and CS-130.

Electronic connections on CS AC generators include a BAT output terminal and either a one- or two-wire connector for the regulator. Figure 8-64 shows the two basic circuits for CS AC generators, but there are a number variations. Refer to the vehicle service manual for complete and accurate circuit diagrams. The use of the P, F, and S terminals is optional.

- The P terminal, connected to the stator, may be connected to a tachometer or other such device.
- The F terminal connects internally to field positive and may be used as a fault indicator.
- The S terminal is externally connected to battery voltage to sense the voltage to be controlled.

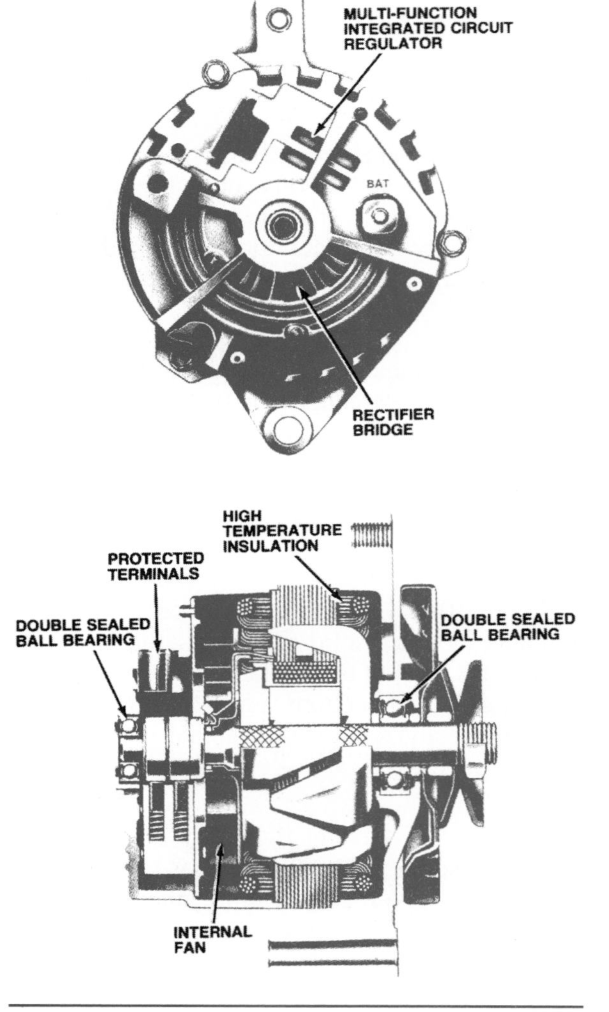

Figure 8-63. Typical Delco-Remy CS series AC generator (alternator) construction. (Delphi Automotive Systems)

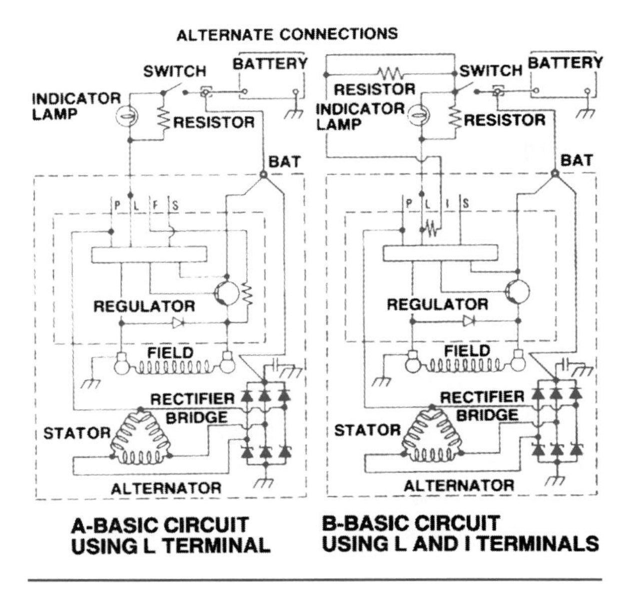

Figure 8-64. Two basic circuits for CS AC generators (alternator). (Delphi Automotive Systems)

- The L terminal connects the regulator to the indicator lamp and battery.

The indicator lamp in a CS charging system works differently than in other Delco-Remy systems: Any defect causes it to light at full brilliance. The lamp also lights if charging voltage is either too low or too high. If the regulator has an I terminal, its wire supplies field current in addition to that applied internally, either directly from the switch or through a resistor.

Motorcraft

Motorcraft, a division of Ford Motor Company, makes most of the AC generators used on domestic Ford vehicles. Model and current rating identifications for later models are stamped on the front housing with a color code. Motorcraft AC generators prior to 1985 are used with either an electromechanical voltage regulator or a remotely mounted solid-state regulator. One exception to this is the 55-ampere model of 1969–1971, which has a solid-state regulator mounted on the rear housing. This model has an A-circuit; all others are B-circuit. The Motorcraft integral alternator/regulator (IAR) model was introduced on some front-wheel drive Ford models in 1985. This AC generator (alternator) has a solid-state regulator mounted on its rear housing. Some Motorcraft charging systems continue to use an external solid-state regulator with either a rear-terminal or side-terminal AC generator. These charging systems are called external voltage regulator (EVR) systems to differentiate them from integral alternator/regulator (IAR) systems.

Integral Alternator/Regulator Models

The Motorcraft IAR AC generators are rated at 40 to 80 amperes. The sealed rectifier assembly is attached to the slipring-end housing. On early models, the connecting terminals (BAT and STA) protruded from the side of the AC generator in a plastic housing. Current models use a single pin stator (STA) connector and separate output stud (BAT). The brushes are attached to, and removed with, the regulator. A Y-type stator is used with a 12-pole rotor. Some applications have an internal cooling fan.

Turning the ignition on sends voltage to the regulator I terminal through a resistor in the circuit. System voltage is sensed and field current is drawn through the regulator A terminal until the ignition is turned off, which shuts off the control circuit.

If the vehicle has a heated windshield, output is switched from the battery to the windshield by an output control relay. This allows output voltage to increase above the normal regulated voltage and vary with engine speed. The regulator I circuit limits the increase to 70 volts, which is controlled by the heated windshield module during the approximate four-minute cycle of heated windshield operation. When the cycle times out, the charging system returns to normal operation.

DaimlerChrysler

DaimlerChrysler Corporation manufactured all of the AC generators for its domestic vehicles until the late 1980s, when it phased in Bosch and Nippondenso AC generators for use on all vehicles.

DaimlerChrysler used two alternator designs from 1972 through 1984. The standard-duty alternator, rated from 50 to 65 amperes, is identified by an internal cooling fan and the stator core extension between the housings. The heavy-duty 100-ampere alternator has an external fan and a totally enclosed stator core. Identification also is stamped on a color-coded tag on the housing.

All models have a 12-pole rotor and use a remotely mounted solid-state regulator. The brushes can be replaced from outside the housing. Individual diodes are mounted in positive and negative heat

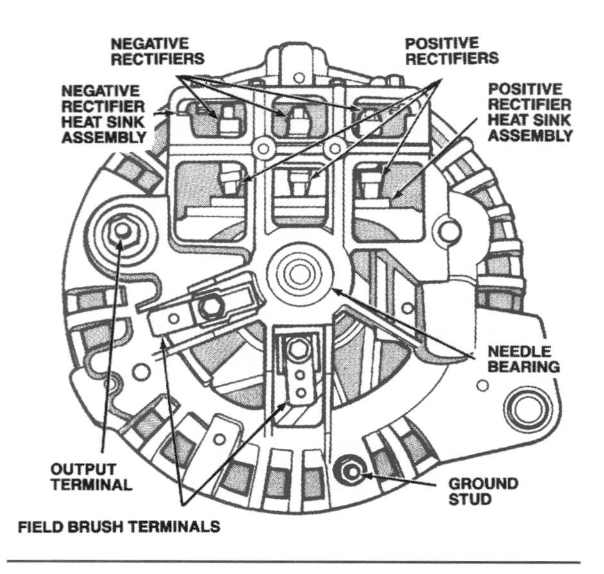

Figure 8-65. The terminals on a DaimlerChrysler standard-duty AC generator (alternator). (DaimlerChrysler Corporation)

sink assemblies, and are protected by a capacitor. The terminals on the standard-duty AC generator are labeled BAT, GRD, and FLD (Figure 8-65).

- The BAT terminal connects AC generator output to the insulated terminal of the battery.
- The GRD terminal is the ground connection.
- The FLD terminals connect to the insulated brushes. On the 100-ampere model, the FLD terminal has two separate prongs that fit into a single connector (Figure 8-66). The additional GRD terminal is a ground path.

DaimlerChrysler standard-duty AC generators have a Y-type stator connected to six diodes. Although both brush holders are insulated from the housing, one is indirectly grounded through the negative diode plate, making it a B-circuit. The 100-ampere AC generator has a delta-type stator. Each of the conductors is attached to two positive and two negative diodes. These 12 diodes create additional parallel circuit branches for high-current output.

DaimlerChrysler eliminated the use of a separate voltage regulator on most 1985 and later fuel injected and turbocharged engines by incorporating the regulator function into the powertrain control module (PCM), as shown in Figure 8-67.

The computer-controlled charging system was introduced with the standard DaimlerChrysler AC generator on GLH and Shelby turbo models. All other four-cylinder engines used either a new DaimlerChrysler 40/90-ampere AC generator or a

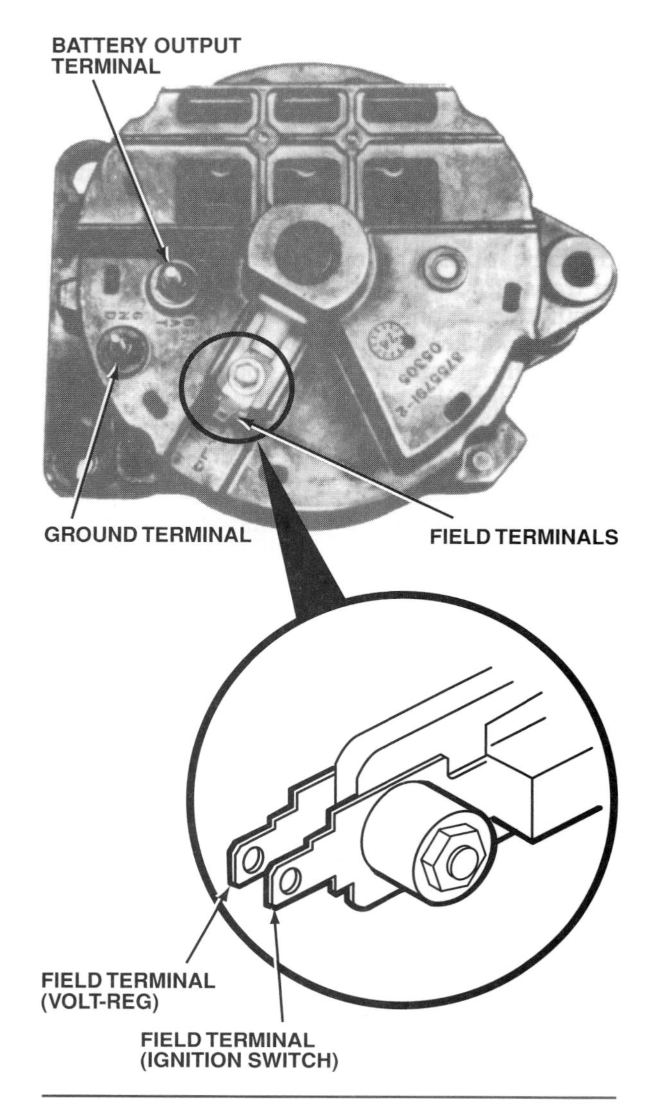

Figure 8-66. The terminals on a 100-amp DaimlerChrysler AC generator (alternator). (DaimlerChrysler Corporation)

modified Bosch 40/90-ampere or 40/100-ampere model.

The DaimlerChrysler-built AC generator uses a delta-type stator. The regulator circuit is basically the isolated-field type, but field current is controlled by integrated circuitry in the logic and power modules (Figure 8-68) or the logic and power circuits of the single-module engine control computer (SMEC) or single-board engine control computer (SBEC). In addition to sensing system voltage, the logic module or circuit senses battery temperature as indicated by system resistance. The computer then switches field current on and off in a duty

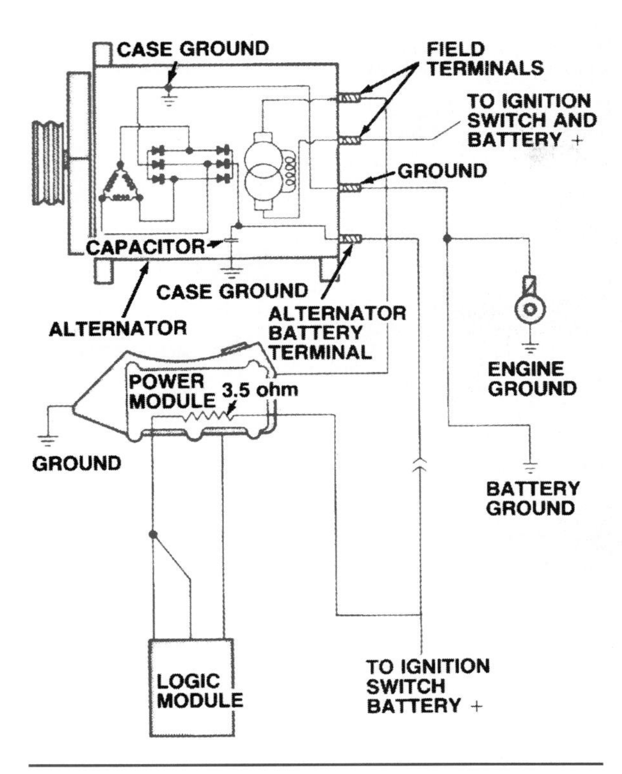

Figure 8-67. Typical DaimlerChrysler computer voltage regulation with DaimlerChrysler 40/90-ampere AC generator (alternator). Circuit connections vary on different models. (DaimlerChrysler Corporation)

cycle that regulates charging voltage, as in any other system. The DaimlerChrysler computer-controlled charging system has the following important features:

- It varies charging voltage relative to ambient temperature and the system voltage requirements.
- A self-diagnostic program can detect charging system problems and record fault codes in system memory. Some codes will light the POWER LOSS, POWER LIMITED, or MALFUNCTION INDICATOR lamp on the instrument panel; others will not.

Turning the ignition on causes the logic circuit to check battery temperature to determine the control voltage. A predriver transistor in the logic module or logic circuit signals a driver translator in the power module or power circuit to turn on the AC generator field current (Figure 8-68). The logic module or logic circuit constantly monitors system voltage and battery temperature and signals the driver in the power module or power circuit when field current adjustment is necessary to keep output voltage within the specified 13.6-to-14.8-volt range.

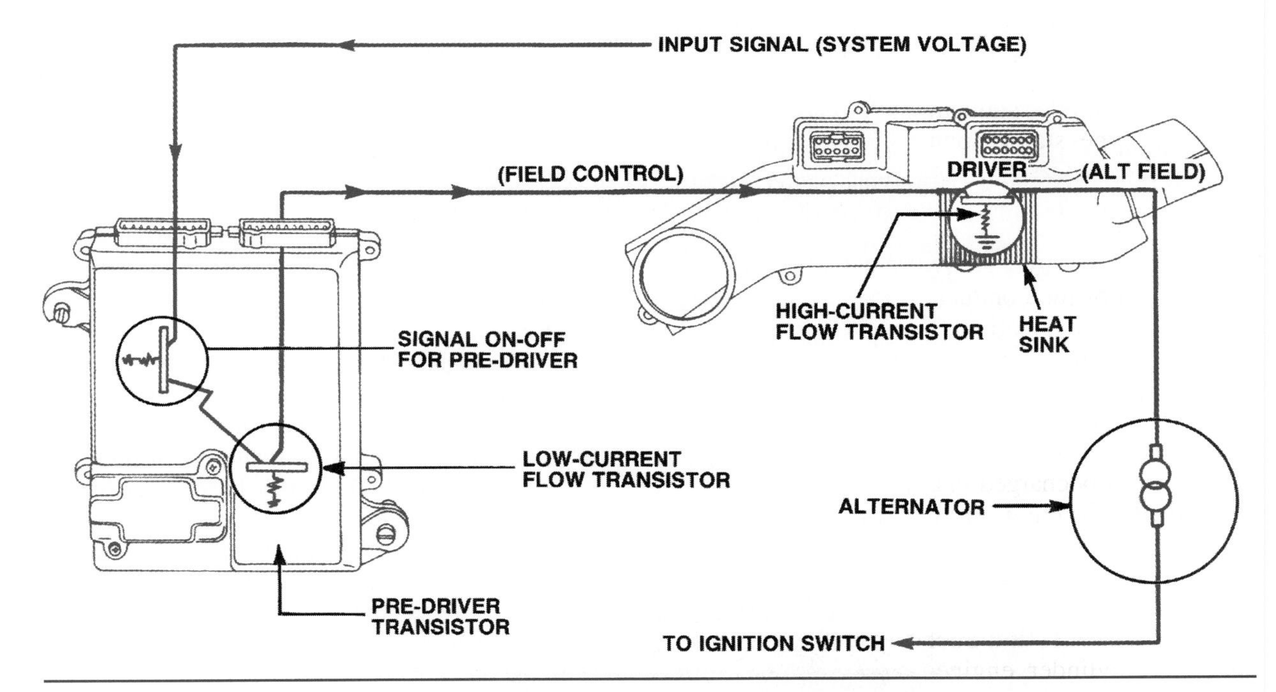

Figure 8-68. DaimlerChrysler computer-regulated charging system internal field control. (DaimlerChrysler Corporation)

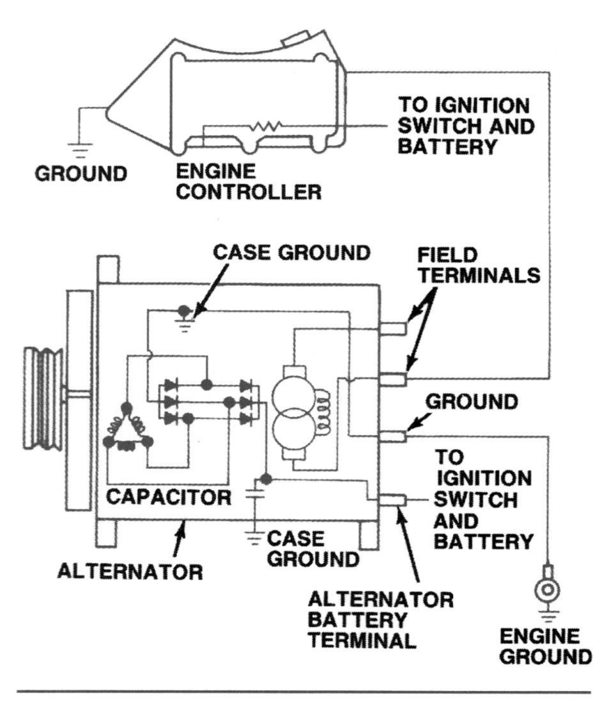

Figure 8-69. Current DaimlerChrysler charging system. (DaimlerChrysler Corporation)

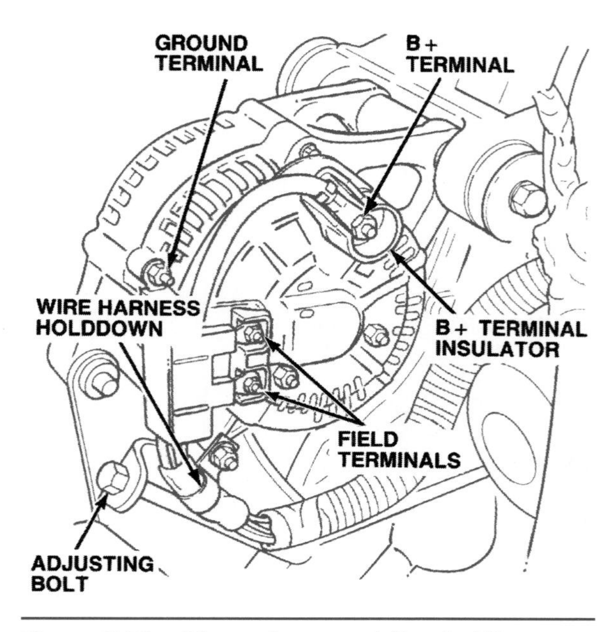

Figure 8-70. Nippondenso and Bosch AC generators used on DaimlerChrysler vehicles have identical connections. (DaimlerChrysler Corporation)

Bosch AC Generators (Alternators)

Modified Bosch 40/90-ampere and 40/100-ampere AC generators (alternators) were introduced in 1985 for use with the DaimlerChrysler computer-controlled charging system. These Bosch dual-output AC generators have a Y-type stator and were modified by removing their internal voltage regulators and changing the external leads. They are fully interchangeable with DaimlerChrysler dual-output AC generators of the same rating. Use of dual-output AC generators was phased out in favor of a single-output Bosch alternator when DaimlerChrysler ceased manufacture of its own alternators in 1989. Current DaimlerChrysler charging systems with a Bosch AC generator (84 or 86 amperes) are essentially the same design as those used with the dual-output AC generators (Figure 8-69). However, an engine controller replaces the separate logic and power modules.

Nippondenso AC Generators

Some current DaimlerChrysler vehicles use Nippondenso AC generators with an output range of 68 to 102 amperes. These are virtual clones of the Bosch design, even to the external wiring connections (Figure 8-70). Charging system circuitry is the same, as are test procedures.

Import Vehicle Charging Systems

Many European vehicles have Bosch AC generators featuring Y-type stators. Bosch models with a remote regulator use six rectifiers and have a threaded battery terminal and two-way spade connector on the rear housing. Those with an integral regulator contain 12 rectifiers and have a threaded battery stud marked B+ and a smaller threaded stud marked D+. This smaller stud is used for voltage from the ignition switch. Models with internal regulators also have a diode trio to supply field current initially and a blocking diode to prevent current from flowing back to the ignition system when the ignition is turned off.

Several manufacturers such as Hitachi, Nippondenso, and Mitsubishi provide AC generators for Japanese vehicles. While all function on the same principles just studied, the design and construction of some units are unique. For example, Figure 8-71 shows a Mitsubishi AC generator that uses an integral regulator with double Y-stator and 12 diodes in a pair of rectifier assemblies to deliver high current with high voltage at low speeds. A diode trio internally supplies the field, and a 50-ohm resistor in the regulator performs the same function as the Bosch blocking diode.

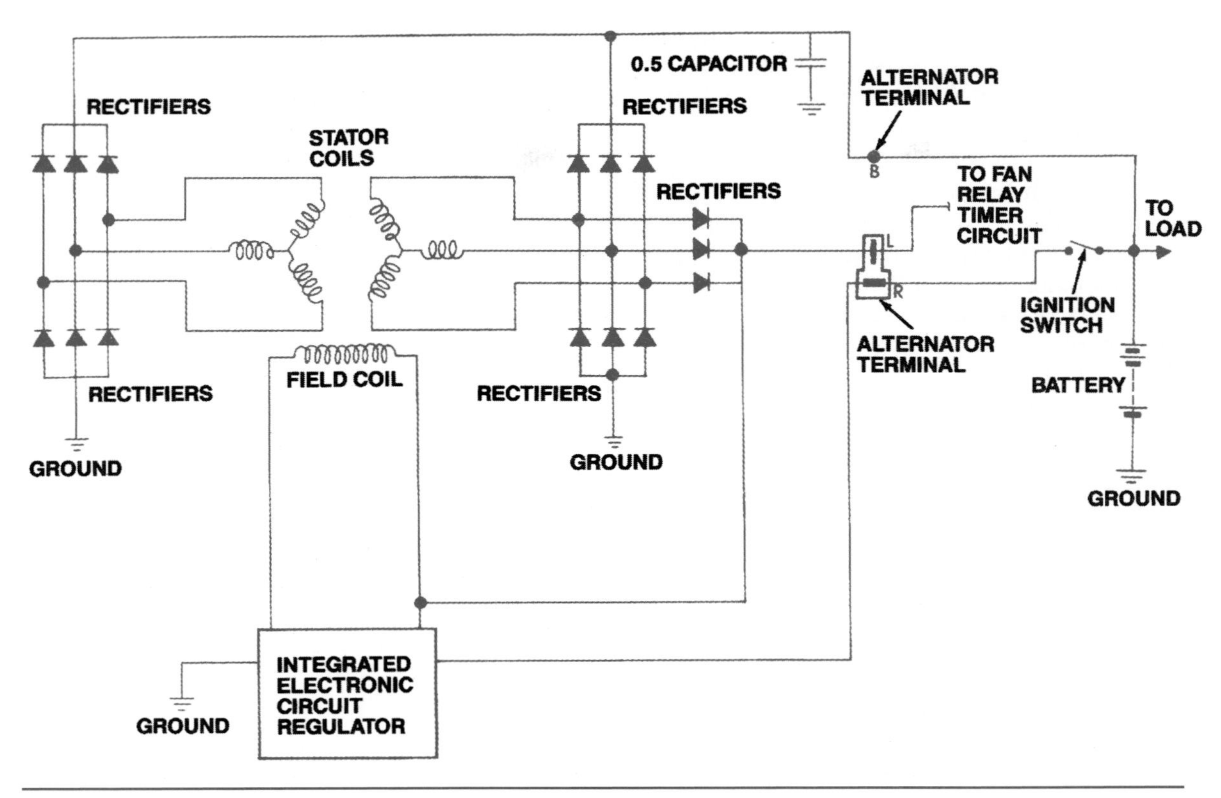

Figure 8-71. Mitsubishi charging system. (Mitsubishi)

SUMMARY

The sine wave voltage induced across one AC generator (alternator) conductor generates single-phase current. By connecting the AC generator conductor diodes, the AC is fully rectified to DC. In an actual automotive AC generator, there are more than two magnetic poles and one conductor. Common AC generators have from 8 to 14 poles on a rotor, and three conductors wound to create a stator. The rotor and stator are held in a two-piece housing. Two brushes attached to the housings, but often insulated from it, ride on sliprings to carry current to the rotor winding. The diodes are installed in the same end housing as the brushes. Three positive diodes are insulated from the housing; three negative diodes are grounded to the housing

Stators with three conductors produce three-phase current. The three conductors may be connected to make a Y-type or a delta-type stator. The rectification process is the same for both types, although the current paths through the stators differ. Rectified output from a multiple-pole AC generator is a rippling DC voltage.

Because the rotor does not retain enough magnetism to begin induction, an excitation circuit must carry battery current to the rotor winding. The rotor winding is part of an externally grounded field, or A-circuit; or an internally grounded field, or B-circuit.

Because a counter-voltage is induced in the stator windings, AC generator current output is self-limiting. Voltage regulation is still needed. Those with integrated circuits have replaced electromagnetic voltage regulators. Voltage regulators now in use are completely solid-state designs. They replaced the electromagnetic type regulators in AC charging systems because they are smaller and have no moving parts. Because of the construction of solid-state regulators, they are non-serviceable and must be replaced if defective. Their function is the same: to control AC generator output by modulating current through the field windings of the AC generator.

Indicators allow the driver to monitor the performance of the charging system: These include ammeters, voltmeters, or warning lamps. An ammeter measures the charging system current into and out of the battery and the entire electrical system. A voltmeter measures electrical pressure and indicates regulated AC generator voltage output or battery voltage. Indicator lamps illuminate on the instrument panel of the vehicle and indicate to the driver general charging system operation status.

Wiring and other components in the charging system may be damaged if the system fails or malfunctions due to excessive current or heat, voltage surges, and other uncontrolled factors. Fusible links are used in these circuits to protect the circuit form high current.

Common AC generators used by domestic manufacturers include the Delco-Remy SI and CS series used by OM; the Motorcraft IAR, rear-terminal, and side-terminal models used by Ford; and the DaimlerChrysler-built, Bosch, and Nippondenso models used by DaimlerChrysler. Most European imports use Bosch AC generators, while Asian imports use alternators made by several manufacturers, including Hitachi, Nippondenso, and Mitsubishi.

Review Questions

1. Alternators induce voltage by rotating:
   a. A magnetic field inside a fixed conductor
   b. A conductor inside a magnetic field
   c. A stator inside a field
   d. A spring past a stator
2. In an alternator, induced voltage is at its maximum value when the angle between the magnetic lines and the looped conductor is:
   a. 0 degrees
   b. 45 degrees
   c. 90 degrees
   d. 180 degrees
3. The sine wave voltage induced across one conductor by one rotor revolution is called:
   a. Single-phase current
   b. Open-circuit voltage
   c. Diode rectification
   d. Half-phase current
4. Alternating current in an alternator is rectified by:
   a. Brushes
   b. Diodes
   c. Slip rings
   d. Transistors
5. An alternator with only one conductor and one diode would show which of the following current output patterns?
   a. Three-phase current
   b. Open-circuit voltage
   c. Half-wave rectification
   d. Full-wave rectification
6. An alternator consists of:
   a. A stator, a rotor, sliprings, brushes, and diodes
   b. A stator, an armature, sliprings, brushes, and diodes
   c. A stator, a rotor, a commutator, brushes, and diodes
   d. A stator, a rotor, a field relay, brushes, and diodes
7. A typical automotive alternator has how many poles?
   a. 2 to 4
   b. 6 to 8
   c. 8 to 14
   d. 12 to 20
8. The three alternator conductors are wound onto a cylindrical, laminated metal-piece called:
   a. Rotor core
   b. Stator core
   c. Armature core
   d. Field core
9. Automotive alternators that have three conductors generally use how many diodes to rectify the output current?
   a. Three
   b. Six
   c. Nine
   d. Twelve
10. Which of the following is *not* true of the positive diodes in an alternator?
    a. They are connected to the insulated terminal of the battery.
    b. They conduct only the current moving from ground into the conductor.
    c. They are mounted in a heat sink.
    d. The bias of the diodes prevents the battery from discharging.
11. A group of three or more like diodes may be called:
    a. Diode wing
    b. Diode triplet
    c. Diode dish
    d. Diode bridge
12. Y-type stators are used in alternators that require:
    a. Low voltage output at high alternator speed
    b. High voltage output at low alternator speed
    c. Low voltage at low alternator speed
    d. High voltage at high alternator speed
13. Which of the following is true of a delta-type stator?
    a. There is no neutral junction.
    b. There is no ground connection.
    c. The windings always form a series circuit.
    d. The circuit diagram looks like a parallelogram.
14. Delta-type stators are used:
    a. When high-voltage output is needed
    b. When low-voltage output is needed

c. When high-current output is needed
d. When low-current output is needed

15. Which of the following is a commonly used type of field circuit in automotive alternators?
    a. X-circuit
    b. Y-circuit
    c. Connected-field circuit
    c. A-circuit

16. Alternator output voltage is directly related to:
    a. Field strength
    b. Rotor speed
    c. Both field strength and rotor speed
    d. Neither field strength nor rotor speed

17. Double-contact voltage regulators contain all of the following except:
    a. An armature
    b. An electromagnet
    c. Two sets of contact points
    d. A solenoid

18. The shorting contacts of a double-contact regulator:
    a. Increase voltage creep
    b. Increase field current
    c. React to battery temperature changes
    d. Short the field circuit to the alternator

19. Which of the following can *not* be used in a totally solid-state regulator?
    a. Zener diodes
    b. Thermistors
    c. Capacitors
    d. Circuit breakers

20. Which of the following is used to smooth out any abrupt voltage surges and protect a regulator?
    a. Transistor
    b. Capacitor
    c. Thermistor
    d. Relays

21. Which of the following is used to monitor the charging system?
    a. Ammeter
    b. Ohmmeter
    c. Dynamometer
    d. Fusible link

22. Warning lamps are installed so that they will not light when the following is true:
    a. The voltage on the battery side of the lamp is higher.
    b. Field current is flowing from the battery to the alternator.
    c. The voltage on both sides of the lamp is equal.
    d. The voltage on the resistor side of the lamp is higher.

23. Maximum current output in an alternator is reached when the following is true:
    a. It reaches maximum designed speed.
    b. Electrical demands from the system are at the minimum.
    c. Induced countervoltage becomes great enough to stop current increase.
    d. Induced countervoltage drops low enough to stop voltage increase.

24. The regulator is a charging system device that controls circuit opening and closing:
    a. Ignition-to-battery
    b. Alternator-to-thermistor
    c. Battery-to-accessory
    d. Voltage source-to-battery

25. Which of the following methods is used to regulate supply current to the alternator field?
    a. Fault codes
    b. Charge indicator lamp
    c. Pulse-width modulation
    d. Shunt resistor

9

Starting System Operation

LEARNING OBJECTIVES

Upon completion and review of this chapter, you should be able to:

- Define the two circuits of the automotive starting system.
- Identify the basic starting systems parts and explain their function in the system.
- Define the different designs of starting systems used by the different automotive manufacturers.
- Identify the internal components of an automotive starter motor and explain their operation.
- Define the term magnetic repulsion and explain how a DC starter motor operates.
- Define the terms *series*, *shunt (parallel),* and *compound (series-parallel)* as they apply to starter motor internal circuitry.
- Explain the operation of the armature and fields.
- Define starter motor drives and explain their operation.
- Define the different designs of starting motors used by the different automotive manufacturers.
- Explain the operation of the overrunning clutch.

KEY TERMS

Armature
Brushes
Clutch Start Switch
Compound Motor
Detented
Ignition Switch
Lap Winding
Magnetic Repulsion
Magnetic Switch
Overrunning Clutch
Pinion Gear
Series Motor
Shunt Motor
Solenoid
Solenoid-Actuated Starter
Starter Drive
Starting Safety Switch
Torque

INTRODUCTION

The engine must be rotated before it will start and run under its own power. The starting system is a combination of mechanical and electrical components that work together to start the engine. The starting system is designed to change electrical energy that is being stored in the battery into mechanical energy. To accomplish this conversion, a starter motor is used. This chapter will explain how the starting system and it components operate.

STARTING SYSTEM CIRCUITS

The starting system draws a large amount of current from the battery to power the starter motor. To handle this current safely and with a minimum voltage loss from resistance, the cables must be the correct size, and all connections must be clean and tight. The driver through the ignition switch controls the starting system. If the heavy cables that carry current to the starter were routed to the instrument panel and the switch, they would be so long that the starter would not get enough current to operate properly. To avoid such a voltage drop, the starting system has the following two circuits, as shown in Figure 9-1:

- Starter circuit
- Control circuit

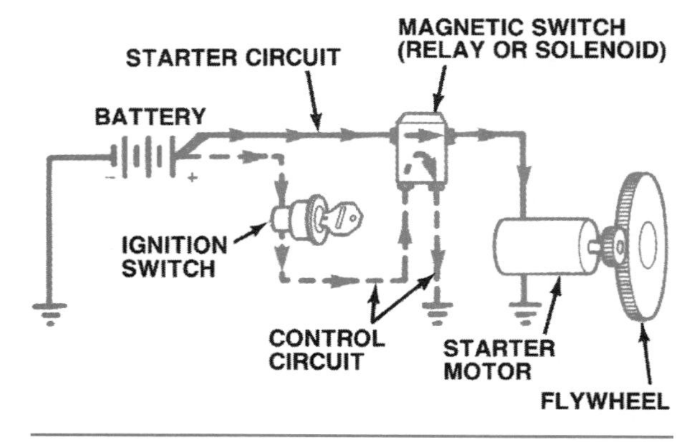

Figure 9-1. In this diagram of the starting system, the starter circuit is shown as a solid line and the control circuit is shown as a dashed line. (Delphi Automotive Systems)

Starter Circuit

The starter circuit, or motor circuit, (shown as the solid lines of Figure 9-1) consists of the following:

- Battery
- Magnetic switch
- Starter motor
- Heavy-gauge cables

The circuit between the battery and the starter motor is controlled by a magnetic switch (a relay or **solenoid**). Switch design and function vary from system to system. A gear on the starter motor armature engages with gear teeth on the engine flywheel. When current reaches the starter motor, it begins to turn. This turns the car's engine, which can quickly fire and run by itself. If the starter motor remained engaged to the engine flywheel, the starter motor would be spun by the engine at a very high speed. This would damage the starter motor. To avoid this, there must be a mechanism to disengage the starter motor from the engine. There are several different designs that will do this, as we will see in this chapter.

Control Circuit

The control circuit is shown by the dashed lines in Figure 9-1. It allows the driver to use a small amount of battery current, about three to five amperes, to control the flow of a large amount of battery current to the starter motor. Control circuits usually consist of an ignition switch connected through normal-gauge wiring to the battery and the magnetic switch. When the ignition switch is in the start position, a small amount of current flows through the coil of the magnetic switch. This closes a set of large contact points within the magnetic switch and allows battery current to flow directly to the starter motor. For more information about control circuits, see the "Starter Control Circuit Devices" section in Chapter 9 of the *Shop Manual*.

BASIC STARTING SYSTEM PARTS

We have already studied the battery, which is an important part of the starting system. The other circuit parts are as follows:

- Ignition switch
- Starting safety switch (on some systems)

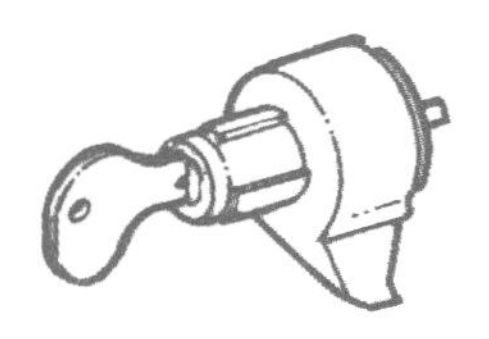

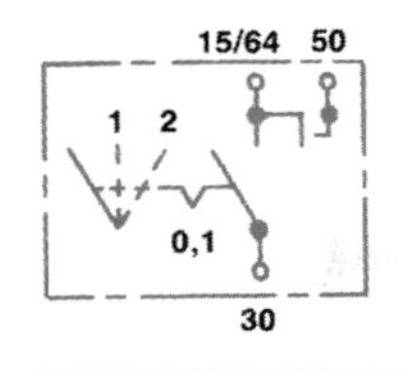

Figure 9-2. This ignition switch acts directly on the contact points. (Reprinted by permission of Robert Bosch GmbH)

- Relays or solenoids (magnetic switches)
- Starter motor
- Wiring

Ignition Switch

The **ignition switch** has jobs other than controlling the starting system. The ignition switch normally has at least four positions:

- ACCESSORIES
- OFF
- ON (RUN)
- START

Switches on late-model cars also have a LOCK position to lock the steering wheel. All positions except START are **detented.** That is, the switch will remain in that position until moved by the driver. When the ignition key is turned to START and released, it will return to the ON (RUN) position. The START position is the actual starter switch part of the ignition switch. It applies battery voltage to the magnetic switch.

There are two types of ignition switches in use. On older cars, the switch is mounted on the instrument panel and contains the contact points (Figure 9-2). The newer type, used on cars with locking steering columns, is usually mounted on the steering column. Many column-mounted switches operate remotely mounted contact points through a rod. Other column-mounted switches operate directly on contact points (Figure 9-3). Older domestic and imported cars sometimes used separate push-button switches or cable-operated switches that controlled the starting system separately from the ignition switch.

Starting Safety Switch

The **starting safety switch** is also called a neutral start switch. It is a normally open switch that prevents the starting system from operating when the automobile's transmission is in gear. If the car has no starting safety switch, it is possible to spin

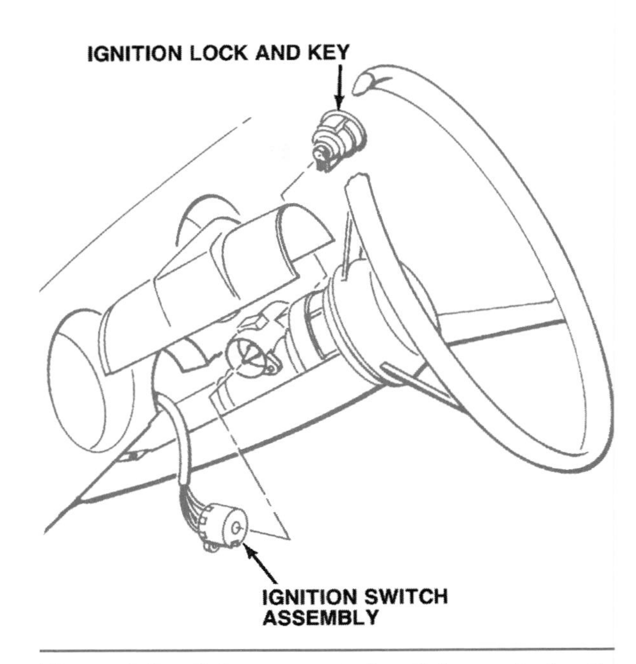

Figure 9-3. Column-mounted switches act directly on the contact points.

the engine with the transmission in gear. This makes the car lurch forward or backward, which could be dangerous. Safety switches or interlock devices are now required by law with all automatic and manual transmissions.

Starting safety switches can be connected in two places within the starting system control circuit. The safety switch can be placed between the ignition switch and the magnetic switch, as shown in Figure 9-4, so that the safety switch must be closed before current can flow to the magnetic switch. The safety switch also can be connected between the magnetic switch and ground (Figure 9-5), so that the switch must be closed before current can flow from the magnetic switch to ground. Where the starting safety switch is installed depends upon the type of transmission used and whether the gearshift lever is column-mounted or floor-mounted.

Automatic Transmissions/Transaxles

The safety switch used with an automatic transmission or transaxle can be either an electrical switch or a mechanical device. Electrical/electronic switches have contact points that are closed only when the gear lever is in PARK or NEUTRAL, as shown in Figure 9-4. The switch can be mounted near the gearshift lever, as in Figures 9-6 and 9-7, or on the transmission-housing, as in Figure 9-8. The

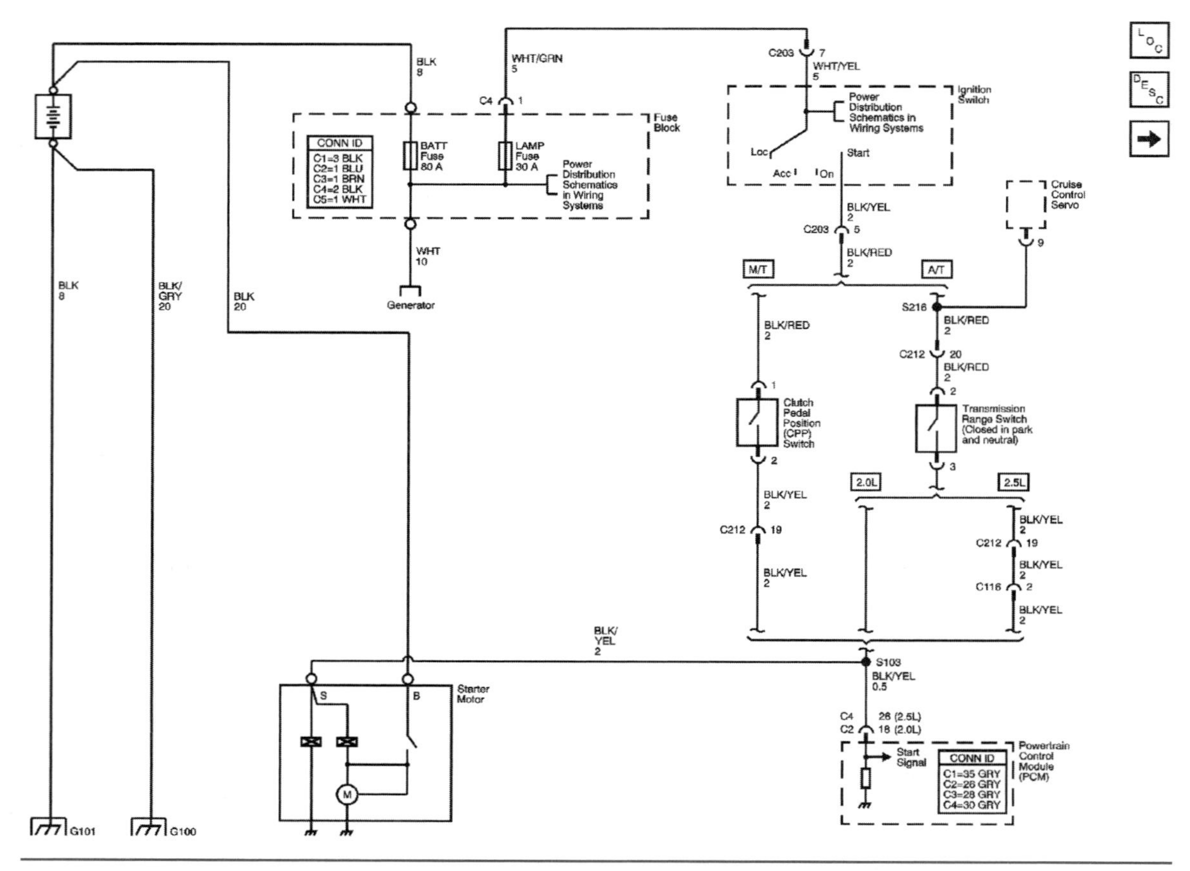

Figure 9-4. This starting safety switch must be closed before battery current can reach the magnetic switch. (GM Service and Parts Operations)

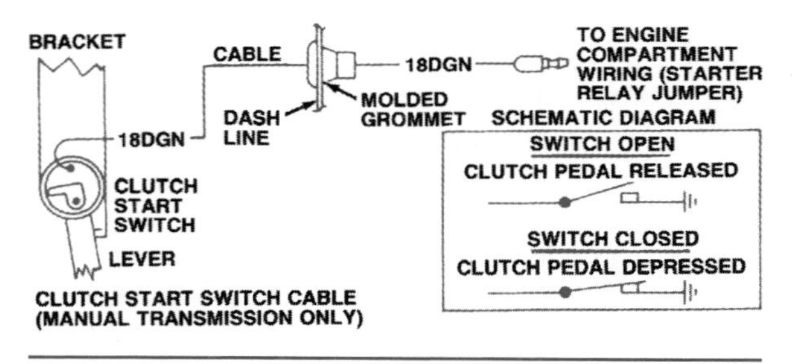

Figure 9-5. The clutch switch must be closed before battery current can flow from the magnetic switch to ground. (DaimlerChrysler Corporation)

contacts are in series with the control circuit, so that no current can flow through the magnetic switch unless the transmission is out of gear.

Mechanical interlock devices physically block the movement of the ignition key when the transmission is in gear, as shown in Figures 9-9 and 9-10. The key can be turned only when the gearshift lever is in PARK or NEUTRAL. Some manufacturers use an additional circuit in the neutral start switch to light the backup lamps

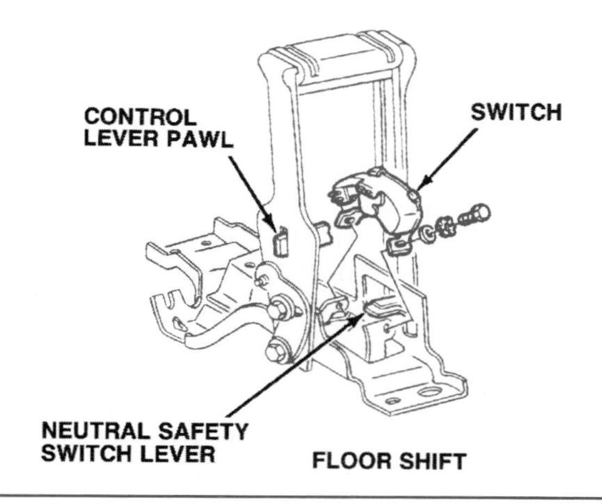

Figure 9-6. An electrical safety switch installed near the floor-mounted gearshift lever. (GM Service and Parts Operations)

when the transmission is placed in REVERSE (Figures 9-7 and 9-8).

Ford vehicles equipped with an electronic automatic transmission or transaxle use an additional circuit in the neutral safety switch to inform

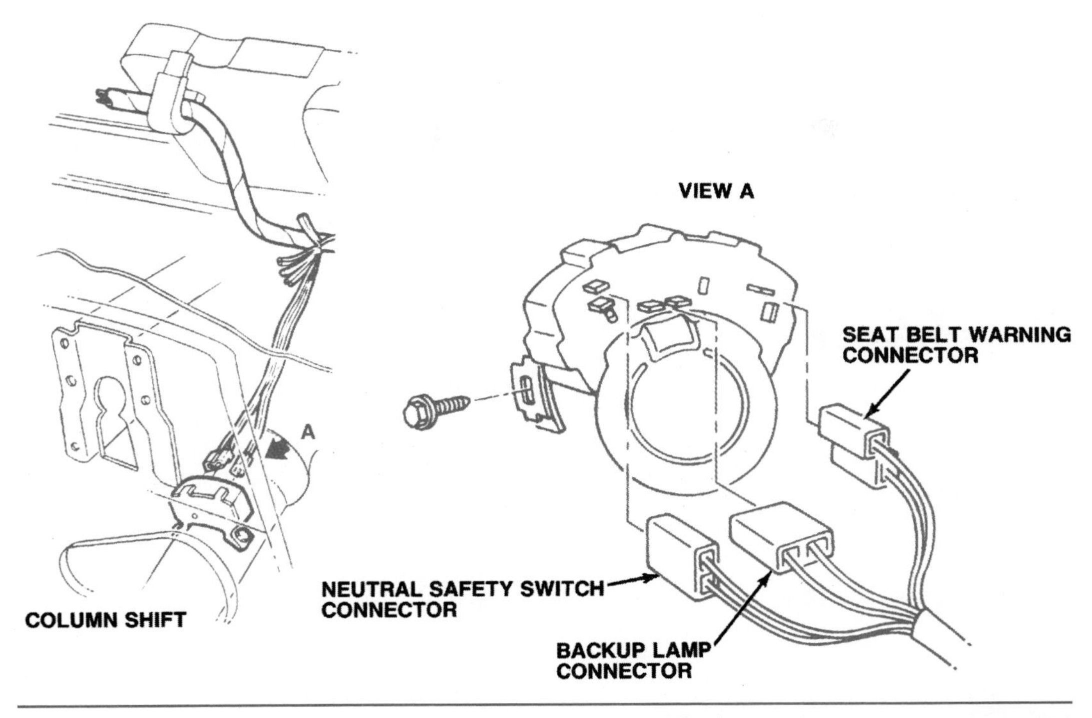

Figure 9-7. Column-mounted neutral safety switch near gearshift tube. (GM Service and Parts Operations)

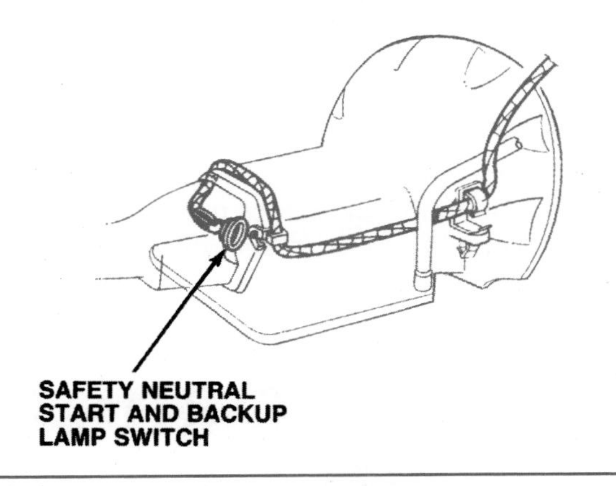

Figure 9-8. Transmission-mounted safety switch. (DaimlerChrysler Corporation)

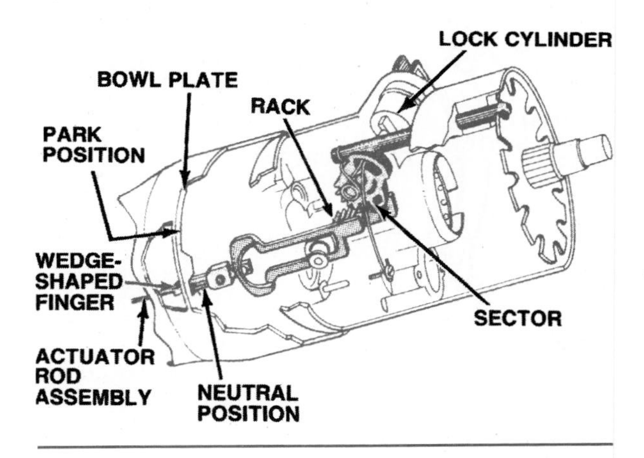

Figure 9-9. A mechanical device within the steering column blocks the movement of the ignition switch.

the microprocessor of the position of the manual lever shaft. This signal is used to determine the desired gear and electronic pressure control. The switch is now called a manual lever position switch (MLPS).

General Motors has done essentially the same as Ford, renaming the PARK/NEUTRAL switch used on its 4T65E and 4T80E transaxles. It now is called either a PRNDL switch or a PARK/NEUTRAL position switch and provides input to the PCM regarding torque converter clutch slip. This input allows the PCM to make the necessary calculations to control clutch apply and release feel.

Manual Transmissions/ Transaxles

The starting safety switch used with a manual transmission on older vehicles is usually an electrical switch similar to those shown in Figures 9-7 and 9-8. A **clutch start switch** (also called an interlock switch) is commonly used with manual

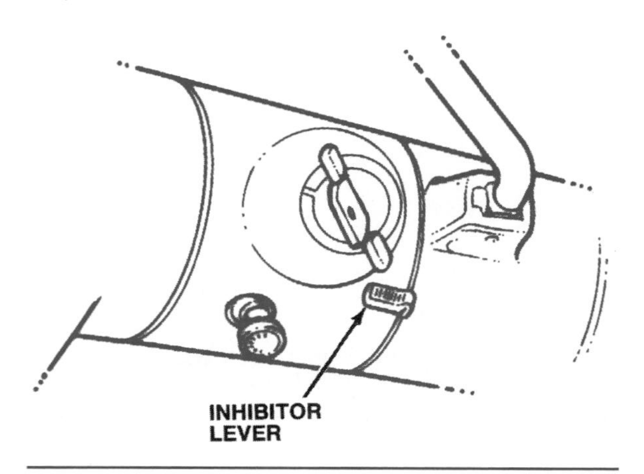

Figure 9-10. A lever on the steering wheel blocks the movement of the ignition key when the transmission is in gear.

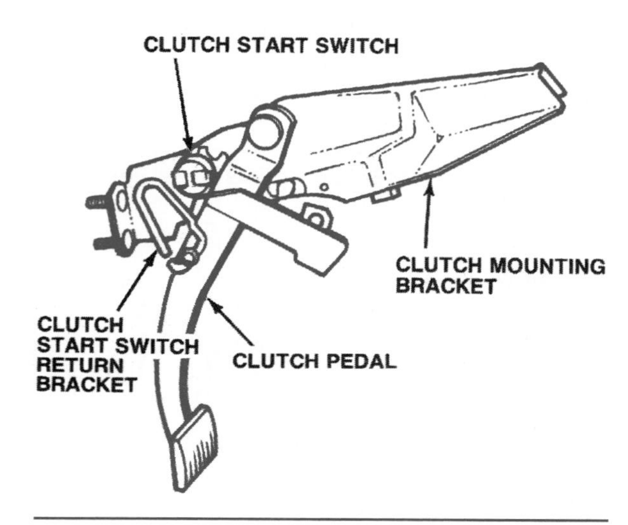

Figure 9-11. The clutch pedal must be fully depressed to close the clutch switch and complete the control circuit.

transmissions and transaxles on late-model vehicles. This is an electric switch mounted on the floor or firewall near the clutch pedal. Its contacts are normally open and close only when the clutch pedal is fully depressed (Figure 9-11).

Relays and Solenoids

A **magnetic switch** in the starting system allows the control circuit to open and close the starter circuit. The switch can be either of the following:

- A relay, which uses the electromagnetic field of a coil to attract an armature and close the contact points
- A solenoid, which uses the electromagnetic field of a coil to pull a plunger into the coil and close the contact points

In addition to closing the contact points, solenoid-equipped circuits often use the movement of the solenoid to engage the starter motor with the engine flywheel. We will explain this in Chapter 10. The terminology used with relays and solenoids is often confusing. Technically, a relay operates with a hinged **armature** and does only an electrical job; a solenoid operates with a movable plunger and usually does a mechanical job. Sometimes, a solenoid is used only to open and close an electric circuit; the movement of the plunger is not used for any mechanical work. Manufacturers sometimes call these solenoids "starter relays." Figure 9-12 shows a commonly used Ford starter relay. We will continue to use the general term *magnetic switch*, and will tell you if the manufacturer uses a different name for the device.

For more information about magnetic switches, see the following sections in Chapter 9 of the *Shop Manual*: "Inspection and Diagnosis," "Starter Control Circuit Devices," and "Unit Removal."

Wiring

The starter motor circuit uses heavy-gauge wiring to carry current to the starter motor. The control circuit carries less current and thus uses lighter-gauge wires.

SPECIFIC STARTING SYSTEMS

Various manufacturers use different starting system components. The following paragraphs briefly describe the circuits used by major manufacturers.

Delco-Remy (Delphi) and Bosch

Delco-Remy and Bosch starter motors are used by General Motors. The most commonly used Delco-Remy and Bosch automotive starter motor depends upon the movement of a solenoid both to control current flow in the starter circuit and to engage the starter motor with the engine flywheel. This is called a **solenoid-actuated starter.** The

Figure 9-12. The Ford starter relay or magnetic switch.

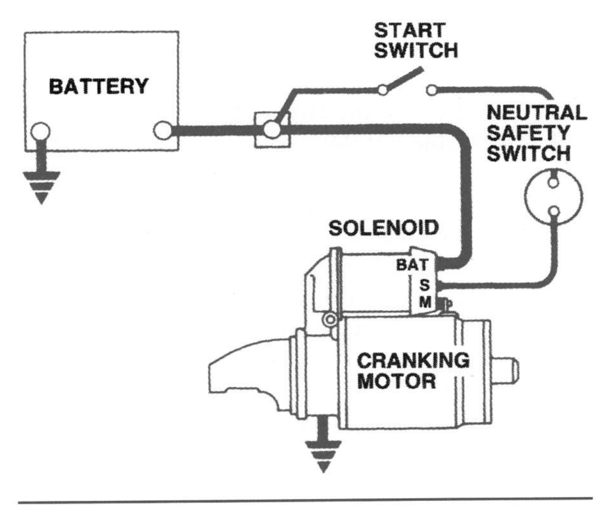

Figure 9-13. GM Starter circuit. (Delphi Automotive Systems)

solenoid is mounted on, or enclosed with, the motor housing (Figure 9-13).

The type and location of starting safety switches vary within the GM vehicle platforms. Larger-size GM cars use a mechanical blocking device in the steering column (Figure 9-9). The intermediate and smaller cars with automatic transmissions have electrical switches mounted near the shift lever. These are either on the column, as shown in Figure 9-7, or on the floor (Figure 9-6). On front-wheel-drive (FWD) cars with automatic transmissions, the PARK/NEUTRAL or PRNDL switch is an electrical switch mounted on the transaxle case manual lever shaft (Figure 9-14). GM cars with floor-shift manual transmissions use a clutch pedal-operated safety switch. With column-shift manual transmissions, an electric switch is mounted on the column.

Ford Motorcraft

Ford has used three types of starter motors, and therefore has several different starting system circuits. The Motorcraft positive engagement starter has a movable-pole shoe that uses electromagnetism to engage the starter motor with the engine. This motor does not use a solenoid to *move* anything, but it uses a solenoid to open and close the starter circuit as a magnetic switch (Figure 9-15). Ford calls this solenoid a starter relay.

The Motorcraft solenoid-actuated starter is very similar to the Delco-Remy unit and depends upon the movement of a solenoid to engage the starter motor with the engine. The solenoid is mounted within the motor housing and receives battery current through the same type of starter relay used in the positive engagement system. Although the motor-mounted solenoid could do the job of this additional starter relay, the second relay is installed

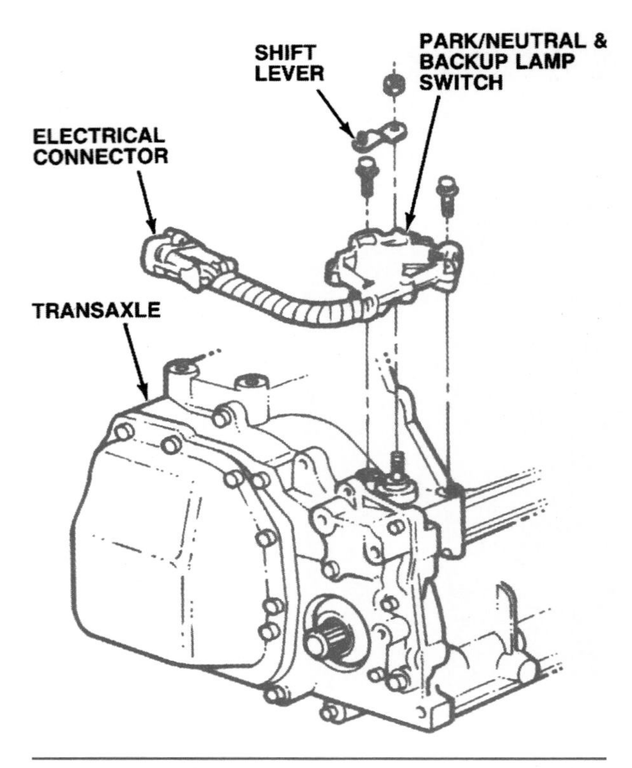

Figure 9-14. GM PRNDL/Park-neutral switch on a GM Transaxle. (GM Service and Parts Operations)

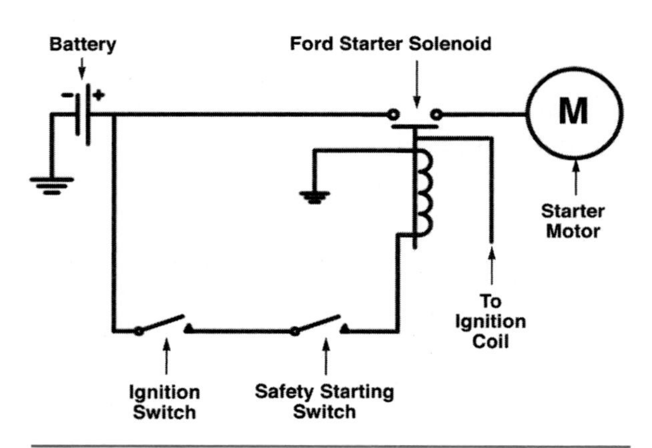

Figure 9-15. The Ford starting system circuit with the positive engagement starter.

on many Ford automobiles to make the cars easier to build. Motorcraft solenoid-actuated starters were used on Ford cars and trucks with large V8 engines. The Motorcraft permanent magnet gear-reduction (PMGR) starter is a solenoid-actuated design that operates much like the Motorcraft solenoid-actuated starter previously described. However, the starter circuit may or may not use a starter relay, depending on the car model.

Rear-wheel-drive (RWD) Ford automobiles with manual transmissions have no starting safety switch. Front-wheel-drive (FWD) models with manual transaxles have a clutch interlock switch. If a Ford car with an automatic transmission has a column-mounted shift lever, a blocking interlock device prevents the ignition key from turning when the transmission is in gear. If the automatic transmission shift lever is mounted on the floor, an electrical switch prevents current from flowing to the starter relay when the transmission is in gear. The switch may be mounted on the transmission case or near the gearshift lever.

DaimlerChrysler

Chrysler uses a solenoid-actuated starter motor. The solenoid is mounted inside the motor housing and receives battery current through a starter relay, as shown in Figure 9-16. Chrysler starter relays used prior to 1977 have four terminals, as shown in Figure 9-17A. In 1977, a second set of contacts and two terminals were added (Figure 9-17B). The extra contacts and terminals allow more current to flow through the relay to the ignition system and to the exhaust gas recirculation (EGR) timer. This has no effect on the operation of the relay within the starting system. These starter relays generally were mounted on the firewall.

Current Chrysler starting systems use a standard five-terminal Bosch relay (Figure 9-18) but only four terminals are used in the circuit (Figure 9-19). The relay is located at the front of the driver's-side strut tower in a power distribution center or cluster.

Chrysler automobiles with manual transmissions have a clutch interlock switch, as shown in Figure 9-20. Current from the starter relay can flow to ground only when the clutch pedal is fully depressed. Cars with automatic transmissions have an electrical neutral start switch mounted on the transmission housing (Figure 9-21). When the transmission is out of gear, the switch provides a ground connection for the starter control circuit.

Toyota and Nissan

Toyota and Nissan use a variety of solenoid-actuated direct drive and reduction-gear starter designs manufactured primarily by Hitachi and Nippondenso, as shown in Figures 9-22 and 9-23. The neutral start switch (called an inhibitor switch by the Japanese automakers) incorporates a relay in its circuit.

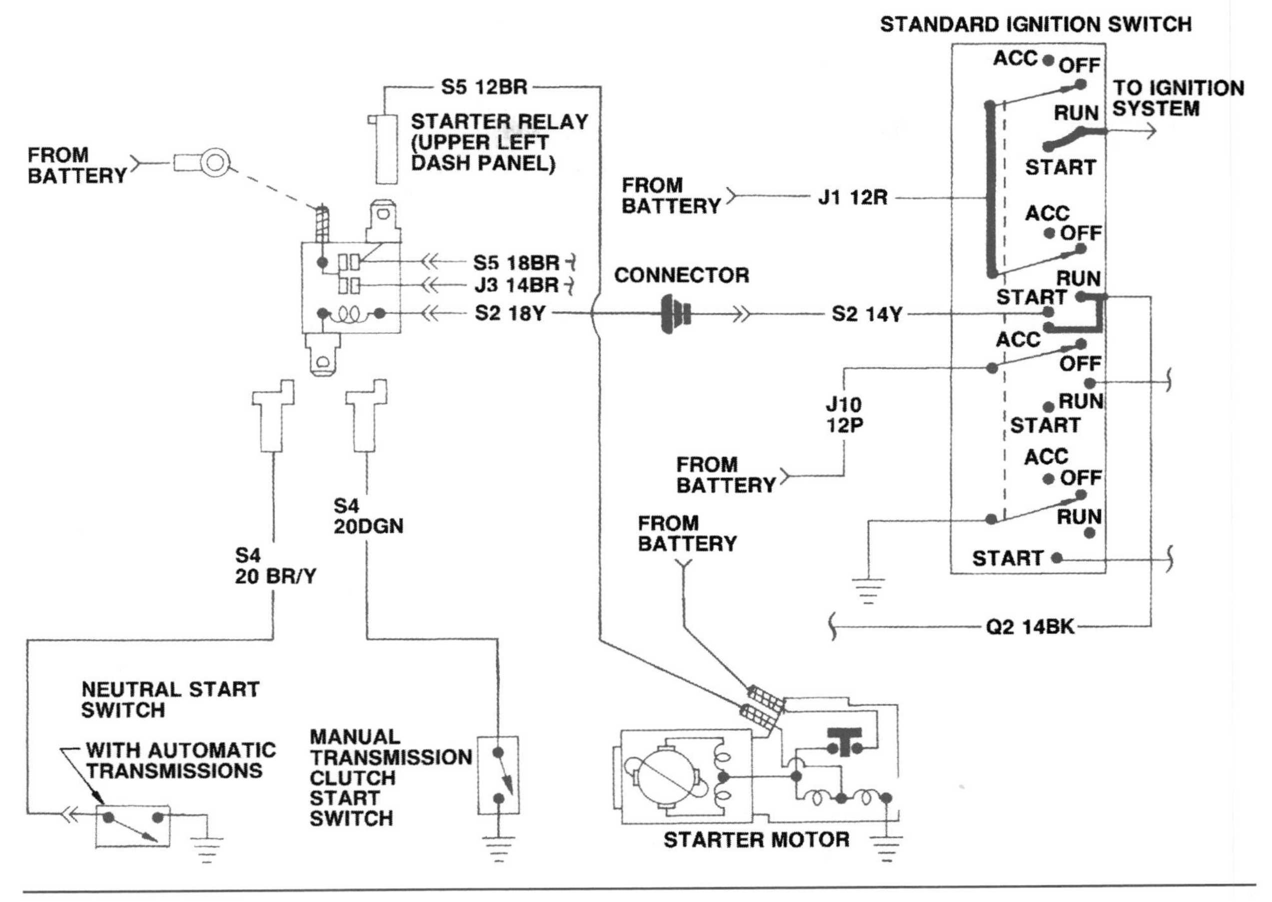

Figure 9-16. Typical DaimlerChrysler starting system. (DaimlerChrysler Corporation)

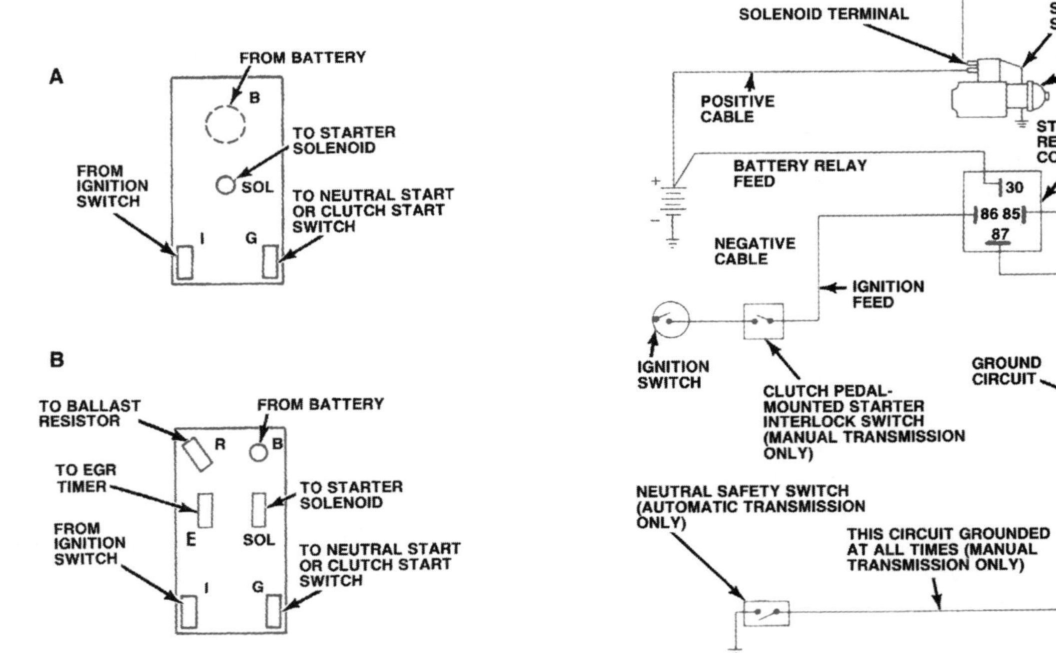

Figure 9-17. Comparison of the terminals on a pre-1977 starter relay (A) and a 1977 or later relay (B). (DaimlerChrysler Corporation)

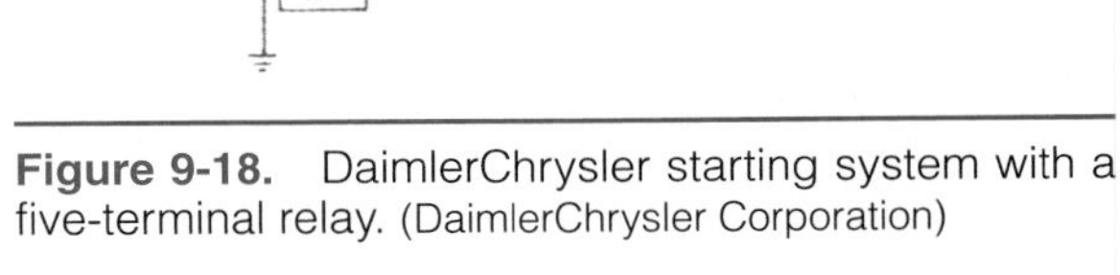

Figure 9-18. DaimlerChrysler starting system with a five-terminal relay. (DaimlerChrysler Corporation)

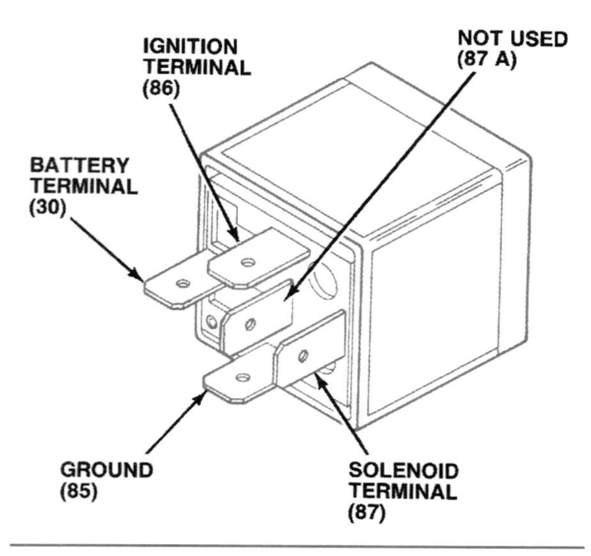

Figure 9-19. Only four of the five relay terminals are used when the Bosch relay is installed. (DaimlerChrysler Corporation)

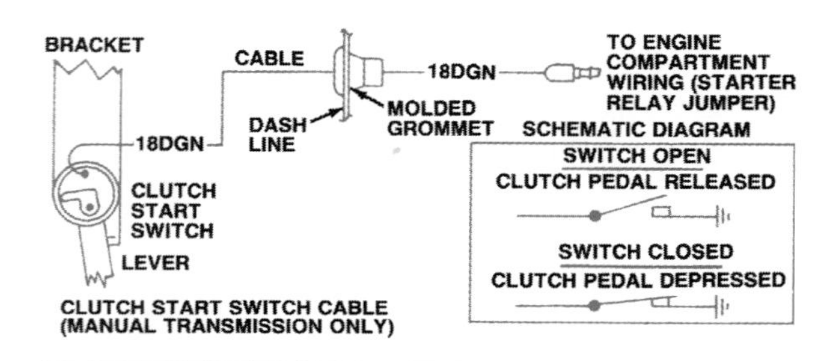

Figure 9-20. DaimlerChrysler clutch switch. (DaimlerChrysler Corporation)

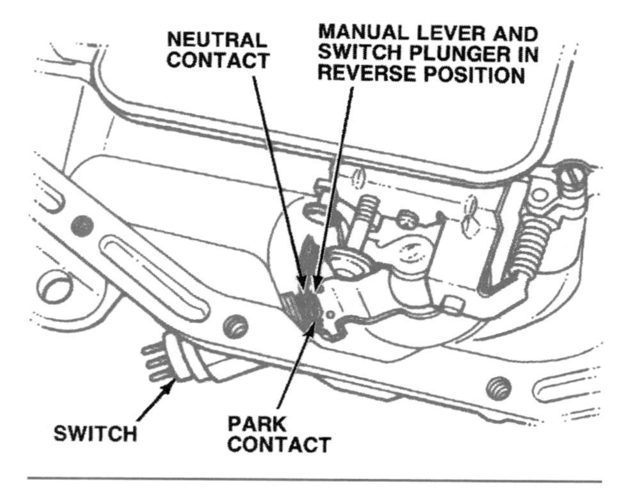

Figure 9-21. When the automatic transmission is in PARK or NEUTRAL, a transmission lever touches the contact and completes the control circuit to ground. (DaimlerChrysler Corporation)

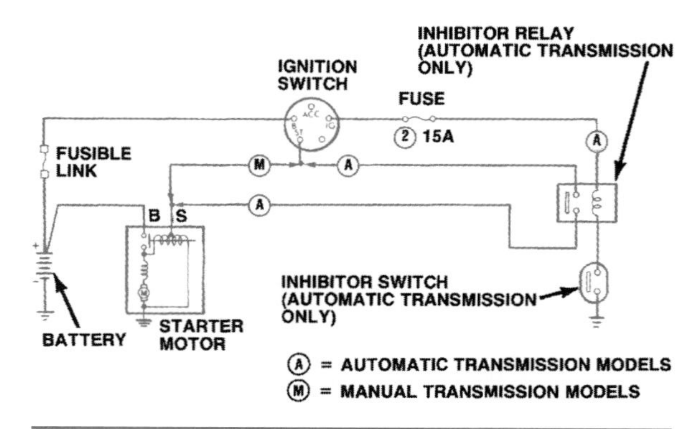

Figure 9-22. Typical Nissan starting system used on gasoline engines. (Courtesy of Nissan North America, Inc.)

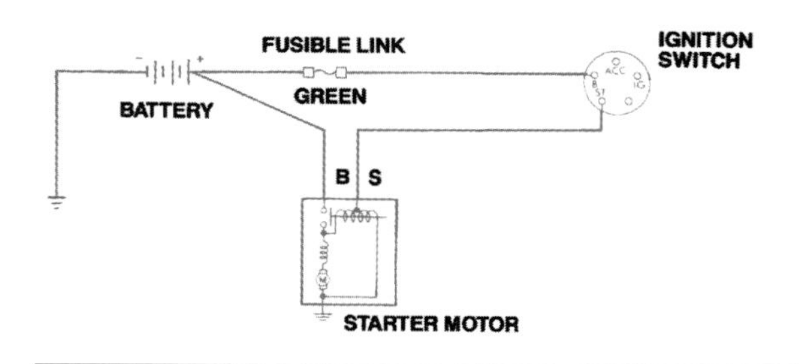

Figure 9-23. A typical Nissan diesel starting system. (Courtesy of Nissan North America, Inc.)

STARTER MOTORS

Starter Motor Purpose

The starter motor converts the electrical energy from the battery into mechanical energy for cranking the engine. The starter is an electric motor designed to operate under great electrical loads and to produce very high horsepower. The starter consists of housing, field coils, an armature, a commutator and brushes, end frames, and a solenoid-operated shift mechanism.

FRAME AND FIELD ASSEMBLY

The frame, or housing, of a starter motor (Figure 9-24) encloses all of the moving motor parts. It supports the parts and protects them from dirt, oil, and other contamination. The part of the frame that encloses the pole shoes and field windings is made of iron to provide a path for magnetic flux lines (Figure 9-25). To reduce weight, other parts of the frame may be made of cast aluminum.

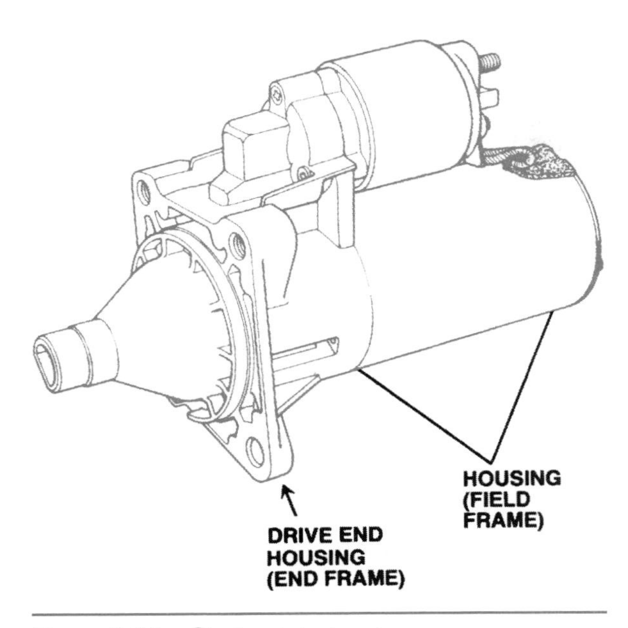

Figure 9-24. Starter motor housing.

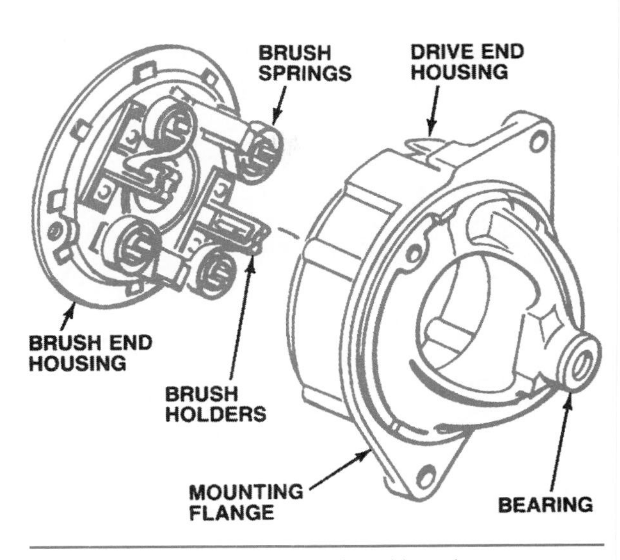

Figure 9-26. Brush end and end housing.

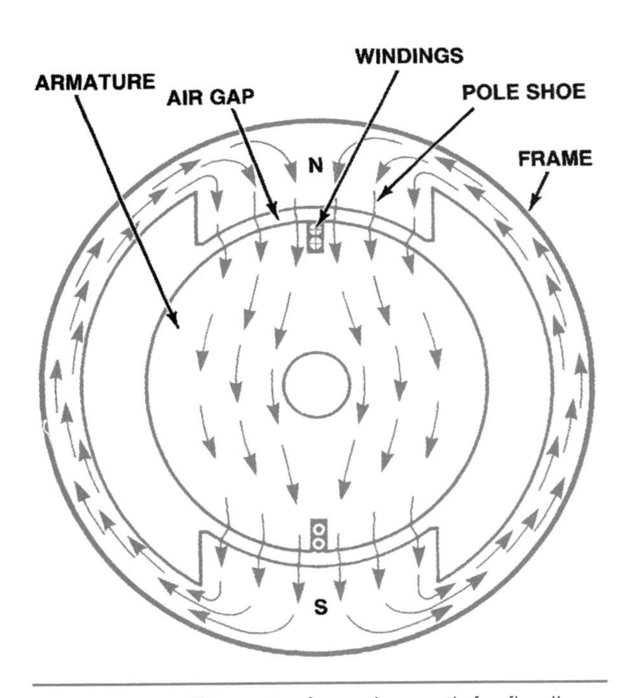

Figure 9-25. The motor frame is a path for flux lines.

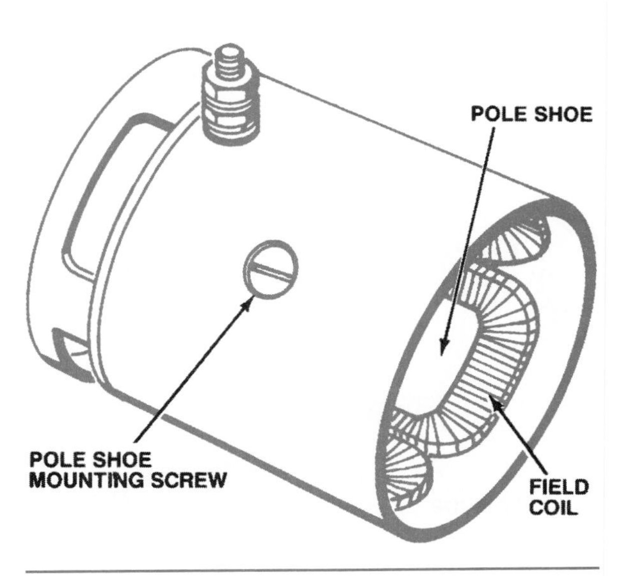

Figure 9-27. Pole shoes and field windings in housing.

One end of the housing holds one of the two bearings or bushings in which the armature shaft turns. On most motors, it also contains the **brushes** that conduct current to the armature (Figure 9-26). This is called the brush, or commutator, end housing. The other end housing holds the second bearing or bushing in which the armature shaft turns. It also encloses the gear that meshes with the engine flywheel. This is called the drive end housing. The drive end housing often provides the engine-to-motor mounting points. These end pieces may be made of aluminum because they do not have to conduct magnetic flux.

The magnetic field of the starter motor is provided by two or more pole shoes and field windings. The pole shoes are made of iron and are attached to the frame with large screws (Figure 9-27). Figure 9-28 shows the paths of magnetic flux lines within a four-pole motor. The field windings are usually made of a heavy

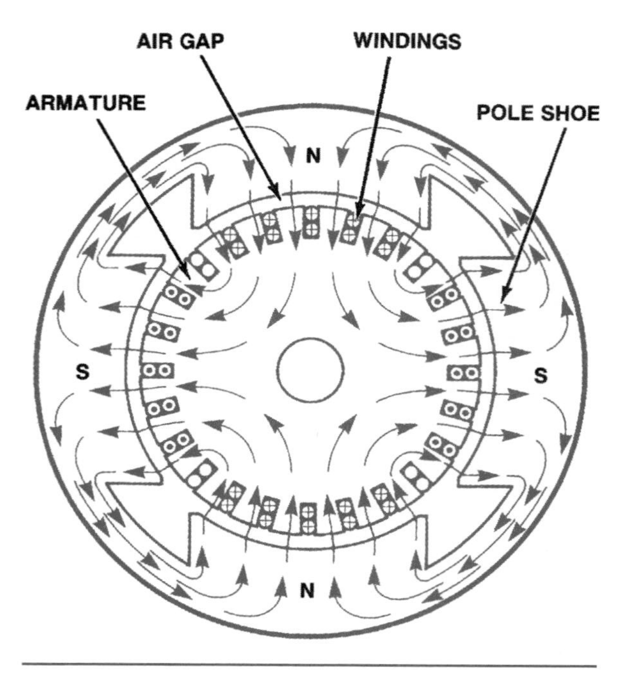

Figure 9-28. Flux path in a four-pole motor.

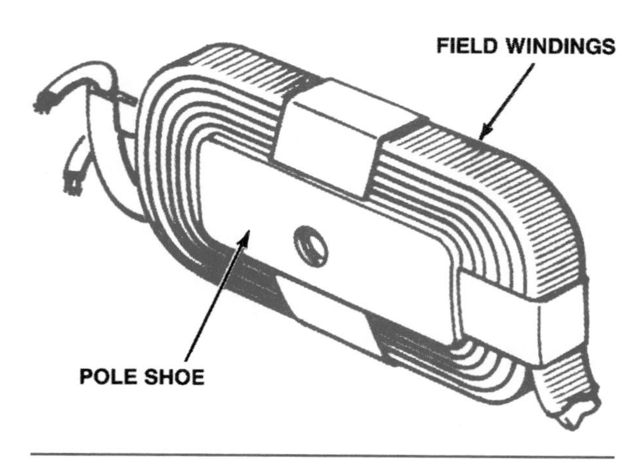

Figure 9-29. Pole shoe and field winding removed from housing.

copper ribbon (Figure 9-29) to increase their current-carrying capacity and electromagnetic field strength. Automotive starter motors usually have four-pole shoes and two to four field windings to provide a strong magnetic field within the motor. Pole shoes that do not have field windings are magnetized by flux lines from the wound poles.

Torque is the force of a starter motor, a force applied in a rotary, or circular direction. The torque, speed, and current draw of a motor are related. As speed increases in most automotive starter motors, torque and current draw decrease. These motors develop maximum torque just before the engine begins to turn. Once the engine begins to turn, the motor speed increases and torque decreases. The maximum amount of torque produced by a motor depends upon the strength of its magnetic fields. As field strength increases, torque increases.

DC STARTER MOTOR OPERATION

DC starter motors (Figure 9-30) work on the principle of magnetic repulsion. This principle states that **magnetic repulsion** occurs when a straight rod conductor composed of the armature, commutator, and brushes is located in a magnetic field (field windings) and current is flowing through the rod. This situation creates two separate magnetic fields: one produced by the magnet (pole shoes of the magnetic field winding) and another produced by the current flowing through the conductor (armature/commutator/brushes).

Figure 9-30 shows the magnet's magnetic field moving from the N pole to the S pole and the conductor's magnetic field flowing around the conductor. The magnetic lines of force have a rubber-band characteristic. That is, they stretch and also try to shorten to minimum length.

Figure 9-30 shows a stronger magnetic field on one side of the rod conductor (armature/commutator/brushes) and a weak magnetic field on the other side. Under these conditions, the conductor (armature) will tend to be repulsed by the strong magnetic field (pole shoes and field winding) and move toward the weak magnetic field. As current in the conductor (armature) and the strength of the magnet (field windings) increases, the following happens:

- More lines of magnetism are created on the strong side.
- More repulsive force is applied to the conductor (armature).
- The conductor tries harder to move toward the weak side in an attempt to reach a balanced neutral state.
- A greater amount of electrical heat is generated.

NOTE: The combination of the U-shaped conductor loop and the split copper ring are called the commutator because they rotate together. Together they become the armature.

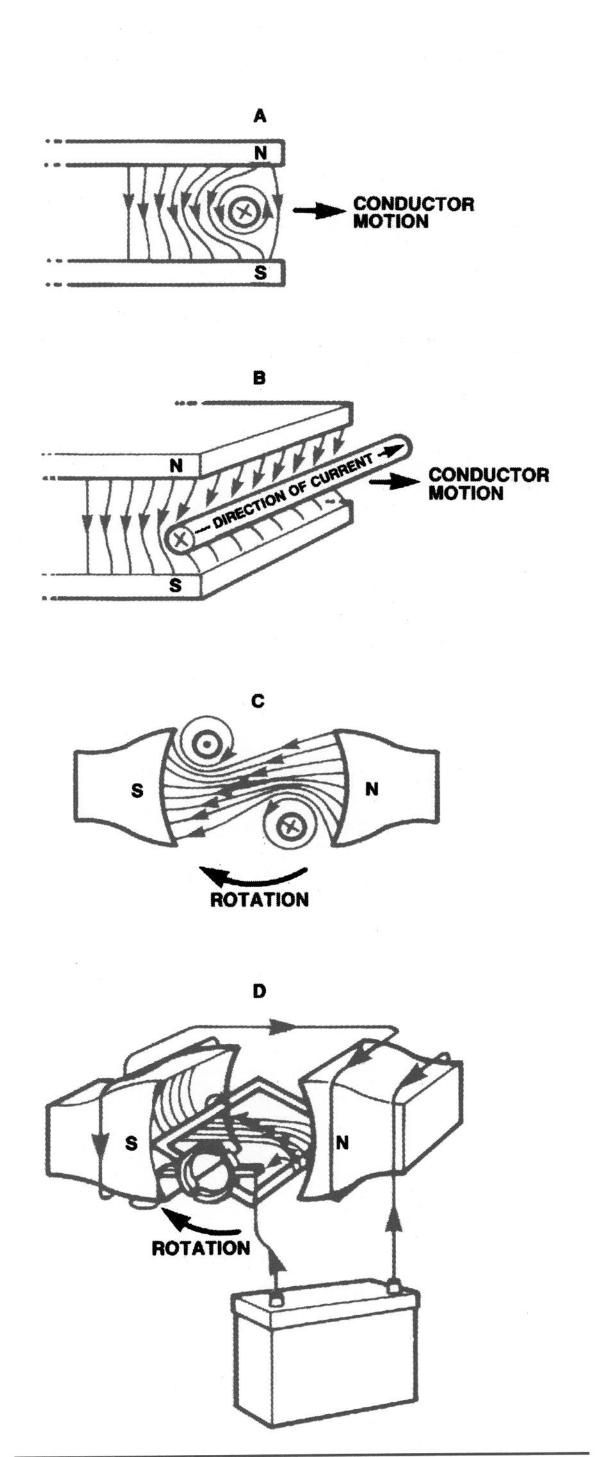

Figure 9-30. Motor principle.

Current flows from the positive (+) battery terminal through the brush and copper ring nearer the N pole, through the conductor (armature) to the copper ring and brush nearer the S pole and back to the negative (−) battery terminal. This electrical flow causes the portion of the loop near the S pole to push downward and the N pole to push upward. With a strong field on one side of the conductor and a weak field on the other side, the conductor will move from the strong to the weak. Put another way, the weaker magnetic field between the S and N poles on one side of the conductor is repulsed by the stronger magnetic field on the other side of the conductor. The commutator then rotates. As it turns, the two sides of the conductor loop reverse positions and the two halves of the split copper ring alternately make contact with the opposite stationary brushes. This causes the flow direction of electrical current to reverse (alternating current) through the commutator and the commutator to continue to rotate in the same direction.

In order to provide smooth rotation and to make the starter powerful enough to start the engine, many armature commutator segments are used. As one segment rotates past the stationary magnetic field pole, another segment immediately takes its place.

When the starter operates, the current passing through the armature produces a magnetic field in each of its conductors. The reaction between the magnetic field of the armature and the magnetic fields produced by the field coils causes the armature to rotate.

Motor Internal Circuitry

Because field current and armature current flow to the motor through one terminal on the housing, the field and armature windings must be connected in a single complete circuit. The internal circuitry of the motor (the way in which the field and armature windings are connected) gives the motor some general operating characteristics.

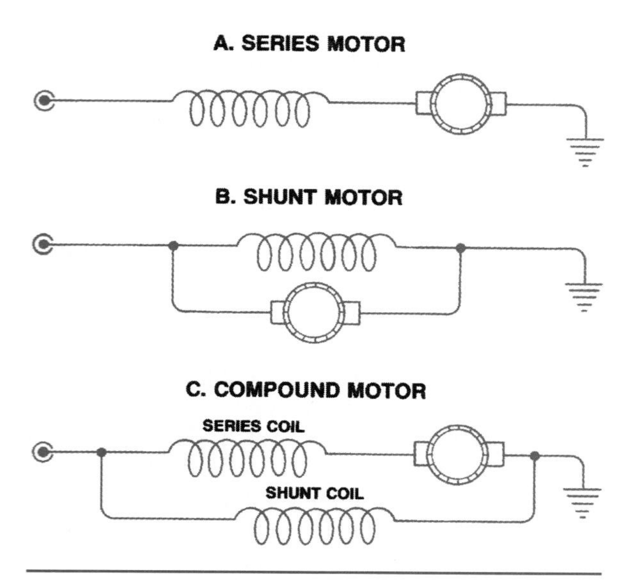

Figure 9-31. Basic motor circuitry.

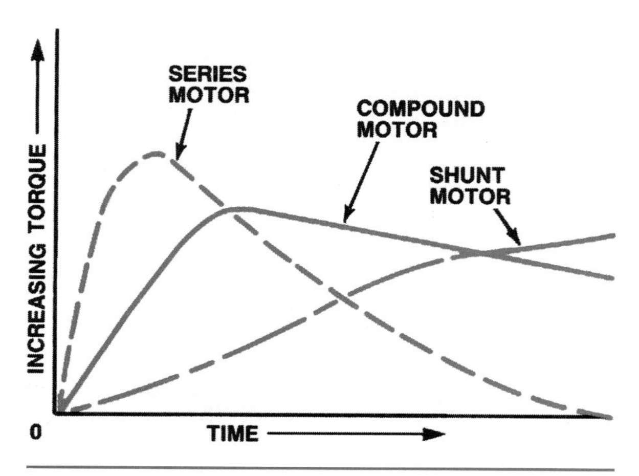

Figure 9-32. Torque output characteristics of series, shunt, and compound motors.

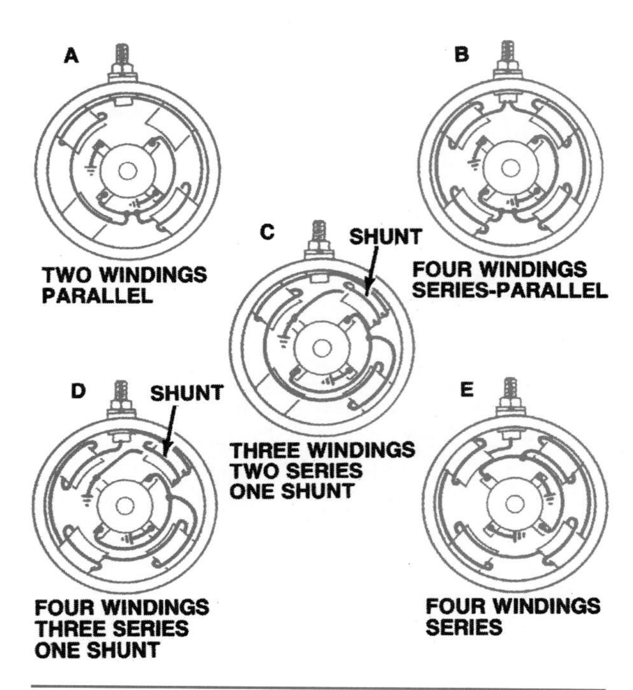

Figure 9-33. Actual relationships of field and armature windings in different types of motors.

Figure 9-31 shows the three general types of motor internal circuitry, as follows:

- Series
- Shunt (parallel)
- Compound (series-parallel)

All automotive starter motors in use today are the series type or the compound type. The **series motor** (Figure 9-31A) has only one path for current. As the armature rotates, its conductors cut magnetic flux lines. A counter-voltage is induced in the armature windings, opposing the original current through them. The counter-voltage decreases the total current through both the field and the armature windings, because they are connected in series. This reduction of current also reduces the magnetic field strength and motor torque. Series motors produce a great amount of torque when they first begin to operate, but torque decreases as the engine begins to turn (Figure 9-32). Series motors work well as automotive starters because cranking an engine requires a great amount of torque at first, and less torque as cranking continues.

The **shunt motor** (Figure 9-31B) does not follow the increasing-speed/decreasing-torque relationship just described. The counter-voltage within the armature does not affect field current, because field current travels through a separate circuit path. A shunt motor, in effect, adjusts its torque output to the imposed load and operates at a constant speed. Shunt motors are not used as automotive starters because of their low initial torque (Figure 9-32), but are used to power other automotive accessories.

The **compound motor,** shown in Figure 9-31C, has both series and shunt field windings. It combines both the good starting torque of the series-type and the relatively constant operating speed of the shunt-type motor (Figure 9-32). A compound motor is often used as an automotive starter. Figure 9-33 shows the actual relationships of field and armature windings in different types of motors.

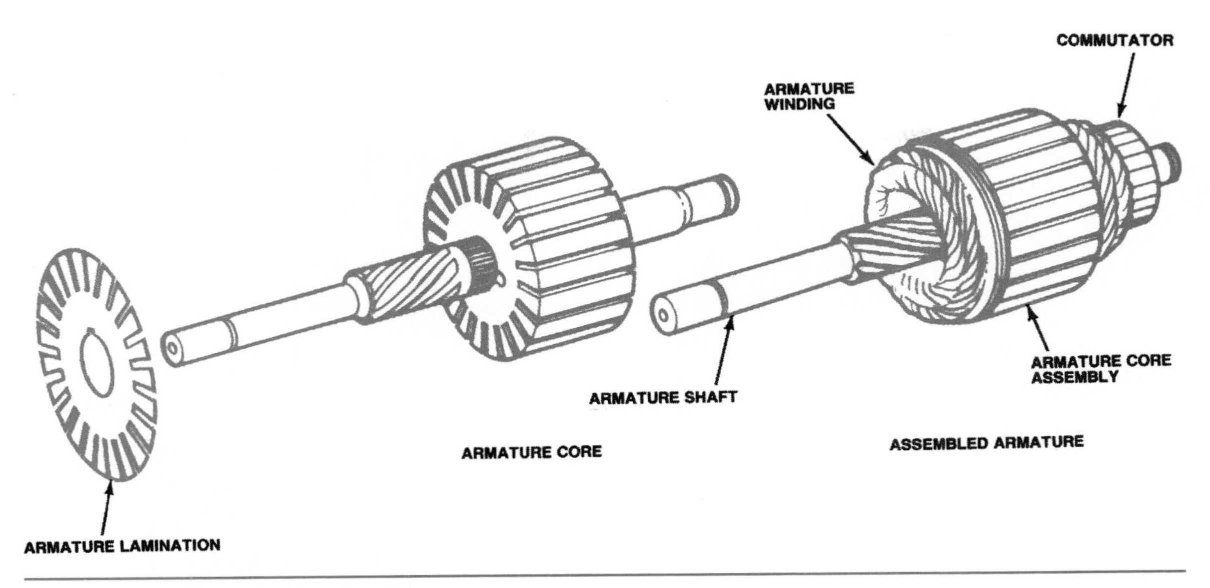

Figure 9-34. Motor armature.

ARMATURE AND COMMUTATOR ASSEMBLY

The motor armature (Figure 9-32) has a laminated core. Insulation between the laminations helps to reduce eddy currents in the core. For reduced resistance, the armature conductors are made of a thick copper wire. Motor armatures are connected to the commutator in one of two ways. In a **lap winding,** the two ends of each conductor are attached to two adjacent commutator bars (Figure 9-35). In a wave winding, the two ends of a conductor are attached to commutator bars that are 180 degrees apart (on opposite sides of the commutator), as shown in Figure 9-36. A lap-wound armature is more commonly used because it offers less resistance.

The commutator is made of copper bars insulated from each other by mica or some other insulating material. The armature core, windings, and commutator are assembled on a long armature shaft. This shaft also carries the **pinion gear** that meshes with the engine flywheel ring gear (Figure 9-37). The shaft is supported by bearings or bushings in the end housings. To supply the proper current to the armature, a four-pole motor must have four brushes riding on the commutator (Figure 9-38). Most automotive starters have two grounded and two insulated brushes. The brushes are held against the commutator by spring force.

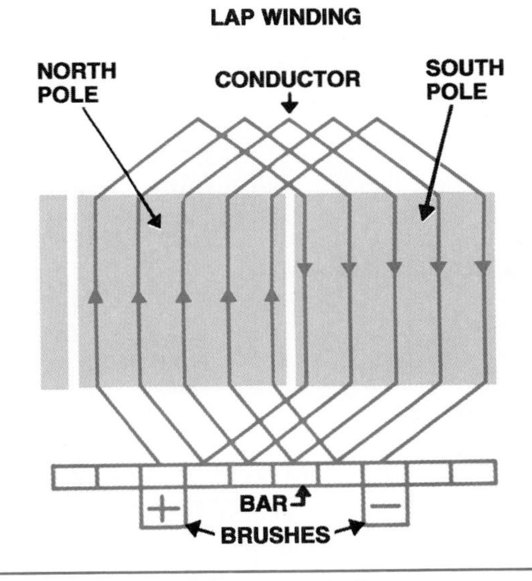

Figure 9-35. Armature lap winding. (Delphi Automotive Systems)

PERMANENT-MAGNET FIELDS

The permanent magnet, planetary-drive starter motor is the first significant advance in starter design in decades. It was first introduced on some 1986 Chrysler and GM models, and in 1989 by Ford on Continental and some Thunderbird models. Permanent magnets are used in place of the electromagnetic field coils and pole shoes. This eliminates

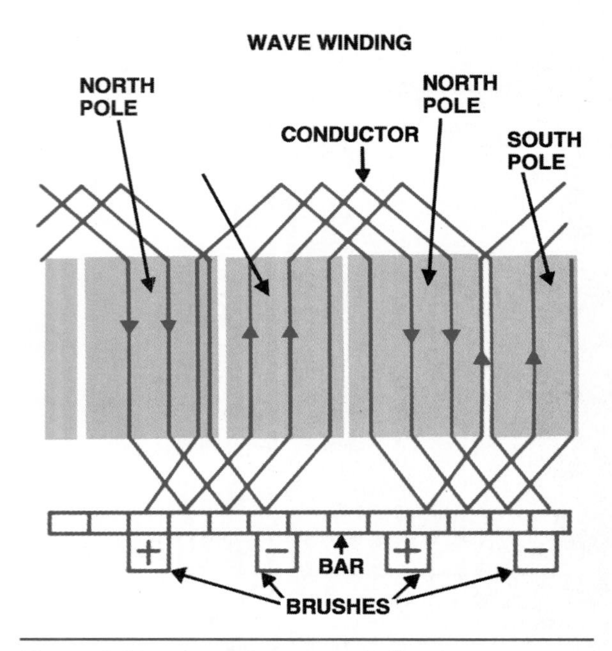

Figure 9-36. Armature wave winding.

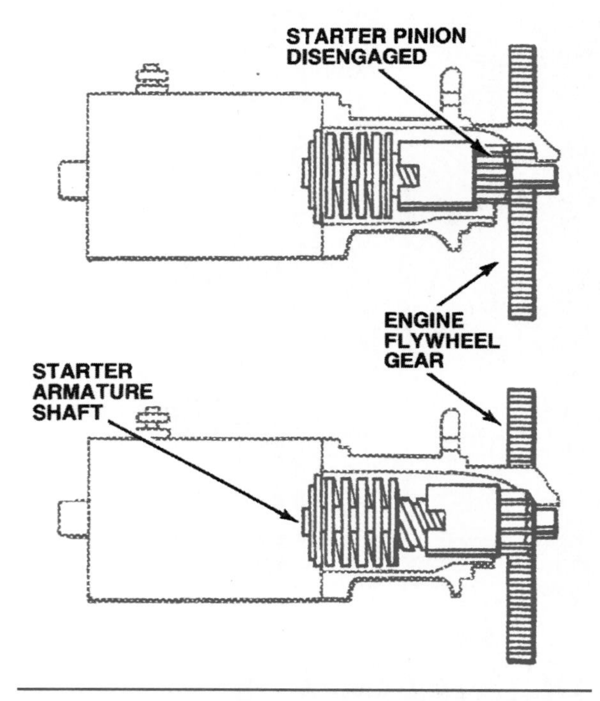

Figure 9-37. The pinion gear meshes with the flywheel ring gear.

the motor field circuit, which in turn eliminates the potential for field wire-to-frame shorts, field coil welding, and other electrical problems. The motor has only an armature circuit. Because the smaller armature in permanent magnet starters uses reinforcement bands, it has a longer life than the armature in wound-field starter motors.

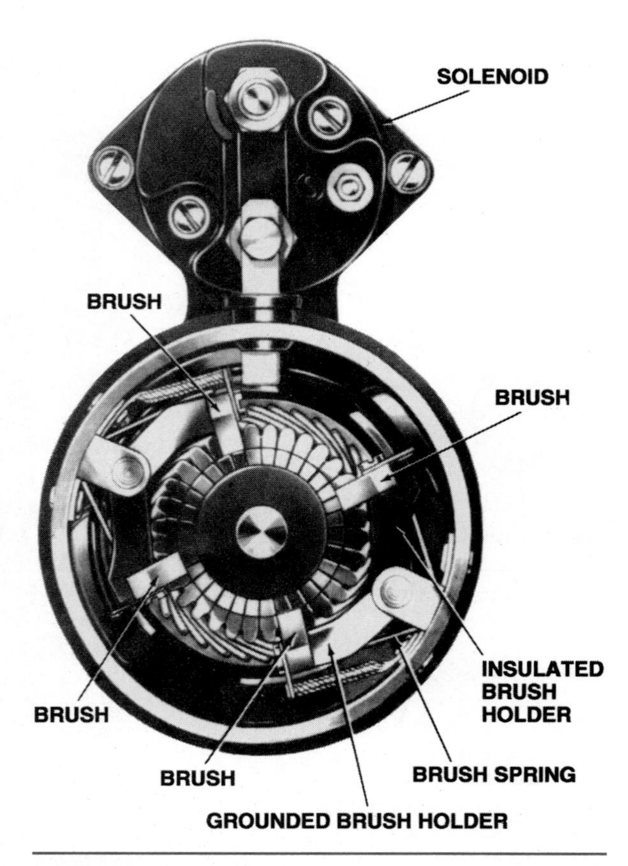

Figure 9-38. A four-brush motor. (Delphi Automotive Systems)

The magnetic field of the starter motor is provided by four or six small permanent magnets. These magnets are made from an alloy of iron and rare-earth materials that produces a magnetic field strong enough to operate the motor without relying on traditional current-carrying field coil windings around iron pole pieces. Removing the field circuit not only minimizes potential electrical problems, the use of permanent-magnet fields allows engineers to design a gear-reduction motor half the size and weight of a conventional wound-field motor without compromising cranking performance.

See Chapter 9 of the *Shop Manual* for service and testing.

STARTER MOTOR AND DRIVE TYPES

Starter motors, as shown in Figure 9-39 are direct-current (DC) motors that use a great amount of current for a short time. The starter motor circuit is a simple one containing just the

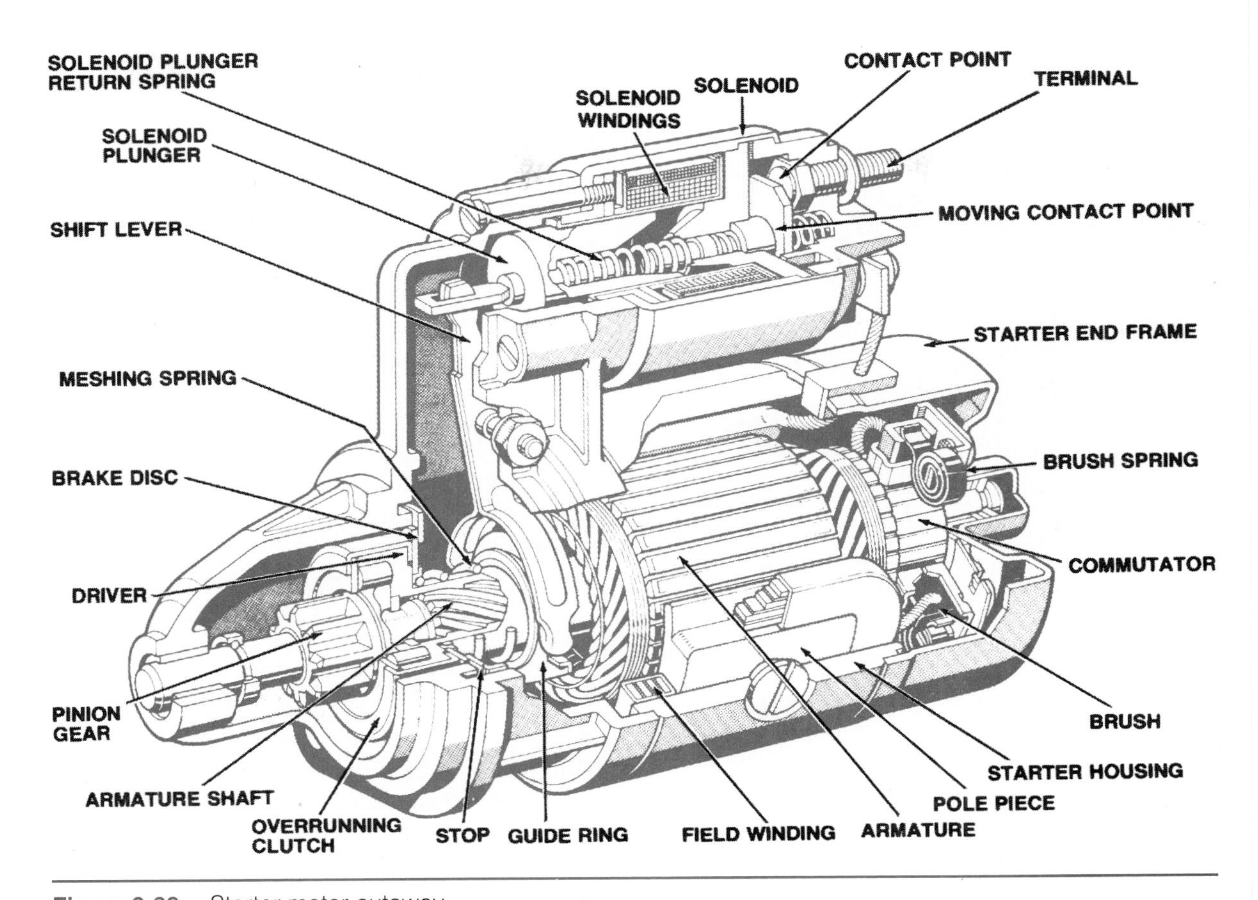

Figure 9-39. Starter motor cutaway.

starter motor and a solenoid or relay. This circuit is a direct path for delivering the momentary high current required by the starter motor from the battery.

The starter motor cranks the engine through a pinion gear that engages a ring gear on the engine flywheel. The pinion gear is driven directly off the starter armature (Figure 9-39) or through a set of reduction gears (Figure 9-40) that provides greater starting torque, although at a lower rpm.

For the starter motor to be able to turn the engine quickly enough, the number of teeth on the flywheel ring gear, relative to the number of teeth on the motor pinion gear, must be between 15 and 20 to 1 (Figure 9-41).

When the engine starts and runs, its speed increases. If the starter motor were permanently engaged to the engine, the motor would be spun at a very high speed. This would throw armature windings off the core. Thus, the motor must be disengaged from the engine as soon as the engine turns more rapidly than the starter motor has cranked it. This job is done by the **starter drive.**

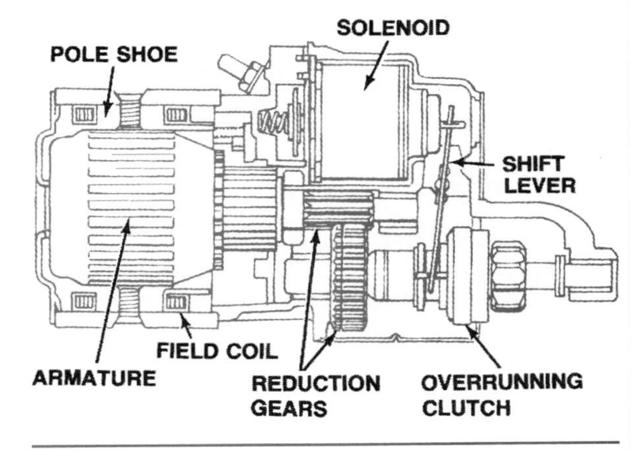

Figure 9-40. The Chrysler reduction-gear starter motor. (DaimlerChrysler Corporation)

Four general kinds of starter motors are used in late-model automobiles:

- Solenoid-actuated, direct drive
- Solenoid-actuated, reduction drive
- Movable-pole shoe
- Permanent-magnet, planetary drive

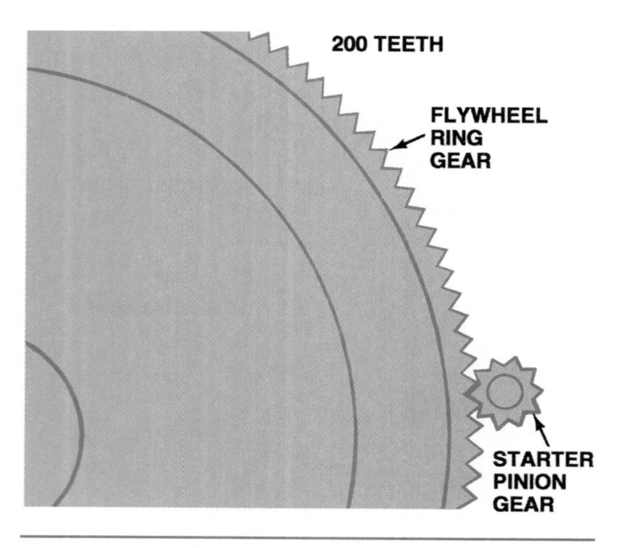

Figure 9-41. The ring-gear-to-pinion-gear ratio is about 20 to 1.

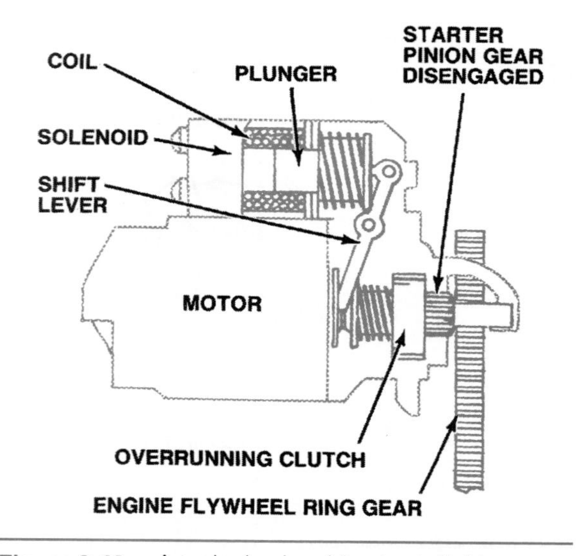

Figure 9-42. A typical solenoid-actuated drive.

Solenoid-Actuated, Direct Drive

The main parts of a solenoid-actuated, direct-drive starter (Figure 9-42), are the solenoid, the shift lever, the overrunning clutch, and the starter pinion gear. The solenoid used to actuate a starter drive has two coils: the pull-in winding and the hold-in, or holding, winding (Figure 9-43). The pull-in winding consists of few turns of a heavy wire. The winding is grounded through the motor armature and grounded brushes. The hold-in winding consists of many turns of a fine wire and is grounded through the solenoid case.

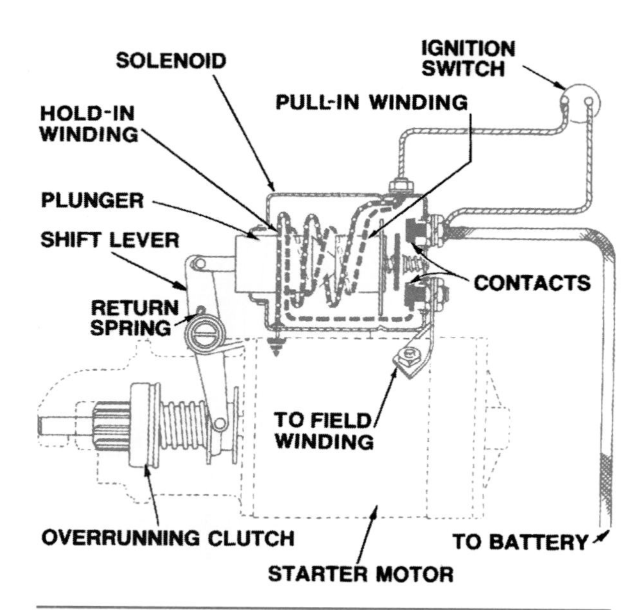

Figure 9-43. The solenoid has a heavy-gauge pull-in winding and a lighter-gauge hold-in winding. (Delphi Automotive Systems)

When the ignition switch is turned to the start position, current flows through both windings. The solenoid plunger is pulled in, and the contacts are closed. This applies battery voltage to both ends of the pull-in winding, and current through it stops. The magnetic field of the hold-in winding is enough to keep the plunger in place. This circuitry reduces the solenoid current draw during cranking, when both the starter motor and the ignition system are drawing current from the battery.

The solenoid plunger action, transferred through the shift lever, pushes the pinion gear into mesh with the flywheel ring gear (Figure 9-44). When the starter motor receives current, its armature begins to turn. This motion is transferred through the overrunning clutch and pinion gear to the engine flywheel.

The teeth on the pinion gear may not immediately mesh with the flywheel ring gear. If this happens, a spring behind the pinion compresses so that the solenoid plunger can complete its stroke. When the motor armature begins to turn, the pinion teeth line up with the flywheel, and spring force pushes the pinion to mesh.

The Delco-Remy MT series, as shown in Figure 9-45, is the most common example of this type of starter motor and has been used for decades on almost all GM cars and light trucks. While this motor is manufactured in different sizes

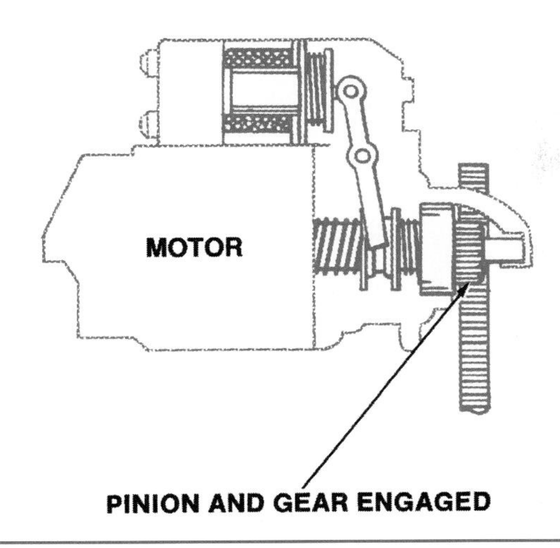

Figure 9-44. The movement of the solenoid plunger meshes the pinion gear and the flywheel ring gear.

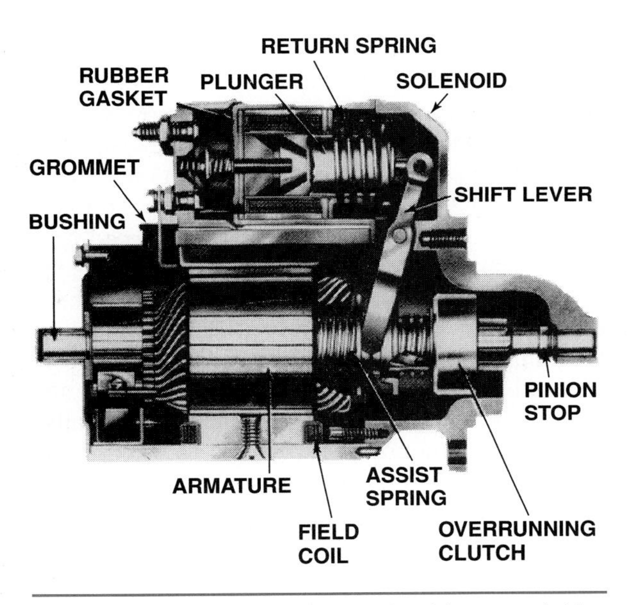

Figure 9-45. The Delco-Remy solenoid-actuated drive motor. (Delphi Automotive Systems)

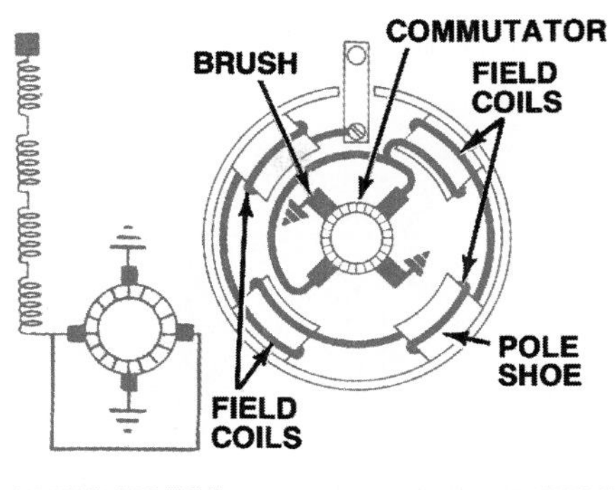

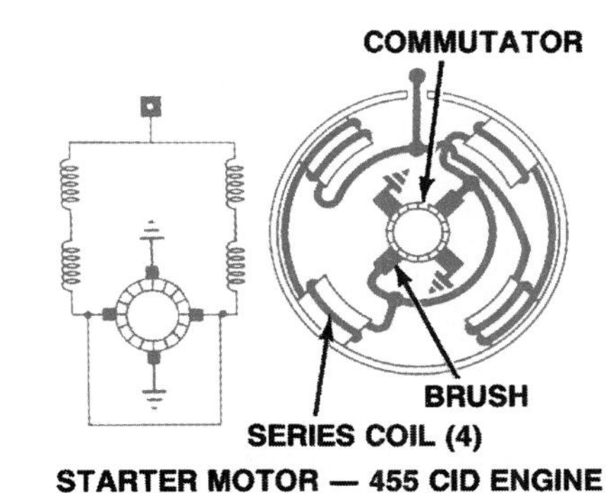

COMMUTATOR
BRUSH
ONE SHUNT COIL
THREE SERIES COILS
SHUNT COIL
POLE SHOE
SERIES COILS
STARTER MOTOR — 260 AND 350 CID ENGINE

Figure 9-46. Delco-Remy provides differently connected starter motors for use with various engines. (GM Service and Parts Operations)

for different engines (Figure 9-46), the most common application is a four-pole, four-brush design.

The solenoid plunger action, in addition to engaging the pinion gear, closes contact points to complete the starter circuit. To avoid closing the contacts before the pinion gear is fully engaged, the solenoid plunger is in two pieces (Figure 9-47). When the solenoid windings are magnetized, the first plunger moves the shift lever. When the pinion gear reaches the flywheel, the first plunger has moved far enough to touch the second plunger. The first plunger continues to move into the solenoid, pushing the second plunger against the contact points.

A similar starter design has been used by Ford on diesel engines and older large-displacement V8 gasoline engines. It operates in the same way as the starter just described. The solenoid action closes a set of contact points.

Because Ford installs a remotely mounted magnetic switch in all of its starting circuits, the solenoid contact points are not required to control the circuit. The solenoid contact points are physically linked, so that they are always "closed."

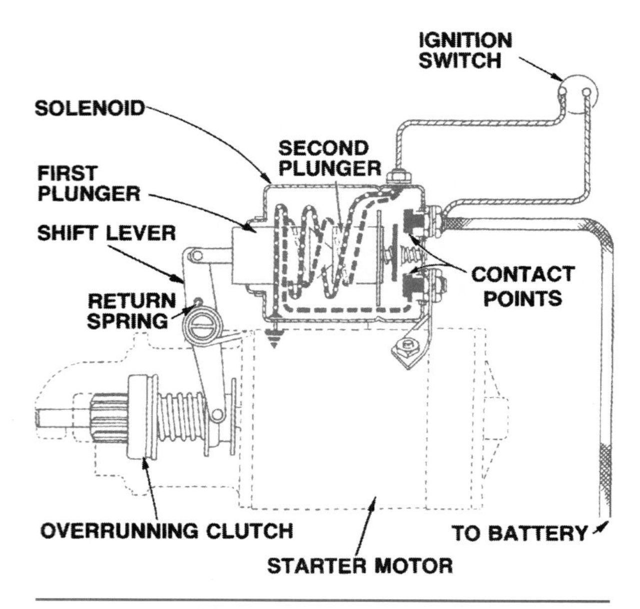

Figure 9-47. The Delco-Remy solenoid plunger is in two pieces. (Delphi Automotive Systems)

In the early 1970s, Chrysler also manufactured fully enclosed direct-drive starter motor. It works in the same way as the solenoid-actuated starters previously described. The solenoid plunger closes contact points to complete the motor circuitry, but the system also has a remotely mounted starter relay. Reduction-drive starters are usually compound motors. Most Bosch and all Japanese starter motors operate on the same principles.

Solenoid-Actuated, Reduction Drive

The Chrysler solenoid-actuated, reduction-drive starter uses a solenoid to engage the pinion with the flywheel and close the motor circuit. The motor armature does not drive the pinion directly, however; it drives a small gear that is permanently meshed with a larger gear. The armature-gear-to-reduction-gear ratio is between 2 and 3.5 to 1, depending upon the engine application. This allows a small, high-speed motor to deliver increased torque at a satisfactory cranking rpm. Solenoid and starter drive operation is basically the same as a solenoid-actuated, direct-drive starter.

Movable-Pole-Shoe Drive

Manufactured by the Motorcraft Division of Ford, the movable-pole-shoe starter motor is used on

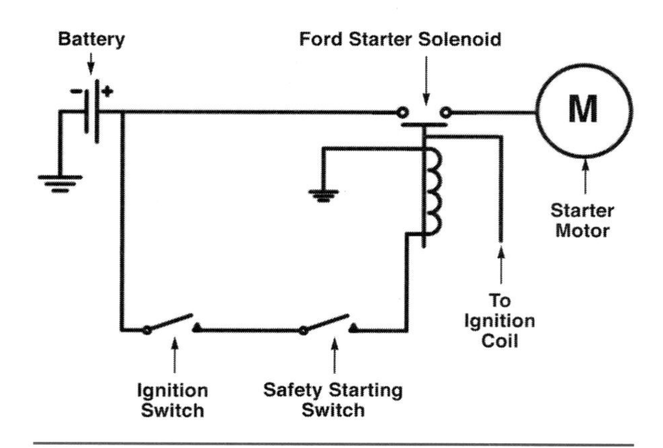

Figure 9-48. Circuit diagram of a system using a movable-pole-shoe starter.

most Ford automobiles (Figure 9-48). One of the motor-pole shoes pivots at the drive end housing. The field winding of this shoe also contains a holding coil, wired in parallel and independently grounded. When the starter relay is closed, battery current flows through the field windings and the holding coil of the pole shoe to ground. This creates a strong magnetic field, and the pole shoe is pulled down into operating position. The motion is transferred through a shift lever, or drive yoke, to mesh the pinion gear with the ring gear.

When the pole shoe is in position, it opens a set of contacts. These contacts break the ground connection of the field windings. Battery current is allowed to flow through the motor's internal circuitry, and the engine is cranked. During cranking, a small amount of current flows through the holding coil directly to ground to keep the shoe and lever assembly engaged.

An overrunning clutch prevents the starter motor from being turned by the engine. When the ignition switch moves out of the start position, current no longer flows through the windings of the movable pole shoe or the rest of the motor. Spring force pulls the shoe up, and the shift lever disengages the pinion from the flywheel.

Permanent-Magnet Planetary Drive

The high-speed, low-torque permanent-magnet planetary-drive motor operates the drive mechanism through gear reduction provided by a simple planetary gearset. Figure 9-49 shows the Bosch gear reduction design, which is similar to that used in Chrysler starters. Figure 9-50 shows the

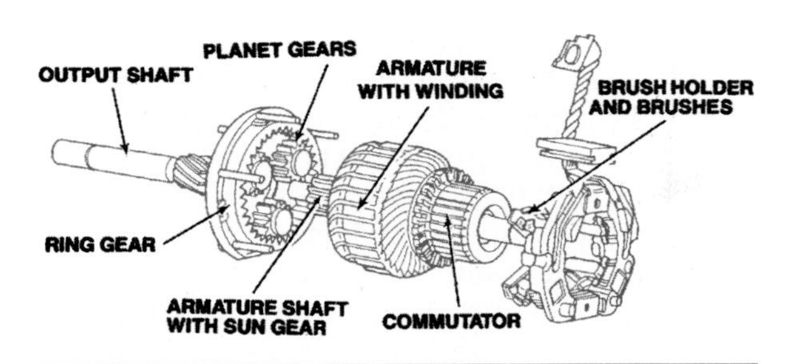

Figure 9-49. Bosch permanent-magnet gear-reduction starter components. (Reprinted by permission of Robert Bosch GmbH)

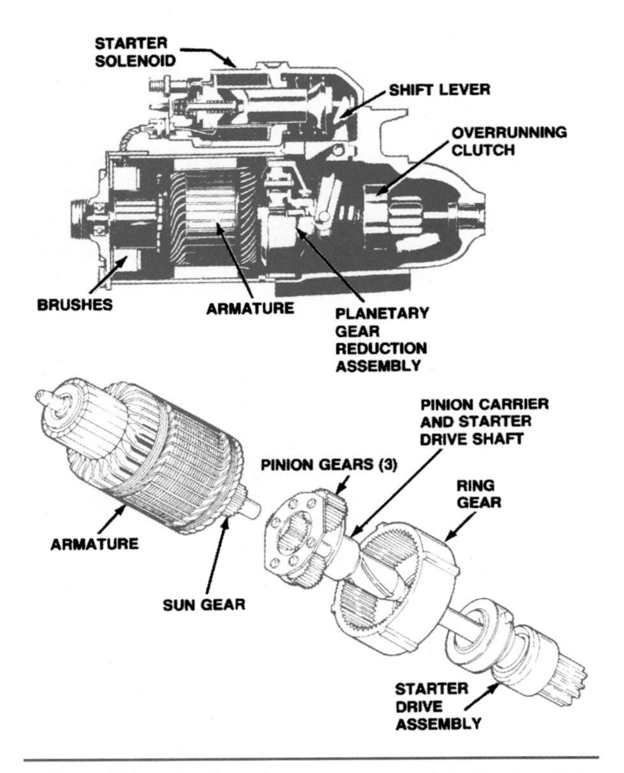

Figure 9-50. Delco-Remy permanent-magnet, gear-reduction starter components. (Delphi Automotive Systems)

gear reduction design used in the Delco-Remy permanent-magnet, gear-reduction (PMGR) starter. All PMGR starter designs use a solenoid to operate the starter drive and close the motor armature circuit. The drive mechanism is identical to that used on other solenoid-actuated starters already described. Some models, however, use lightweight plastic shift levers.

The planetary gearset between the motor armature and the starter drive reduces the speed and increases the torque at the drive pinion. The compact gearset is only 1/2 to 3/4 inch (13 to 19 mm) deep and is mounted inline with the armature and

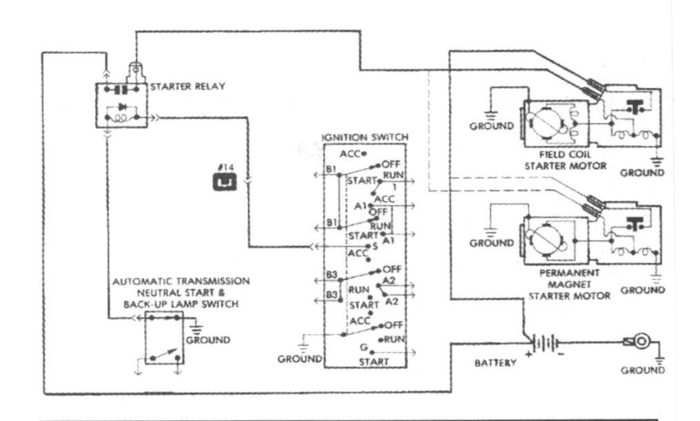

Figure 9-51. Field coils and permanent-magnet starters use the same electrical wiring.

drive pinion. An internal ring gear is keyed to the field frame and held stationary in the motor. The armature shaft drives the sun gear for the planetary gearset. The sun gear meshes with three planetary pinions, which drive the pinion carrier in reduction as they rotate around the ring gear. The starter driveshaft is mounted on the carrier and driven at reduced speed and increased torque. This application of internal gear reduction through planetary gears delivers armature speeds in the 7,000-rpm range. The armature and driveshaft ride on roller or ball bearings rather than bushings.

Permanent-magnet, planetary-drive starters differ mechanically in how they do their job, but their electrical wiring is the same as that used in the field-coil designs (Figure 9-51).

Although PMGR motors are lighter in weight and simpler to service than traditional designs, they do require special handling precautions. The material used for the permanent magnet fields is quite brittle. A sharp impact caused by hitting or dropping the starter can destroy the fields.

OVERRUNNING CLUTCH

Regardless of the type of starter motor used, when the engine starts and runs, its speed increases. The motor must be disengaged from the engine as soon as the engine is turning more rapidly than the starter motor that has cranked it. With a movable-pole-shoe or solenoid-actuated drive, however, the pinion remains engaged until power stops flowing to the starter. In these applications, the starter is protected by an over-running clutch (Figure 9-52).

The **overrunning clutch** consists of rollers that ride between a collar on the pinion gear and an outer shell. The outer shell has tapered slots for the rollers so that the rollers either ride freely or wedge tightly between the collar and the shell. Figure 9-53 shows the operation of an overrunning clutch. In Figure 9-53A, the armature is turning, cranking the engine. The rollers are wedged against spring force into their slots. In Figure 9-53B, the engine has started and is turning faster than the motor armature. Spring force pushes the rollers so that they float freely. The engine's motion is not transferred to the motor armature. These devices are sometimes called one-way clutches because they transmit motion in one direction only.

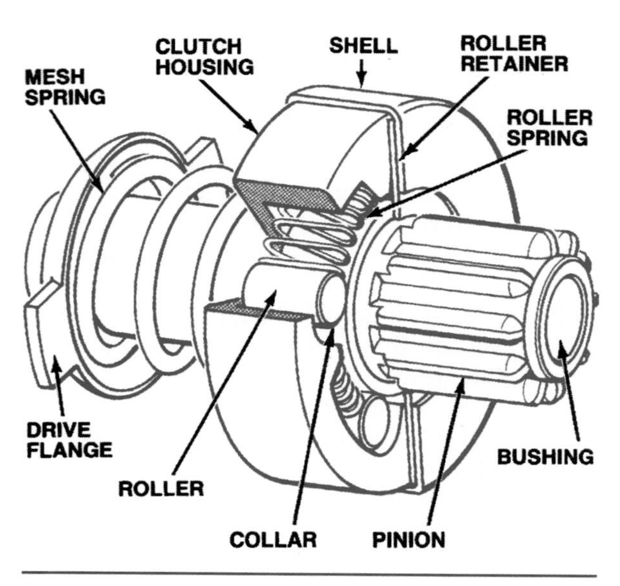

Figure 9-52. Cutaway view of an overrunning clutch.

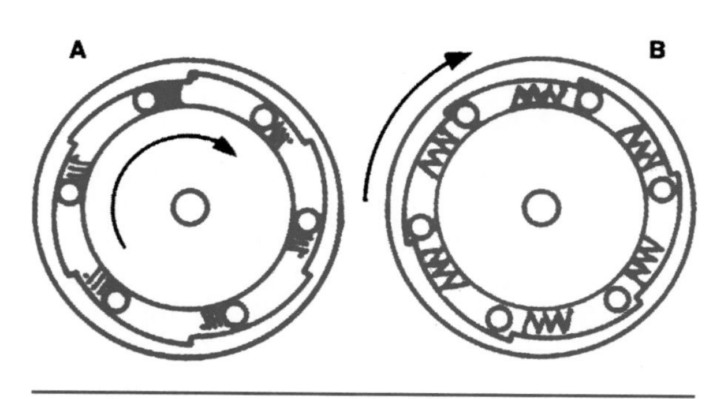

Figure 9-53. The operation of an overrunning clutch.

Once the engine starts, the ignition switch is to be released from the start position. The solenoid hold-in winding is demagnetized, and a return spring moves the plunger out of the solenoid. This moves the shift lever back so that the overrunning clutch and pinion gear slide away from the flywheel. For more information about overrunning clutches, see the following sections of Chapter 9 in the *Shop Manual*, "Bench Tests" and "Starter Motor Overhaul Procedure."

SUMMARY

Electrical starting systems consist of a high-current starter circuit controlled by a low-current control circuit. The ignition switch includes contacts that conduct battery current to the magnetic switch. The magnetic switch may be a relay or a solenoid and may have other jobs besides controlling the starter circuit current flow. The starter motor and connecting wires are also included in the system. Variations are common among the starting systems used by the various manufacturers. Magnetic repulsion occurs when a straight-rod conductor composed of the armature, commutator, and brushes is located in a magnetic field (field windings) and current is flowing through the rod.

When the starter operates, the current passing through the armature produces a magnetic field in each of its conductors. The reaction between the magnetic field of the armature and the magnetic fields produced by the field coils causes the armature to rotate.

Traditional starter motors have pole pieces wound with heavy copper field windings attached to the housing. A new design, the permanent-magnet planetary drive, uses small permanent magnets to create a magnetic field instead of pole pieces and field windings.

One end housing holds the brushes; the other end housing shields the pinion gear. The motor armature windings are installed on a laminated core and mounted on a shaft. The commutator bars are mounted on, but insulated from, the shaft.

The solenoid-actuated drive uses the movement of a solenoid to engage the pinion gear with the ring gear. Delco-Remy, Chrysler, Motorcraft, and many foreign manufacturers use this type of starter drive. The movable-pole-shoe drive, used by Ford, has a pivoting pole piece that is moved by electromagnetism to engage the pinion gear with the ring gear. In the planetary-gear drive used by Chrysler, Ford, and GM, an armature-shaft sun gear meshes with the planetary pinions, which drive the pinion carrier in reduction as they rotate around the ring gear. The starter driveshaft is mounted on the carrier and driven at reduced speed and increased torque. An overrunning clutch is used with all starter designs to prevent the engine from spinning the motor and damaging it.

Review Questions

1. All of these are part of the control circuit *except*:
   a. A starting switch
   b. An OCP thermostat
   c. A starting safety switch
   d. A magnetic switch
2. Which of the following is a component of a starting circuit?
   a. Magnetic switch
   b. Ballast resistor
   c. Voltage regulator
   d. Powertrain control module (PCM)
3. Two technicians are discussing the operation of a DC automotive starter. Technician A says the principle of magnetic repulsion causes the motor to turn. Technician B says the starter uses a mechanical connection to the engine that turns the armature. Who is right?
   a. A only
   b. B only
   c. Both A and B
   d. Neither A nor B
4. All of these are part of a starter motor *except*:
   a. An armature
   b. A commutator
   c. Field coils
   d. A regulator
5. The starting system has ________ circuits to avoid excessive voltage drop.
   a. Two
   b. Three
   c. Four
   d. Six
6. The starter circuit consists of which of the following?
   a. Battery, ignition switch, starter motor, large cables
   b. Battery, ignition switch, relays or solenoids, large cables
   c. Battery, magnetic switch, starter motor, primary wiring
   d. Battery, magnetic switch, starter motor, large cables
7. Which of the following is *not* part of the starter control circuit?
   a. The ignition switch
   b. The starting safety switch
   c. The starter relay
   d. The starter motor
8. The ignition switch will *not* remain in which of the following positions?
   a. ACCESSORIES
   b. OFF
   c. ON (RUN)
   d. START
9. The starting safety switch is also called a:
   a. Remote-operated switch
   b. Manual-override switch
   c. Neutral-start switch
   d. Single-pole, double-throw switch
10. Safety switches are most commonly used with:
    a. Automatic transmissions
    b. Imported automobiles
    c. Domestic automobiles
    d. Manual transmissions
11. Starting safety switches used with manual transmissions are usually:
    a. Electrical
    b. Mechanical
    c. Floor-mounted
    d. Column-mounted
12. Which of the following is not true of solenoids?
    a. They use the electromagnetic field of a coil to pull a plunger into the coil.
    b. They are generally used to engage the starter motor with the engine flywheel.
    c. They operate with a movable plunger and usually do a mechanical job.
    d. They send electronic signals to the control module and have no moving parts.
13. Starter motors usually have how many pole shoes?
    a. Two
    b. Four
    c. Six
    d. Eight
14. The rotational force of a starter motor is:
    a. Polarized
    b. Rectified

c. Torque
d. Current

15. Which of the following is true of a shunt motor?
 a. It has high initial torque.
 b. It operates at variable speed.
 c. It has only one path for current flow.
 d. It is not often used as a starting motor.

16. Which of the following is true of a compound motor?
 a. It has low initial torque.
 b. It operates at variable speeds.
 c. It has only one path for current flow.
 d. It is often used as a starting motor.

17. In a lap-wound motor armature, the two ends of each conductor are attached to commutator segments that are:
 a. Adjacent
 b. 45 degrees apart
 c. 90 degrees apart
 d. 180 degrees apart

18. Most automotive starters have ________ grounded and ________ insulated brushes.
 a. 2, 2
 b. 2, 4
 c. 4, 4
 d. 4, 8

19. The ratio between the number of teeth on the flywheel and the motor pinion gear is about:
 a. 1:1
 b. 5:1
 c. 20:1
 d. 50:1

20. The overrunning clutch accomplishes which of the following?
 a. Separates the starter motor from the starter solenoid
 b. Brings the starter motor into contact with the ignition circuit
 c. Lets the starter motor rotate in either direction
 d. Protects the starter motor from spinning too rapidly

21. A starting motor must have ___________ brushes as poles.
 a. The same number of
 b. Twice as many
 c. One-half as many
 d. Three times as many

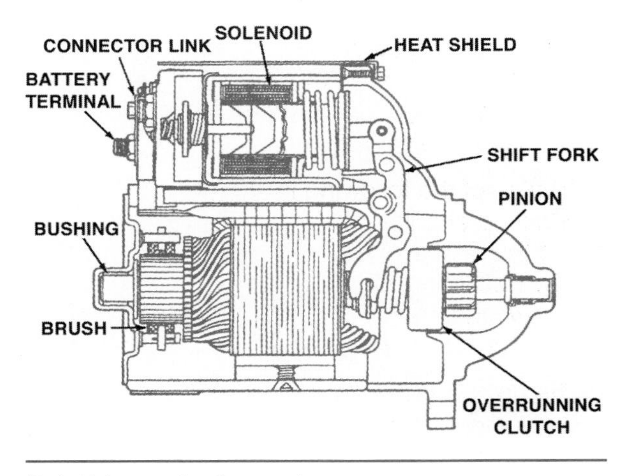

(Delphi Automotive Systems)

22. The preceding illustration shows a:
 a. Permanent-magnet planetary-gear starter
 b. Movable-pole-shoe starter
 c. Direct-drive, solenoid-actuated starter
 d. Reduction-gear drive, solenoid-actuated starter

23. Which type of starter drive is *not* used on late-model cars?
 a. Direct drive
 b. Bendix drive
 c. Reduction drive
 d. Planetary drive

24. A solenoid uses two coils. Their windings are called:
 a. Push-in and pull-out
 b. Pull-in and push-out
 c. Push-in and hold-out
 d. Pull-in and hold-in

25. Which of the following is true of a reduction drive?
 a. The motor armature drives the pinion directly.
 b. The sun gear is mounted on the armature shaft.
 c. The overrunning clutch reduces battery current.
 d. The small gear driven by the armature is permanently meshed with a larger gear.

26. The planetary drive starter uses:
    a. Permanent magnets
    b. Field coils
    c. Both A and B
    d. Neither A nor B

27. Which of the following is *not* required of a permanent magnet starter?
    a. Brush testing
    b. Commutator testing
    c. Field circuit testing
    d. Armature testing

10

Automotive Electronics

LEARNING OBJECTIVES

Upon completion and review of this chapter, you should be able to:

- Identify the basic types and construction of solid-state devices used in automotive electronic circuits.
- Identify an ESD symbol and explain its use.
- Explain the use and function of diodes in an automotive circuit.
- Explain forward-biased diodes and reverse-biased diodes.
- Explain the use and function of transistors and the different types used in automotive circuits.
- Explain the operation of a silicon-controlled rectifier (SCR).
- Identify and explain the operation of photonic semiconductors.
- Identify and explain the term integrated circuit and how one uses electronic signals.

KEY TERMS

Anode
Cathode
Diode
Doping
Electrostatic Discharge (ESD)
Field-Effect Transistor (FET)
Forward Bias
Integrated Circuit (ICs)
Light-Emitting Diode (LED)
N-Type Material
P-Type Material
PN Junction
Photonic Semiconductors
Rectifier
Reverse Bias
Semiconductors
Silicon-Controlled Rectifier (SCR)
Transistor
Zener Diode

INTRODUCTION

This chapter will review the basic semiconductors and the electronic principles required to understand how electronic and computer-controlled systems manage the various systems in today's vehicles. A technician must have a thorough grasp of the basis of electronics, it has become the single most important subject area, and the days when technicians could avoid working on an electronic circuit throughout an entire career are long past. This course of study will begin with semiconductor components.

Chapter 3 examined the atom's valence ring, the outermost electron shell. We learned that elements whose atoms have three or fewer electrons in their valence rings are good conductors because the free electrons in the valence ring readily join with the valence electrons of other, similar atoms. We also learned that elements whose atoms have five or more electrons in their valence rings are good insulators (poor conductors) because the valence electrons do not readily join with those of other atoms.

SEMICONDUCTORS

Elements whose atoms have four electrons in their valence rings are neither good insulators nor good conductors. Their four valence electrons cause special electrical properties, which give them the name *semiconductors*. Germanium and silicon are two widely used semiconductor elements. **Semiconductors** are materials with exactly four electrons in their outer shell, so they cannot be classified as insulators or conductors. If an element falls into this group but can be changed into a useful conductor, it is a semiconductor.

When semiconductor elements are in the form of a crystal, they bond together so that each atom has eight electrons in its valence ring: It has its own four electrons and shares four with surrounding atoms (Figure 10-1). In this form it is an excellent insulator, because there are no free electrons to carry current flow. Selenium, copper oxide, and gallium arsenide are all semiconductors, but silicon and germanium are the most commonly used. Pure semiconductors have tight electron bonding; there's no place for electrons to move. In this natural state, the elements aren't useful for conducting electricity.

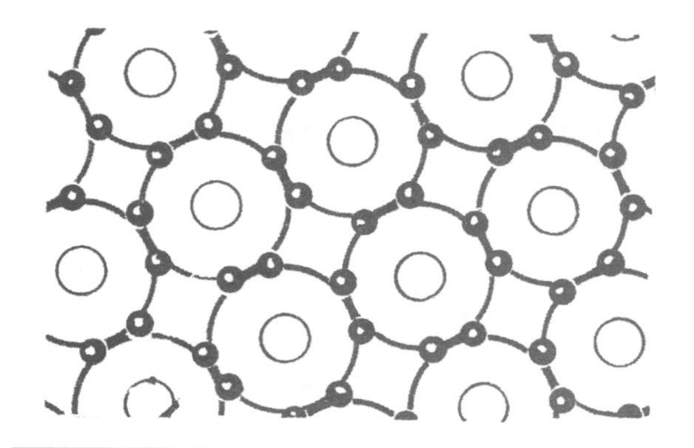

Figure 10-1. Crystalline silicon is an excellent insulator. (Delphi Automotive Systems)

Other elements can be added to silicon and germanium to change this crystalline structure. This is called **doping** the semiconductor. The ratio of doping elements to silicon or germanium is about 1 to 10,000,000. The doping elements are often called impurities because their addition to the silicon or germanium makes the semiconductor materials impure.

Semiconductor Doping

Silicon has four electrons in its outer shell. Silicon can be grown into large crystals by applying heat to melt the silicon followed by a period of cooling. Pure silicon is not very useful in electronics components. For silicon to be useful it must be doped, that is, to have small quantities of impurities added to it. The impurities affect how many free electrons the semiconductor has. Depending on which impurity is added, the resulting material will have either an excess of free electrons or a shortage of free electrons. The doping agents are usually phosphorus and boron. The doping intensity will define the electrical behavior of the crystal. After doping, you can slice the silicon crystals into thin sections known as wafers.

The type of doping agent used to produce silicon crystals will define the electrical properties of the crystals produced. A boron atom has three electrons in its outer shell. The outer shell is known as the valence ring and an atom with three electrons in its outer shell is known as *trivalent*. A boron atom in a crystallized cluster of silicon atoms will produce an outer shell with seven electrons instead of eight: this "vacant" electron opening is known as a *hole*. The hole makes it possible for an electron from a nearby atom to fall into the hole. In other words, the holes can move, permitting a flow of electrons. Silicon crystals doped with boron (or other elemental atoms with three electrons in the outer shell, that is, trivalent) form *P-type silicon*.

A phosphorus atom has five electrons in its outer shell. It is *pentavalent*. In the bonding between the semiconductors and the doping material, there is room for only eight electrons in the center shell. Even when the material is in an electrically neutral state, the extra electron can move through the crystal. When a silicon crystal is manufactured using a doping material with five electrons in the outer shell (pentavalent), it forms *N-type silicon* or semiconductor material. Like silicon, germanium has four electrons in its outer shell: Figure 10-2 shows a germanium atom in a crystallized cluster with shared electrons.

P- and N-Type Material

If there is an excess of free electrons, the semiconductor is **N-type material,** where *N* stands for *negative*. If there is a shortage of free electrons, the semiconductor is **P-type material,** where *P* stands for *positive*. Figure 10-3 shows P- and N-type materials together. In a conductor, we describe current flow as the movement of electrons, in a semiconductor, something else is going on. Just as in a conductor, there is a movement of free electrons, but at the same time, there is a movement of "holes."

A pure crystal of silicon or germanium is of no use by itself. To provide a semiconductor crystal with useful characteristics, it must be doped with an impurity. If silicon or germanium is doped with an element such as phosphorus, arsenic, or antimony, each of which has five electrons in its valence ring, there will not be enough space for the ninth electron in any of the shared valence rings. This extra electron is free (Figure 10-4). This type of doped material is called negative or *N-type* material because it already has excess electrons and will repel additional negative charges. The doping of a semiconductor crystal will always define its electrical characteristics.

Germanium is used once again to show how the element forms a P-type semiconductor. If silicon or germanium is doped with an element such as boron or indium, each of which has only three electrons in its outer shell, some of the atoms will have only seven electrons in their valence rings. There will be a hole in these valence rings (Figure 10-5). This type of doped material is called positive or *P-type* material, because it will attract a negative charge

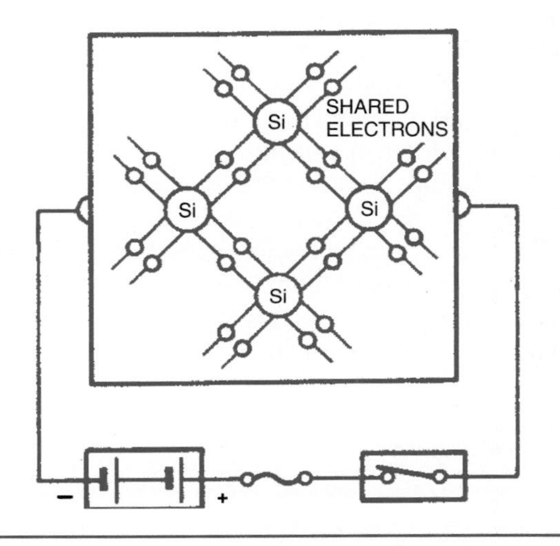

Figure 10-2. Germanium atom in a crystallized cluster with shared electrons.

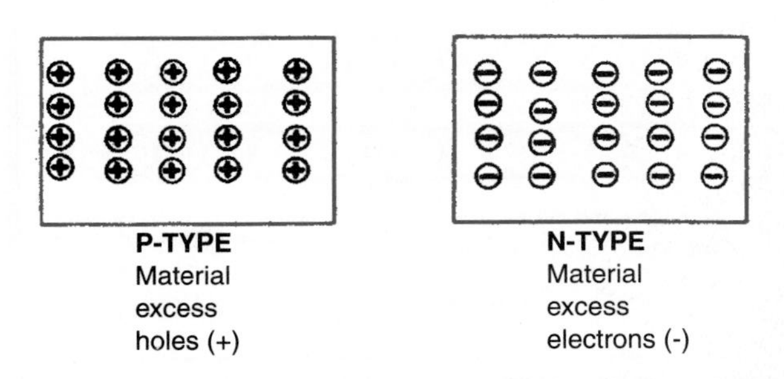

Figure 10-3. P- and N-type material.

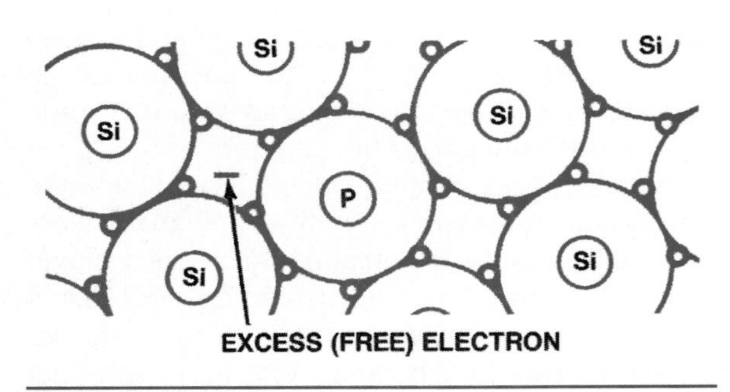

Figure 10-4. N-type material has an extra, or free, electron. (Delphi Automotive Systems)

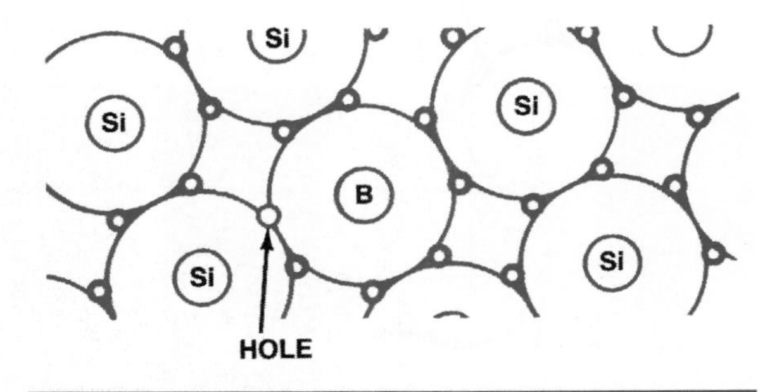

Figure 10-5. P-type material has a "hole" in some of its valence rings. (Delphi Automotive Systems)

(an electron). In different ways, P-type and N-type silicon crystals may permit an electrical current flow: In the P-type semiconductor, current flow is occasioned by a deficit of electrons, while in the N-type semiconductor, current flow is occasioned by an excess of electrons.

PN-Junction

When the two semiconductor materials are brought together, as shown in Figure 10-6, something happens at the area where the two touch. This area is called the **PN junction.** You chemically join P- and N-type materials either by growing them together or fusing them with some type of heat process. Either way they join together as a single crystal structure. The excess electrons in the N-type material are attracted to the holes in the P-type material. Some electrons drift across the junction to combine with holes. As an electron combines with a hole, both effectively disappear. Whenever a voltage is applied to a semiconductor, electrons will flow towards the positive terminal and the holes will move towards the negative terminal. The electron is no longer free and the hole no longer exists. Because of the cancellation of charges, the material at the junction assumes a positive charge in the N-type material and a negative charge in the P-type material; PN-junctions become diodes.

As long as no external voltage is applied to the semiconductors, there is a limit to how many electrons will cross the PN junction. Each electron that crosses the junction leaves behind an atom that is missing a negative charge. Such an atom is called a *positive ion*. In the same way, each hole that crosses the junction leaves behind a *negative ion*. As positive ions accumulate in the N-type material, they exert a force (a potential) that prevents any more electrons from leaving. As negative ions accumulate in the P-type material, they exert a potential that keeps any more holes from leaving.

Eventually, this results in a stable condition that leaves a deficiency of both holes and electrons at the PN junction. This zone is called the *depletion region*. The potentials exerted by the negative and positive ions on opposite sides of the depletion region are two unlike charges. Combining unlike charges creates voltage potential. The voltage potential across the PN junction is called the *barrier voltage*. Doped germanium has a barrier voltage of about 0.3 volt. Doped silicon has a barrier voltage of about 0.7 volt.

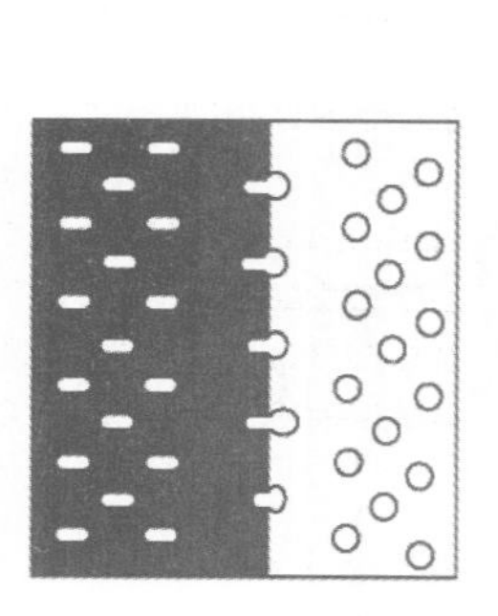

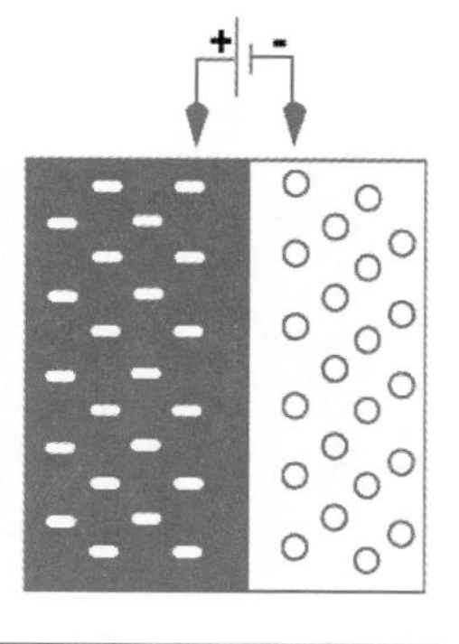

Figure 10-6. PN junction.

Free Electrons and Movement of Holes

In P-type semiconductor material, there is a predominance of *acceptor atoms* and holes. In N-type material, there is a predominance of *donor atoms* and free electrons. When the two semiconductors are kept separate, holes and electrons are distributed randomly throughout the respective materials. A hole is a location where an electron normally resides but is currently absent. A hole is not a particle, but it behaves like one (Figure 10-7).

Because a hole is the absence of an electron, it represents a missing negative charge. As a result, a hole acts like a positively charged particle. Electrons and holes occur in both types of semi-

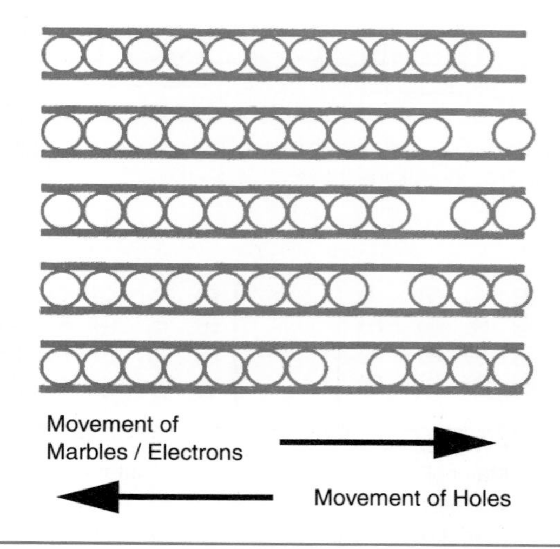

Figure 10-7. Movement of holes. (GM Service and Parts Operations)

conductor material. In P-type material, we describe current flow as holes moving. In N-type material, we describe current flow as electrons moving. Holes cannot actually carry current. When we talk about current flow in terms of holes moving in one direction, we're actually describing the movement of the absent electrons moving in the opposite direction.

ELECTROSTATIC DISCHARGE (ESD)

An electrostatic charge can build up on the surface of your body. If you touch something your charge can be discharged to the other surface. This is called **electrostatic discharge (ESD).** Figure 10-8 shows the ESD symbol indicating that the component is solid-state. Some service manuals use the words *solid-state* instead of the ESD symbol. Look for these indicators and take the suggested ESD precautions when you work on sensitive components.

DIODES

Now that you've learned what a semiconductor is, let's look at some basic examples. The simplest kind of semiconductor is a **diode** (Figure 10-9). It's made of one layer of P-type material and one of N-type material. The diode is the simplest semiconductor device that allows current to flow in only one direction. A diode can function as a switch, acting as either conductor or insulator, depending on the direction of current flow. Diodes turn on when the polarity of the current is correct and turn off when the flow has the wrong polarity. The suffix *-ode* literally means *terminal*. For instance, it is used as the suffix for *cathode* and *anode*. The word *diode* means literally, having two terminals.

Previous sections discussed how both P-type and N-type semiconductor crystals can conduct electricity. Either the proportion of holes or surplus of electrons determines the actual resistance of each type. When a chip is manufactured using both P- and N-type semiconductors, electrons will flow in only one direction. The diode is used in electronic circuitry as a sort of one-way check valve that will conduct electricity in one direction (forward) and block it in the other (reverse).

Figure 10-8. ESD symbol. (GM Service and Parts Operations)

Anode/Cathode

If current flows from left to right in Figure 10-9, it's correct to place a positive plus sign to the left and a negative minus sign to the right of the diode. The positive side of the diode is the **anode** and the negative side is the **cathode.** Associate *anode* with A+ (it's the positive side) and *cathode* with C− (the negative side). The cathode is the end with the stripe. Basically, the following types of diodes are used in automotive applications:

- Regular diodes (used for rectification, or changing AC to DC)
- Clamping diodes (to control voltage spikes and surges that could damage solid-state circuits)
- LEDs (light emitting diodes, used as indicators)
- Zener diodes (voltage regulation)
- Photodiodes

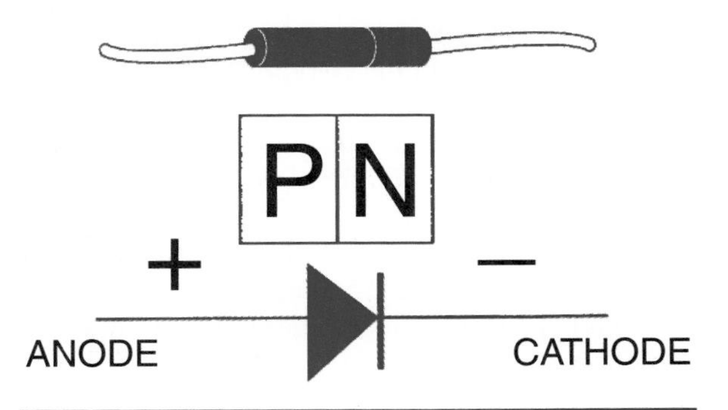

Figure 10-9. Diode.

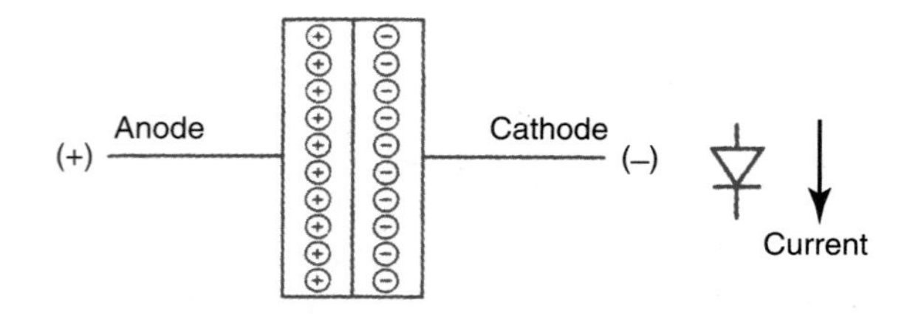

Figure 10-10. Diode triangle.

Diodes allow current flow in only one direction. In Figure 10-10, the triangle in the diode symbol points in the direction current is permitted to flow using conventional current flow theory. Inside a diode are small positive (P) and negative (N) areas, which are separated by the thin boundary area called the PN Junction. When a diode is placed in a circuit with the positive side of the circuit connected to the positive side of the diode, and the negative side of the circuit connected to the negative side of the diode, the diode is said to have *forward bias*. As an electrical one-way check valve, diodes will permit current flow only when correctly polarized. Diodes are used in AC generators (alternators) to produce a DC characteristic from AC and are also used extensively in electronic circuits.

Forward-Biased Diodes

A diode that's connected to voltage so that current is able to flow has **forward bias.** Bias refers to how the voltage is applied. In Figure 10-11, the negative voltage terminal is connected to the N side of the diode, and the positive voltage terminal is connected to the P side.

If you cover up the P-type material in the diode, you can think of this as any other circuit. Electrons are repelled from the negative voltage terminal through the conductor and towards the diode. The electrons at the end of the conductor repel the electrons in the N-type material. If there weren't a barrier voltage, the movement of electrons would continue through the conductor towards the positive terminal.

Now think about the P-type material. The positive voltage terminal repels the electron holes in the P side of the diode. This means that both electrons and electron holes are being forced into the depletion zone.

Unlike electrical charges are attracted to each other and like charges repel each other. Therefore,

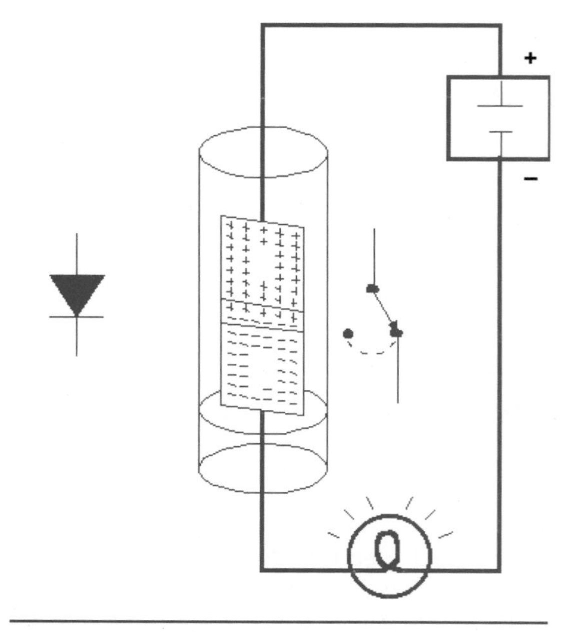

Figure 10-11. Forward-biased diode. (GM Service and Parts Operations)

the positive charge from the circuit's power supply is attracted to the negative side of the circuit. The voltage in the circuit is much stronger than the charges inside the diode and causes the charges inside the diode to move. The diode's P conductive material is repelled by the positive charge of the circuit and is pushed toward the N material, and the N material is pushed toward the P. This causes the PN junction to become a conductor, allowing the circuit's current to flow. Diodes may be destroyed when subjected to voltage or current values that exceed their rated capacity. Excessive reverse current may cause a diode to conduct in the wrong direction and excessive heat can melt the semiconductor material.

Reverse-Biased Diode

A diode that's connected to voltage so that current cannot flow has **reverse bias** (Figure 10-12). This means the negative voltage terminal is connected to the P side of the diode, and the positive voltage terminal is connected to the N side. Let's see what happens when voltage is applied to this circuit. The electrons from the negative voltage terminal combine with the electron holes in the P-type material. The electrons in the N-type material are attracted towards the positive voltage terminal. This enlarges the depletion area. Since the holes

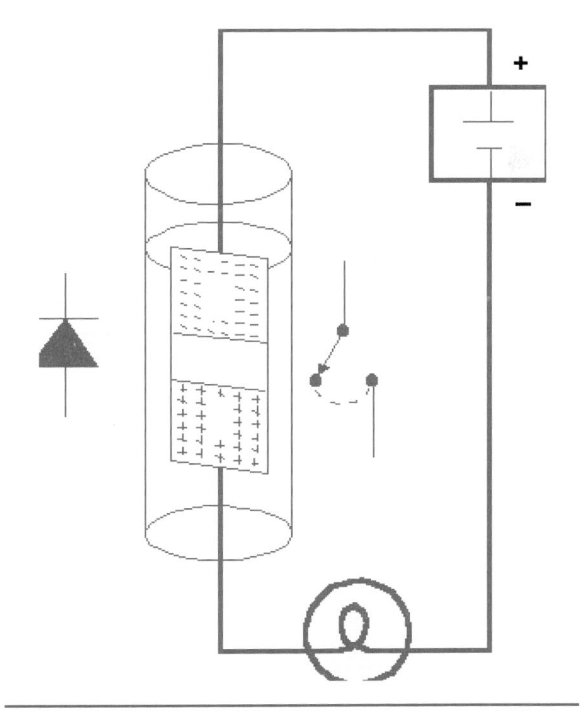

Figure 10-12. Reverse-biased diode. (GM Service and Parts Operations)

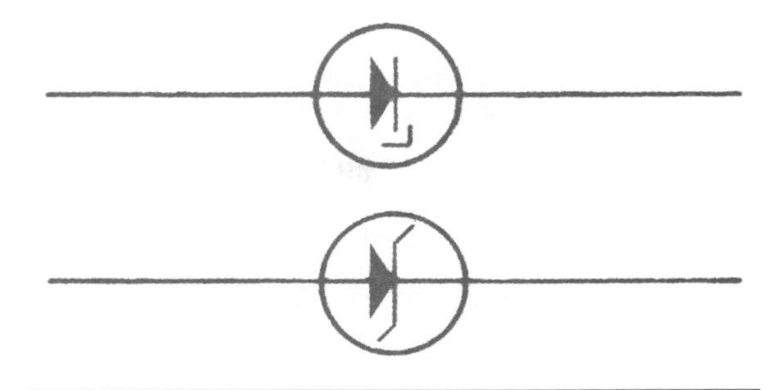

Figure 10-13. Zener diode.

and electrons in the depletion area don't combine, current can't flow.

When reverse bias is applied to the diode, the P and N areas of the diode are connected to opposite charges. Since opposites attract, the P material moves toward the negative part of the circuit whereas the N material moves toward the positive part of the circuit. This empties the PN junction and current flow stops.

Types of Diodes

There are numerous types of diodes and they play a variety of roles in electronic circuits. The following are some of the more common types:

Small-Signal Diodes

Small-signal diodes are used to transform low-current AC to DC (rectification), perform logic data flows, and absorb voltage spikes.

Power Rectifiers

Power rectifiers function in the same manner as small-signal diodes but are designed to permit much greater current flow. They are used in multiples and often mounted on a heat sink to dissipate excess heat caused by high current flow.

Zener Diodes

The Zener diode, invented by Clarence Zener in 1934, functions as a voltage-sensitive switch. **Zener diodes** (Figure 10-13) work the same way as regular diodes when forward biased, but they are placed backwards in a circuit. When the Zener voltage is reached, the Zener diode begins to allow current flow but maintains a voltage drop across itself. The Zener diode is designed to block reverse-bias current but only up to a specific voltage value between 2 and 200 volts. When this reverse breakdown voltage (V_2) is attained, it will conduct the reverse-bias current flow without damage to the semiconductor material. Zener diodes are manufactured from heavily doped semiconductor crystals. They are used in electronic voltage regulators and in other automotive electronic circuitry.

PHOTONIC SEMICONDUCTORS

Photonic semiconductors, like the photodiode in Figure 10-14, emit and detect light or *photons*. Photons are produced when certain electrons excited to a higher-than-normal energy level return to a more normal level. Photons act like waves, and the distance between the wave nodes and anti-nodes (wave crests and valleys) is known as wavelength. Electrons are excited to higher energy levels and photons with shorter wavelengths than electrons are excited to lower levels. Photons are not necessarily visible, and it is important to note that they may

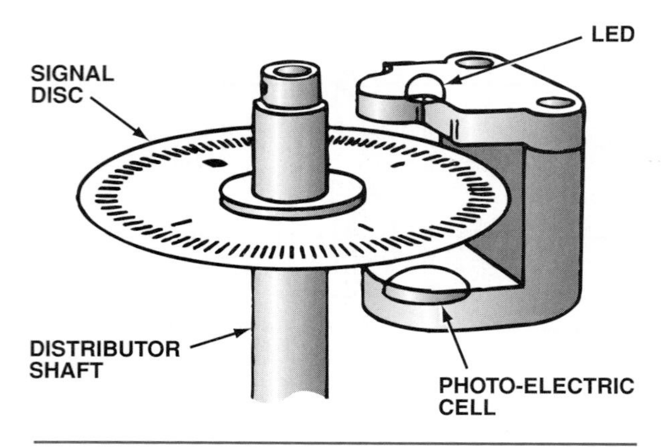

Figure 10-14. Nissan optical signal generator works by interrupting a beam of light passing from the LED to a photodiode. (Courtesy of Nissan North America, Inc.)

only truly be described as *light* when they are visible. Photodiodes are designed specifically to detect light. These diodes are constructed with a glass or plastic window through which the light enters. Often, photodiodes have a large, exposed PN junction region. These diodes are often used in automatic headlamp control systems.

The Optical Spectrum

All visible light is classified as electromagnetic radiation. The specific wavelength of light rays will define its characteristics. Light wavelengths are specified in nanometers, which are billionths of a meter. The optical light spectrum includes ultraviolet, visible, and infrared radiation. Photonic semiconductors can emit or detect near-infrared radiation, so near-infrared is usually referred to as light.

Solar Cells

A solar cell consists of a PN or NP silicon semiconductor junction built onto contact plates. A single silicon solar cell may generate up to 0.5 V in ideal light conditions (bright sunlight), but output values are usually lower. Like battery cells, solar cells are normally arranged in series groups, in which case the output voltage would be the sum of cell voltages, or in parallel, where the output current would be the sum of the cell currents. They are sometimes used as battery chargers on vehicles.

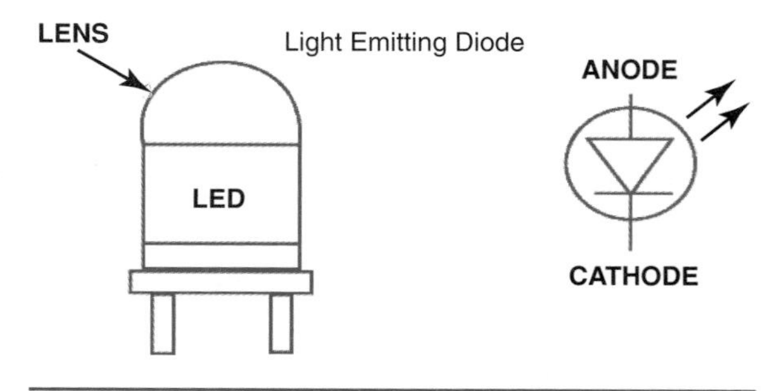

Figure 10-15. Light-emitting diode (LED).

Light-Emitting Diodes (LEDs)

All diodes emit electromagnetic radiation when forward biased, but diodes manufactured from some semiconductors (such as gallium arsenide phosphide) emit it at much higher levels. These are known as **light-emitting diodes (LEDs)** (Figure 10-15). They are much the same as regular diodes except that they emit light when they are forward biased. LEDs are very current-sensitive and can be damaged if they are subjected to more than 50 milliamps. LEDs also require higher voltages to turn on than do regular diodes; normally, 1.5 to 2.5 volts are required to forward-bias an LED to cause it to light up. LEDs also offer much less resistance to reverse-bias voltages. High reverse-bias voltages may cause the LED to light or cause it to burn up.

LEDs may be constructed to produce a variety of colors and are commonly used for digital data displays. For instance, a digital display with seven linear LED bars arranged in a bridged rectangle could display any single numeric digit by energizing or not energizing each of the seven LEDs. LED arrangements constructed to display alpha characters are only slightly more complex. LEDs convert electrical current directly into light (photons) and therefore are highly efficient, as there are no heat losses. In complex electrical circuits, LEDs are an excellent alternative to incandescent lamps. They produce much less heat and need less current to operate. They also turn on and off more quickly. LEDs are also used in some steering wheel controls.

Photodiode

All diodes produce some electrical response when subjected to light. A photodiode is designed to detect light and therefore has a clear window through which light can enter. Silicon is the usual semiconductor medium used in photodiodes.

RECTIFIER CIRCUITS

A **rectifier** (Figure 10-16) converts an undulating (alternating current voltage) signal into a single-polarity (direct current voltage) signal. Rectifiers change alternating current (AC) to direct current (DC). Several diodes are combined to build a diode rectifier, which is also called a bridge rectifier. Bridge rectifiers comprise a network of four diodes that converts both halves of an AC signal.

AC Generator (Alternator)

The most common use of a rectifier in today's automotive systems is in the AC generator. The AC generator produces alternating current (AC). Because automotive electrical systems use direct current, the generator must somehow convert the AC to DC. The DC is provided at the alternator's output terminal. Figure 10-17 shows how a diode rectifier works inside a generator.

Full-Wave Rectifier Bridge Operation

Think about this example in terms of conventional theory (Figure 10-16). First you must understand that the stator voltage is AC. That means the voltage at *A* alternates between positive and negative. When the voltage at *A* is positive, current flows from *A* to the junction between diodes D1 and D2. Notice the direction of the diode arrows. Current can't flow through D1, only through D2. The current reaches another junction. It can't flow through D4, so current must pass through the circuit load. The current continues along the circuit until it reaches the junction of D1 and D3.

Even though the voltage applied to D1 is forward biased, current can't flow because there's positive voltage on the other side of the diode. Current flows through D3 to ground at B. When the stator voltage reverses so that point *B* is positive, current flows along the opposite path. Whether the stator voltage at point *A* is positive or negative, current always flows from top to bottom through the load (R1). This means the current is DC.

The rectifiers in generators are designed to have an output (positive) and an input (negative) diode for each alternation of current. This type of rectifier is called a full-wave rectifier. In this type of rectifier, there is one pulse of DC for each pulse of AC. The DC that's generated is called full-wave pulsating DC. Figure 10-16 is an illustration of full-wave pulsating DC.

This example is simplified to include only one stator and four diodes. In a real AC generator, there are three stator coils and six diodes. The diodes are mounted inside two heat sinks. The heat sinks are cooled by air to dissipate the heat generated in the six diodes. The combination of the six diodes and the heat sink is called a rectifier bridge.

TRANSISTORS

Transistors (Figure 10-18) are semiconductor devices with three leads. A very small current or voltage at one lead can control a much larger current flowing through the other two leads. This means that transistors can be used as amplifiers and switches. There are two main families of transistors: bipolar transistors and field-effect transistors. Many of their functions are either directly or

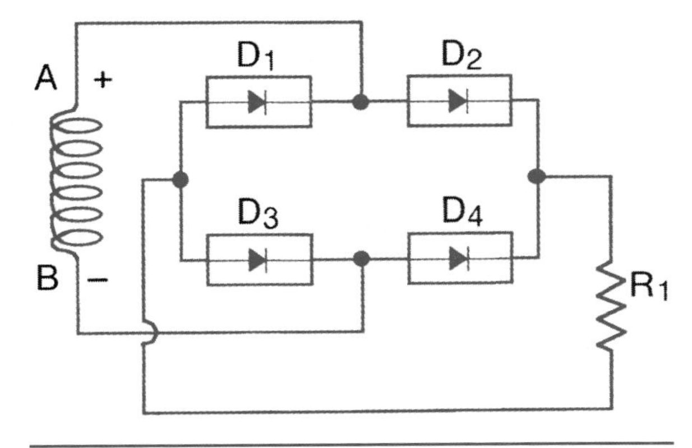

Figure 10-16. Rectifier bridge. (GM Service and Parts Operations)

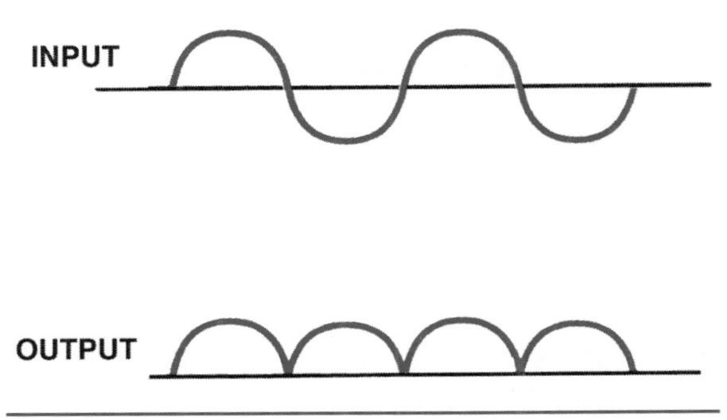

Figure 10-17. AC to DC rectification. (GM Service and Parts Operations)

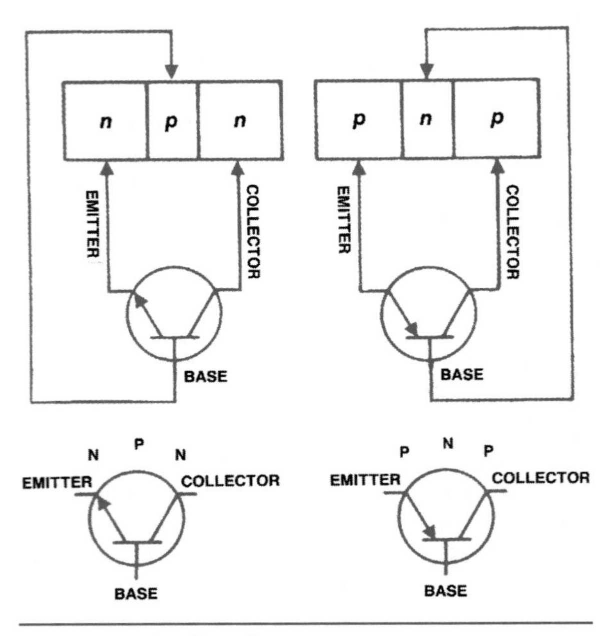

Figure 10-18. Transistors.

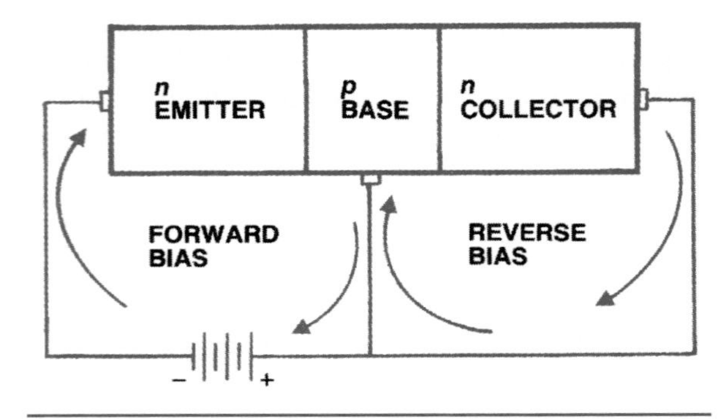

Figure 10-19. Transistor operation.

indirectly associated with circuit switching, and in this capacity they can be likened to *relays* in electrical circuits.

Bipolar Transistors

Add a second junction to a PN junction diode and you get a three-layer silicon sandwich (Figure 10-19). The sandwich can be NPN or PNP. Either way, the middle layer acts like a faucet or gate that controls the current moving through the three layers. The three layers of a bipolar transistor are the emitter, base, and collector. The base (middle) of the sandwich always acts as a *gate* capable of controlling the current flow through all three layers. This base is fairly thin and it has comparatively fewer doping atoms than that of the semiconductor material either side of it. The base is designed so that a small emitter-base current will "ungate" the transistor and permit a larger emitter-collector current flow.

When the base of an NPN Transistor is grounded (0 volts), no current flows from the emitter to the collector—the transistor is off. If the base is forward-biased by at least 0.6 volt, the current will flow from the emitter to the collector—the transistor is on. When operated in only these two modes, the transistor functions as a *switch* (Figure 10-19). If the base is forward biased, the emitter-collector current will follow variations in a much smaller base current. The transistor then functions as an amplifier. This discussion applies to a transistor in which the emitter is the ground connection for both the input and output and is called the common-emitter circuit. Diodes and transistors share several key features:

- The base-emitter junction (or diode) will not conduct until the forward voltage exceeds 0.6 volt.
- Too much current will cause a transistor to become hot and operate improperly. If a transistor is hot when touched, disconnect the power to it.
- Too much current or voltage may damage or permanently destroy the "chip" that forms a transistor. If the chip isn't harmed, its tiny connection wires may melt or separate from the chip. Never connect a transistor backwards!

Examples of Bipolar Transistors

A few types of the more common bipolar transistors are listed in this section, including a brief description of their application.

Small-Signal Switching Transistors

These are used to amplify signals. They may be designed to fully gate current flow in their off position and others may both amplify and switch.

Power Transistors

These are used in power supply circuits. Power transistors may conduct high current loads and may be mounted on heat sinks to enable them to dissipate heat. They are sometimes known as drivers because they serve as the final or output switch in an electronic circuit used to control a component such as a solenoid or pilot switch.

High Frequency Transistors

These operate at radio and microwave frequencies and are designed with a very thin base.

Transistor Operation

The input circuit for a NPN transistor is the emitter-base circuit (Figure 10-20). Because the base is thinner and doped less than the emitter, it has fewer holes than the emitter has free electrons. Therefore, when forward bias is applied, the numerous free electrons from the emitter do not find enough holes to combine with in the base. In this condition, the free electrons accumulate in the base and eventually restrict further current flow.

The output circuit for this NPN transistor is shown in Figure 10-21 with reverse bias applied. The base is thinner and doped less than the collector. Therefore, under reverse bias, it has few minority carriers (free electrons) to combine with many minority carriers (holes in the collector). When the collector-base output circuit is reverse-biased, very little reverse current will flow. This is similar to the effect of forward-biasing the emitter-base input circuit, in which very little forward current will flow.

When the input and output circuits are connected, with the forward and reverse biases maintained, the overall operation of the NPN transistor changes (Figure 10-22). Now, the majority of free electrons from the emitter, which could not combine with the base, are attracted through the base to the holes in the collector. The free electrons from the emitter are moved by the negative forward bias toward the base, but most pass through the base to the holes in the collector, where they are attracted by a positive bias. Reverse current flows in the collector-base output circuit, but the overall current flow in the transistor is forward.

A slight change in the emitter-base bias causes a large change in emitter-to-collector current flow. This is similar to a small amount of current flow controlling a large current flow through a relay. As the emitter-base bias changes, either more or fewer free electrons are moved toward the base. This causes the collector current to increase or decrease. If the emitter-base circuit is opened or the bias is removed, no forward current will flow through the transistor because the base-collector junction acts like a PN diode with reverse bias applied.

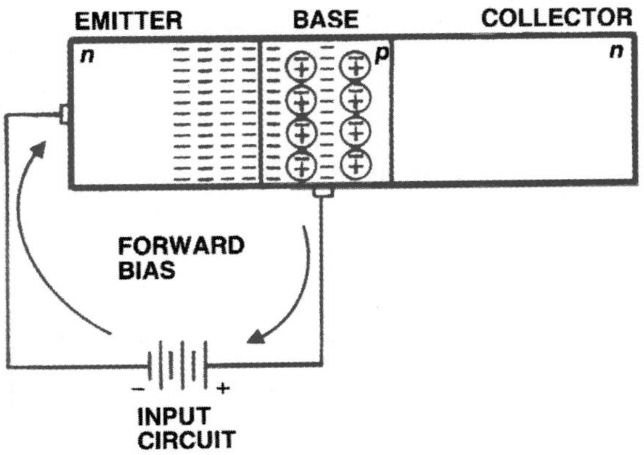

Figure 10-20. Forward-biased transistor operation.

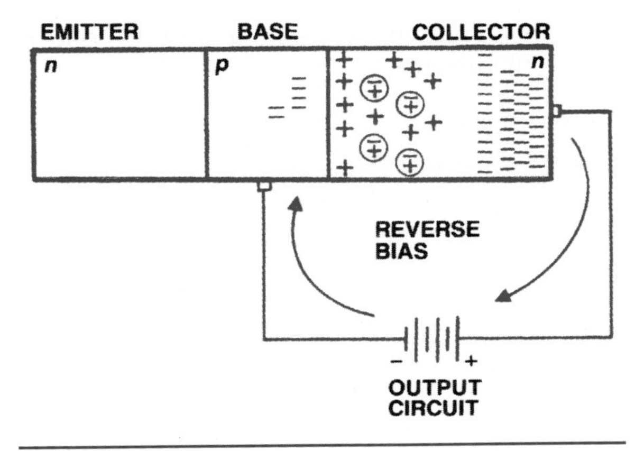

Figure 10-21. Reverse-biased transistor operation.

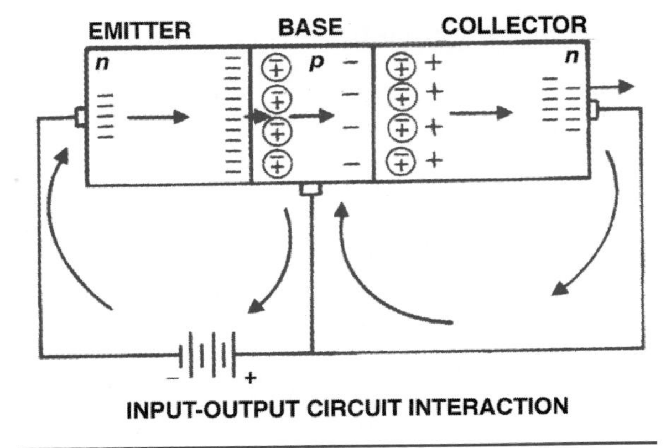

Figure 10-22. Transistor input/output current interaction.

■ **Tran(sfer) + (Re)sistor**

The word *transistor* was originally a business trademark used as a name for an electrical part that transferred electric signals across a resistor. Two scientists, John Bardeen and Walter H. Brattain, at Bell Telephone Laboratories in 1948, invented the point-contact transistor. In 1951, their colleague, William Shockley, invented the junction transistor. As a result of their work, these three men received the 1956 Nobel Prize for physics.

Field-Effect Transistors (FETs)

Field-effect transistors (FET) have become more important than bipolar transistors. They are easier to make and require less silicon (Figure 10-23). There are two major FET families, junction FETs and metal-oxide-semiconductor FETs. In both kinds, a small input voltage and practically no input current control an output current.

Junction FETs (JFETs)

There are two types of junction FETs: N-channel and P-channel. The channel behaves as a resistor that conducts current from the source side to the drain side: Voltage at the gate will act to increase the channel resistance, thereby reducing drain to source current flow. This enables the FET to be used either as an amplifier or a switch: when gate voltage is high, high resistance fields are created, narrowing the channel available for conductivity current to the drain. In fact, if the gate voltage is high enough, the fields created in the channel can join and completely block current flow.

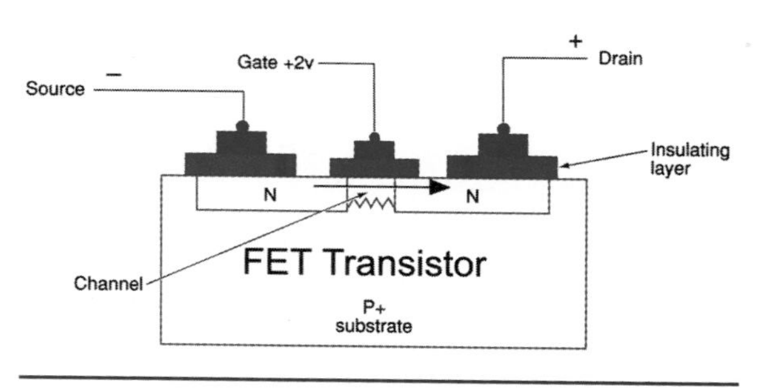

Figure 10-23. Field-effect transistor (FET).

Since JFETs are voltage-controlled, they have the following important advantages over current-controlled bipolar transistors:

- JFET gate-channel resistance is very high, so the device has almost no effect on external components connected to the gate.
- Almost no current flows in the gate circuit because the gate-channel resistance is high. The gate and channel form a "diode" and so long as the input signal reverse-biases this diode, the gate will show high resistance.

Like bipolar transistors, JFETs can be damaged or destroyed by excessive current or voltage.

Metal-Oxide Semiconductor Field Effect Transistors (MOSFETs)

The Metal-Oxide-Semiconductor FET (MOSFET) has become the most important transistor. Most microcomputer and memory integrated circuits are arrays of thousands of MOSFETs on a small sliver of silicon. MOSFETs are easy to make, they can be very small, and some MOSFET circuits consume negligible power. New kinds of power MOSFETs are also very useful. The input resistance of the MOSFET is the highest of any transistor. This and other factors give MOSFETs the following important advantages:

- The gate-channel resistance is almost infinite. This means the gate pulls no current from external circuits.
- MOSFETs can function as voltage-controlled variable resistors. The gate voltage controls channel resistance.

MOSFETs are classified as P-type or N-type. However, in a MOSFET there is no direct electrical contact between the gate with the source and the drain. A silicon oxide insulator from the remainder of the transistor material separates the gate's aluminum contact. When the gate voltage is positive, electrons are attracted to the region around the insulation in the P-type semiconductor medium: This produces a path between the source and the drain in the N-type semiconductor material, permitting current flow. The gate voltage will define the resistance of the path or channel created through the transistor, permitting them to both switch and act as variable resistors. MOSFETs can be switched at very high speeds and because the gate-channel resistance is so high, almost no current is drawn from external circuits. Newer types of MOSFETs can switch very high currents in a few billionths of a second.

Unijunction Transistor (UJT)

The unijunction transistor functions more like a three-terminal diode and perhaps is best described as a voltage-controlled switch: It has no ability to amplify. The UJT will permit a small current flow from base I to base II; however, when the voltage applied to the either terminal reaches a specific trigger value, the UJT is switched to its *on* position, enabling a high current flow between base I and base II.

Darlington Pairs

A Darlington pair (Figure 10-24) consists of a pair of transistors wired so that the emitter of one supplies the base signal to a second, through which a large current flows. The objective once again is to use a very small current to switch a much larger current. This action is better known as amplification. Darlington pairs are used extensively in vehicle computer control systems.

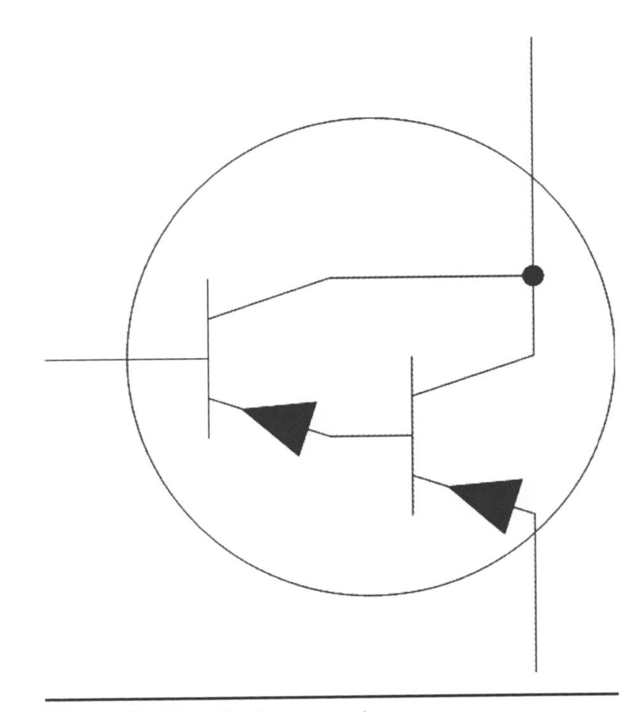

Figure 10-24. Darlington pairs.

SILICON-CONTROLLED RECTIFIERS (SCRs)

A **silicon-controlled rectifier (SCR)** (Figure 10-25) is similar to a bipolar transistor with a fourth semiconductor layer added. SCRs are used to switch DC. When the anode of an SCR is made more positive than the cathode, the outer two PN junctions become forward-biased: The middle PN junction is reverse-biased and will block current flow. However, a small gate current will forward-bias the middle PN junction, enabling a large current to flow through the thyristor. SCRs will remain on even when the gate current is removed: the on condition will remain until the anode-cathode circuit is opened or reverse-biased. SCRs are used for switching circuits in vehicle electronic and ignition systems.

TRIACs are essentially two SCRs connected in parallel: they will switch both DC and AC circuits. TRIACs have five P-N layers plus an extra N region at the gate terminal. The three terminals each make contact with two semiconductor P-N layers. TRIACs enable low gate current to switch high current flow through the device; current flow through the TRIAC will occur only while gate current is present.

Thyristor

Thyristors are a type of SCR that has three terminals and acts as a solid-state switch. A small current flow through one of the terminals will

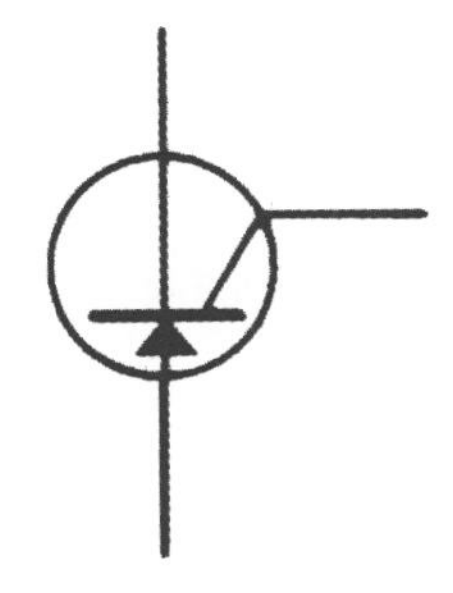

Figure 10-25. Silicon-controlled rectifiers (SCR).

switch the thyristor *on* and permit a larger current flow between the other two terminals. Thyristors are switches so they are either in an *on* or *off* condition: they fall into two classifications based on whether they switch AC or DC current. Some thyristors are designed with two terminals only: they will conduct current when a specific trigger or breakdown voltage is achieved.

INTEGRATED CIRCUITS

Integrated circuits (ICs) (Figure 10-26) fall into two general categories. Analog integrated circuits operate on variable voltage values: electronic voltage regulators are a good vehicle example of an analog IC. Digital integrated circuits operate on two voltage values only, usually *presence of voltage* and *no voltage*. Digital ICs are the basis of most computer hardware, including processing units, main memory, and data retention chips. Integrated circuit chips can be soldered onto motherboard (main circuit) or socketed; the latter has the advantage of easy removal and replacement. A common chip package used in computers and vehicle ECMs is the DIP (dual in-line package), a rectangular plastic enclosed IC with usually 14 to 16 pins arranged evenly on either side. DIPs may be soldered (not removable) or socketed to the motherboard.

USING ELECTRONIC SIGNALS

Simple electronic circuits can be designed to transmit relatively complex data by switching the circuit on or off, resulting in the presence of voltage or no voltage. The term *pulse width modulation (PWM)* is used extensively in digital electronics and the technician should develop an understanding of this concept.

If a circuit consisting of a power source, light bulb and switch is constructed, the switch can used be to "pulse" the on/off time of the bulb. This pulsing can be coded into many types of data, such as alphabetic or numeric values. Pulses are controlled immediate variations in current flow in a circuit where the increase or decrease in current would ideally be instantaneous. However, true pulse-shaping results in a graduated rise when the circuit is switched to the *on* state and a graduated fall when the circuit is switched to the *off* state. Waves are rhythmic fluctuations in circuit current or voltage.

The term *signals* is used to describe pulses and waveforms that are shaped to transmit data. The mechanisms and processes used to shape data signals is modulation. In vehicle electronics, the term *modulation* is more commonly used in reference to digital signaling. Electronic noise is unwanted pulse or waveform interference that can scramble signals. All electrical and electronic components produce electromagnetic fields, which may generate noise. Note that all electronic circuits are valuable to magnetic and electromagnetic field effects.

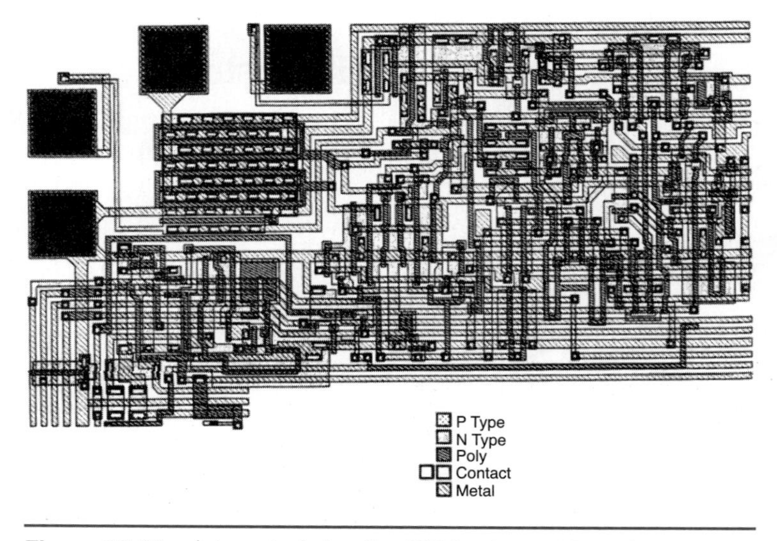

Figure 10-26. Integrated circuits. (GM Service and Parts Operations)

SUMMARY

Semiconductors are, by definition, elemental materials with four electrons in their outer shells. Silicon is the most commonly used semiconductor material. Semiconductors must be doped to provide them with the electrical properties that can make them useful as electronic components. After doping, semiconductor crystals may be classified as having N or P electrical properties. Diodes are two-terminal semiconductors that often function as a sort of electrical one-way check valve. Zener diodes are commonly used in vehicle electronic systems; they act as voltage-sensitive switches in a circuit.

Transistors are three-terminal semiconductor chips. Transistors can be generally grouped into bipolar and field effect types. Essentially a transistor is a semiconductor sandwich with the middle layer acting as a control gate, a small current flow through the base-emitter will ungate the transistor and permit a much larger emitter-collector current flow. Many different types of transistor are used in vehicle electronic circuits, but their roles are primarily concerned with switching and amplification. The optical spectrum includes ultraviolet, visible, and infrared radiation. Optical components conduct, reflect, refract, or modify light. Vehicle electronics are using increased amounts of fiber optics components.

Integrated circuits consist of resistors, diodes, and transistors arranged in a circuit on a chip of silicon. A common integrated circuit chip package used in computer and vehicle electronic systems is a DIP with either 14 or 16 terminals. Many different chips with different functions are often arranged on a primary circuit board also known as a motherboard.

Review Questions

1. Semiconductors are made into good conductors through which of these processes:
   a. Electrolysis
   b. Purification
   c. Ionization
   d. Doping
2. In an electrical circuit, a diode functions as a:
   a. Timing device
   b. Switch
   c. Resistor
   d. Energy storage device
3. What is the purpose of a diode rectifier circuit?
   a. To convert AC voltage to DC
   b. To convert DC voltage to AC
   c. To smooth out voltage pulsations
   d. To convert DC voltage to AC
4. The positive terminal of a diode is correctly called an:
   a. Electrode
   b. Cathode
   c. Anode
   d. Emitter
5. Semiconductors are elements that have how many electrons in their valence rings?
   a. Two
   b. Four
   c. Six
   d. Eight
6. A diode is a simple device which joins:
   a. P-material and N-material
   b. P-material and P-material
   c. N-material and N-material
   d. P-material and a conductor
7. Which of the following is *not* true of forward bias in a diode?
   a. Free electrons in the N-material and holes in the P-material both move toward the junction.
   b. N-material electrons move across the junction to fill the holes in the P-material.
   c. Negatively charged holes left behind in the N-material attract electrons from the negative voltage source.
   d. The free electrons which moved into the P-material continue to move toward the positive voltage source.
8. When reverse bias is applied to a simple diode, which of the following will result?
   a. The free electrons will move toward the junction.
   b. The holes of the P-material move toward the junction.
   c. No current will flow across the junction.
   d. The voltage increases.
9. Under normal use, a simple diode acts to:
   a. Allow current to flow in one direction only
   b. Allow current to flow in alternating directions
   c. Block the flow of current from any direction
   d. Allow current to flow from either direction at once
10. "Breakdown voltage" is the voltage at which a Zener diode will:
    a. Allow reverse current to flow
    b. Stop the flow of reverse current
    c. Sustain damage as a result of current overload
    d. Stop the flow of either forward or reverse current
11. Which of the following combinations of materials can exist in the composition of a transistor?
    a. NPN
    b. PNN
    c. ABS
    d. BSA
12. To use a transistor as a simple solid-state relay:
    a. The emitter-base junction must be reverse-biased, and the collector-base junction must be forward-biased.
    b. The emitter-base junction must be forward-biased, and the collector-base junction must be reverse-biased.
    c. The emitter-base junction and the collector-base junction both must be forward-biased.
    d. The emitter-base junction and the collector-base junction both must be reverse-biased.
13. Which of the following is *not* commonly used as a doping element?
    a. Arsenic
    b. Antimony

c. Phosphorus
d. Silicon

14. The display faces of digital instruments are often made of:
    a. Silicon-controlled rectifiers
    b. Integrated circuits
    c. Light-emitting diodes
    d. Thyristors

15. All of the following are parts of a field-effect transistor (FET), *except* the:
    a. Drain
    b. Source
    c. Base
    d. Gate

16. All of the following are characteristics of an integrated circuit (IC), except:
    a. It is extremely small.
    b. It contains thousands of individual components.
    c. It is manufactured from silicon.
    d. It is larger than a hybrid circuit.

11

Computers and Data Communication

LEARNING OBJECTIVES

Upon completion and review of this chapter, you should be able to:

- Identify the components in a basic personal computer system and explain their operation.
- Define personal computer inputs and outputs.
- Outline the four stages of a personal computer processing cycle.
- Describe the role played by the Central Processing Unit (CPU) in the processing cycle.
- Describe the role played by the various memory components of a computer.
- Identify the monitoring automotive computer input devices and explain their operation.
- Identify the automotive computer output devices and explain their operation.
- Define the four basic functions of the microprocessor processing cycle in the PCM.
- Demonstrate a basic understanding of the Internet.

KEY TERMS

Analog
CPU (Central Processing Unit)
Digital
ECM (Engine/Electronic Control Unit)
EEPROM (Electronically Erasable PROM)
EPROM (Erasable PROM)
Handshake
Indicator Lights
Input
Internet
Memory
Network
Output
PCM (Powertrain Control Module)
PROM (Programmable Read Only Memory)
RAM (Random Access Memory)
ROM (Read Only Memory)
Sensor

INTRODUCTION

Today's automotive technician is required to have a working knowledge of personal computers (PCs) as well as the computers that manage vehicle systems and their process of communication. The PC has played an ever-increasing role on the shop floor, where it has been used to obtain service information, guide sequential troubleshooting, connect to the Internet, and assist in on-board vehicle computer reprogramming. There is no doubt that automotive technicians with basic computer skills learn the technology of vehicle electronic management systems much more quickly than those without these skills. This fact has opened up many opportunities for younger technicians to advance at an accelerated rate in the automotive industry. By definition, a computer is a device used for information processing, arithmetic calculation, and information storage. This chapter deals with computer technology in a general sense, attempting to provide a broad introduction to automotive computer technology.

On-board vehicle computers are referred to as one of the following: **ECM (engine/electronic control module),** ECU (electronic/engine control unit), or **PCM (powertrain control module).** This text will use the term PCM because it is the most common type used in OBD II (on-board diagnostic systems second design). The term PCM refers to the fact that this computer manages both the engine and the drivetrain.

The PCM normally contains a microprocessor, data retention media, and often the output or switching apparatus. Some manufacturers use multiple computers to manage the engine, transmission, and ABS/traction control. One central computer manages the three systems for synchronized performance, or each separate system computer can be bussed (electronically connected) to the vehicle management computer through a controller area network (CAN). Some manufacturers use the term *multiplexing* to refer to a vehicle management system using multiple, interconnected computers or controllers.

This chapter will also introduce some of the communications technology that a technician should have an awareness of to work effectively. Technicians will need to know how to download PCM programming from a communications system, sometimes using a modem or through a PC. A majority of today's computer systems are networked via the *Internet* (global multimedia communications network) and various in-house, manufacturer or dealer *intranet* systems (internal networks), which facilitate communications within a company. This technology will merely be introduced in this section, with a reminder that its use is projected to expand rapidly in the next few years and with it, the technician's requirement to understand it.

A technician is required to have a basic understanding of vehicle and personal computers and computer communications to interact effectively with today's technology. This section will introduce the essentials of electronic management of vehicle systems. The terminology presented in this chapter introduces basic personal computers, vehicle computers, and the computer communicating process.

HISTORY OF COMPUTERS

The first computer is generally believed to be a simple counting and calculation tool called an *abacus.* The abacus is believed to have originated in Babylonia 2,500 years ago. It survives as a calculating instrument in remote parts of East Asia and the Middle East and as a child's toy in the West. The abacus would be classified as an analog computer. An **analog** device operates on data represented by numbers or voltage values, which it processes to output a physical analogy of a mathematical problem to be solved. A slide rule, much used until the widespread acceptance of the electronic multitasking calculator in the 1960s, is another example of an analog computer.

Most modern computers are **digital.** The digital computer works with data in discreet form, that is, expressed as digits of binary codes. The digital computer processes data by counting, listing, and rearranging bits of data in accordance with detailed program instructions. The results of processing are then translated into outputs that may display on a screen, print onto paper, or actuate other peripherals. Peripherals are input and output devices that are part of the computer system but not integral to the main housing.

After the abacus, the next significant events in the development of the computer took place in Europe during the Seventeenth century. In 1620, Edmund Gunter, an English mathematician, invented the slide rule. In 1642 the French scientist Blaise Pascal introduced the first mechanical digital calculating machine. These significant innovations, plus the ideas of many scientists and mathematicians who followed, paved the road to the development of the modern computer.

The invention of the first electronic digital computer is attributed to Presper Eckert and John Mauchly of the University of Pennsylvania in 1946. By the 1960s, the computer had developed into a business tool used for data storage and arithmetic processing. The computers of the 1960s were physically vast, power-hungry machines full of vacuum tubes that required considerable maintenance.

THE PERSONAL COMPUTER

Automotive technicians are required to have a working knowledge of personal computers (Figure 11-1) as well as the computers that manage vehicle systems. Most of the automotive manufacturers use PCs for electronic service information and in the diagnostic process of repair to connect to their scan tools. Since the early 1980s, PCs have filed and organized the business activity of the automotive industry in service/repair operations. And from the early 1990s on, the PC has played an increasing role on the shop floor, where it is used to read, diagnose, guide sequential troubleshooting, and reprogram on-board computers. By definition, a computer is a device used for information processing, arithmetic calculation, and information storage. A computer system is made up of a number of pieces of hardware. Figure 11-1 shows a typical personal computer.

Figure 11-1. Personal computer.

PC INPUT/OUTPUT DEVICES

A personal computer, like any computer, has inputs and outputs (Figure 11-2). The computer processes input information and provides outputs that are used by the end user. **Input** devices are what allow users to control the computer. They provide a means for inputting data and telling the computer what to do. **Output** devices are what let us see what the computer has produced. They translate the work of the computer into a form that we can understand.

Output devices turn the computer's work into a usable form. Strictly speaking, any outcome the computer produces as a result of processing data is an output, even if that outcome is in programming code and never seen by human eyes. In this sense, the computer is continually producing outputs that it uses as inputs to produce other outputs, and so on. For example, the Automotive Personal Computer Module (PCM) PCM does this when it uses the crankshaft position sensor (CKP) signal (an input) to help calculate engine speed (an output), which in turn is used as a factor (an input) in fuel injector firing. Many devices of the computer serve as both an input and an output. The following paragraphs contain descriptions of inputs and outputs.

Mouse

The mouse input device (Figure 11-3) is most effective when the computer is running software that uses a graphical user interface (GUI). GUI software uses pictures and icons, instead of just text, to allow users to control the software (Microsoft Windows® is an example of GUI software). The mouse consists of a trackball in a housing. As the ball rotates, two rollers in the housing pick up their movements that read left-to-right (X) and up-to-down (Y) movements. The movement of the rollers sends signals to the computer. These signals are used to move the cursor on the screen.

The mouse has two buttons. The left button is used for making selections (e.g., clicking on an icon, highlighting text, selecting a file). The right button is used to access other features.

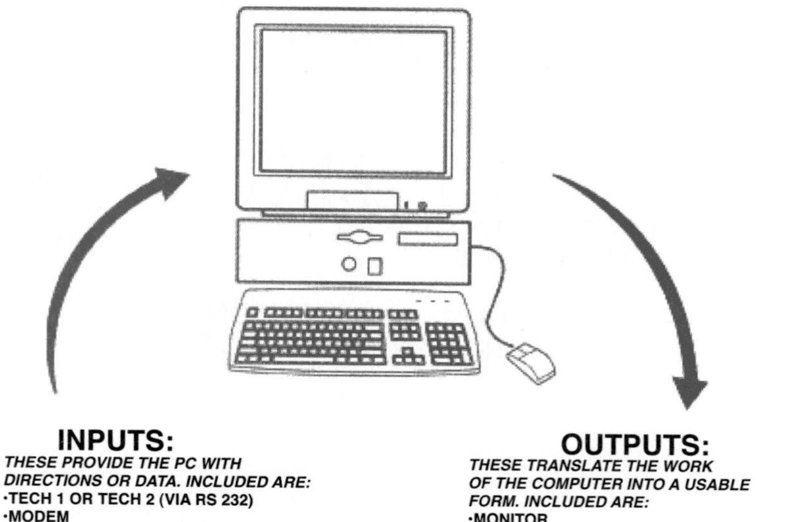

Figure 11-2. Inputs and outputs. (GM Service and Parts Operations)

Figure 11-3. Mouse.

Keyboard

The keyboard (Figure 11-4) is another input device. It allows the user to enter letters, numbers, and other characters that tell the computer what to do. The most obvious keyboard use is for typing information. It is also required to operate non-GUI software. The keyboard can also be used to maneuver GUI software through predefined keyboard commands that initiate the same kind of activity as a mouse click.

System Unit

The system unit (Figure 11-5) is the processing portion of the computer. It consists of several hardware devices housed in a tower or desktop case. The processing components of the computer (e.g., the CPU, RAM, video graphics card) are stored or attached to the motherboard that is housed in the unit. Other devices are attached to the system unit and connected to the motherboard directly or through expansion cards and slots. Figure 11-5 shows internal and external views of a typical personal computer.

Motherboard

The main circuit board or deck to which the internal chips, cards, and other components are attached is called the motherboard (Figure 11-6).

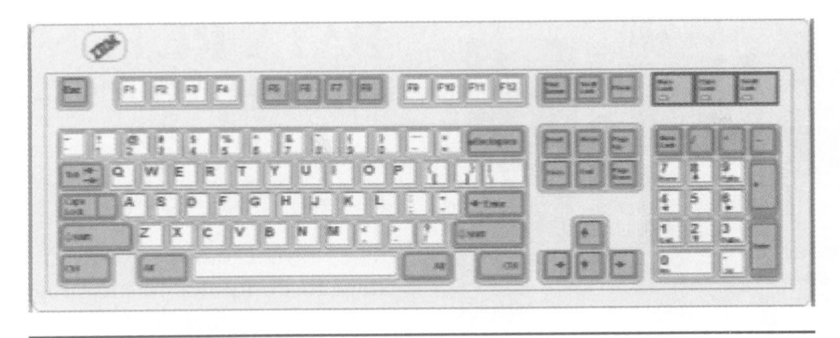

Figure 11-4. Keyboard.

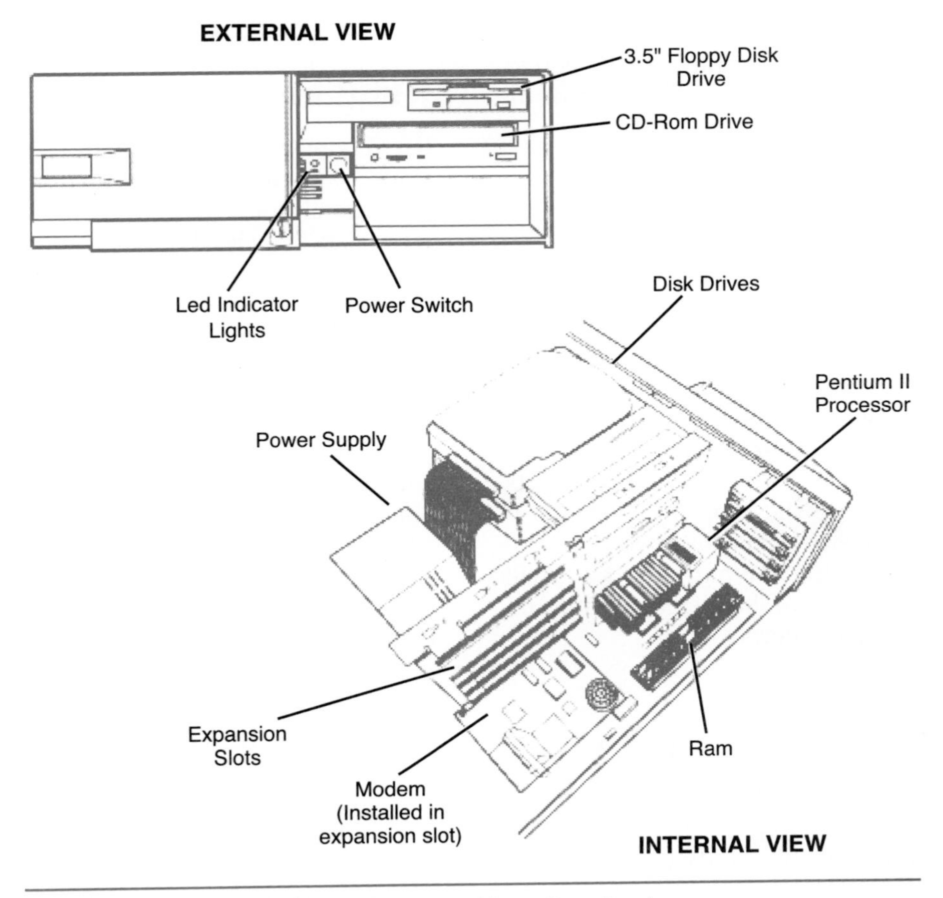

Figure 11-5. System unit. (GM Service and Parts Operations)

Central Processing Unit

The CPU (central processing unit) is also known as the MPU (microprocessor unit). It is the "brain" of the computer system and the primary indicator of its processing speed. It is attached to the motherboard by a multipin connector. The function of the CPU is decision making, addressing and data transfer, timing and logic control, arithmetic and logic operations, fetching data and program instructions from memory, decoding instructions, and responding to control signals from input and output (I/O) devices. Microprocessors contain a complex of many electronic circuits photo-infused on a silicon chip. The CPU's individual

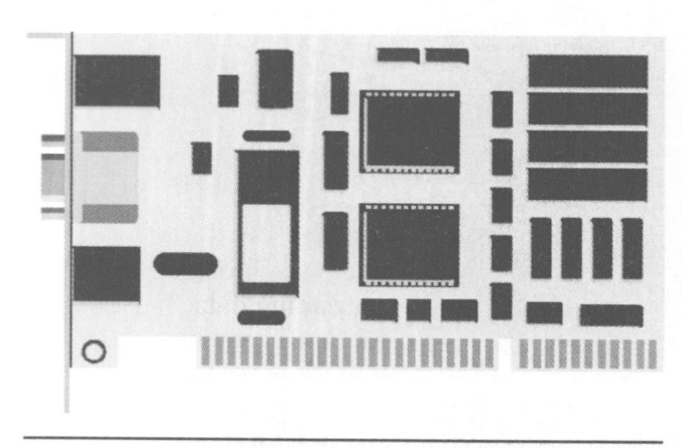

Figure 11-6. Motherboard.

subcircuits are referred to as blocks, and collectively are referred to as the architecture of the microprocessor.

System Internal Buses

System internal buses are simply connection and conduction devices. System internal buses link the main modules in a system as follows:

> **Control bus—bi-directional** transmits and receives operational commands to all parts of the computer system.
> **Address bus—unidirectional** connects the CPU to the memory banks and I/O ports. The CPU selects the address or location in memory to which data is targeted.
> **Data bus—bi-directional** acts as a data conduit to and from the memory banks, the CPU, and the I/O ports.

ELECTRONICALLY REPRESENTED DATA

In a digital computer, binary digits are used to represent data. The digits "O" and "1" are the only numbers used to represent billions of data possibilities. These digits represent the on (1) or off (O) status. Each on or off digital value is called a *bit*. A bit is the smallest unit of data that can be handled by a computer. The word *bit* is derived from the words *binary digit*.

A byte comprises 8 bits. A byte is capable of representing 256 data possibilities. ASCII (American Standard Code for Information Interchange) is a commonly used secondary data-storage code. ASCII is an 8-bit storage code used on PC systems.

DATA RETENTION HARDWARE—MEMORY CATEGORIES

The term **memory** is used to describe digital data in computer terminology. Memory or data can be retained using a number of different means, determined by factors such as whether the information must be accessed at high speeds by the processor and whether it needs to retained permanently.

Random Access Memory

RAM (random access memory) is often referred to as *main memory* or *primary storage*. It is retained electronically in semiconductor chips. RAM data includes the computer operating system software (a set of protocols that coordinate computer operations), program application software (direct procedure[s] for the specific task[s] performed), and the data currently being processed. RAM data is volatile. Because it is electronically retained, it is lost when the electrical circuit supplying it is opened. Some vehicles have on-board computers with RAM capability described as both volatile RAM and non-volatile RAM. Vehicle volatile RAM can be likened to main memory in a PC. The RAM is the computer's scratch pad.

In the PC processing cycle, RAM is loaded using a SIMM (single in-line memory module) or DIMM (double in-line memory module), cards containing multiple chips. The SIMM or DIMM is directly connected to the motherboard of the computer. SIMM chips come in 30-, 72-, 144-, and 168-pin designs. Using high-speed RAM located between the CPU and main memory may increase processing efficiency. This high-speed RAM is called *cache memory*. Cache is used to retain frequently consulted program instructions or data. When the CPU requires data, it will check cache first, enabling faster execution than retrieval from (slower) main memory.

RAM data on PCs can be categorized as follows:

- *Conventional memory* is the first logged in RAM at boot-up (when the computer is turned on) and it consists of operating system data and programs.

- *Upper memory* is used to control the computer's integral hardware, input and output devices, and some peripherals.
- *Extended memory* is used to retain programs and data. Multiple programs may be logged in extended memory, permitting multitasking, or the driving of more than one separate operations simultaneously.
- *Expanded memory* is built into a memory expansion board and was used on older PCs. It is driven by software called EMM (expanded memory manager) and results in slower processing speeds. Current PC systems tend to use increased extended memory capability and eliminate expanded memory.

Read Only Memory

ROM (read only memory) chips retain data, protocols (rules and regulations), and instructions that once written, cannot be changed. In other words, the information can be read and used but never altered (though it can be destroyed or corrupted). For example, the boot-up procedure in a PC system is used repeatedly and does not change, so it makes sense that it is written to ROM. Automotive computers (ECM/PCM) have major portions of the total requisite management data written to ROM. This data in itself may not be sufficient to run the engine without further data qualification. However, it would contain all of the common running characteristics required to run the engine in each of the very different applications. ROM describes any non-volatile data retention media; it is retained magnetically. Volatile memory is RAM data that is only retained when a circuit is switched on.

Memory Speed

The speed of memory is usually measured in nanoseconds (billionths of a second). Dynamic RAM (DRAM) chips may have access speeds of 50–100 nanoseconds while cache (static RAM or SRAM) may have access times of as little as 10 nanoseconds. Registers designed into CPU chips have access times of as little as a single nanosecond and represent the fastest type of memory. A vehicle ROM has access times in the 100–250-nanosecond range while a PC hard disk produces access times in the 10–20-millisecond range. A nanosecond is one-billionth of a second or 10^{-9}. Light travels one foot in one nanosecond.

SECONDARY DATA-STORAGE MEDIA

Floppy Disk Construction

Secondary data-storage or auxiliary storage devices, such as diskettes, are those device used for writing, storing, and reading data. The earliest home PCs used cassette tapes on which data was magnetically encoded; these presented some limitations imposed by the speed at which the tape could be driven by the reading heads and the sequential storage characteristic of cassette tape. Sequential data storage works fine to retain an album of music where the notes and songs are replayed in sequence, but severely limits access speed when specific units of data logged on different locations on the tape are required. Cassette tapes were replaced by 5 1/4" floppy diskettes and then by 3.5" disks. The 3.5" floppy diskette is encased in rigid plastic and only the internal circular disk is "floppy": it is nylon plastic coated with metal oxide. It is encoded with data magnetically in the same way an audiotape is encoded. It is read by a rotating drive mechanism and read heads.

Hard Drive

Hard disks (Figure 11-7) also retain data magnetically. They consist of multiple rigid platters coated with metal oxide that permit data to be magnetically written to the surface. On a hard disk assembly, the read-write heads, the drive mechanism, and the data retention platters are encased in an airtight, sealed cover. Hard disks are usually a permanent fixture within a PC housing. They rotate at high speeds (7,200 rpm) to receive and unload data.

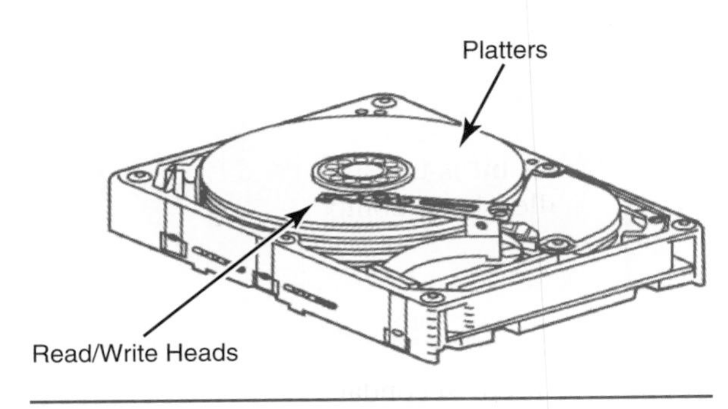

Figure 11-7. Hard drive. (GM Service and Parts Operations)

CD-ROM

Optical disks, or CD-ROMs (compact disk, read only memory), are being used increasingly to store data (Figure 11-8). A laser is used to burn microscopic holes on the surface of a rigid plastic disk. The disk can then be scanned by a low-power laser, which reflects light off the encoded disk surface to read digital data. This read-only format is being widely used in the automotive industry. Because of the amount of data that a CD can retain, many computer programs are now sold in this format. Optical disks have the additional advantage that the data written to them cannot be damaged by magnetic force. The CD-ROM drive is used to read CD-ROMs. CD-ROMs are used for transporting very large amounts of data—they can contain up to 650 MB of data, which is the equivalent of over 450 floppy disks. CDs are also used for transporting data that users will not change—for example, EPROM (erasable programmable read only memory) calibration files.

Unlike hard disks or floppy disks, CD-ROMs usually cannot be written to; they are read from. The data on a CD is usually permanent. Thus, you wouldn't be able to save a file to a CD-ROM disk unless you are using a CD-RW drive—the CD-RW drive that reads, writes, and rewrites to the CD.

Physically, CD-ROMs are made of plastic. During the production process, pits are stamped onto the plastic disk in a very controlled pattern. These pits represent the data that is stored on the CD. After stamping, the disk is covered with a very thin reflective coating. The coating is then covered with a scratch-resistant layer that also protects the disk from oxygen, which can oxidize the reflective coating. Finally, a label may be silk-screened onto the disk.

Figure 11-8. CD-ROM.

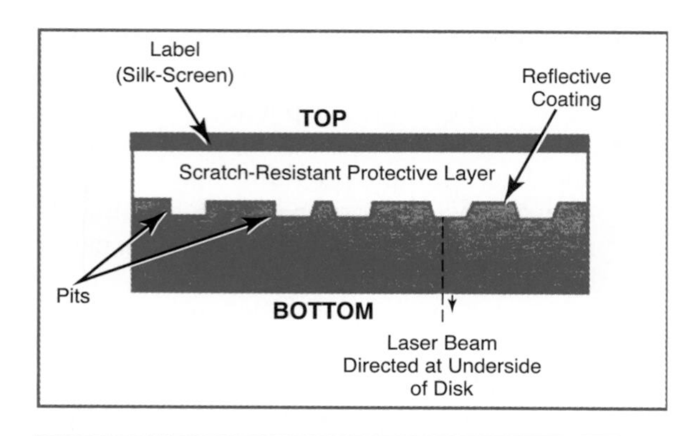

Figure 11-9. CD-ROM architecture. (GM Service and Parts Operations)

When the disk spins in the drive, a laser beam is directed at the bottom surface of the disk (Figure 11-9). Read heads pick up the light as it is either reflected or not reflected on a pitted or nonpitted area. This reflection is turned into code that the computer can read. CD-ROM drives are rated according to their speed. Common CD-ROM drive speeds range from 4× (which transfers data at a rate of 614,000 bytes per second) up to 32× (which transfers at eight times the rate of 4× drives).

DVD-ROM

Another recently developed media is the DVD or digital versatile disk. The DVD operates in the same way as a CD-ROM, but is smaller, being about 3 inches, and holds approximately 5.4 gigabytes of data and graphics.

PCMCIA Card

Optical memory cards (Figure 11-10) such as the PCMCIA (Personal Computer Memory Card International Association) card are about the size of a credit card and may contain data equivalent to 1,600 pages of text. Their use is increasing. They retain data for read-out and can be written to with the correct equipment. The GM Tech 2 scan tool currently uses a PCMCIA card.

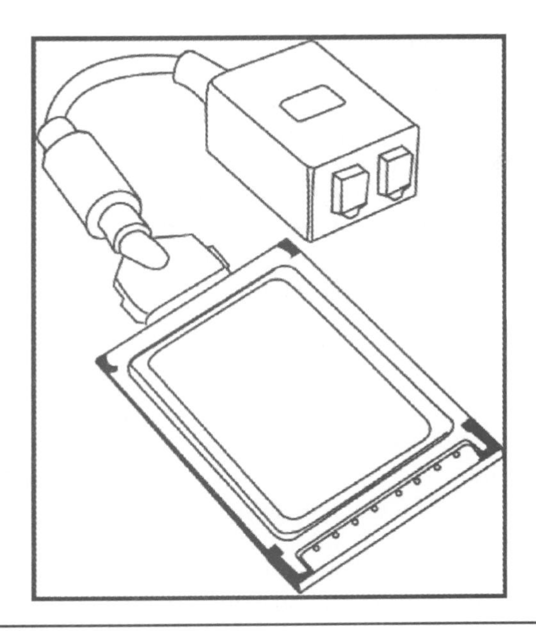

Figure 11-10. PCMCIA card.

Figure 11-11. Monitor/display.

Monitor/Display

The monitor (Figure 11-11) displays information in color or black-and-white. Monitors come in a variety of dimensional sizes, ranging from 13 inches to 21 inches. In addition to a physical size, monitors have a display size. This is expressed in pixels. For example, 14-inch or 15-inch monitors work well displaying a picture 640 pixels wide and 480 pixels in height. Larger monitors can display at 1,024 by 768 pixels or more. Many monitors can display at different resolution settings. Color monitors are classified as either VGA (Video Graphics Array) or SVGA (Super Video Graphics Array) types. VGA monitors display images at 72 or 75 dpi or ppi (dots per inch or pixels per inch). VGA provides a 640 by 480 pixel display. SVGA displays at 96 dpi. SVGA monitors can display more pixels for a larger desktop area—thus creating a larger "viewing window" to see an application.

In order to function correctly, most monitors require that the computer contain a video graphics card. This may be a separate card that plugs into the computer's motherboard, or it may be a built-in function of the motherboard. The monitor connects directly to the card. The card acts as an interpreter that lets the monitor display what the computer is doing. The number of colors the monitor can display and its display resolution are determined by the capabilities of the graphics card, which is largely a function of the card's VRAM (video RAM). The ability of a graphics card to display colors is expressed as "bit depth." The greater the bit depth, the more colors the card/monitor can display. Many newer monitors like laptop computers are flat-panel displays. These displays take up much less room and have greater resolution. They currently use multiple technologies to operate, the most common being LCD (liquid crystal display technology) and the newest being TFT (thin film transistor).

The monitor has controls for brightness and contrast that allow users to create the best picture for viewing. Newer monitors sometimes come with antiglare screens that reduce the reflections caused by lighting. Antiglare attachments are also available.

Printer

The printer (Figure 11-12) is an output device that produces a paper copy of whatever the user has instructed the computer to print. Printers range from black-and-white dot matrix all the way up to color laser. Printers are speed rated according to how many pages per minute (PPM) they produce. They are quality rated by their resolution, which is measured in dots-per-inch (dpi). In either case the higher the numbers, the better the printer's performance.

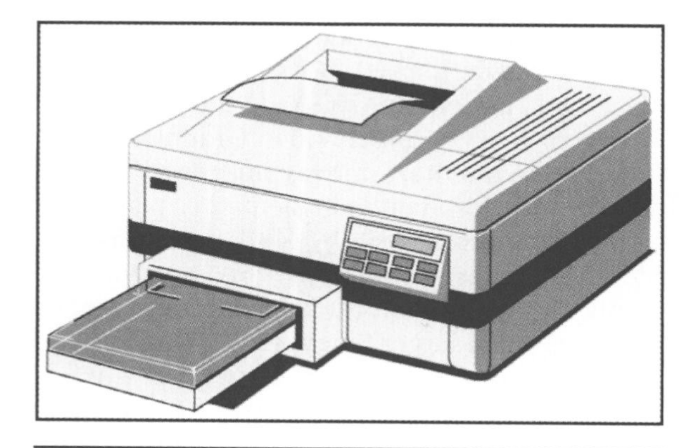

Figure 11-12. Printer.

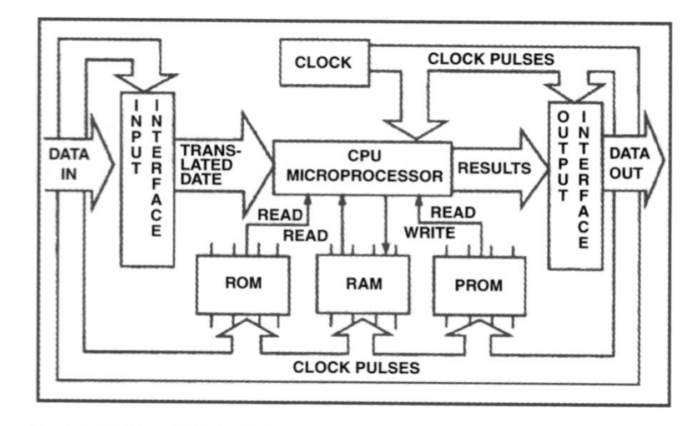

Figure 11-13. Central processing unit (CPU).

COMPUTER OPERATION AND PROCESSING

The system unit in a typical computer consists of a housing containing a motherboard, central processing unit (CPU) (Figure 11-13), electronic and magnetic data retention media, coprocessors, connection slots and ports, and a power supply. The motherboard is a circuit board; most of the system unit components are either permanently connected or plugged in socket to the motherboard. In a typical PC, the CPU is contained in a single integrated circuit called a microprocessor. The CPU contains the control/command unit and the ALU (arithmetic/logic unit). Figure 11-13 shows the processing cycle of the CPU. The control/command unit in the CPU operates by sequencing a four part processing cycle:

- Fetching
- Decoding
- Executing
- Storing

System Clock

The control unit uses a *system clock* to synchronize all computer operations; the system clock generates electronic pulses at a fixed rate specified in megahertz (MH_z).

CPU Operation

The CPU's (Figure 11-13) ALU incorporates the electronic circuitry required to perform arithmetic and logic data manipulation. Arithmetic operations include addition, subtraction, multiplication, and division. Logical operations consist primarily of data comparison and sequential strategizing. The CPU contains *registers,* memory storage locations that may temporarily store certain types of data. A critical factor of computer processing speed is work size: this indicates the number of bits that a CPU can process simultaneously. Electronically retained data is volatile; when the computer is switched off, it is dumped. After booting up a computer, RAM is loaded with the following data:

- The Operating System
- The Application Program Instructions
- The Data Currently Being Processed

The speed of memory is measured in nanoseconds. Registers in the CPU chip have access times as fast as a single nanosecond; ROM has access times between 50 and 250 nanoseconds, while accessing data on a typical PC hard disk drive takes between 10 and 20 milliseconds.

Using a *coprocessor,* a chip or card that performs specific tasks, may increase computer-processing efficiency. The computer *power supply* depends on the computer's functions. A PC typically receives an electrical supply at 110–120 volts AC and transforms this to an operating value ranging from 3 to 12 volts DC. A vehicle PCM is usually supplied with 12 volts DC and transforms this to a lower voltage value, usually 5 volts within the system. Operating voltage values tend to be lower in more recent computers, the objective being to reduce heat. PCs normally incorporate a cooling fan to remove heat from system components.

Expansion slots are used in PCs to add in devices that increase computing power or capability. *Ports* are used to connect peripherals such as printers, modems, and keyboards. *Parallel ports* are capable of transmitting data 8 bits (1 byte) at a time using cable that has 8 data lines. Printers and disk drives use parallel port connections. *Serial ports* transmit data 1 bit at a time and therefore function more slowly than parallel ports. A keyboard, a mouse, and communication devices (e.g., modems) use serial ports. A *bay* is simply a location within the system unit used to store hardware such as disk and CD drives. Two or more bays stacked vertically can be called a *cage*.

A machine language instruction is the driving command in binary data that the CPU interprets and generates outputs from. Computers may use serial processing in which the CPU processes a single instruction at a given moment until the instruction is completed, it then begins the execution of the next instruction and sequences processing one instruction at a time until the program is completed. *Parallel processing* uses multiple CPUs, enabling the execution of several sets of instructions simultaneously. Massively parallel processors (MPPs) are used in supercomputers where thousands of CPUs function simultaneously.

AUTOMOTIVE COMPUTERS

Vehicle computerized system management is summarized as a set of electronically connected components that enable an information processing cycle comprising three distinct stages:

- Data input
- Data processing
- Outputs

At the heart of the electronic engine control system is a computer or controller. This computer is given different names, depending on the manufacturer. DaimlerChrysler, Ford, and General Motors call it the powertrain control module, or PCM. These manufacturers began using *PCM* in the early 1990s when the computer started to also control the electronic transmission, or transaxle. The powertrain includes the transmission, hence the term *powertrain control module*. VW and Toyota may call it the engine control module, or ECM. The fact is that practically all vehicles today have some type of engine controller. The following is a list of some typical fuel-related systems that this microprocessor controls:

- Fuel delivery
- Emissions controls
- Charging
- Idle
- Radiator fan
- Air conditioning
- Speed control system

The microprocessor controls these systems by receiving information from a variety of sources such as input sensors, switches, and the data bus (communication circuit). The microprocessor then processes this information in order to control output devices that regulate engine performance. Outputs include the following:

- Ignition system
- Fuel injectors
- Generator field
- Air conditioning compressor
- Radiator fans
- Speed control servos

Data Input

Figure 11-14 shows, on the left, the systems that provide operating parameters or *inputs*, which is the data or information that the PCM (microprocessor) will use to make its decisions to provide *output* to the systems controlled, shown on the right in Figure 11-14. The microprocessor should deliver the correct ignition advance and fuel to the engine under all operating conditions. Data is simply raw information. Most of the data to be inputted to an automotive PCM comes from monitoring sensors, such as rpm signals, and command sensors, such as the throttle position sensor. Most of this data is in analog format, such as voltage values, and it must be converted to a digital format, or digitized, to enable processing by the PCM. The cycle of input/processing/output, as shown in Figure 11-14, ensures that the engine meets emission, performance, fuel economy, driveability, and customer expectations.

There are differences in the design, function, and terminology used from one manufacturer to another. For example, the microprocessor is known by many different names, such as:

- Engine controller
- Electronic control module (ECM)
- Engine control module (ECM)
- Powertrain control module (PCM)

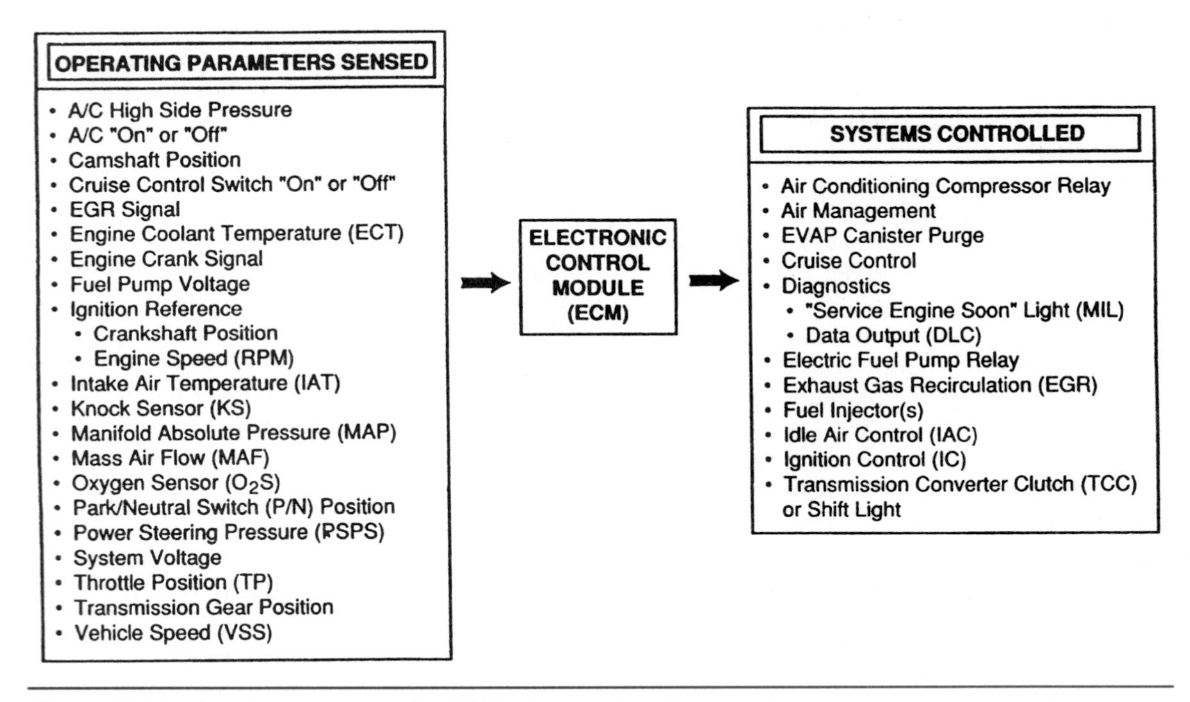

Figure 11-14. Inputs versus outputs. (GM Service and Parts Operations)

- Vehicle control module (VCM)
- Body control module (BCM)

Engine computers such as the VCM control the ABS, transmission, and various body electrical systems such as air conditioning and lighting functions. As always, refer to the appropriate service or diagnostic publication for accurate, up-to-date information. For more information on this topic, see the section on "On-Board Diagnostic Systems" in Chapter 11 of the *Shop Manual*.

Multiplexing

When a vehicle uses multiple electronically managed circuits, bussing (electronically connecting) the system PCMs is necessary to synergize the operation of each system. For instance, both the engine management and transmission computers require accelerator position input to function. If you allow the two computers to communicate with each other this would permit them to share input data and reduce the amount of input circuit hardware. The terms used to describe two or more ECMs/PCMs connected to operate multiple systems are *multiprocessing* and *multiplexing*. These systems are also referred to as controller area networks (CAN). To connect two ECMs/PCMs, they must use common operating protocols; that is, they must be speaking the same language. However, when two electronic systems are required to operate with each other and their communication and operating protocols do not match, an electronic translator called an interface module is required.

PCM (ECM) DATA RETENTION

Data is retained both electronically and magnetically in current-generation PCMs. The current PCM uses the data categories discussed in the following sections.

RAM

The amount of data that may be retained in RAM is a primary factor in quantifying the computing power of a system. It is also known as *main memory* because the CPU can only manipulate data when it is retained electronically. At start-up, RAM is electronically loaded with the vehicle management operating system instructions and all necessary running data retained in other data categories that the CPU must access at high speed.

RAM data is electronically retained, which means that it is always volatile. Data storage could be described as temporary—if the circuit is opened, RAM data is lost. It should be noted that most PCMs use only fully volatile RAM. In other words, when the ignition circuit is opened, all RAM data is dumped.

Data such as coolant temperature, oil pressure, and boost pressure are also logged into RAM. These data are within a threshold window defined in the ROM, PROM, EEPROM, or EPROM storage device. They only have significance at that specific moment in time and therefore can be safely discarded when the ignition key is turned off. A process known as *sampling* continually monitors this category of input data. Most RAM in engine management systems is volatile RAM; nothing is retained when the ignition switch is opened.

However, a second type of RAM is used in some systems—this is non-volatile RAM (NV-RAM) in which data is retained until either the battery is disconnected or the PCM is reset, usually by depressing a computer reset button which temporarily opens the circuit. Codes and failure strategy (action sequence) would be written to NV-RAM and retained until reset.

Some vehicles, especially those with first-generation computerized system management electronics, have a non-volatile RAM memory component: this data is electronically retained but supplied by a circuit directly from the vehicle battery after the ignition circuit is opened. Fault codes and failure mode strategy are written to non-volatile RAM and *kept alive* until either the battery is disconnected or the PCM reset. In fact, one manufacturer uses the acronym *KAM* (keep alive memory) to describe non-volatile RAM.

ROM

ROM data is magnetically retained and cannot be overwritten. It is *permanent,* though it can be corrupted and it is susceptible to damage when exposed to powerful magnetic fields. Low-level radiation such as police radar and high-tension electrical wiring will not affect any current PCMs. A majority of the data retained in the PCM is logged in ROM. The master program for the system management is loaded into ROM. Constructing ROM architecture so that it contains common requisite data for a number of different systems permits production standardization.

The ROM contains all the protocols ("rules and regulations") to master engine (or system) management, including all the hard threshold values (limits). The term *hard parameter* is often used to describe parameters/values that are interpreted by the PCM in a fixed manner. For instance, the temperature at which the PCM is programmed to identify an engine overheat condition is always a fixed value; at this specific value, a fault code and failure strategy are effected. Engine overspeed would be another example of a hard parameter. A *soft parameter* value is one that is interpreted by the PCM with a cushion around the value, according to the input of sensors monitoring the engine and the conditions it is running under.

Programmable Read Only Memory

PROM (Programmable Read Only Memory) is magnetically retained data—usually a chip, set of chips, or card socketed into the PCM motherboard. PROM can be removed and replaced. PROM's function is to qualify ROM to a specific chassis application. In the earliest engine management systems, replacing the PROM chip could only alter programming options such as idle shutdown time. Customer programmable options are written to an EEPROM (discussed in the next section) in current systems where they can be easily altered. Some OEMs describe the PROM chip as a *personality module,* an appropriate description of its actual function of trimming or fine-tuning the ROM data to a specific application.

Electronically Erasable PROM

The **EEPROM (electronically erasable PROM)** data category contains data programmable options and data that can be altered and modified using a variety of tools, ranging from a generic reader programmer to a mainframe computer. A scan tool can be used to adjust programming in EEPROM.

EEPROM is used in some current engines and body computers. It can be written to by the on-board software, receive customer and proprietary data programming, log driving and performance analyses, and retain codes manifestly and covertly. The type of data programmed to EEPROM includes cruise parameters, transmission ratios, final drive ratios, tire size, and failure mode strategy. It would also log data such as the fuel map profiles and covert documentation of failure conditions. A piece of mylar tape, a common type

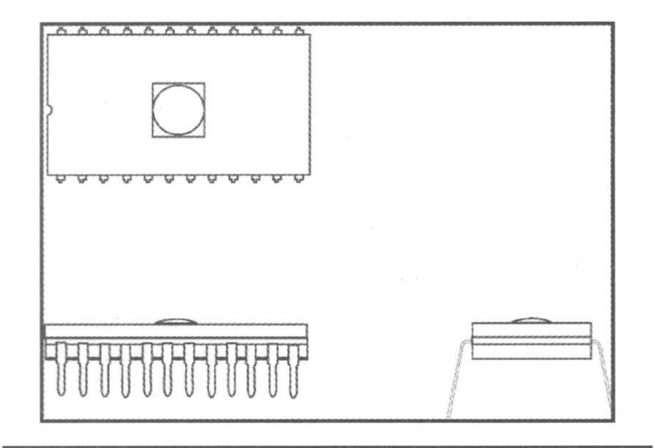

Figure 11-15. EPROM (erasable programmable read only memory).

of tape used in electronics, covers the window on the top of the EEPROM. If the tape is removed, the memory circuit is exposed to ultraviolet light that erases its memory.

Erasable PROM

The newest form of rewritable ROM is **erasable PROM** or **EPROM.** (See Figure 11-15.) It is similar to EEPROM except that its contents can be completely erased to allow new data to be installed, instead of partially erased in the case of the EEPROM. The EPROM is also known as a flash PROM.

Data Processing

A CPU microprocessor contains a control unit that executes program instructions and an ALU to perform numeric calculations and logic processing such as comparing data. As noted, RAM is data that is electronically retained in the PCM—the CPU can manipulate this data. Input data and magnetically retained data in ROM, PROM, EEPROM, and EPROM are transferred to RAM for processing.

INPUT CIRCUITS

PCM inputs can be divided into sensor device inputs and switched inputs. There is a high degree of input component commonality among automotive manufacturers. These inputs can also be further described by grouping them into system monitoring and system command inputs.

Sensor Definition

Computers need input information to make decisions. This information is called input. In the previous module, you learned that computers monitor changing voltage levels and interpret different levels to mean different things. In addition, a computer might make decisions based on:

- How fast the voltage level changes
- Changes in the voltage shape or wave form
- Whether or not a terminal is grounded

The components that collect information and send it to the computer are called **sensors.** Other names for a sensor are *sender* or *sending unit*. These names are often used for sensors that send information to a device other than a computer. Sensors may be simple switches that an operator toggles open or closed. They may ground a reference voltage, modulate a reference voltage, or be powered up either by voltage reference (V-ref) or outside of the V-ref circuit. Some sensors use a reference voltage to produce an analog or digital signal. These sensors receive a constant reference voltage (5 volts on OBD I and 7 volts on OBD II). The PCM compares the return signals with values logged in its non-volatile memory banks.

Signals

Computers use voltage as an input signal that tells them how a circuit is operating. However, as we learned in the last module, computers "think" in binary. This means that computers do not understand every kind of signal that is sent. Before discussing this in detail, analog and digital signals must be explained.

Analog Signals

Analog signals are continuous between the highest and lowest possible value. For example, if the lowest possible measurement were 0 volts and the highest were 5 volts, a measurement could be 0, 5, or any number in between, such as 0.5, 2.5, 0.99, and so on. Analog signals are shown as curved waves (Figure 11-16). A 5-volt value is above or below the "0" centerline of the sine wave graph. This value is the V-ref signal. All input voltage signals must be converted to digital binary so that the microprocessor can understand them. The conversion is done through an interface.

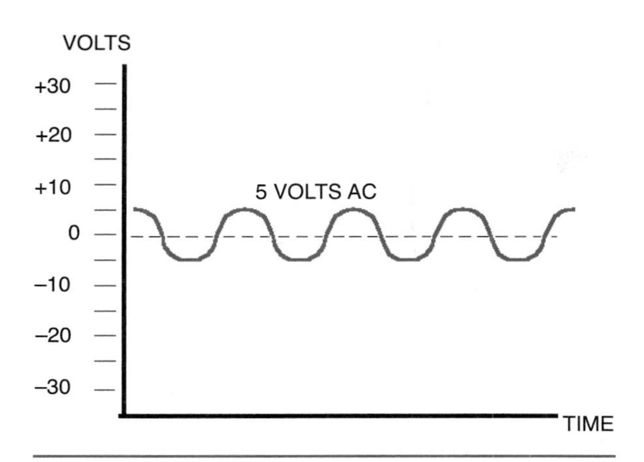

Figure 11-16. Analog signal. (GM Service and Parts Operations)

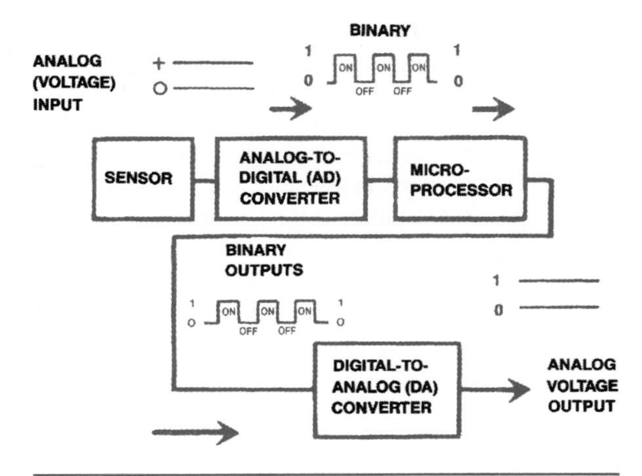

Figure 11-17. Analog to digital converter changes input signals into a binary system that the microprocessor can understand. (DaimlerChrysler Corporation)

Interfaces

Interfaces have several possible functions. They can translate an incoming signal into binary code, and they can translate output from binary code into an analog signal, as shown by the analog to digital converter in Figure 11-17. They can be used to protect a computer's sensitive solid-state components. In addition, an interface sometimes has a small amount of its own memory, called a *buffer*.

Digital Signals

Digital signals have only two possible measurements:—0/1 (or ON/OFF or YES/NO). This is called a digital/binary signal, and can be understood by a microprocessor. Digital binary signals are represented as square waves, where the low measurement is 0 volts and the high measurement is a positive voltage (Figure 11-18). The low measurement would be assigned a "0" by a microprocessor and the high measurement would be assigned a "1."

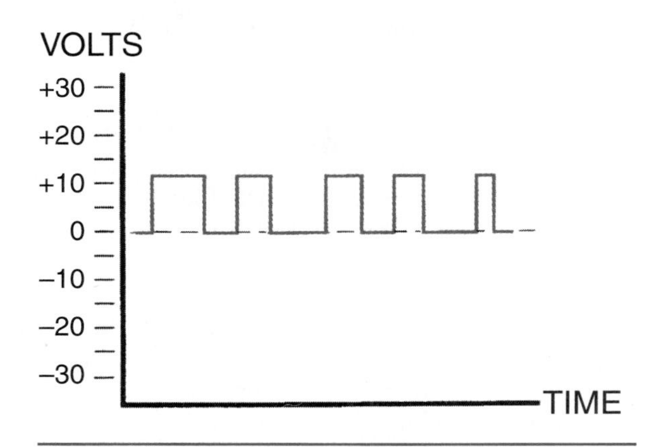

Figure 11-18. Digital signal. (GM Service and Parts Operations)

Switch Signals

In general, switches are the simplest sensors. Usually, a switch has a closed position that offers almost no resistance and an open position that offers close to infinite resistance. This means that the output signal is either 0 volts or it is equal to the voltage applied to the switch. The computer receiving this signal will know how to interpret the two different voltages. For example, one voltage might mean oil pressure is too low while the other means it is OK. Or, one voltage might mean the brake pedal is fully depressed while the other voltage means the brake pedal is not fully depressed. Switches in the PCM system, for example, are used to signal the module of the condition of certain systems, such as power steering pressure, A/C status, transmission temperature, and so on. For this reason, the switches will be discussed according to how they signal the module. Signaling is done in one of two ways: either by grounding a signal or by closing to allow a voltage to signal the PCM. The two types of switched inputs, depending on the source of power to the circuit, are as follows:

- Grounding or pull-down circuit
- Voltage or pull-up circuit

Grounding or Pull-Down Circuit

Grounding switches always produce a digital signal to the module. A grounding switch must always be used in series with a fixed, current-limiting resistor. The current-limiting resistor reduces current in the circuit to prevent damage to the control module. In Figure 11-19, the signal is zero (low) with the switch closed. When the switch is open, the voltage will be signal voltage (usually 5 or 12 volts) high. A "pull-down" circuit (Figure 11-19) is provided with a reference voltage signal from the computer. The power source for the circuit is internal to the computer. When the switch is closed, source voltage is pulled low to an external ground. The computer registers a low voltage reference signal. When the switch is open, the computer registers a high voltage reference signal.

Voltage or Pull-Up Circuit

Voltage input signal switches allow a 12-volt signal to the module when the switches are closed. When they are opened, there is no voltage sent to the module. This provides a digital input B+ (high) or zero (low) signal to the module. A "pull-up" circuit (Figure 11-20) has a power source outside the computer, like fused battery or ignition voltage. In a pull-up circuit, the computer is not providing the reference voltage signal. When the switch is closed, external source voltage is applied to the computer.

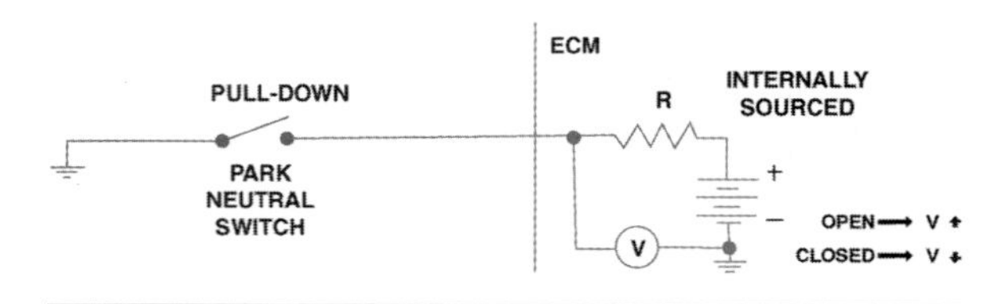

Figure 11-19. Grounding (Pull-Down) Switch. (GM Service and Parts Operations).

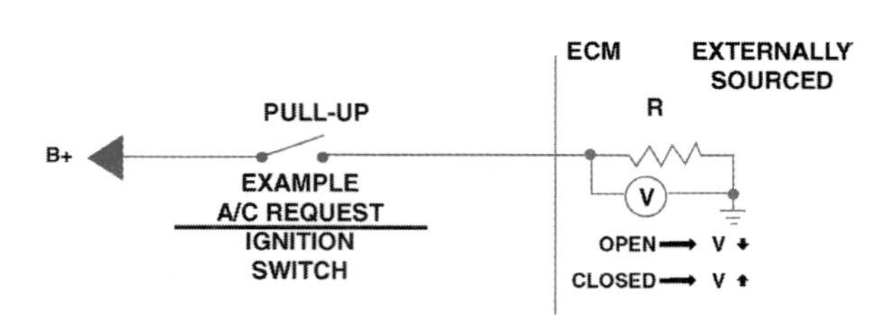

Figure 11-20. Voltage or pull-up circuit. (GM Service and Parts Operations)

An open switch, on the other hand, generates a low, or zero, reference signal.

Photo Diodes

Photo diodes are designed specifically to detect light. These diodes are constructed with a glass or plastic window through which the light enters. These diodes are often used in automatic headlamp control systems.

Thermistors

A *thermistor* is a device in which resistance changes with temperature. The electrical resistance decreases as temperature increases, and increases as temperature decreases. The hotter a thermistor gets the lower the resistance. The varying resistance affects the voltage drop across the sensor terminals and provides electrical signals corresponding to temperature. In a two-wire sensor, one of the wires is the reference voltage from the computer. The other is the ground wire, which goes to the computer or to an external ground. The computer gets the input signal by monitoring the reference voltage line. An example of this would be the Engine Coolant Temperature (ECT) sensor, as shown in Figure 11-21.

Piezoelectric Devices

A piezoelectric device utilizes a wafer of a crystalline material that produces a voltage when bent by the pressure of sound waves. *Knock sensors* determine when the engine is pinging (pre-ignition) and signals the PCM to retard timing. Additionally, microphones and speakers sometimes use a piezo

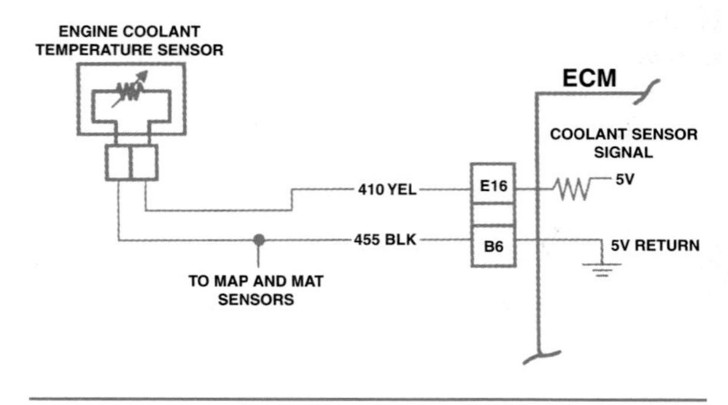

Figure 11-21. Thermistor. (GM Service and Parts Operations)

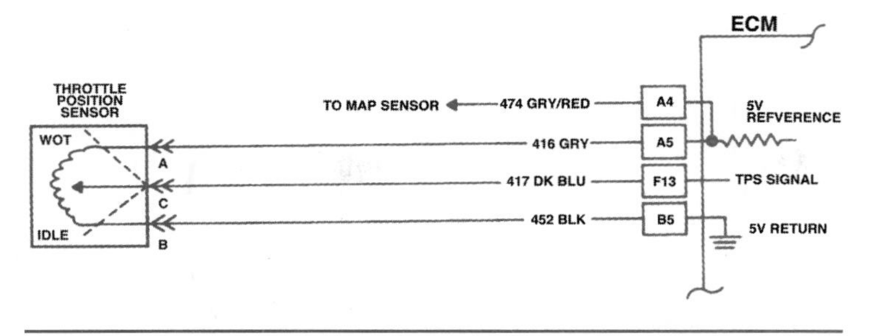

Figure 11-22. Potentiometer. (GM Service and Parts Operations)

crystal. Microphones act the same as the knock sensor—electric current is produced by sound waves applying pressure to the crystal/diaphragm. In piezo speakers, sound is produced when an electric current is applied to the piezo crystal.

Potentiometers

A *potentiometer* is a three-wire, variable resistor that acts as a voltage divider to produce a continuously variable output signal proportional to a mechanical position. The throttle position sensor is a common example of a potentiometer (Figure 11-22).

Hall-Effect Devices

The Hall-Effect sensor (switch) shown in Figure 11-23 is the most commonly used engine-position sensor. Unlike a magnetic pulse generator, the Hall-Effect sensor produces an accurate voltage signal throughout the entire rpm range of the engine. The Hall-Effect, which is a voltage dependant upon a magnetic fields, is also known as the *Lorenz* theory. Furthermore, a Hall-Effect switch produces a square wave signal that is more compatible with the digital signals required by on-board computers. The Hall-Effect principal states: If a current is allowed to flow through a thin conducting material, and that material is exposed to a magnetic field, voltage is produced.

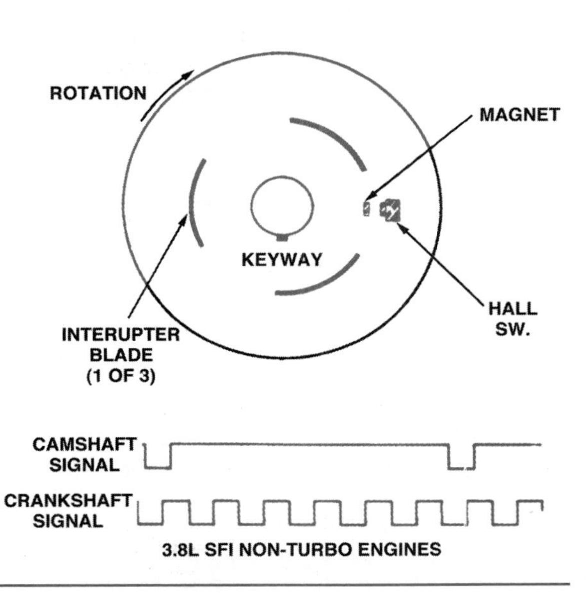

Figure 11-23. Hall-effect sensor (switch). (GM Service and Parts Operations)

Radio Frequency Signals

Remote Keyless Entry (RKE) systems use a hand-held transmitter to send radio frequency signals to a receiver located in the vehicle. With a press of a button, the user can lock or unlock the doors, arm or disarm the alarm system, and in some cases, start the engine from a distance of 25 feet or more. In some systems, the RKE module communicates through the data bus to the body control module to activate the driver memory functions and adjusts the seats, mirrors, and radio to a preset condition.

Shutter Wheel

A shutter wheel (Figure 11-24) is a device consisting of a series of alternating windows and vanes that pass between the Hall layer and a magnet. When a vane of the shutter wheel is positioned between the magnet and the Hall element, the metallic vane blocks the magnetic field from the Hall layer and Hall voltage is low. When a shutter wheel window rotates into the air gap, the magnetic field flows to the Hall layer and pushes

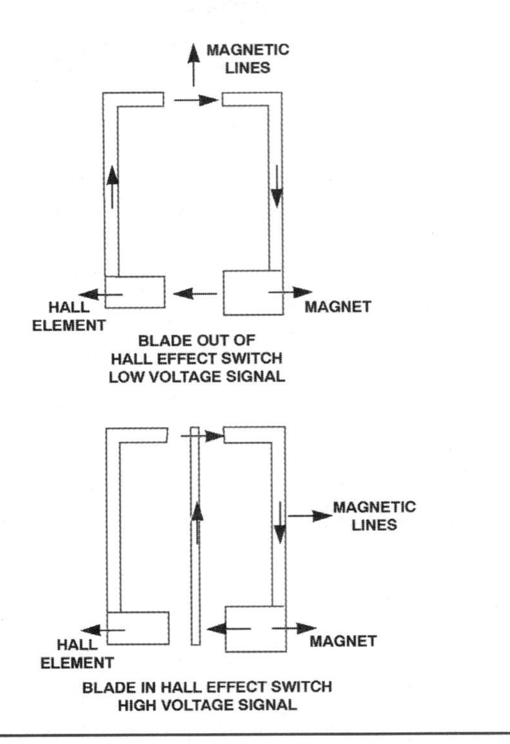

Figure 11-24. Shutter wheel.

the Hall voltage to its maximum range. For more information on this topic, see the section on "Circuit Tests" in Chapter 11 of the *Shop Manual*.

OUTPUT DEVICES

We have looked at several examples of input signals a computer might receive. Now we'll learn a little more about output signals. We know that input devices send information to the module, which is processed and checked against the module's memory. The module performs its internal computation and sends commands to various output devices. These output devices may be the instrument panel display or a system actuator, for example. The output of one module can also be used as an input to another module.

Types of Output Controls

A computer can control an output in one of three different ways. It can:

- Apply voltage
- Apply ground
- Alternately apply voltage and ground

On/Off (Switched Power)

Terminals related to the solenoids of an electronic brake control module (EBCM) are examples of ON/OFF switched power. Under certain circumstances, the EBCM will close one of its internal switches. This, in turn, applies battery voltage to an ABS solenoid, and the solenoid is actuated. In this case, the computer is applying voltage as an output control.

On/Off (Switched Ground)

The terminal related to the brake switch input of EBCM is an example of ON/OFF switched ground. Normally, the switch inside the EBCM is open, and the brake indicator circuit is not grounded. Under certain circumstances, however, the EBCM closes the switch and applies ground to the circuit. The indicator lights up. Of course, there are other conditions that could close this circuit and cause the brake indicator to light.

Alternating Ground and Power

Depending on the circumstances, some ECM/PCM controllers apply either battery voltage or ground to a stepper motor through several different terminals. For example, when the switch at terminal "A" is at battery voltage, the switch at "B" is in the ground position. When the switch at terminal "A" is at ground, the switch at "B" is at the battery voltage position. This means that the ECM/PCM can pulse voltage to a stepper motor in two different polarities.

The information in the OEM service manual states that the computer decides when to apply ground or voltage. The computer makes these decisions based on the input signals it receives and the computer's programming.

Actuators

An *actuator* is a component that performs a task based on the output signal it receives from a computer. For example, the stepper motor is an actuator. It performs the task of moving a valve or mechanical arm.

Variable Control (Pulse-Width Modulation)

So far, we've discussed output signals that turn another device on and off. Sometimes, more complicated signals need to be sent. For example, it is often necessary to control a motor's speed as well as to turn it fully on or fully off. We know that a digital signal has only two possible values, which we usually call ON and OFF. If the signal is OFF, the actuator doesn't operate. If the signal is ON, the actuator does operate. Now think about what happens if the signal fluctuates quickly between ON and OFF.

In Figure 11-25, each digital signal is shown over time, and the total time has been divided into periods of equal lengths. During each period, the signal is on for a certain percentage of the total time. The ratio of ON time to total time is called the *duty cycle*. The higher the duty cycle percentage, the longer the actuator operates. If the actuator is a motor, it runs faster as the duty cycle increases. If the actuator is a light, it gets brighter as the duty cycle increases. If the duty cycle percentage is 100 percent, the actuator operates at its fastest speed or greatest intensity.

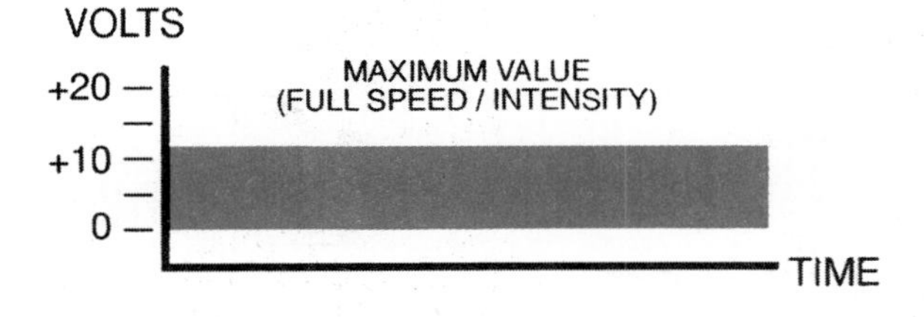

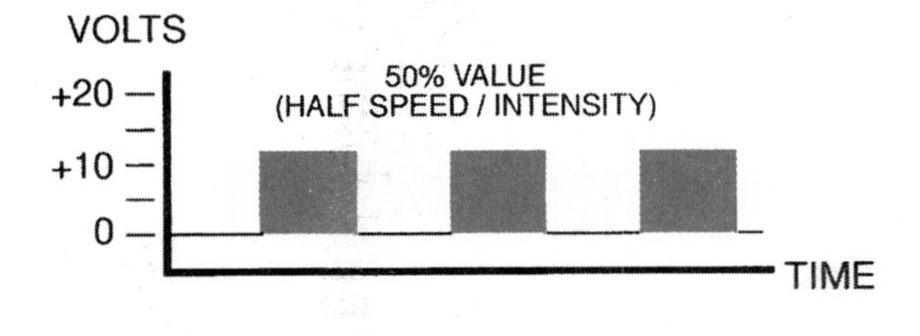

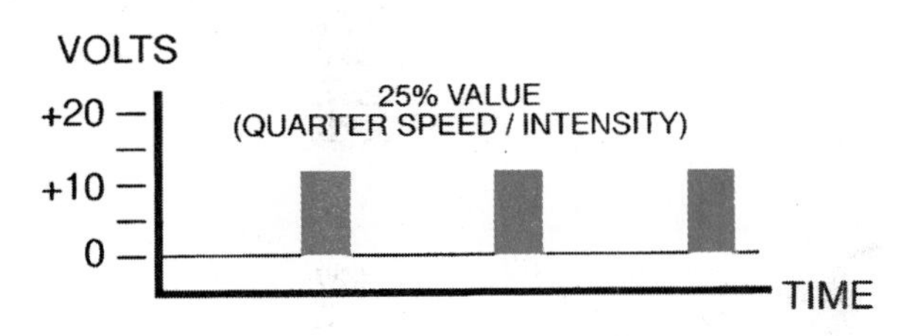

Figure 11-25. Variable control (pulse-width modulation). (GM Service and Parts Operations)

Relays

As previously covered in Chapter 4, a *relay* is a switching device operated by a low-current circuit that controls the opening and closing of another circuit of higher current capacity. Relays can be either "normally open" or "normally closed." The normal position of a relay refers to its condition at rest, without a signal from the control module to energize. When the normally open relay is energized, the internal contacts are closed and the high-current circuit is completed. When the normally closed relay is energized, the internal contacts are opened and the high-current circuit is disconnected.

Solenoids

As discussed in Chapter 4, a solenoid is a device that has a moveable core surrounded by a wire coil. When an electrical current flows through the coil, the force of electromagnetism moves the solenoid core. This process of converting electrical energy to mechanical movement can be applied to many of the control valves found in today's engines. These valves can control air flow, fuel, vacuum, or hydraulic fluids.

Solenoids can be cycled on and off very rapidly. This is called duty cycle and is expressed as a percentage in relation to a period of time. A duty cycle of 100 percent means that the solenoid is energized all of the time. A duty cycle of 50 percent means that the solenoid is energized half of the time.

Indicator Lights

Indicator lights illuminate to warn the driver that something in the system is not functioning properly or that a situation exists that must be corrected. Control modules continuously monitor the various components of a vehicle system and many modules have the ability to trigger a warning light. The "Check Engine" light, technically referred to as the *Malfunction Indicator Lamp* (MIL), alerts the driver of certain malfunctions in the Engine Control System directly related to exhaust emissions. The Inflatable Restraint (Air Bag) and Antilock Brake System (ABS) are two safety-related vehicle systems monitored by individual control modules.

Motors

Electric motors are used throughout the vehicle. They are used to run hydraulic pumps in ABS systems, fuel pumps, radiator fans, heater blower motors, power windows and seats, air suspension compressors, and so on. Like any other electrically powered device in the vehicle, a control module can easily turn a motor on or off to satisfy a computer request, and can do so without the driver's input.

MICROPROCESSOR PROCESSING CYCLE

The microprocessor processing cycle in the PCM has four basic functions:

- Regulating reference voltage (V-ref)
- Input conditioning (amplification, ADC)
- Processing inputs and outputs
- Managing output drivers

PCM is the generic term for the unit housing the complete computer assembly, and often, some or all of the switching apparatus. The engine management (or vehicle management) PCM generally receives command and monitoring data from an electronic subcircuit that operates at 5 volts. The PCM also manages reference voltage (V-ref) at either 5 volts or in the case of OBD II, 7 volts, to power-up and to benchmark input signals. The electronic subcircuit consists primarily of sensors located variously on the engine and vehicle chassis. Because the data they input is mostly in the form of analog voltage values, these must be converted to digital codes for processing by an ADC (analog to digital converter) integral with the PCM. Weak signals may be conditioned or strengthened prior to processing. Preparing input signals for processing are PCM functions categorized as *input conditioning*. Figure 11-26 shows the input sensors on the left and the output actuators on the right. The voltage regulator provides the V-ref and signal conditioning in a PCM.

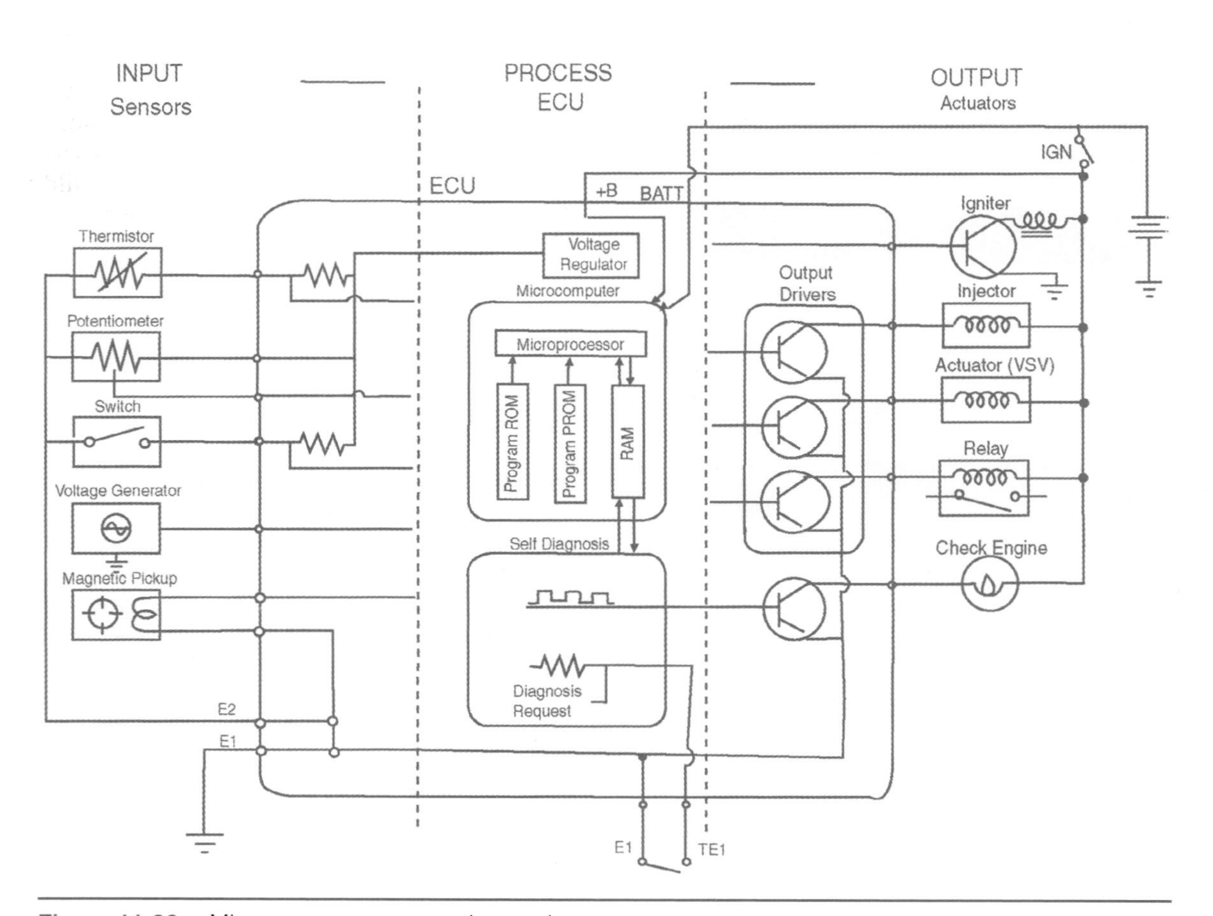

Figure 11-26. Microprocessor processing cycle.

Processing in the PCM involves scanning the programmed fuel map data (in ROM, complemented; finally defined in PROM, EEPROM, or EPROM), engine and chassis monitoring sensors (such as Engine Coolant Temperature sensors, Vehicle Speed sensors [VSS], etc.) and command inputs (Throttle Position sensor, cruise control, Oxygen sensor, etc.), and subsequently plotting an actual fuel quantity to be delivered.

NOTE: SAE Standard J1930 pertains to electrical/electronic systems diagnostic terms, definitions, and acronyms. As of June 1993, it defines ECM specifically as an engine control module and PCM as powertrain control module. Automotive manufacturers are required to abide by the terms in J1930.

The results of CPU logic and arithmetic processing must next be converted to action by means of outputs. PCM processing may occur at different frequencies classified as foreground and background computations. Foreground operations would include response to a critical command input such as the TPS (Throttle Position sensor) or Oxygen sensor whose signal must be factored immediately to generate the appropriate outcome. The monitoring of engine-intake air temperature is often classified as a background operation; though this is obviously valid, this signal does not require as immediate adjustment to operating strategy as that from the Oxygen sensor. PCM outputs are switching functions. When the engine-management PCM is bussed into other system PCMs (transmission, ABS, BCM) or there is a single vehicle management PCM managing multiple systems, switching mechanisms increase in complexity. For more information on this subject, see the section on "PCM Replacement" in Chapter 11 of the *Shop Manual.*

THE INTERNET AND WORLD WIDE WEB

Networks

A **network** is a term that describes the connection of several different devices for the purpose of communication. In automobiles, many of the various systems rely on computerized control modules communicating with sensors, actuators, and motors to achieve proper operation. This communication is accomplished through a series of wires that allow electrical signals or current to flow back and forth between the components of a system or between several systems. However, the more complex a system is, the larger the network needs to be, resulting in more and more wiring. This wiring system is also limited to the communication of one signal at a time for any given wire.

A network in its simplest form would be a *handshake* connection between two computers. **Handshake** is the term used to describe the establishing of a communications link. More often, a network refers to a series of interconnected computers, usually PCs, for purposes of sharing data, software, and hard resources. A large office with several computer stations may network them to permit data exchange and may perhaps have a single printer to serve all the stations. The means of connecting the computers in a network vary, but if the network extends beyond a single building, it usually involves a *modem* connection into the telephone system.

Modem

A *modem* (Figure 11-27) is a device for converting the digital output of the computer to the analog signals required for transmission by the phone system. The speed at which a modem can transmit data is specified by its baud rate. Baud rate indicates the number of times per second one

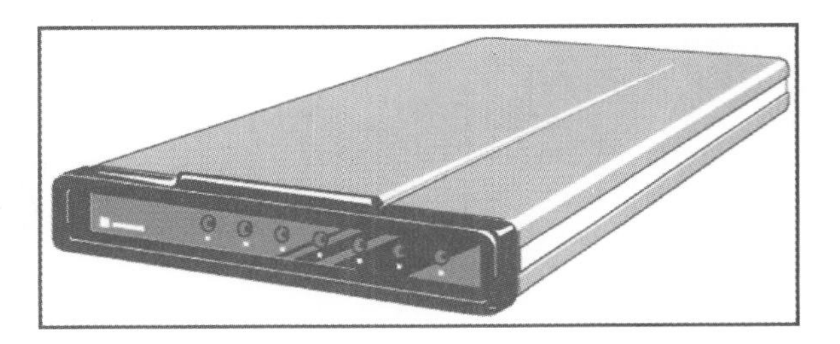

Figure 11-27. Modem.

bit can be transmitted. The baud rate specification is critical when considering the purchase of any equipment that must be networked. Engine OEMs will always specify a minimum baud rate for networking computers with their mainframe systems.

A *star network* is constructed around a central computer with multiple terminals or nodes. *Node* is a term used to describe each user station. The central computer may be a mainframe. Often a star network will use *dumb nodes*—stations that consist of a screen and keyboard but no processing or data storage capability independent of the central computer.

INTERNET

The **Internet** (Figure 11-28) is a network of networks. It now connects most of the countries in the world and is used by over 50 million people world wide, a number that grows at an astonishing 5 percent per month. It is very appropriately referred to as the World Wide Web or WWW.

The following paragraphs discuss some of the uses of the Internet system.

Sending E-Mail (Electronic Mail)

E-Mail is a highly efficient means of messaging users served by the Internet. It has the advantage of nearly immediate transfer and is facilitated without the protocols (and delay) of hard mail.

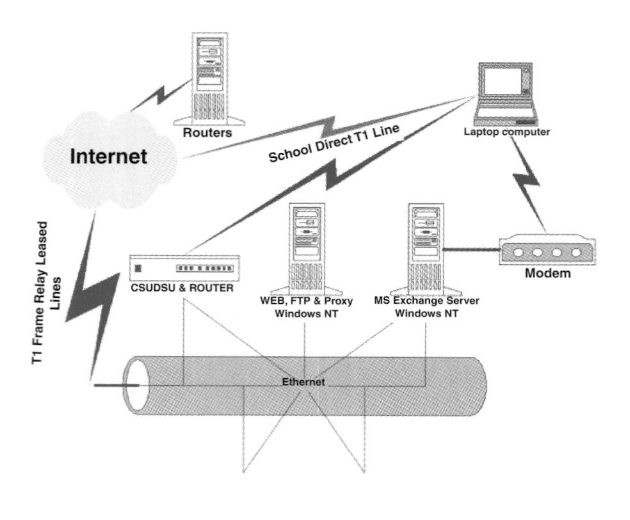

Figure 11-28. The Internet.

Users must have an e-mail address and a modicum of computer skills, which should include some keyboarding ability.

File Transfer

Files may be transferred from the Internet to a user PC system using a GUI (graphical user interface) and dragging the file title from the Internet menu to the user menu.

Search Databases

Most government agencies and educational institutions make extensive quantities of information available on the Internet. Automotive manufacturers now offer electronic service information via the Internet and Web-based training is also offered. This makes accessing information, especially detailed, subject-specific data, easy, certainly much easier than attempting to locate it by phoning bureaucrats or searching a public library with its limited hard resources. A variety of search engines are used on the Internet. Search engines will hunt down data on the Web using a key word or phrase.

Research a Company

Most U.S. companies have a Web page to maintain a profile on the Internet and to offer products and services. The extent of information accessible varies with each company.

Go Shopping

Thousands of products and services are advertised on the Internet and offered for sale using a credit card number. The Internet term for this type of business is called *e-commerce.* The growth rate for e-commerce rivals the growth of the industrial revolution in the early twentieth century.

Find a Job

Many companies list job vacancies on the Internet. There are search engines dedicated exclusively to job searching on the Internet. Web sites like www.monster.com are examples of this type of Internet use.

Receive On-Line News

Magazines, newspapers, and other news services make information available on the Web, and depending on the service, it may be up-to-date information. Sports scores throughout the world are an example of data that is almost constantly updated. The scores are usually accompanied by stories, which can be read or downloaded for printing.

PR (Public Relations)

Almost every company, government agency, and organization in North America and Western Europe realizes the importance of the Internet as a communications tool. Connected to the White House, for instance, the user may select anything from a pictorial tour to messages from the President and Vice-President. A presence on the Internet is essential for any major company transacting business in North America.

LANs and Intranet Systems

LANs (local area networks) and *intranet* are terms that usually describe private communication networks run by most OEM automotive corporations and government agencies to handle the specific communications needs of their operation. Access is usually restricted, with passwords required for system entry and fields within the system. The protocols used are Internet derived and while general access *from* the Internet may be to some extent blocked, access *to* it is usually enabled. A LAN could be housed in a single room and consist of two or three nodes, or it may be spread over the country and incorporate thousands of nodes.

Telecommunications Systems

The phone systems are responsible for both the hard (wiring/switching/decoding) and soft (microwave/radio waves, etc.) linkages that enable networking. Of the telecommunications systems in use in the world today, the telephone system is probably the most used. The telephone system is continually being expanded, largely to handle the immense growth in the data-communications traffic over the past decade.

A telephone consists of a transmitter, a receiver, and a dial or push-button mechanism. The telephone transmits voice or sound waves by having them vibrate a thin metal diaphragm backed by carbon granules through which current flows; when a sound wave acts to pulse the diaphragm inwards, the carbon granules pack closely, permitting higher current flow. So, the current flow depends on sound waves. This fluctuating current flows to a receiver. The receiver earpiece consists of an annular armature located between a permanent magnet and a coil; attached to the armature is a plastic diaphragm. When fluctuating current flows into the coil it becomes an electromagnet. This acts to attract the armature, causing it to vibrate and creating sound waves that replicate speech.

The telephone must also have a means of targeting calls. A dial or touch-tone device is used. A dial operates by sending a series of pulses to the switching network; when the number 5 is dialled, current from the source set is interrupted five times. The dialing creates a pattern of interruptions, which are used to target the phone signal to the correct location. With a touch-tone device, each numeric button produces a different tone or frequency, which is electronically decoded to target the phone call.

Most modern telephone switching is performed digitally. Incoming signals from a telephone set are first converted into digital signals, patterns of on-off pulses. A computer processes these pulses to search the shortest (and least overloaded) path to connect the target phone.

TRANSMISSION MEDIA

Early telephone systems used exclusively hard-wire transmission media to carry signals from one point to another. Today a telephone call to a foreign country can involve the use of several different types of transmission media.

Copper Wire

Most local calls are carried over pairs of copper wires. Coaxial cable is also used and can carry signals of higher frequency and can allow more transmissions on a single line using multiple coaxial conductors.

Integrated Services Digital Network

ISDN stands for Integrated Services Digital Network (ISDN). ISDN is an ITU (International Telecommunications Union) defined digital telephone standard that can be used to provide noise-free phone service and Internet access. The ITU is the leading publisher of telecommunication technology, regulatory, and standards information. What's neat about ISDN is its ability to work with ordinary phone wiring, called *twisted pair,* that's found in most homes and offices. Unlike ordinary telephone service, ISDN is fully digital, so you don't need a modem. (You will need some special equipment, though, as this section describes.) Another welcome ISDN feature is near-instantaneous connection; there's no tedious dial-up or logon process to go through before you're connected to the Internet.

In order to obtain ISDN service for Internet access, you need a local telephone company that offers ISDN at a reasonable price. (In some areas, ISDN service is not yet available.) You also need an Internet service provider that offers an ISDN connection that is accessible by means of a local telephone call. Never sign up for ISDN telephone service without first making sure that your local ISP supports ISDN connections. An ISDN connection consists of two different channels: a 16 Kbps channel used for control signals and a 56 Kbps or 64 Kbps channel (called a *B channel*) for voice or data.

DSL (Digital Subscriber Line)

DSL is a blanket term for a group of related technologies, including *Asymmetric Digital Subscriber Line (ADSL)*. Already available in major metropolitan markets, ADSL is akin to ISDN in that it uses existing copper wiring. But ADSL is much faster than ISDN, typically, up to 1.5 Mbps when downloading data, and up to 256 Kbps when uploading data. In ADSL, the downloading speeds are much faster than uploading speeds; the discrepancy explains why the service is called "asymmetric." It is assumed that you will be consuming information more often than publishing it. How fast is DSL? With a 56 Kbps modem, you will need about 2.5 minutes to download a 1 MB video clip. With DSL, you can download the same video in less than 5 seconds.

To connect to a DSL line, your computer will need a DSL "modem," as well as a DSL phone line and an Internet service subscription that includes DSL service. One potential problem is that DSL service is not standardized, so you will need a DSL modem that is compatible with your telephone company's particular type of DSL service. (Generally, you get the ADSL modem from the telephone company when you sign up for the service.) Standardization efforts are underway, however, and these efforts will reduce the cost and complexity of DSL installations. Don't expect DSL service if you are located in an out-of-the-way area; DSL service is available only within a distance of 2 or 3 miles from a local telephone office.

Microwave

Microwave (Figure 11-29) is the modulation of high-frequency radio signals. These radio signals are transmitted between antennae located on towers that are within sight of each other. Microwaves travel only in straight lines and will not refract from the ionosphere layer of the atmosphere. So, because they will only travel short distances, multiple relays must be used. Microwaves have frequencies of many gigahertz, permitting large numbers of simultaneous transmissions on each frequency band. Microwave currently accounts for the major portion of long distance transmissions transacted over land in North America.

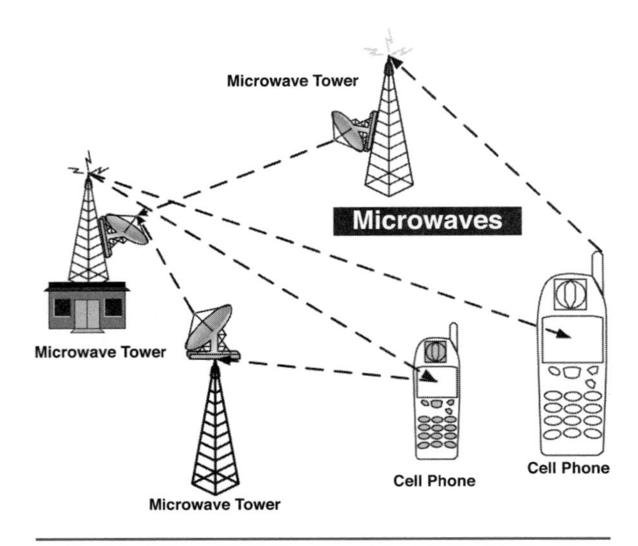

Figure 11-29. Microwave.

Cellular Phones

The term *cellular phone* (Figure 11-30) originates from the fact that each cell phone geographic region is divided into cells, each with its own transmitter-receiver. The cells are arranged so that the frequencies used in adjacent cell locations are different—though the same frequencies are reused many times over throughout the whole system. If a call is made from a cell phone in a car traveling over the boundary of one cell to another, the phone automatically switches frequencies, permitting a continuous phone conversation.

Fiber Optics

Fiber optic technology will dominate the telecommunications industry in the near future because of the vast volume of transmissions it can handle. Optical fibers use light as the transmitting medium rather than radio waves. Since light has a frequency around 100 TH_3 (100 trillion hertz), it has the potential to handle (theoretically) almost infinite numbers of transmissions. Telephone signals are first digitized then fed to semiconductor lasers, which produce pulses of light to a conduit of extremely thin glass fiber. At the receiving end of the optical fiber, photo detectors convert the optical signal to an electrical signal for local transmission. The fibers used to transmit the optical signals can be made almost transparent, and simply upgrading the transmission and receiving electronics and lasers can increase the transmission volume. Fiber optics already handle a growing percentage of the world's transcontinental telecommunications—up-front expense is really the only thing hindering its rate of growth. In time, first continental long-distance services then local systems will adopt the technology.

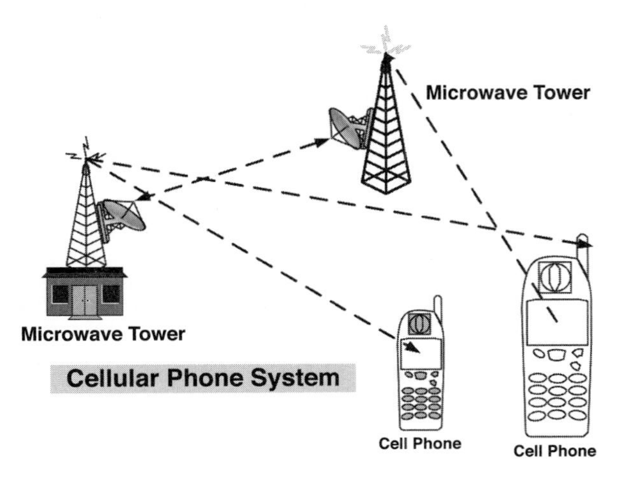

Figure 11-30. Cellular phone.

Satellites

A satellite is a small body that orbits a larger astronomical object; by this definition, the Earth's moon is a natural satellite. Telecommunications satellites are parked in a geosynchronous orbit (Figure 11-31) that acts as a relay station for microwave signals receiving a signal from a ground base and retransmitting it to another Earth-based station. Communications satellites are often placed in geosynchronous orbit. The time that it takes for a satellite to orbit the Earth once correlates exactly to its altitude. A satellite parked exactly 22,300 miles (35,900 km) above the Earth's surface on the equator will be positioned at a fixed point above the Earth's surface throughout the rotation. Such a satellite is said to be in

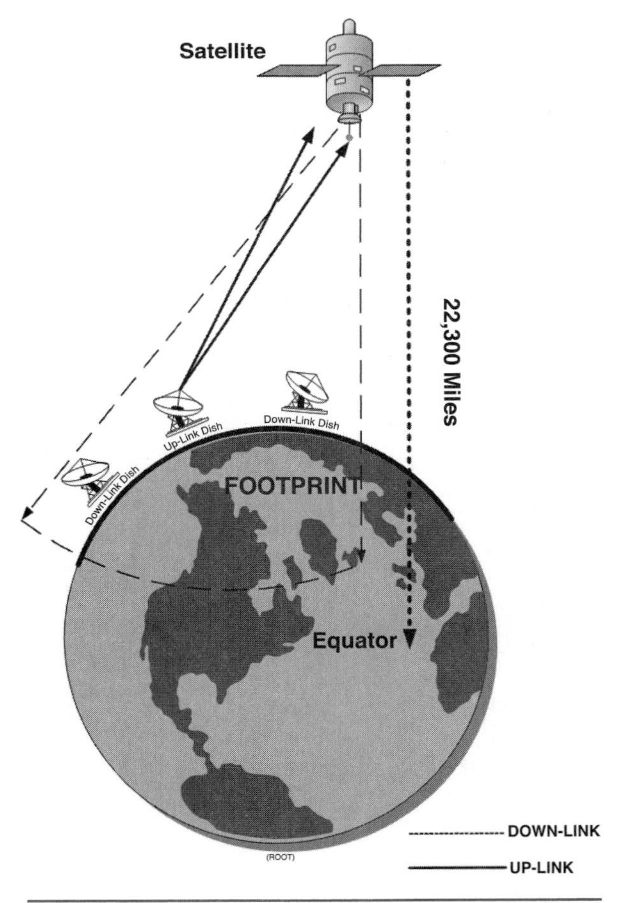

Figure 11-31. Geosynchronous orbit.

geostationary orbit. If its altitude were to increase slightly, its orbit would cause it to fall behind the fixed point—if it descended below the geosynchronous altitude, the reverse would occur.

NAVIGATION SYSTEMS

Luxury model cars and trucks may contain an optional navigation (NAV) system (Figure 11-32) that allows the driver to get directions and find map locations using the global positioning system via a geosynchronous orbiting satellite as shown in Figure 11-31. The navigation system contains the following components:

- Navigation radio
- Global positioning system (GPS) antenna

The export navigation system contains the following additional components:

- Vehicle information communication system (VICS) microprocessor card
- VICS optical/microwave beacon antenna
- Four or more TV (television) antennae integrated into the rear window glass
- VICS FM antenna integrated into the rear glass
- TV antenna module to control antenna selection
- Two TV antenna amplifiers on some models
- Auxiliary RCA video jacks

Navigation Radio

The radio acts as the operator interface for the navigation system, provides the data input from the operator to the navigation system, and provides navigation information to the operator via the display screen. The navigation radio is located in the center of the instrument panel (Figure 11-33). The navigation radio provides the following:

- A display screen—all navigation, audio and TV functions are displayed on this screen.
- Soft key buttons on the display to allow selection from menus and to operate the navigation system, the audio system, and the export TV system.

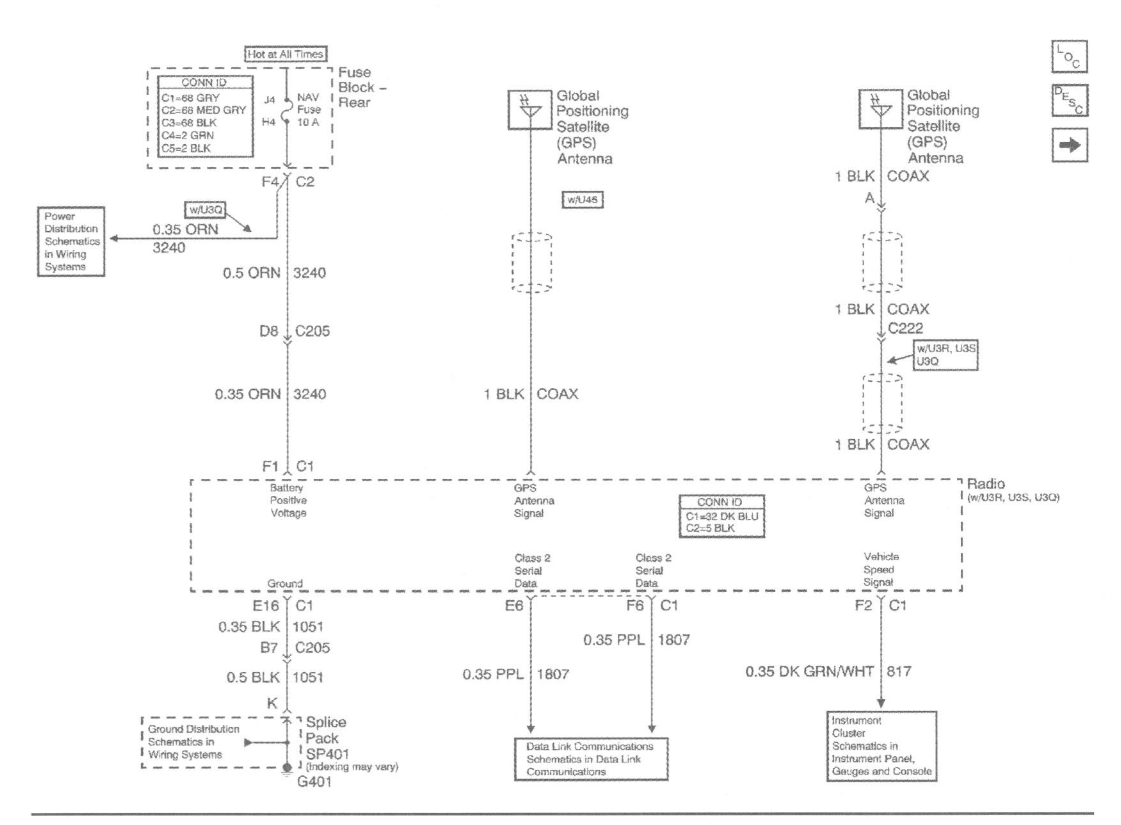

Figure 11-32. GM navigation system. (Courtesy of GM Service and Parts Operations)

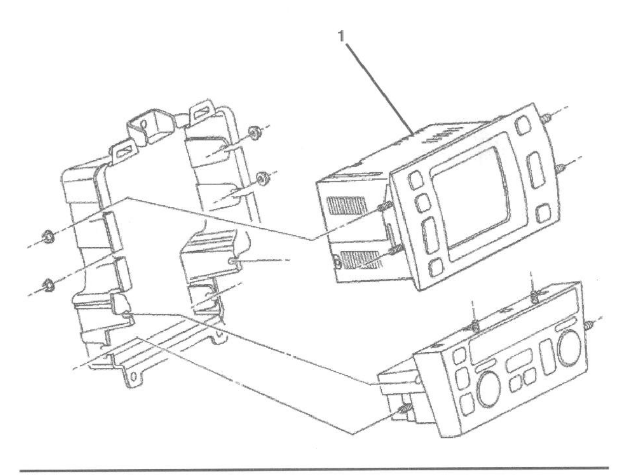

Figure 11-33. Navigation system radio. (GM Service and Parts Operations)

- The navigation system map with routing information displayed on the navigation radio screen.
- Verbal guidance to the operator.

Audio System Interface

The navigation radio controls the audio system by sending key-pressed messages to the remote radio and displaying audio information.

Global Positioning System (GPS) Antenna

The GPS antenna (shown in Figure 11-32) is attached to the bottom of the rear shelf and can be accessed through the rear compartment. The GPS antenna is powered through the same coaxial cable used to send the signals to the NAV system.

VICS Card

The *vehicle information communication system (VICS) card,* which is usually an export-only, dealer-installed option, is installed inside the navigation radio. The VICS card uses signals from the optical/microwave beacon antenna and the rear glass VICS antenna via the TV antenna module to convey routing information to the NAV. The NAV either revises planned routing or relays the VICS information to the operator through the screen of the navigation radio and the voice guidance system.

TV and VICS Antenna

The antennae array circuit (Figure 11-34) consists of conductive traces placed on the inside of the rear window glass.

TV Antenna Module

The TV antenna module is located behind the trim on the right rear side of the passenger compartment. The TV antenna module is used by the navigation radio to automatically select the antenna combination that provides the strongest TV signal to the navigation radio. The TV antenna module also provides the VICS FM signal to the VICS card.

TV Antenna Amplifiers

The TV antenna amplifiers are mounted behind the trim on each side of the rear window. The TV antenna module powers the TV antenna amplifiers through the same coaxial cable that it used to carry signals. They are attached to two of the four TV antenna trace patterns in the rear window glass.

Auxiliary RCA Video Jacks

The auxiliary RCA video jacks, used to attach a remote video device, are located at the rear of the center console. These connections may be used to provide audio and video input from a video player or camera to the navigation radio.

VICS Optical/Microwave Beacon Antenna

The VICS optical/microwave beacon antenna is a dealer-installed option. The antenna, which is mounted on the right side of the instrument panel, picks up signals through the front window glass. A coaxial cable carries the power to and signals from this antenna to the navigation radio.

SUMMARY

A computer system requires hardware and software to function. In a PC system, the system housing encases most of the components in either a desktop horizontal case or a vertical tower.

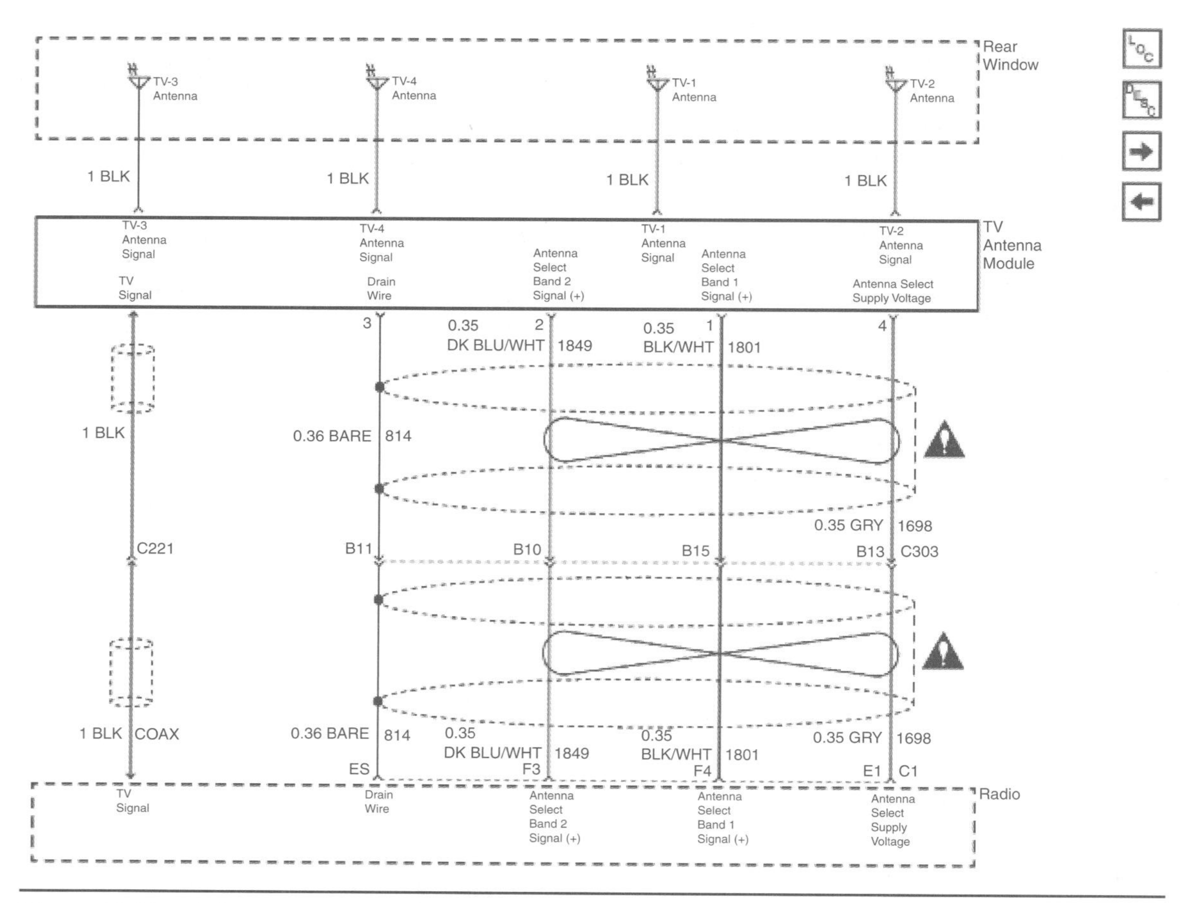

Figure 11-34. Antenna array circuit. (GM Service and Parts Operations)

Within the system housing, most of the internal components are connected to a main circuit board called a motherboard. The CPU is often referred to as the "brain" of a computer system. RAM is electronically retained and can be called main memory. A cache is a type of high-speed RAM used to store frequently consulted operating data requiring fast access. Commonly used data-storage media include cassette tapes, floppy disks, hard drive disk platters, and optically encoded disks (CDs).

Data is represented electronically in computer systems in the form of bits and bytes. A bit is the smallest unit of data that can be handled by a computer: it is simply an on or off condition used to represent digital values. There are 8 bits to a byte. A byte is capable of representing 256 data possibilities. The four-part processing cycle consists of fetching, decoding, executing, and storing. A vehicle ECM or PCM information-processing cycle comprises three stages: data input, data processing, and outputs.

RAM, or main memory, is electronically retained and therefore volatile. The master program for engine management in an engine controller PCM is usually written to ROM. PROM data is used to qualify the ROM data to a specific chassis application.

Some OEMs describe their PROM component as a personality module. EEPROM gives the PCM a read/write/erase memory component used for logging fault codes, masked codes, failure strategy management, and customer/proprietary data programming.

Input data may be categorized as command data and system monitoring data. Thermistors precisely measure temperature and may operate on either a NTC or PTC principle. Variable-capacitance-type sensors are often used to measure pressure values. Throttle position sensors are usually of the potentiometer type. Hall-Effect sensors generate a digital signal and are used to signal shaft-position data. Induction pulse generators are used to input shaft rotational speed data.

Engine management PCMs can be mounted on the engine itself or in a remote location such as under the dash.

The connection of a computer either to another single computer or to a network of computers can be referred to as a handshake. A network is a series of interconnected computers for purposes of data sharing, communication, and hardware sharing. The Internet is a network of networks with millions of users spread all over the world and is used by private individuals, government agencies, and corporations. The phone system is responsible for both the hard and soft links that enable networking. Most local telephone system transactions use pairs of copper wires. Most long distance telecommunications traffic currently uses microwave transmission with multiple relays. Fiber optics technology is positioned to replace both the local and long distance transmission media. It already handles a growing percentage of the world's transcontinental telecommunications transmissions.

Most telecommunications satellites are parked in a geosynchronous orbit. Telecommunications satellites generally use microwave transmissions. An up-link is an Earth-based signal broadcast to a satellite. A down-link is satellite signal received signal. The use of GPS technology in the automotive industry has brought about the widened use of on-board navigation systems in luxury model cars and trucks.

Review Questions

1. In which of the following can the data not be overwritten?
   a. RAM
   b. ROM
   c. PROM
   d. EEPROM
2. Which of the following components is sometimes referred to as the "brain" of a computer system?
   a. CRT
   b. VDT
   c. CPU
   d. CD-ROM
3. Which of the following data-retention categories is often referred to as *main memory?*
   a. RAM
   b. ROM
   c. PROM
   d. EEPROM
4. The main circuit board in computer is often referred to as a:
   a. Motherboard
   b. ROM card
   c. CPU chip
   d. BIOS
5. A PC keyboard would normally be connected to the system housing using which type of port?
   a. Com port
   b. Serial port
   c. Parallel port
   d. PCMCIA slot
6. In which of the following memory categories would customer data programming be written to from a scan tool?
   a. RAM
   b. ROM
   c. PROM
   d. EEPROM
7. Which of the following components conditions V-ref?
   a. PCM
   b. Voltage regulator
   c. Mouse
   d. Keyboard
8. A Hall-Effect sensor is usually responsible for signaling data concerning:
   a. Pressure
   b. Temperature
   c. Rotational speed
   d. Voltage
9. Which of the following could be described as being a computer input?
   a. Ground
   b. Actuator
   c. Relay
   d. Solenoid
10. Technician A says the speed at which a modem can transmit data is specified by its voltage rate. Technician B says that a modem is a device for converting the digital output of the computer to analog signals used by the phone company. Who is right?
    a. A only
    b. B only
    c. Both A and B
    d. Neither A nor B

12

Automotive Lighting Systems

LEARNING OBJECTIVES

Upon completion and review of this chapter, you should be able to:

- Identify and explain the operation of most automotive headlight systems.
- Identify the common bulbs used in automotive lighting systems.
- Define the taillamp, license plate lamp, and parking lamp circuits.
- Identify the components and define the stop lamp and turn signal circuits.
- Define the hazard warning lamp (emergency flasher) circuit.
- Identify the components and explain the operation of the backup light, side marker, and clearance lamp circuits.
- Identify the components and explain the operation of the instrument panel lighting.

KEY TERMS

Asymmetrical
Backup Lamps
Clearance Markers
Daytime Running Lights (DRL)
Dimmer Switch
Flasher Units
Halogen Sealed-Beam Headlamps
Hazard Warning Lamp
Headlamp Circuit
High-Intensity Discharge (HID) Lamp
Potentiometers
Rheostats
Sealed-Beam Headlamps
Side Marker Lamps
Stop Lamps
Symmetrical
Turn Signal Switch

INTRODUCTION

Automotive lighting circuits include important safety features, so they must be properly understood and serviced. Lighting circuits follow general patterns, according to the devices they serve, although slight variations appear from manufacturer to manufacturer.

HEADLAMP CIRCUITS

The **headlamp circuit** is one of the most standardized automotive circuits, because headlamp use is regulated by laws that until recently had seen little change since the 1940s. There are two basic types of headlamp circuits, as follows:

- Two-lamp circuit
- Four-lamp circuit

Manufacturers select the type of circuit on the basis of automotive body styling. Each circuit must provide a high-beam and a low-beam light, a switch or switches to control the beams, and a high-beam indicator.

Circuit Diagrams

Most often, the headlamps are grounded and switches are installed between the lamps and the power source, as shown in Figure 12-1. Some circuits have insulated bulbs and a grounded switch, as shown in Figure 12-2. In both cases, all lamp filaments are connected in parallel. The failure of one filament will not affect current flow through the others.

A two-lamp circuit (Figure 12-3) uses lamps that contain both a high-beam and a low-beam filament. A four-lamp circuit (Figure 12-4) has two double-filament lamps and two lamps with single, high-beam filaments. Lamp types are explained in more detail later in this chapter.

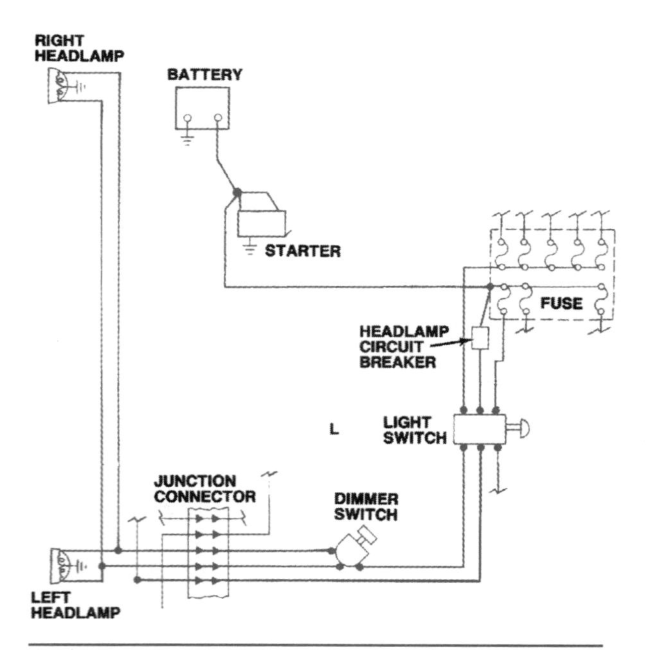

Figure 12-1. Most headlamp circuits have insulated switches and grounded bulbs.

Switches and Circuit Breakers

The three operating conditions of a headlamp circuit are as follows:

- **Off**—No current
- **Low-beam**—Current through low-beam filaments
- **High-beam**—Current through both the low-beam and the high-beam filaments

One or two switches control these current paths; the switches may control other lighting circuits as well. Most domestic cars have a main headlamp switch with three positions, as shown in Figure 12-5.

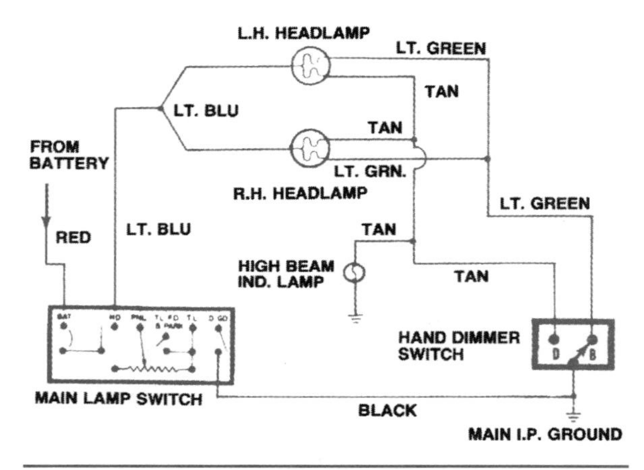

Figure 12-2. Some headlamp circuits use grounded switches and insulated bulbs. (GM Service and Parts Operations)

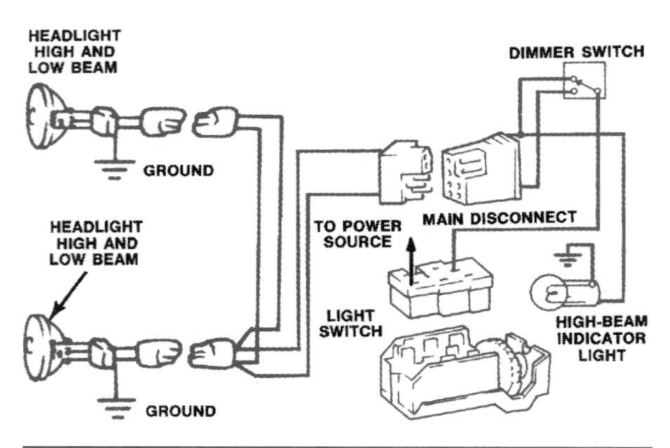

Figure 12-3. A two-lamp headlamp circuit uses two double-filament bulbs.

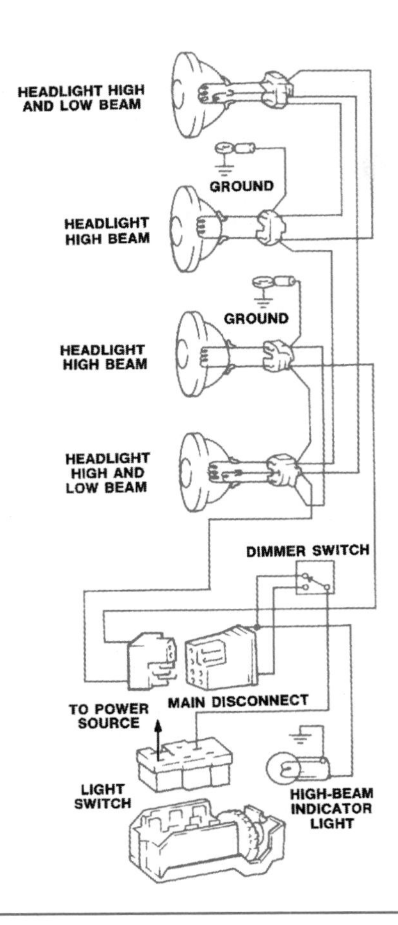

Figure 12-4. A four-lamp headlamp circuit uses two double-filament bulbs and two single-filament bulbs.

- **First position**—Off, no current.
- **Second position**—Current flows to parking lamps, taillamps, and other circuits.
- **Third position**—Current flows to both the second-position circuits and to the headlamp circuit.

The headlamp switch is connected to the battery whether the ignition switch is on or off. A two-position **dimmer switch** is connected in series with the headlamp switch. The dimmer switch controls the high- and low-beam current paths. If the headlamps are grounded at the bulb, as shown in Figure 12-6, the dimmer switch is installed between the main headlamp switch and the bulbs. If the headlamps are remotely grounded, as shown in Figure 12-2, the dimmer switch is installed between the bulbs and ground.

The dimmer switch on older cars and most light-duty trucks is foot operated and mounted near the pedals. On late-model cars, it generally is mounted on the steering column and operated by a multifunction stalk or lever, as shown in Figure 12-7. Some imported and late-model domestic cars use a

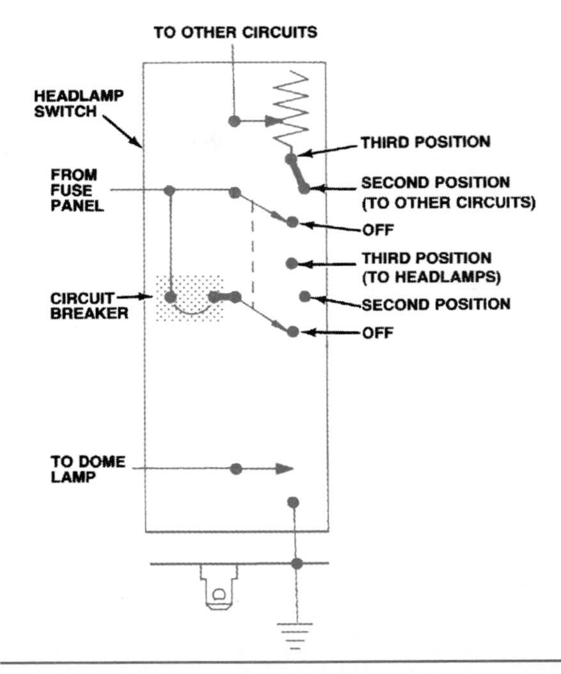

Figure 12-5. The main headlamp switch controls both the headlamp circuit and various other lighting circuits. (DaimlerChrysler Corporation)

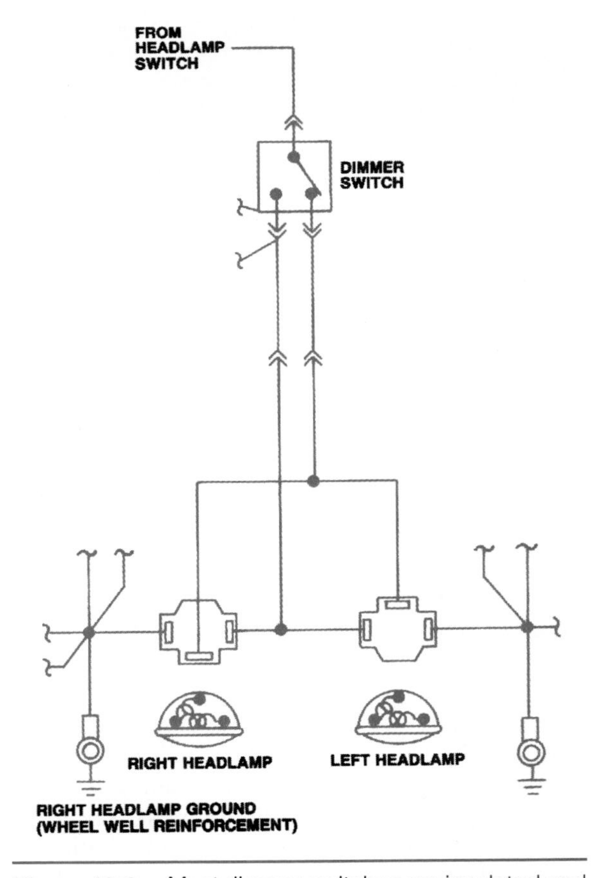

Figure 12-6. Most dimmer switches are insulated and control current flow to grounded bulbs. (DaimlerChrysler Corporation)

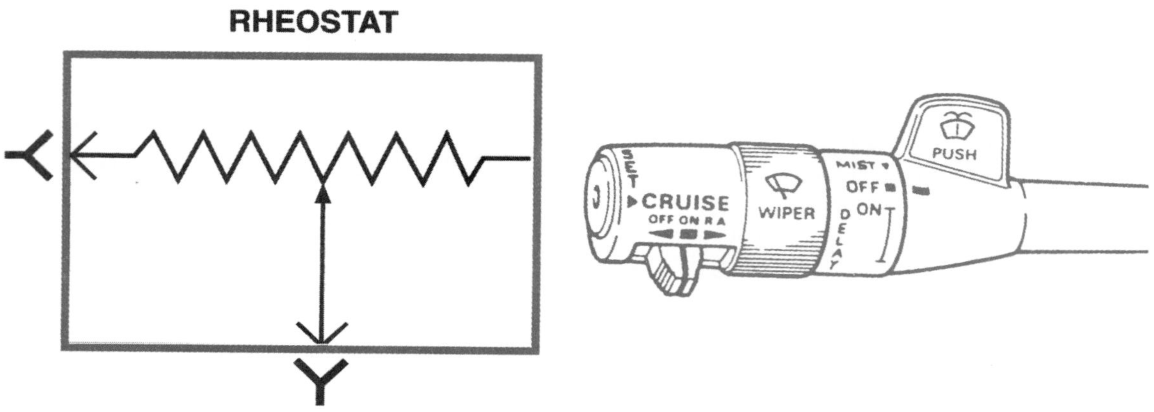

Figure 12-7. Late-model dimmer switches are operated by a steering column-mounted multifunction stalk or lever and control other lamp circuits. (GM Service and Parts Operations)

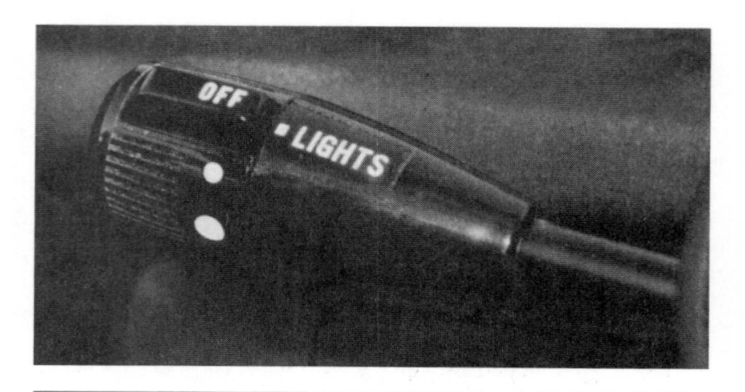

Figure 12-8. Headlamp switches may be mounted on the steering column and operated by a stalk or lever.

single switch to control all of the headlamp circuit operations. The following basic types of headlamp switches are used:

- Mounted on the steering column and operated by a lever (Figure 12-8)
- A push-pull switch mounted on the instrument panel (Figure 12-9)
- A rocker-type switch mounted on the instrument panel (Figure 12-10)

All systems must have an indicator lamp for high-beam operation. The indicator lamp is mounted on the instrument panel. It forms a parallel path to ground for a small amount of high-beam current and lights when the high-beam filaments light.

Because headlamps are an important safety feature, a Type 1, self-setting circuit breaker protects the circuitry. The circuit breaker can be built into the headlamp switch, as shown in Figure 12-5, or it can be a separate unit, as shown in Figure 12-1.

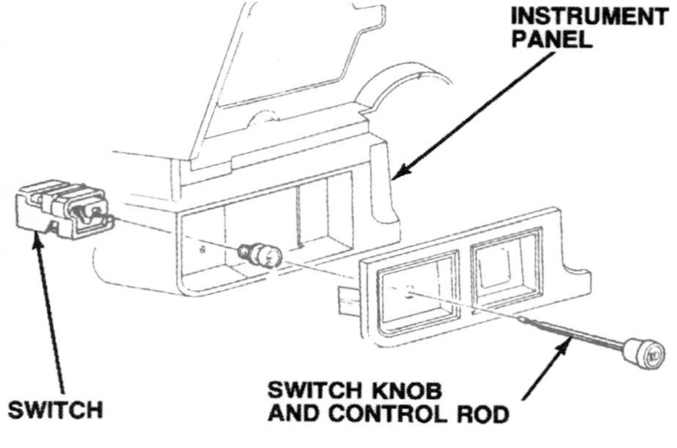

Figure 12-9. Push-pull headlamp switches are mounted on the instrument panel.

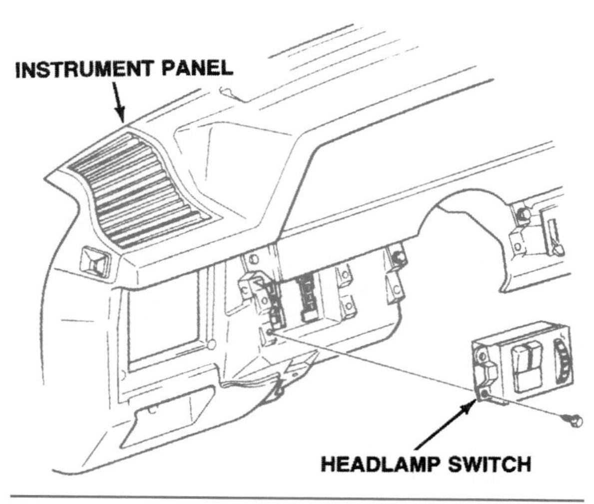

Figure 12-10. Rocker-type headlamp switches usually have a separate rotary rheostat control. (GM Service and Parts Operations)

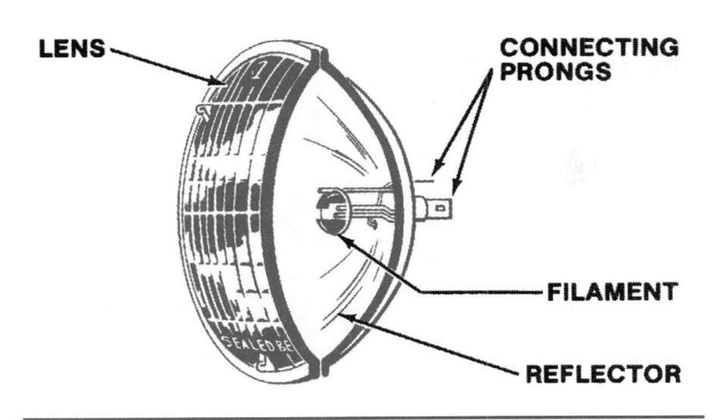

Figure 12-11. A cutaway view of a conventional sealed-beam headlamp.

Headlamps

Until 1940, a small replaceable bulb mounted behind a glass lens was used to provide light for night driving. Safety standards established in the United States in 1940 made round, sealed-beam units mandatory on domestic cars. Repeated attempts to modify the standards after World War II were only partially successful, beginning with the introduction of rectangular sealed-beam units in the mid-1970s. The first major change in headlight design came with the rectangular halogen headlamp, which appeared on some 1980 models. Since that time, considerable progress has been made in establishing other types, such as composite headlamps that use replaceable halogen bulbs.

Conventional Sealed-Beam Headlamps

A **sealed-beam headlamp** is a one-piece, replaceable unit containing a tungsten filament, a reflector, a lens, and connecting terminals, as shown in Figure 12-12. The position of the filament in front of the reflector determines whether the filament will cast a high or a low beam. The glass lens is designed to spread this beam in a specific way. Headlamps have **symmetrical** or **asymmetrical** beams, as shown in Figure 12-12. All high beams are spread symmetrically; all low beams are spread asymmetrically.

Halogen Sealed-Beam Headlamps

Halogen sealed-beam headlamps (Figure 12-13) first appeared as options on some 1980 domestic cars. Their illumination comes from passing current

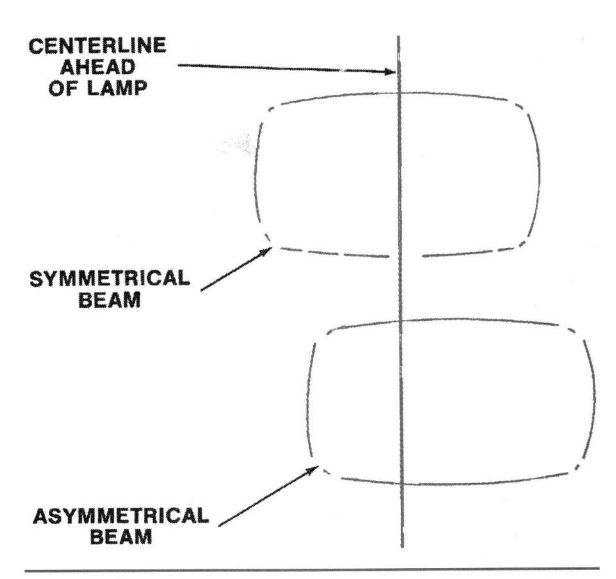

Figure 12-12. The glass lens design determines whether the beam is symmetrical or asymmetrical.

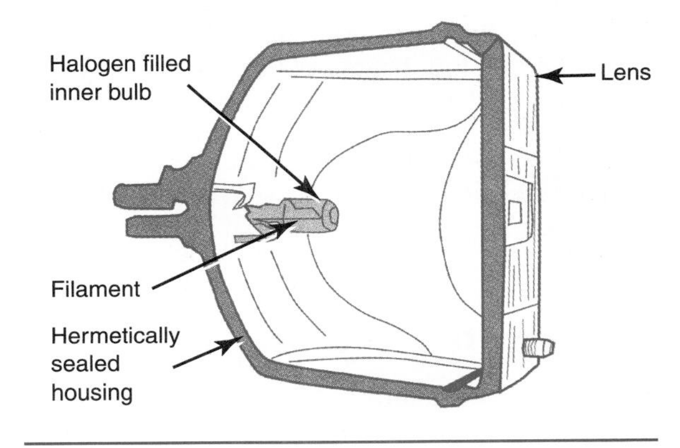

Figure 12-13. Halogen sealed-beam headlamps. (DaimlerChrysler Corporation)

through a filament in a pressure-filled halogen capsule, instead of through a filament in a conventional evacuated sealed-beam bulb. Halogen lamps provide brighter, whiter light than conventional headlamps.

Service and adjustment procedures are the same for halogen sealed-beam lamps as they are for conventional sealed-beams. Early bulb-type halogen lamps were not interchangeable with conventional sealed-beam headlamps, but today halogen sealed-beam lamps can often be used as direct replacements for their conventional counterpart. Halogen lamps are manufactured of glass or plastic. Glass lamps carry an "H" prefix; plastic lamps have an "HP" prefix. Plastic lamps are less susceptible to stone damage and also weigh considerably less than glass lamps.

Historical Headlamp Control Levers

Late-model cars are not the first to have a column-mounted lever controlling the headlamp circuit. The headlamps on the 1929 REO Wolverine Model B were turned on and off by a lever mounted to the left of the horn button on the steering wheel. This lever also controlled the high-low beam switching. The REO instruction book pointed out that, because each headlamp filament produced twenty-one candlepower, the headlamps should not be used when the car was standing still, to avoid draining the battery.

Halogen sealed-beam lamps are manufactured in the same sizes and types as conventional sealed-beams, with one additional type, as follows:

- Type 2E lamps contain both a high-beam and a low-beam inside a rectangular, 4 × 6 1/2-inch (102 × 165-mm) housing.

Like conventional sealed-beams, the type code and aiming pads are molded into the lens of the bulb.

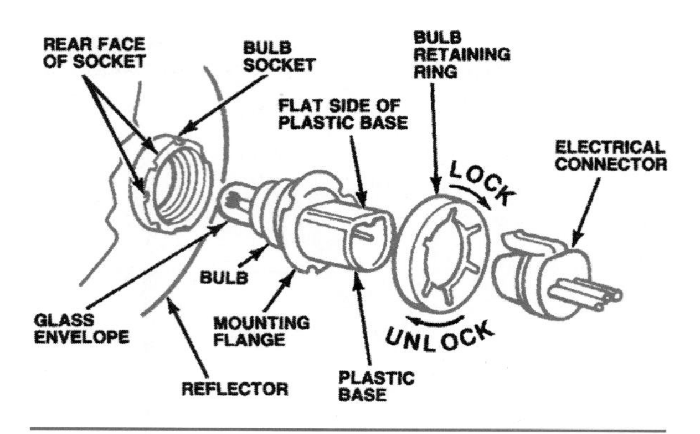

Figure 12-14. A replaceable halogen bulb is installed through the rear of the reflector and held in place with a retaining ring. (GM Service and Parts Operations)

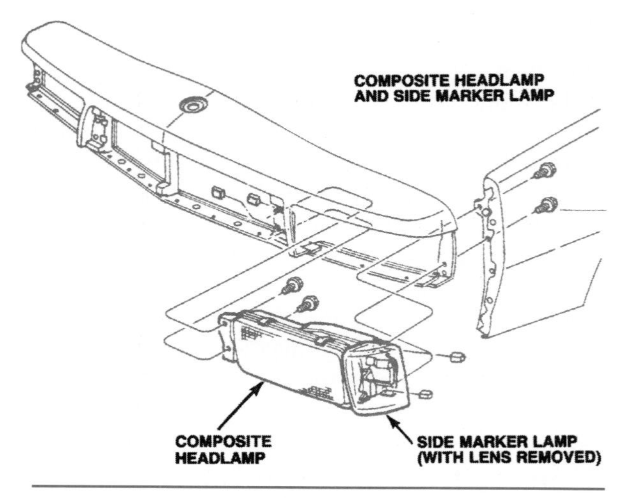

Figure 12-15. Composite headlights use a polycarbonate lens and form a permanent part of the car's styling. (GM Service and Parts Operations)

Composite Headlamps

Composite headlamps first appeared on some 1984 models as a part of the aerodynamic styling concept that has characterized car design since the mid-1980s. Composite headlamp design uses a replaceable halogen headlamp bulb that fits into a socket at the rear of the reflector, as shown in Figure 12-14. Since the headlamp housing does not require replacement unless damaged, it can be incorporated as a permanent feature of automotive styling. The housing can be designed to accept a single bulb or dual bulbs.

Composite headlamps can be manufactured by two different methods. In one, polycarbonate plastic is used to form the lens portion of the headlamp housing, as shown Figure 12-15, and the inside of the housing is completely sealed. In the other, a glass lens cover is permanently bonded with a reflector housing to form a single unit. Because this type of composite headlamp is vented to the atmosphere, water droplets may form on the inside of the glass lens cover when the headlamps are off. Such condensation disappears rapidly when the lights are turned on and does not affect headlamp performance.

Replacement halogen bulbs may contain both high- and low-beam filaments for use in two-headlamp systems, and individual high- or low-beam filaments for use in four-headlamp systems. The halogen bulbs have a quartz surface that can be easily stained when handled. For this reason, the bulbs are furnished in protective plastic covers that should not be removed until the bulb has been installed. If the quartz surface is accidentally touched with bare hands, it should be cleaned immediately with a soft cloth moistened with alcohol.

AUTOMOTIVE HEADLAMPS

HEADLAMP TYPE AND SIZE	ID CODE①	TRADE NO.②	DESIGN WATTS @ 12.8 VOLTS	
			HIGH BEAM	LOW BEAM
5-3/4 INCH CIRCULAR SEALED-BEAM	2C1	4000	37.5	60
	2C1	4040③	37.5	60
	2C1	H5006	35	35
	1C1	4001	37.5	
	1C1	H4001	37.5	
	1C1	5001	50	
	1C1	H5001	50	
7 INCH CIRCULAR SEALED-BEAM	2D1	6014	50	50
	2D1	H6014	60	50
	2D1	6015③	60	50
	2D1	6016④	60	50
	2D1	H6017	60	35
4 X 6-12 INCH RECTANGULAR SEALED-BEAM	1A1	4651	50	
	1A1	H4651	50	
	2A1	4652	40	60
	2A1	H4656	35	35
	2A1	H4662	40	45
	2A1	H4739	40	50
	2E1	H4666	65	45
5-1/2 X 8 INCH RECTANGULAR SEALED-BEAM	2B1	6052	65	
	2B1	H6052	65	
	2B1	H6054	65	55
3 X 5 INCH RECTANGULAR	H1	H4701	65	
	H3	H4703	65	
	H4	H4704	60	55④
REPLACEMENT HALOGEN BULB		9094	65	45

NOTES:
① THE FIRST CHARACTER INDICATES THE NUMBER OF BEAMS IN THE BULB, THE SECOND INDICATES THE SIZE AND NUMBER OF BULBS USIED ON THE CAR, AND THE THIRD IS AN SAE PHOTOMETRIC SPECIFICATION. ②H INDICATES A HALOGEN SEALED-BEAM. ③ HEAVY DUTY. ④ AT 13.2 VOLTS.

Figure 12-16. Common sealed-beam headlamps and replaceable halogen bulbs. (GM Service and Parts Operations)

Replacing the halogen bulb in a composite headlamp does not normally disturb the alignment of the headlamp assembly. There should be no need for headlamp alignment unless the composite headlamp assembly is removed or replaced. If alignment is required, however, special adapters must be used with the alignment devices. Figure 12-16 describes the automotive headlamps currently in use.

High-Intensity Discharge (HID) Lamps

The latest headlight development is the **high-intensity discharge (HID) lamp** (Figure 12-17). These headlamps put out three times more light and twice the light spread on the road than conventional halogen headlamps. They also use about two-thirds less power to operate and will last two to three times as long. HID lamps produce light in both ultraviolet and visible wave-lengths, causing highway signs and other reflective materials to glow.

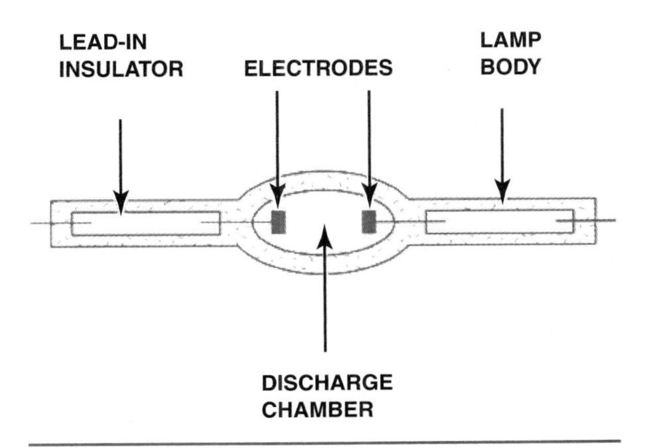

Figure 12-17. High-intensity discharge (HID) lamps.

These lamps do not rely on a glowing element for light. Instead the HID lamp contains a pair of electrodes, similar to spark plug electrodes, surrounded by gas. The electrode is the end of an electrical conductor that produces a spark. Light is produced by an arc that jumps from one electrode to another, like a welder's arc. The presence of an inert gas amplifies the light given off by the arcing. More than 15,000 volts are needed to jump the gap in the electrodes. To provide this voltage, a voltage booster and controller are required. Once the gap is bridged, only 80 volts is needed to keep the current flow across the gap. The large light output of the HID allows them to be smaller in size. HIDs will usually show signs of failure before they burn out.

Headlamp Location and Mounting

State and federal laws control the installation of headlamps. Automotive designers must place headlamps within certain height and width ranges. In addition, two- or four-lamp systems must follow one of the patterns shown in Figure 12-18.

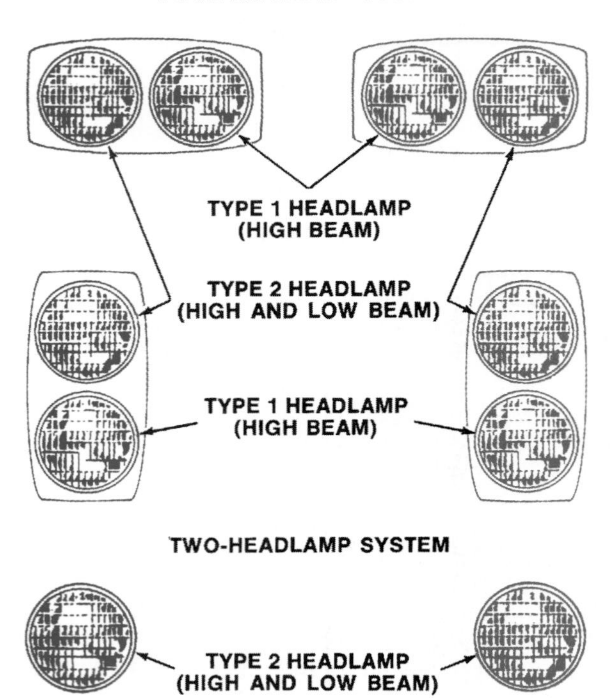

Figure 12-18. The law requires that headlamps be arranged in one of these patterns. The same requirements apply to rectangular lamps.

Headlamps are mounted so that their aim can be adjusted. Most circular and rectangular lamps have three adjustment points, as shown in Figure 12-19. The sealed-beam unit is placed in an adjustable mounting, which is retained by a stationary mounting. Many cars have a decorative bezel that hides this hardware while still allowing lamp adjustment, as shown in Figure 12-20. Composite headlamps use a similar two-point adjustment system, as in Figure 12-21, but require the use of special adapters with the alignment devices.

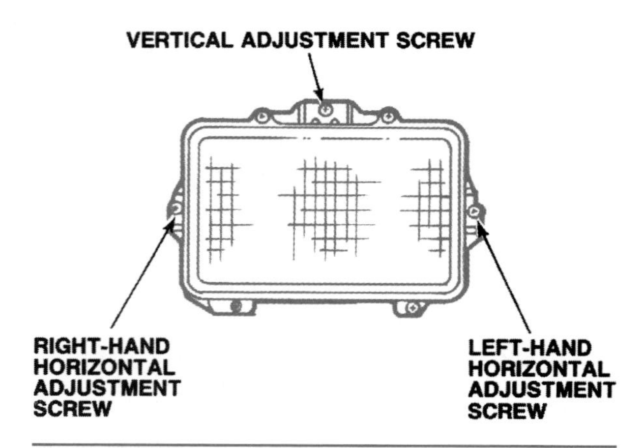

Figure 12-19. Most sealed-beam headlamps have vertical and horizontal adjusting screws. (GM Service and Parts Operations)

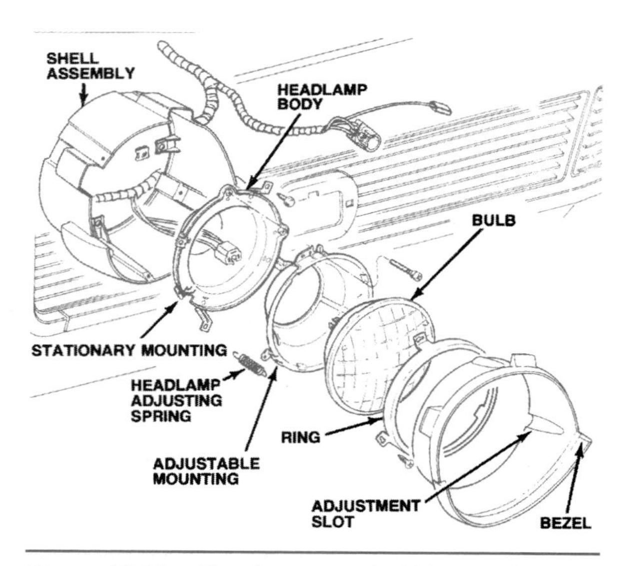

Figure 12-20. Headlamps are held in an adjustable mounting which is generally concealed by a decorative bezel.

Concealed Headlamps

Another automotive styling feature is concealed headlamps, either stationary lamps behind movable doors, as in Figure 12-22, or lamps that move in and out of the car's bodywork as in, Figure 12-23. The doors can be metal or clear plastic.

Electric motors or vacuum actuators operate headlamp-concealing mechanisms. Electrically operated systems usually have a relay controlling current flow to the motor. Vacuum-actuated systems work with engine vacuum stored in a reservoir.

Federal law requires that the main headlamp switch control the concealing mechanisms on late-model cars and that "pop-up" headlamps that rise out of the hood must not come on until they have completed 75 percent of their travel. Switches used with electrically operated headlamp doors have additional contacts to activate the relay (Figure 12-24). Vacuum-actuated systems usually have a vacuum switch attached to the headlamp switch. Some older cars may have a separate switch to control the door. All concealed headlamp systems also must have a manual opening method, such as a crank or a lever, as a backup system.

Some 1967 and earlier cars have a clear plastic lens cover, or fairing, over the sealed-beam unit. These are not legal on later-model cars.

Automatic Headlamp Systems

Photocells and solid-state circuitry are used to control headlamp operation in many vehicles today. A system can turn the lamps on and off; on past models they controlled the high- and low-beam switching. Some parts can be adjusted, but defective parts cannot be repaired. All automatic

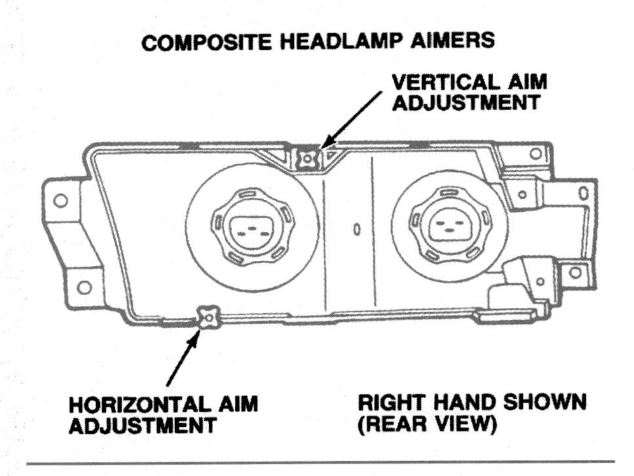

Figure 12-21. Composite headlamps also have vertical and horizontal adjustments. (GM Service and Parts Operations)

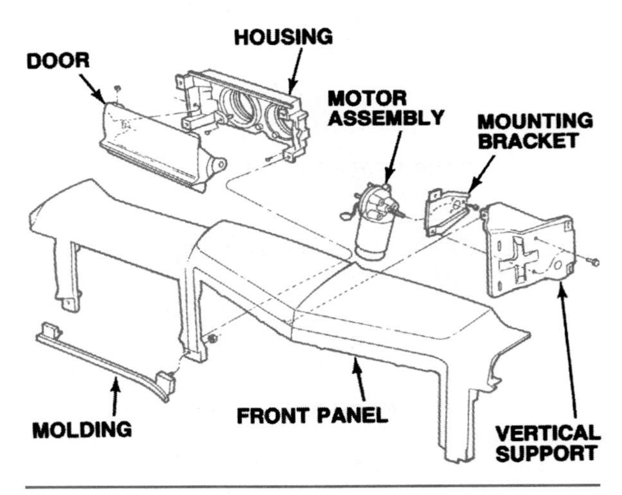

Figure 12-22. Headlamps can be concealed by a movable door. (DaimlerChrysler Corporation)

Figure 12-23. Headlamps can be concealed by moving them into and out of the car's bodywork.

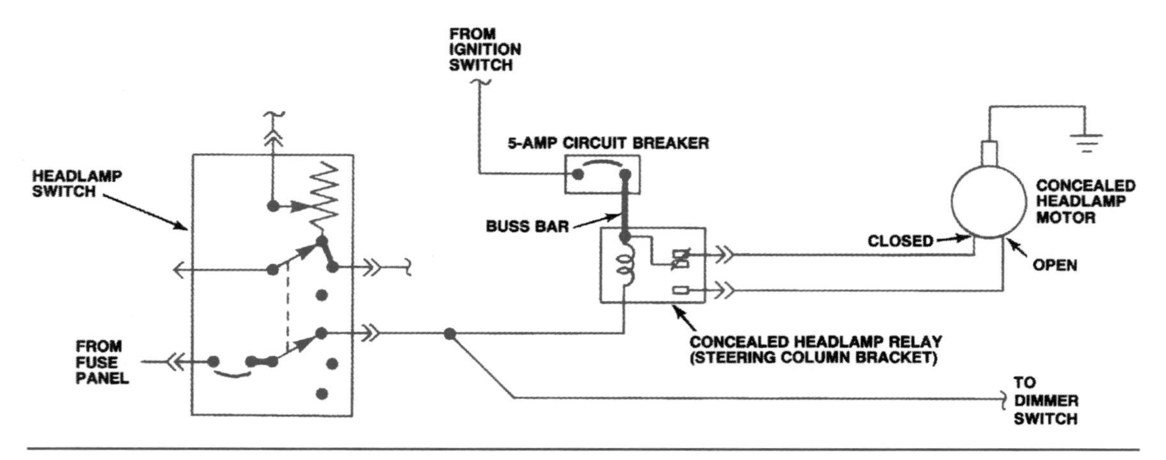

Figure 12-24. The main headlamp switch must operate the concealing mechanism. (DaimlerChrysler Corporation)

systems have manual switches to override the automatic functions.

On-Off Control

The photocell or ambient light sensor used in this system may be mounted on top of the instrument panel (Figure 12-25) facing upward so that it is exposed to natural outside light. In some older applications it may be mounted to the rearview mirror assembly facing outward for exposure to outside light. The photocell voltage is amplified and applied to a solid-state control module. Photocell voltage decreases as outside light decreases. Most photocells are adjustable for earlier or later turn-on. At a predetermined low light and voltage level, the module turns the headlamps on. The module often contains time-delay circuitry, so that:

- When the vehicle is momentarily in dark or light, such as when passing under a bridge or a streetlamp, the headlights do not flash on or off.
- When the automobile's ignition is turned off, the headlights remain on for a specified length of time and then are turned off.

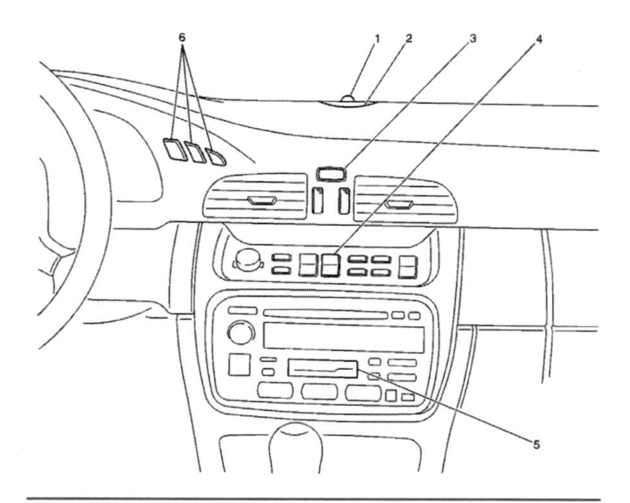

Figure 12-25. This photocell or ambient light sensor is mounted near the center of the dash panel and reacts to outside light to control the headlight on-off operation. The instrument panel integration module (IPM), which is the system amplifier is also shown. (GM Service and Parts Operations)

Twilight Sentinel

GM luxury vehicles use a system called Twilight Sentinel. The twilight delay switch in the headlamp switch assembly is supplied a 5 volt reference from the instrument panel integration module (IPM) as shown in Figure 12-26. The IPM also provides ground to the twilight delay switch. The switch is a potentiometer in which the resistance varies as the switch is moved. The IPM receives an input voltage proportional to the resistance of the potentiometer through the twilight delay signal circuit. The IPM sends a class 2 message to the dash integration module (DIM) indicating the on/off status and delay length for the twilight sentinel. With the twilight sentinel switch in any position other than OFF, the DIM will turn the headlamps on or off according to the daytime/nighttime status sent by the IPM. The DIM uses the twilight delay signal in order to keep the headlamps and park lamps on after the

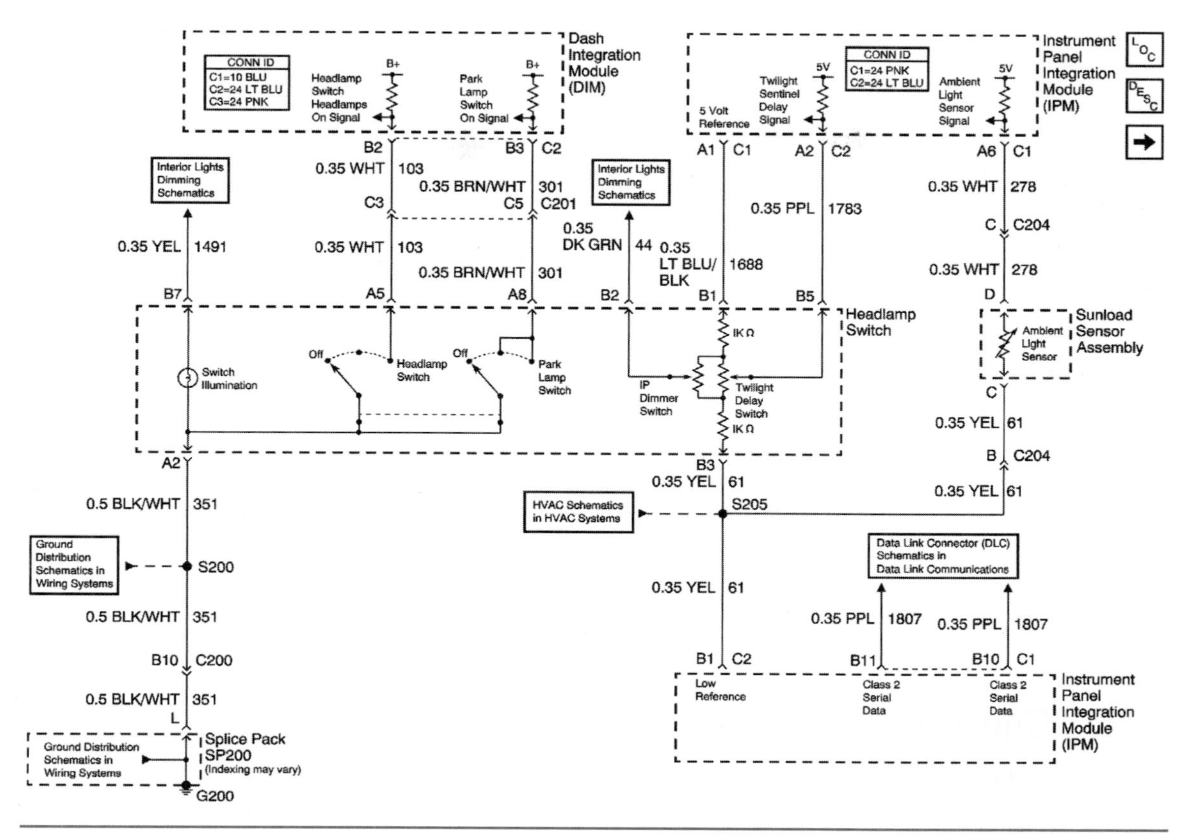

Figure 12-26. GM twilight sentinel.

ignition switch transitions from ON to OFF during nighttime conditions.

Daytime Running Lights

All late-model Canadian vehicles and GM vehicles after 1996 use **daytime running lights (DRL)** as shown in Figure 12-27. The basic idea behind these lights is that dimly lit headlights during the day make the vehicle more visible to other drivers, especially when the sun is behind the vehicle during sunset or after dawn. Generally, when the ignition is on and it is not dusk or dark, the daytime running lights will be on. When it is dusk, the system operates like an automatic headlamp system.

The DRL systems use an ambient light sensor, a light-sensitive transistor that varies its voltage signal to the body control module (BCM) in response to changes to the outside (ambient) light level. When the BCM receives this signal it will either turn on the DRL or the headlight relay for auto headlamp operation. Any function or condition that turns on the headlights will cancel the daytime running lamps operation. The DRL are separate lamps independent of the headlamps. With the headlight switch in the OFF position, the DRL will either be turned on or off after an approximately 8-second delay, depending on whether daylight or low light conditions are sensed. The DRL 10-amp fuse in the engine wiring harness junction block supplies battery positive voltage to the DRL relay switch contacts. The ignition 10-amp fuse in the engine wiring harness junction block supplies ignition positive voltage to the DRL relay coil. When the BCM energizes the DRL relay in daylight conditions, the current flows to both DRL lamps and to ground. The DRL will operate when the ignition switch is in the RUN position, the gear selector is not in the PARK position, and the parking brake is released. When these conditions have been met and the ambient light sensor indicates daytime conditions, the DRL will illuminate.

Some systems channel the headlamp current through a resistor and reduce the current and power to the lights to reduce their daytime intensity. Others use pulse-width modulation (PWM) through a separate control module that modulates

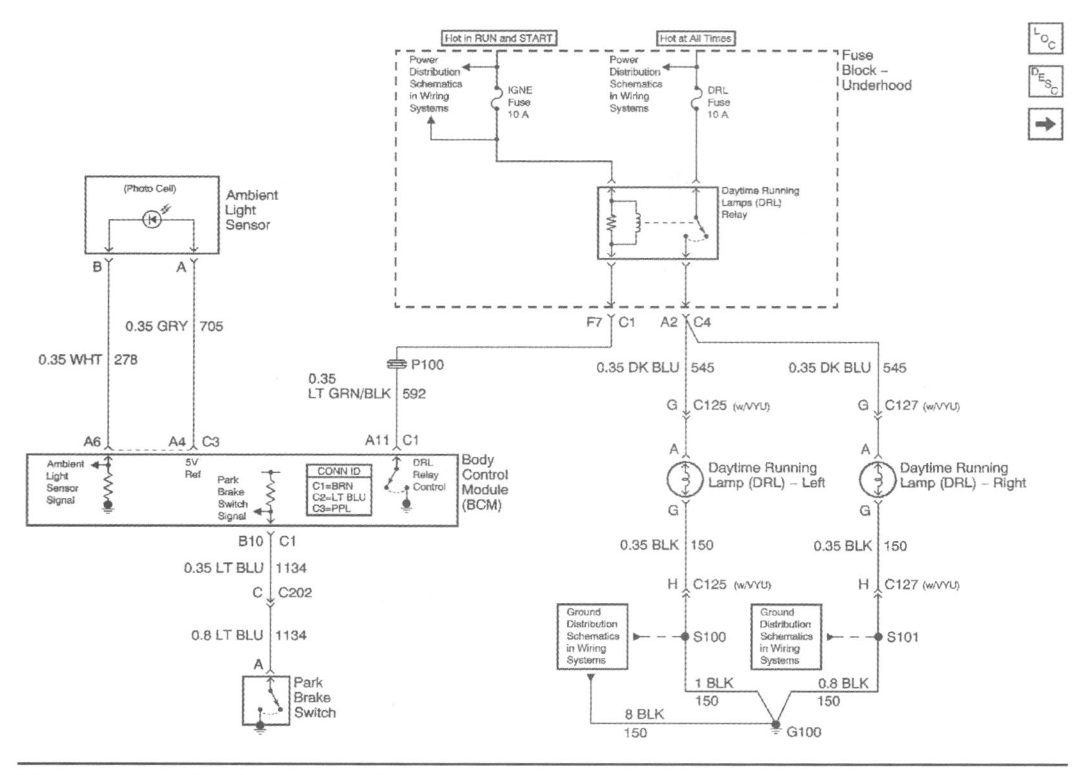

Figure 12-27. Daytime running lights. (GM Service and Parts Operations)

the voltage to the headlights and reduces the daytime intensity.

COMMON AUTOMOTIVE BULBS

Sealed-beam and composite headlamps are very specialized types of lamp bulbs. The other bulbs used in automotive lighting circuits are much smaller and less standardized. Each specific bulb has a unique trade number that is used consistently by all manufacturers.

Most small automotive bulbs are clear and are mounted behind colored lenses. Some applications, however, may call for a red (R) or an amber (NA) bulb.

Small automotive bulbs use either a brass or a glass wedge base. Bulbs with a brass base fit into a matching socket. The single or double contacts on the base of the bulb are the insulated contacts for the bulb's filaments. A matching contact in the socket supplies current to the bulb filament (Figure 12-28). A single-contact bulb contains one filament; a double-contact bulb has two filaments. The ground end of the bulb filament is connected directly to the base of the bulb, which is grounded through contact with the socket. In many cases, a separate ground wire leads from the socket to a ground connection. All double-contact bulbs are indexed so that they will fit into the socket in only one way. This is called an indexed base.

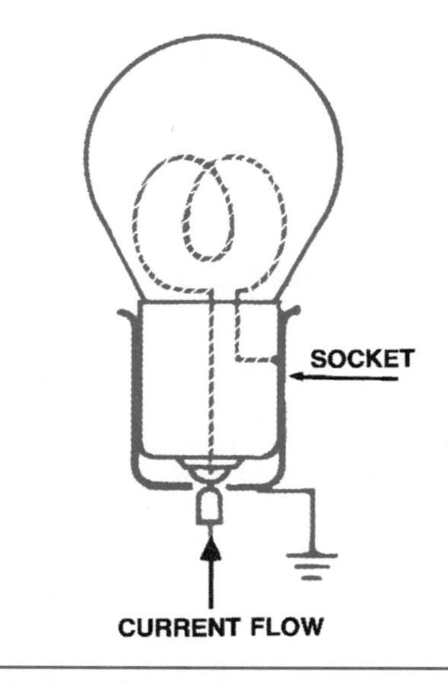

Figure 12-28. Automotive bulbs and sockets must be matched.

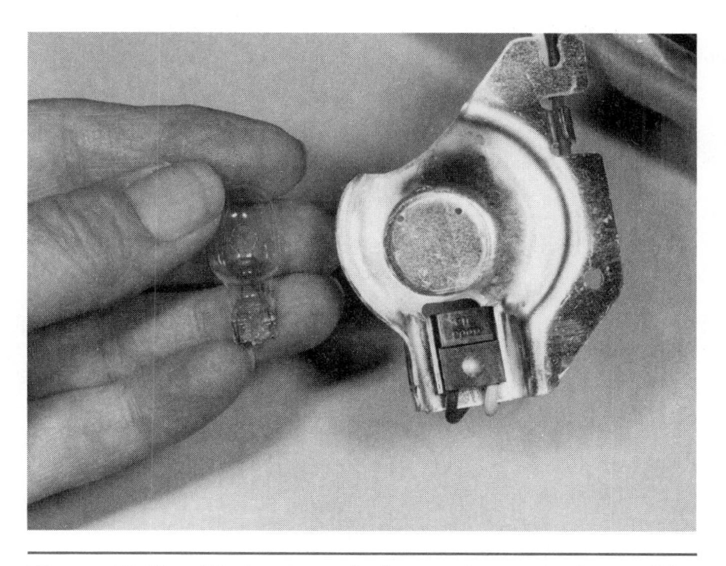

Figure 12-29. Wedge-base bulbs are increasingly used for interior lighting applications.

Historical Fact: Gas Lighting

Headlamps that burned acetylene gas were used on early cars, trucks, and motorcycles. The acetylene gas came from a prefilled pressurized container or from a "gas generator."

One type of acetylene gas generator used a drip method. A tank filled with water was mounted above another tank containing calcium carbide. A valve controlled the dripping of water onto the calcium carbide. When water was allowed to drip onto the calcium carbide, acetylene gas formed. The gas was routed through a small pipe to the headlamps. The headlamps were lit with a match or by an electric spark across a special lighting attachment.

Wedge-base bulbs generally have been used for instrument cluster and other interior lighting applications. The base and optical part of the bulb are a one-piece, formed-glass shell with four filament wires extending through the base and crimped around it to form the external contacts (Figure 12-29). The design locates the contacts accurately, permitting direct electrical contact with the socket, which contains shoulders to hold the bulb in place. The bulb is installed by pushing it straight into its socket, with no indexing required.

Wedge-base 2358 bulbs with a new socket design were introduced in 1987 as replacements for the brass-base 1157 and 2057 bulbs for exterior lighting applications. The wires of the low-profile plastic socket exit from the side instead of the rear (Figure 12-30). This reduces the possibility of wire damage and permits the socket to be used in more confined areas. Since the introduction of this base-socket design, a series of these bulbs has been made available in both clear and amber versions.

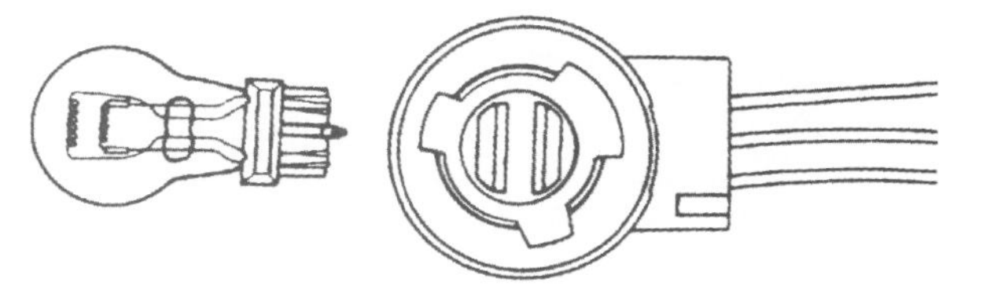

Figure 12-30. Wedge-base bulbs with plastic sockets are used for some external lighting applications. (DaimlerChrysler Corporation)

TAILLAMP, LICENSE PLATE LAMP, AND PARKING LAMP CIRCUITS

The taillamps, license plate lamps, and parking lamps illuminate the car for other drivers to see.

Circuit Diagram

These lamps usually share a single circuit because the laws of some states require that they be lit at the same time. Figure 12-31 shows a typical circuit diagram. Since the main headlamp switch controls the lamps, they can be lit whether the ignition switch is on or off.

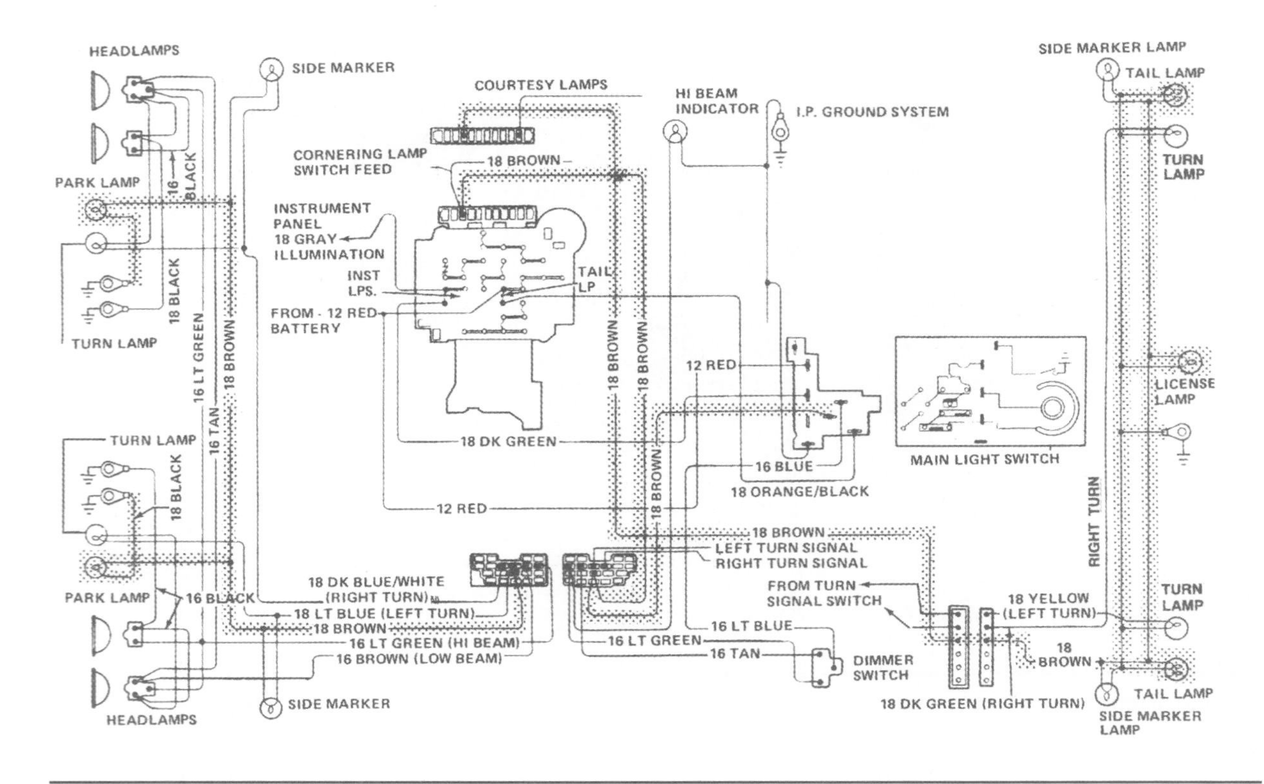

Figure 12-31. A taillamp, license plate lamp, and parking lamp circuit diagram. (GM Service and Parts Operations)

Switches and Fuses

These lamps are controlled by contacts within the main headlamp switch. They can be lit when the headlamps are off (Figure 12-32). A fuse (usually 20 amperes) protects the circuit.

Bulbs

The bulb designs most commonly used as taillamps and parking lamps are the G-6 single-contact bayonet and the S-8 double-contact bayonet. The tail and parking lamps may each be one filament of a double-filament bulb. License plate lamps are usually G-6 single-contact bayonet or T-3 1/4-wedge bulbs.

STOP LAMP AND TURN SIGNAL CIRCUITS

Stop lamps, also called brake lamps, are always red. Federal law requires a red center high-mounted stop lamp (CHMSL), on 1986 and later

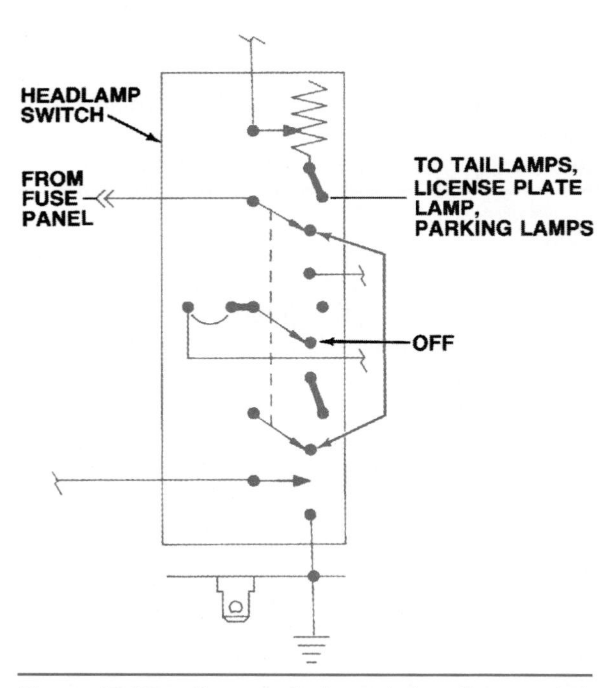

Figure 12-32. Contacts in the main headlamp switch provide current to the taillamps, license plate lamps, and parking lamps. (DaimlerChrysler Corporation)

models. Turn signals, or directionals, are either amber or white on the front of the car and either red or amber on the rear.

Circuit Diagram

A typical circuit diagram with stop and turn lamps as separate bulbs is shown in figure 12-38A. When the brakes are applied, the brake switch is closed and the stop lamps light. The brake switch receives current from the fuse panel and is not affected by the ignition switch.

When the turn signal switch is moved in either direction, the corresponding turn signal lamps receive current through the flasher unit. The flasher unit causes the current to start and stop very rapidly, as we will see later. The turn signal lamp flashes on and off with the interrupted current. The turn signal switch receives current through the ignition switch, so that the signals will light only if the ignition switch is on.

In many cars, the stop and turn signals are both provided by one filament, as shown in Figure 12-33B and Figure 12-34. When the turn signal switch is closed, the filament receives interrupted current through the flasher unit. When the brakes are applied, the filament receives a steady flow of current through the brake switch and special contacts in the turn signal switch. If both switches are closed at once, brake switch current is not allowed through the turn signal switch to the filament on the signaling side. The signaling filament receives interrupted current through the flasher unit, so it flashes on and off. The filament on the opposite side of the car receives a steady flow of current through the brake switch and the turn signal switch, so it is continuously lit. Figure 12-34 shows the integration of the single-filament CHMSL in a typical stop-and-turn signal circuit.

Switches, Fuses, and Flashers

Several units affect current flow through the stop lamp and turn signal circuits. The ignition switch is located between the battery and the turn signal switch (Figure 12-35), so the current cannot flow through the turn signal switch if the ignition switch is off. The ignition switch does not control the brake switch; it is connected directly to battery voltage through the fuse panel (Figure 12-35).

Before the mid-1960s, the brake switch was often located within the brake hydraulic system and operated by hydraulic pressure. Because of changes in braking system design, this type of switch is no longer commonly used. On late-model cars, the brake switch is usually mounted

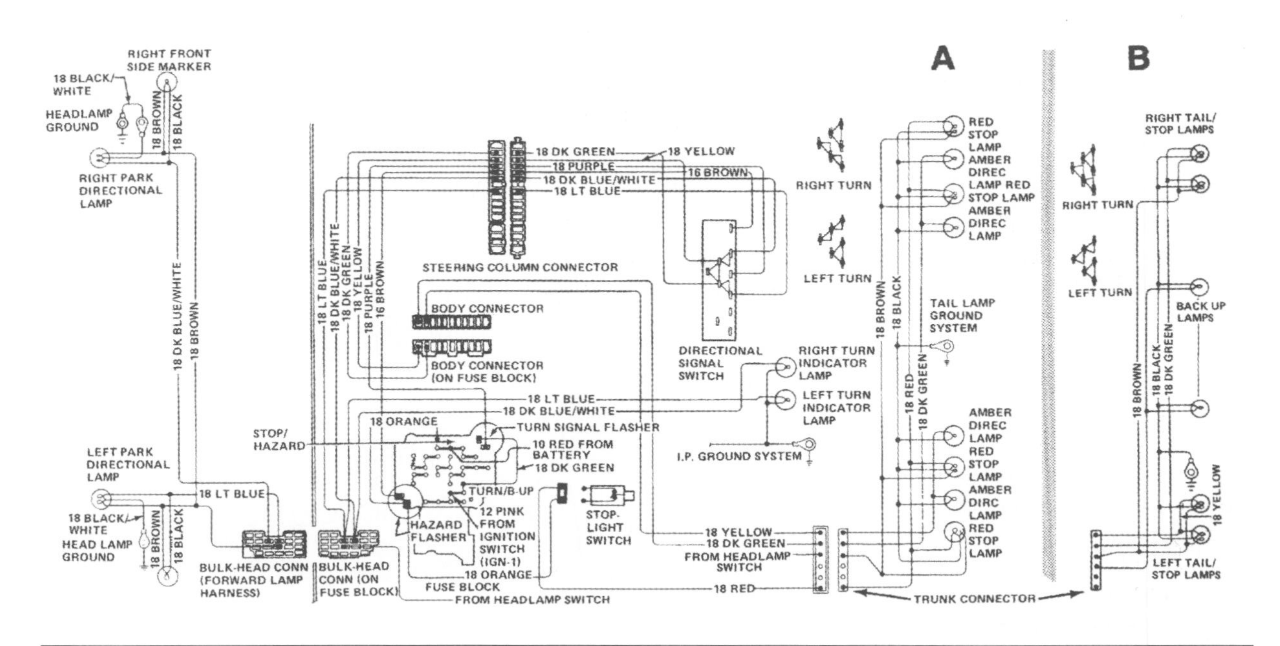

Figure 12-33. Stop lamp and turn signal circuits. The basic drawing (A) has separate bulbs for each function. The alternate view of the rear lamps (B) has single bulbs with double filaments. One filament of each bulb works for stop lamps and for turn signals. (GM Service and Parts Operations)

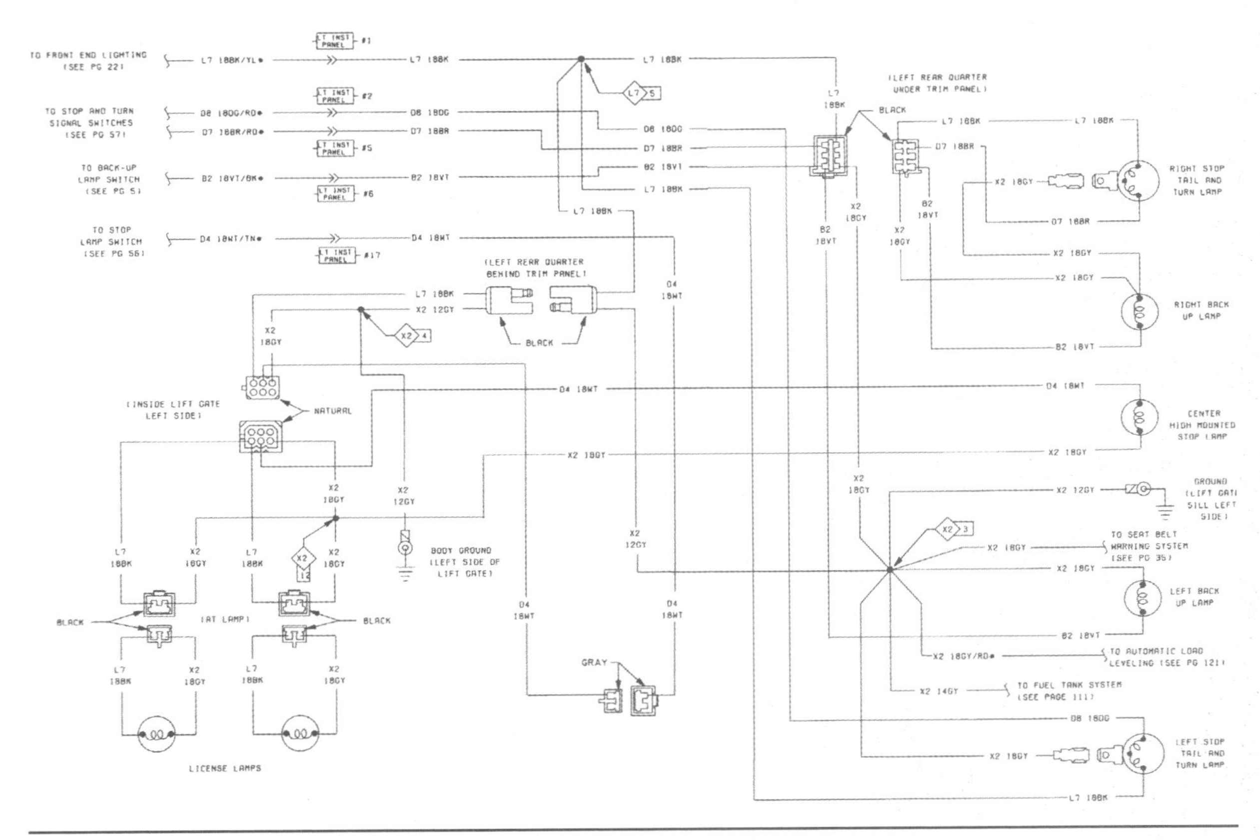

Figure 12-34. A typical rear lighting circuit diagram showing the inclusion of the center high-mounted stop lamp (CHMSL) mandated by law on 1986 and later models. (DaimlerChrysler Corporation)

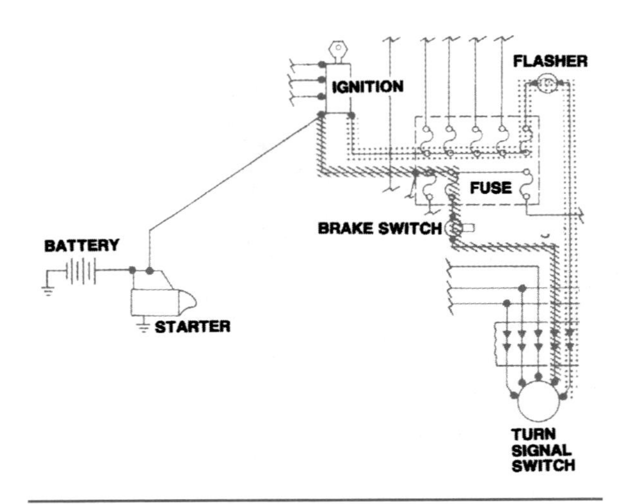

Figure 12-35. The ignition switch controls current to the turn signal switch, but does not affect current to the brake switch.

on the bracket that holds the brake pedal. When the pedal is pressed, the switch is closed.

The **turn signal switch** is mounted within the steering column and operated by a lever (Figure 12-36). Moving the lever up or down

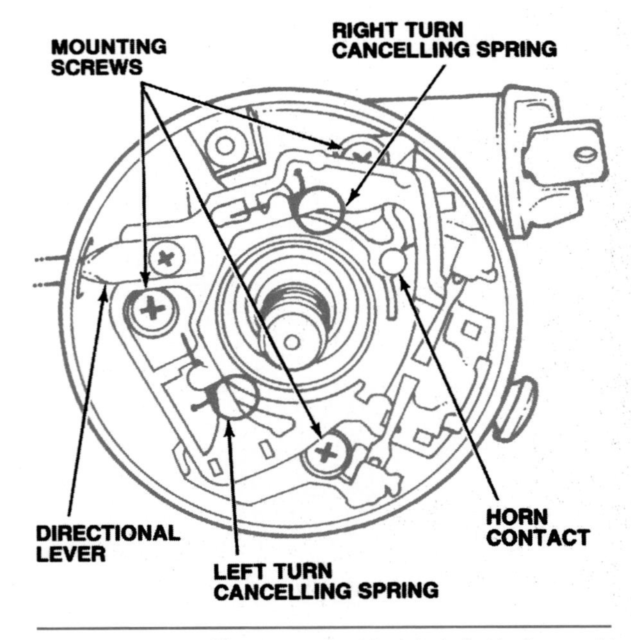

Figure 12-36. The turn signal switch includes various springs and cams to control the contact points.

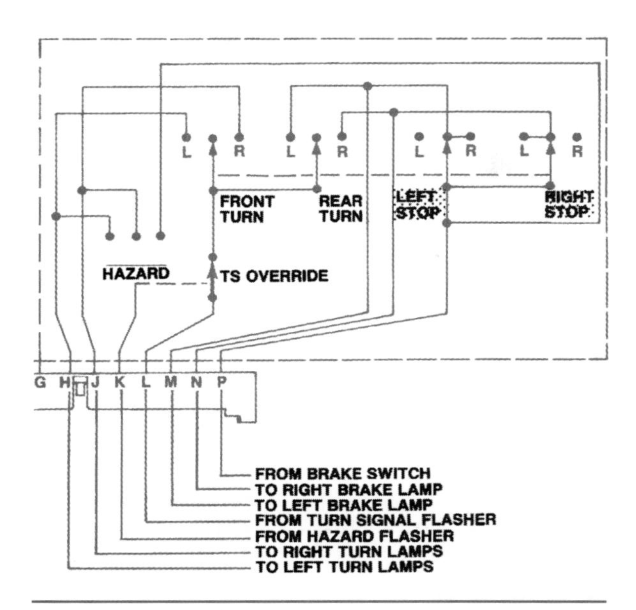

Figure 12-37. When the stop lamps and turn signals share a common filament, stop lamp current flows through the turn signal switch.

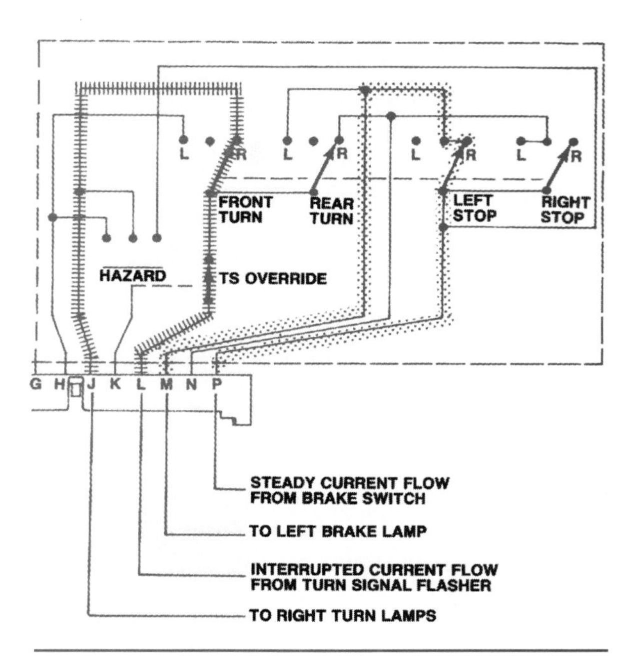

Figure 12-38. When a right turn is signaled, the turn signal switch contacts send flasher current to the right-hand filament and brake switch current to the left-hand filament.

closes contacts to supply current to the flasher unit and to the appropriate turn signal lamp. A turn signal switch includes cams and springs that cancel the signal after the turn has been completed. That is, as the steering wheel is turned in the signaled direction and then returns to its normal position, the cams and springs separate the turn signal switch contacts.

In systems using separate filaments for the stop and turn lamps, the brake and turn signal switches are not connected. If the car uses the same filament for both purposes, there must be a way for the turn signal switch to interrupt the brake switch current and allow only flasher unit current to the filament on the side being signaled. To do this, brake switch current is routed through contacts within the turn signal switch (Figure 12-37). By linking certain contacts, the bulbs can receive either brake switch current or flasher current, depending upon which direction is being signaled.

For example, Figure 12-38 shows current flow through the switch when the brake switch is closed and a right turn is signaled. Steady current through the brake switch is sent to the left brake lamp. Interrupted current from the turn signal is sent to the right turn lamps.

Flasher units supply a rapid on-off-on current to the turn signal lamps. To do this, they act very much like Type 1 self-setting circuit breakers. Current flows through a bimetallic arm (Figure 12-39), heating the arm until it bends and opens a set of

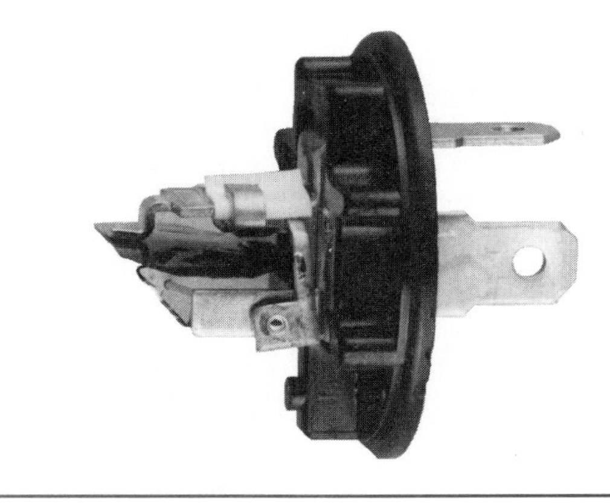

Figure 12-39. The internal components of a turn signal flasher.

contact points. When the current stops, the arm cools and the contact points close again. This cycle occurs rapidly so that the turn signal lamps flash on and off about once every second. Flasher units usually are installed in the wiring harness beneath the instrument panel or in the fuse panel.

Some manufacturers use two flashers, one for the turn signals and one for the hazard warning lamps. Other manufacturers use a single flasher that controls both the turn signals and the hazard warning lamps. This type of flasher is called a *combination flasher*.

■ Switch the Bulbs, Not the Switch

Have you ever been stumped by a turn signal problem where the lamps on one side flashed properly, but those on the other side lit and burned steadily without flashing? The flasher checks out okay, and the panel indicator lights but doesn't flash. Both bulbs, front and rear, light; power is getting to the sockets. Sounds like trouble with the switch? Maybe it is. However, before you tear into the steering column, try swapping the front and rear bulbs from one side to the other. Sometimes, a little corrosion on a socket and the resistance of an individual bulb can add up to cumulative resistance that unbalances the circuit and prevents flashing. Swapping the bulbs or cleaning the contacts can reduce the resistance to within limits, restore equilibrium, and get the system working correctly again.

The turn signal circuit must include one or more indicators to show the driver that the turn signals are operating. These indicators are small bulbs in the instrument panel that provide a parallel path to ground for some of the flasher unit current. Most systems have separate indicators for the right and left sides, although some cars use only one indicator bulb for both sides. On some models, additional indicators are mounted on the front fenders facing the driver. Two separate fuses, rated at about 20 amperes, usually protect the stop lamp and turn signal lamp circuits.

Bulbs

The bulb types traditionally used as stop or turn signal lamps are the S-8 single- and double-contact bayonet base, although the 2358 wedge-base bulbs are being used more frequently. The stop and turn filaments may be part of a double-filament bulb.

HAZARD WARNING LAMP (EMERGENCY FLASHER) CIRCUITS

All motor vehicles sold in the United States since 1967 have a **hazard warning lamp** circuit. It is designed to warn other drivers of possible danger in emergencies.

Circuit Diagram

The hazard warning lamp circuit uses the turn signal lamp circuitry, a special switch, and a heavy-duty flasher unit. The switch receives battery current through the fuse panel. When the switch is closed, all of the car's turn signal lamps receive current through the hazard flasher unit. An indicator bulb in the instrument panel provides a parallel path to ground for some of the flasher current.

Switches, Fuses, and Flashers

The hazard warning switch can be a separate unit or it can be part of the turn signal switch (Figure 12-40). In both cases, the switch contacts route battery current from the fuse panel through the hazard flasher unit to all of the turn signal lamps at once. In most systems, the hazard warning switch overrides the operation of the turn signal switch. A 15- to 20-ampere fuse protects the hazard warning circuit.

The hazard warning flasher looks like a turn signal flasher when assembled, but it is constructed differently and operates differently, in order to control the large amount of current required to flash all of the turn signal lamps at once.

The flasher consists of a stationary contact, a movable contact mounted on a bimetallic arm, and a high-resistance coil (Figure 12-41). The coil is connected in parallel with the contact points, which are normally open. When the hazard warning switch

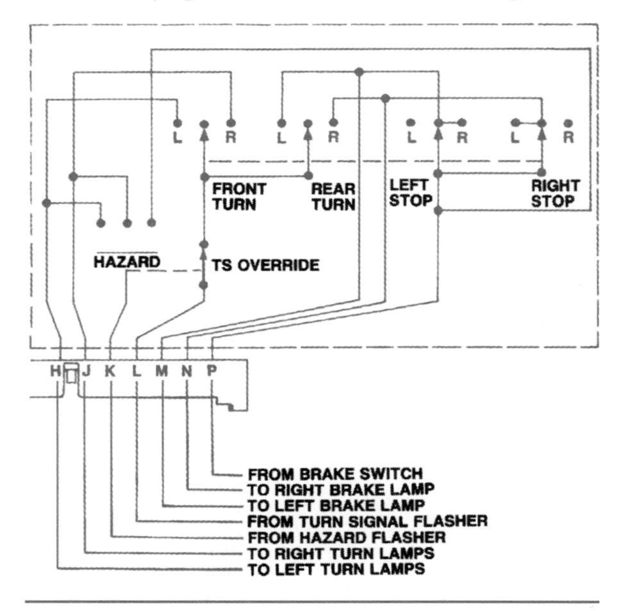

Figure 12-40. The hazard warning switch is often a part of the turn signal switch. (DaimlerChrysler Corporation)

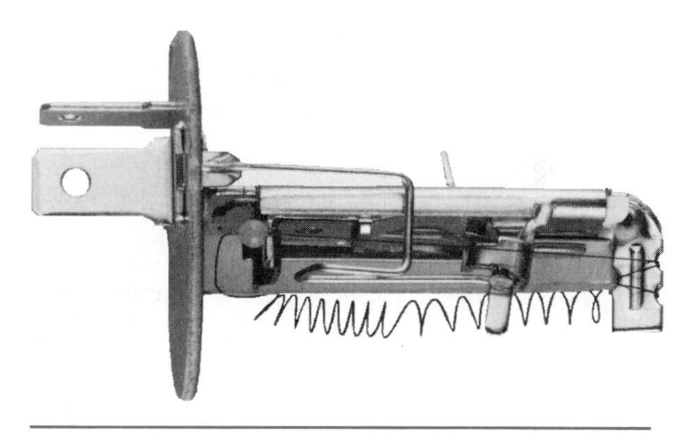

Figure 12-41. The hazard flasher is constructed to control a large amount of current.

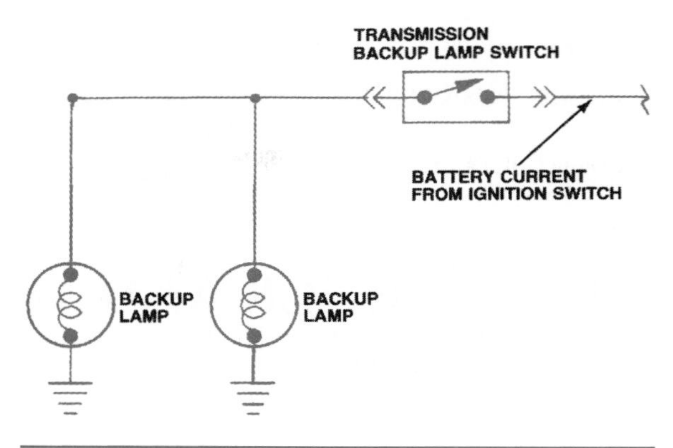

Figure 12-42. A typical backup lamp circuit. (DaimlerChrysler Corporation)

is closed and current flows to the flasher, the high resistance of the coil does not allow enough current to light the lamps. However, the coil heats up and causes the bimetallic strip to close the contacts. The contacts form a parallel circuit branch and conduct current to the lamps. Decreased current flow through the coil allows it to cool and the bimetallic strip opens the contacts again. This cycle repeats about 30 times per minute.

BACKUP LAMP CIRCUITS

The white **backup lamps** light when the car's transmission or transaxle is in reverse. The lamps have been used for decades, but have been required by law since 1971. Backup lamps and license plate lamps are the only white lamps allowed on the rear of a car.

Circuit Diagram

A typical backup lamp circuit diagram is shown in Figure 12-42. Figure 12-43 shows integration of the backup lamp with the stop, taillamp, and turn signals in a typical rear lighting diagram. When the transmission switch is closed, the backup lamps receive current through the ignition switch. The lamps will not light when the ignition switch is off.

Switches and Fuses

The backup lamp switch generally is installed on the transmission or transaxle housing (Figure 12-43), but it may be mounted near the gear selector lever (Figure 12-42) on some vehicles. The backup switch may be combined with the neutral safety switch. A 15- to 20-ampere fuse, which often is shared with other circuits, protects the circuit.

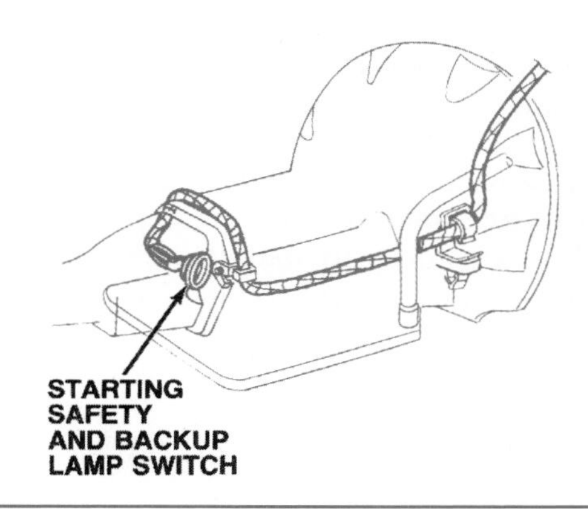

Figure 12-43. The backup lamp switch can be mounted on the transmission housing. (DaimlerChrysler Corporation)

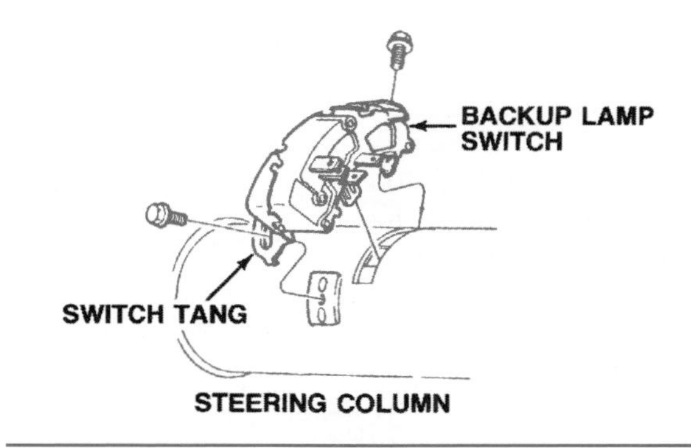

Figure 12-44. The backup lamp switch can be mounted near the gearshift lever. (GM Service and Parts Operations)

Bulbs

The bulb designs most commonly used for backup lamps are S-8 single-contact bayonet and double-contact indexed.

SIDE MARKER AND CLEARANCE LAMP CIRCUITS

Side marker lamps are mounted on the right and left sides toward the front and rear of the vehicle to indicate its length. Side marker lamps are required on all cars built since 1969 and are found on many earlier models as well. Front side markers are amber; rear side markers are red. On some vehicles, the parking lamp or taillamp bulbs are used to provide the side marker function.

Clearance markers are required on some vehicles, according to their height and width. Clearance lamps, if used, are included in the side marker lamp circuitry. Clearance lamps face forward or rearward. Like side markers, front clearance lamps are amber; rear lamps are red.

Circuit Diagrams

Side marker lamps can be either grounded or insulated. Figures 12-45 shows a GM Cadillac Seville circuit with independently grounded side markers. Figure 12-46 shows a circuit that has the ground path for current through the turn signal bulb filaments. When the headlamp switch is off and the turn signal switch is on, both the turn signal and the side marker on the side being signaled will flash. When the headlamp switch is on, only enough current flows to light the side marker. The turn signal lamp does not light until the turn signal switch and flasher are closed; then the turn signal lamp will light. The side marker lamp will not light, because 12 volts are applied to each end of the filament. There is no voltage drop and no current flow. When the turn signal flasher opens, the turn signal lamp goes out.

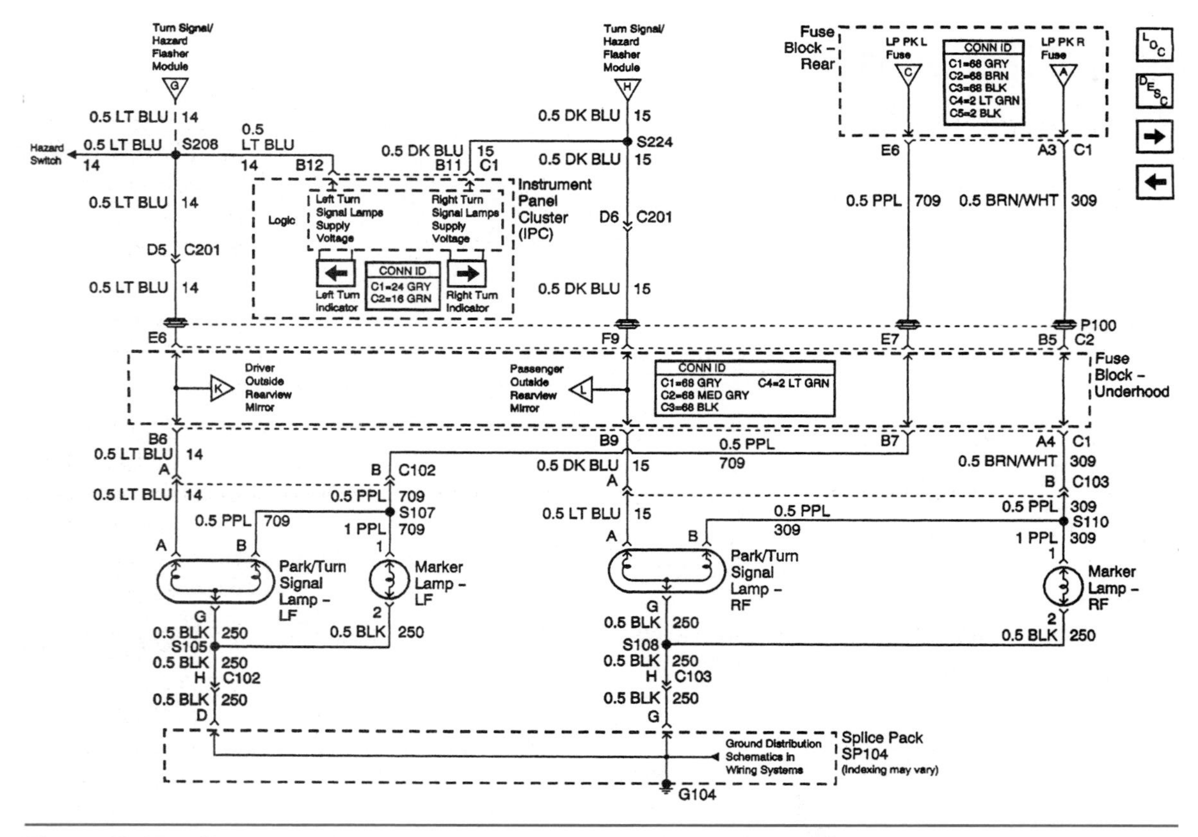

Figure 12-45. Side marker lamps can be independently grounded. (GM Service and Parts Operations)

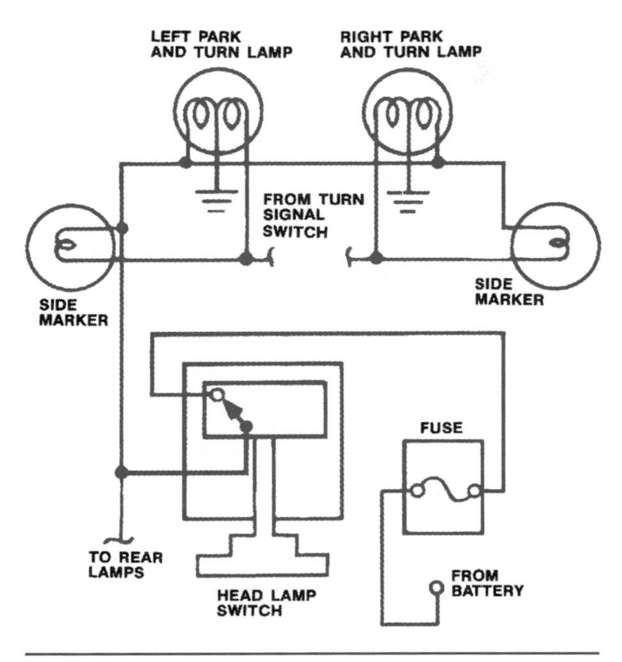

Figure 12-46. Side marker lamps can be grounded through the turn signal filaments.

Normal headlamp circuit current lights the side marker lamp. This sequence makes the two lamps flash alternately—one is lit while the other is not.

Switches and Fuses

Side marker lamps are controlled by contacts within the main headlamp switch. Their circuit is protected by a 20-ampere fuse, which usually is shared with other circuits.

Bulbs

The G-6 and S-8 single- and double-contact bayonet base bulbs commonly are used for side marker lamps.

■ Installing New Bulbs

If you replace a bulb in a parking light, turn signal, stop lamp, or taillamp, you may find that the bulb will not light unless you hold it against the socket. This may be due to weakened springs or flattened contacts. To solve the problem, apply a drop of solder to the contact points at the base of the bulb. Add more solder if necessary or file off the excess. The result will be a good solid connection. Also, for a weak spring, if the wires going to the socket are given slack, you may be able to gently stretch the spring.

INSTRUMENT PANEL AND INTERIOR LAMP CIRCUITS

Instrument Panel

The lamps within the instrument panel can be divided into three categories: indicator lamps, warning lamps, and illumination lamps. We have seen that some circuits, such as high-beam headlamps and turn signal lamps, include an indicator mounted on the instrument panel. Warning lamps, which alert the driver to vehicle operating conditions, are discussed in Chapter 13. Lamps that simply illuminate the instrument panel are explained in the following section.

Circuit Diagram

All late-model automobiles use a printed circuit behind the instrument panel to simplify connections and conserve space. The connections can also be made with conventional wiring (Figure 12-47).

Switches and Rheostats

The rheostat may be a separate unit in the panel lamp circuit. Current to the panel lamps is controlled by contacts within the main headlamp switch (Figure 12-48). The instrument panel lamps receive current when the parking and taillamps are lit and when the headlamps are lit.

Rheostats and **potentiometers** are variable resistors that allow the driver to control the brightness of the panel lamps. The rheostat or the potentiometer for the panel lamps can be a separate unit (Figure 12-46) or it can be a part of the main headlamp switch.

Bulbs

The T-3 1/4 bulb with a wedge or miniature bayonet base is a design commonly used in instrument panel illumination.

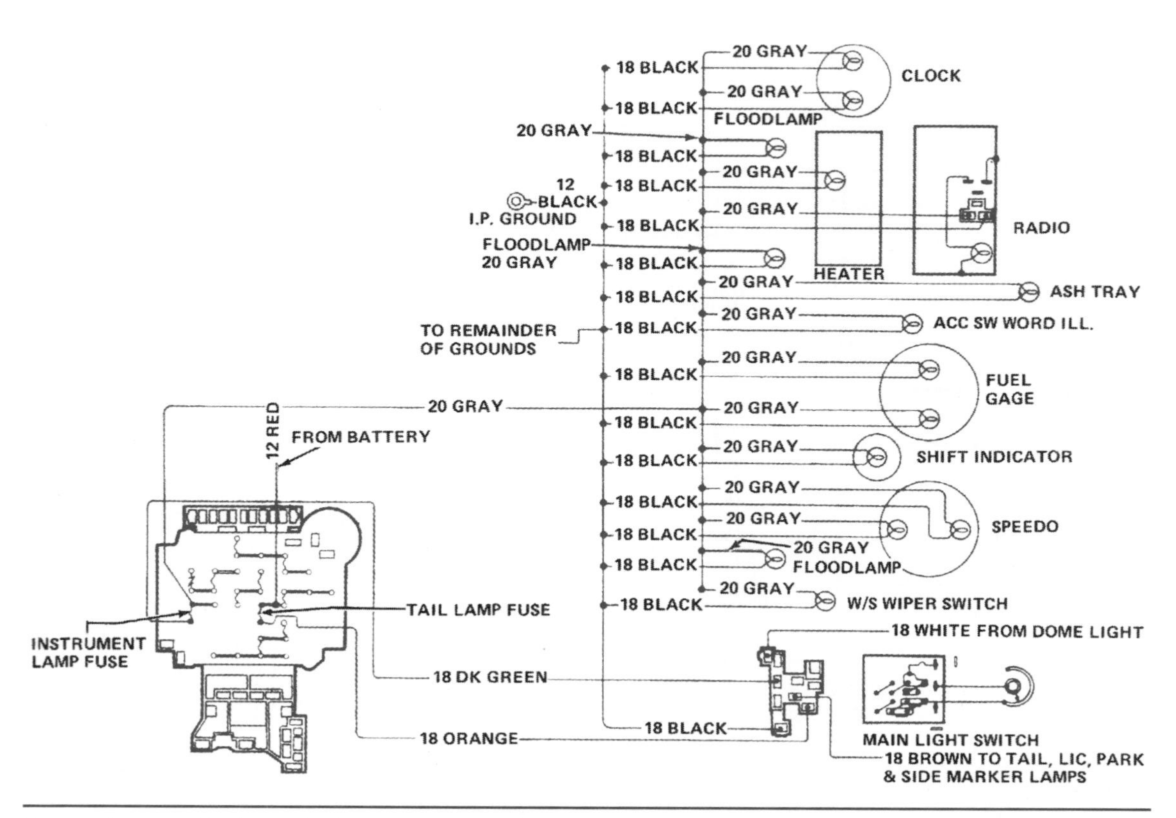

Figure 12-47. Multistrand wiring can be used behind the instrument panel. (GM Service and Parts Operations)

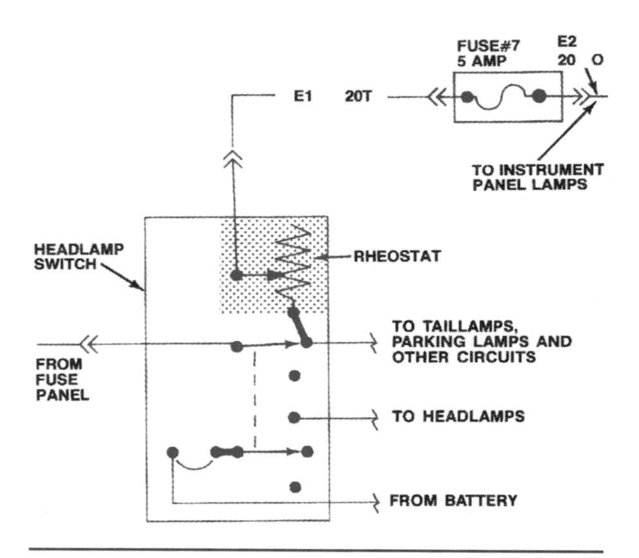

Figure 12-48. The panel lamps receive current through the main headlamp switch, which may also contain a rheostat to control the current. (DaimlerChrysler Corporation)

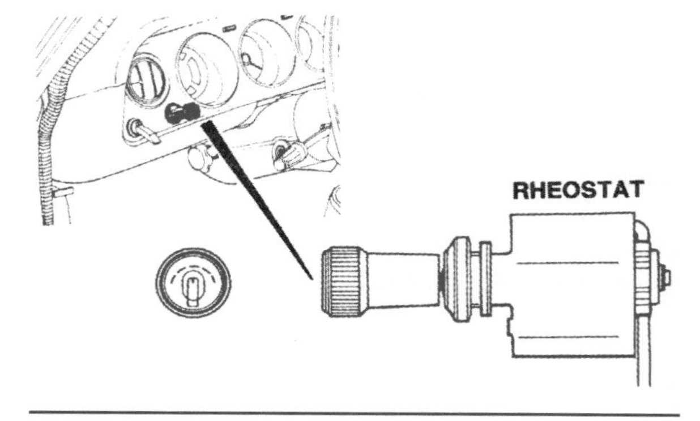

Figure 12-49. The rheostat may be a separate unit in the panel lamp circuit. (DaimlerChrysler Corporation)

Interior Lamps

Interior (courtesy) lamps light the interior of the car for the convenience of the driver and passengers.

Circuit diagram

Interior lamps receive current from the battery through the fuse panel. Switches at the doors control this current and light the lamps when one of the doors is opened. Many manufacturers install the bulbs between the power source and the grounded switch (Figure 12-50). Others, including Ford and Chrysler, install the switches between the power source and the grounded bulb (Figure 12-51).

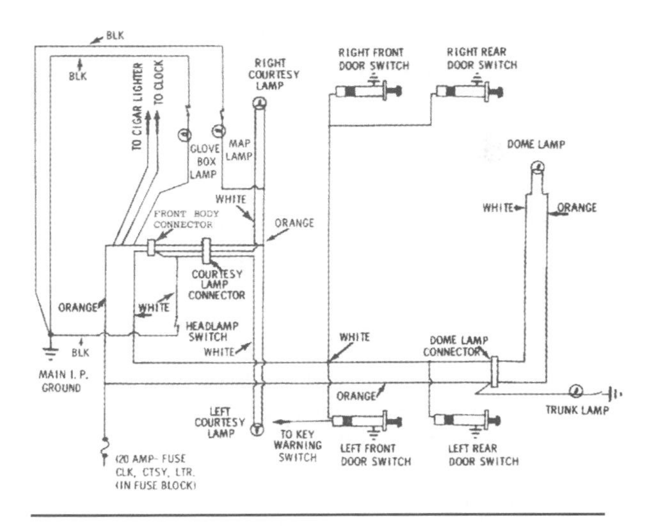

Figure 12-50. The interior (courtesy) lamp circuit may have insulated bulbs and grounded switches. (GM Service and Parts Operations)

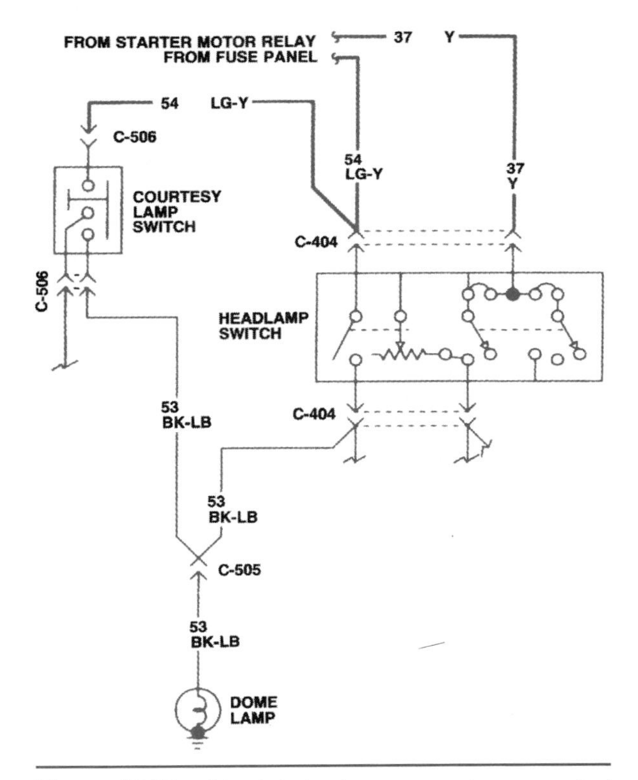

Figure 12-51. The interior lamps may be grounded and also have insulated switches.

Courtesy lamp circuits also may contain lamps to illuminate the glove box, trunk, and engine compartment. Additional switches that react to glove box door, trunk lid, or hood opening control current through these bulbs.

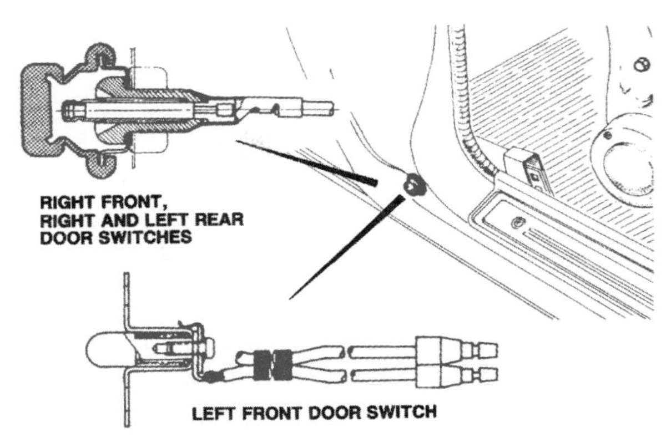

Figure 12-52. Interior lamps are controlled by switches at the door jambs. (DaimlerChrysler Corporation)

Switches

The switches used in courtesy lamp circuits are push-pull types (Figure 12-52). Spring tension closes the contacts when a door is opened. When a door is closed, it pushes the contacts apart to stop current flow. When any one switch is closed, the circuit is complete and all lamps are lit.

Accessory Lighting

Every car manufacturer offers unique accessory lighting circuits. These range from hand-controlled spotlights to driving and fog lamps. Each additional accessory circuit requires more bulbs, more wiring, and possibly an additional switch.

For example, cornering lamps can be mounted on the front sides of the car to provide more light in the direction of a turn. When the turn signal switch is operated while the headlamps are on, special contacts in the turn signal switch conduct a steady flow of current to the cornering lamp on the side being signaled.

One or more of the interior lamps may have a manually controlled switch to complete the circuit, (Figure 12-53). This switch allows the driver or passengers to light the bulb even when all the doors are closed.

Bulbs

The S-8 bulbs are used for trunk and engine compartment lamps, with T-3 1/4 wedge and T-3 3/4 double-end-cap bulbs used as courtesy lamps.

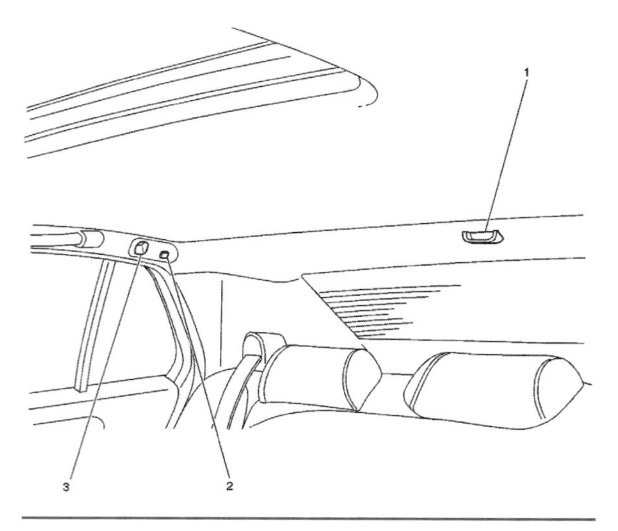

Figure 12-53. Interior lamps often have a manual switch to override the automatic operation. (GM Service and Parts Operations)

SUMMARY

Headlamp circuits must provide low- and high-beam lights for driving, and a high-beam indicator lamp for driver use. Two or four lamps may be used. Most often, the lamps are grounded, but some circuits have insulated lamps and grounded switches. The main headlamp switch also controls other lamp circuits. The main switch sometimes controls high-low beam switching, but a separate dimmer switch usually controls this. A circuit breaker protects the headlamp circuit.

Conventional sealed-beam headlamps use a tungsten filament; halogen sealed-beam lamps pass current through a pressure-filled halogen capsule. Halogen headlamps provide a brighter light with less current. Sealed-beam headlamps have either a high-beam filament (Type 1) or both a high- and low-beam filament (Type 2). Headlamps are always connected with multiple-plug connectors.

Changes in federal lighting standards have permitted sealed-beam headlamps made of plastic instead of glass. Plastic headlamps weigh considerably less and are more damage-resistant than glass. The changes also have resulted in the use of a composite headlamp in place of sealed-beams. The composite headlamp consists of a polycarbonate lens housing and a replaceable halogen bulb that contains a dual filament. Since the lens housing is not replaced, it has been integrated into vehicle styling.

Headlights are mounted so that their aim can be adjusted vertically and horizontally. Some cars have concealed headlamps with doors or mountings that are operated by vacuum or by electric motors. The main headlamp switch controls these mechanisms, and there also is a manual control provided to open and close the mechanisms if necessary.

Photocells and solid-state modules are used to control headlamp on-off switching, beam switching operation, and daytime running lights (DRL) on many late-model vehicles. Bulbs used in other lighting circuits are smaller than sealed-beam units and must be installed in matching sockets. These other lighting circuits include the following:

- Taillamp, license plate lamp, and parking lamp circuits
- Stop lamp and turn signal circuits
- Hazard warning (emergency flasher) circuit
- Backup lamp circuit
- Side marker and clearance lamp circuit
- Instrument panel and interior lamp circuits

Review Questions

Choose the single most correct answer.

1. Which of the following is true concerning headlamp circuits?
   a. The circuits can have totally insulated bulbs and grounded switches.
   b. The lamp filaments are connected in series.
   c. All circuits use lamps that contain both a high beam and a low beam.
   d. The headlamps receive power through the ignition switch.
2. Because headlamps are an important safety feature, they are protected by:
   a. Heavy fuses
   b. Fusible links
   c. Type-I circuit breakers
   d. Type-II circuit breakers
3. All high beams are spread:
   a. Symmetrically
   b. Asymmetrically
   c. Either A or B
   d. Neither A nor B
4. All types of headlamps have _____ that are used when adjusting the beam.
   a. Connecting prongs
   b. Aiming pads
   c. Filaments
   d. Reflectors
5. Concealed headlamps can be operated by all of the following methods, *except*:
   a. Electric motor
   b. Vacuum actuator
   c. Manually
   d. Accessory belt
6. Which of the following is not true of automatic headlamp systems?
   a. Can turn headlamps on or off
   b. Can control high- and low-beam switching
   c. Are easily repaired when defective
   d. Can be adjusted to fit various conditions
7. On small automobile bulbs that have only one contact, the contact is:
   a. Insulated
   b. Indexed
   c. Grounded
   d. Festooned
8. Double-contact bulbs that are designed to fit into the socket only one way are called:
   a. Miniature bayonet base
   b. Single-contact bayonet
   c. Double-contact indexed
   d. Double-contact bayonet
9. The taillamps, license plate lamps, and parking lamps are generally protected by:
   a. A Type-I circuit breaker
   b. A Type-II circuit breaker
   c. A 20-amp fuse
   d. Three fusible links
10. Turn signal flasher units supply a rapid on-off-on current flow to the turn signal lamps by acting very much like:
    a. Circuit breakers
    b. Fuses
    c. Zener diodes
    d. Transistorized regulators
11. Which of the following is *not* part of the hazard warning lamp circuit?
    a. Turn signal lamps
    b. Brake lamps
    c. Flasher unit
    d. Instrument panel
12. The only white lamps allowed on the rear of a car are:
    a. Backup lamps and turn signal lamps
    b. Backup lamps and license plate lamps
    c. License plate lamps and turn signal lamps
    d. Turn signal lamps and backup lamps
13. Brightness of the instrument panel lamps is *not* controlled by:
    a. Diodes
    b. Rheostats
    c. Potentiometers
    d. Variable resistors
14. The switches used in courtesy lamp circuits are:
    a. Compound switches
    b. Push-pull switches
    c. Three-way switches
    d. Rheostats

15. Which of the following is *not* true of composite headlamps?
    a. Use replaceable bulbs
    b. Are part of the car's styling
    c. Have a polycarbonate lens
    d. May be red or amber

16. Halogen sealed-beam lamps:
    a. Are not as damage-resistant as glass
    b. Produce 30 percent less light
    c. May be made of plastic
    d. Have been used since the 1940s

13

Gauges, Warning Devices, and Driver Information System Operation

LEARNING OBJECTIVES

Upon completion and review of this chapter, you should be able to:

- Identify and explain the operation of electromagnetic instrument circuits, including gauges and sending units.
- Explain the operation of malfunction indicator lamps (MIL).
- Explain the operation of a simple mechanical speedometer.
- Identify and explain the operation of driver information systems (electronic instrument circuits).
- Explain the operation of the head-up display (HUD).

KEY TERMS

Air Core Gauge
Bimetallic Gauge
D'Arsonval Movement
Driver Information Center (DIC)
Driver Information System (DIS)
Gauges
Ground Sensor
Head-up Display (HUD)
Instrument Panel Cluster (IPC)
Malfunction Indicator Lamp (MIL)
Menu Driven
Nonvolatile RAM
Three-Coil Movement
Vacuum Fluorescent Display (VFD)
Warning Lamps

INTRODUCTION

This chapter covers the operation and common circuits of gauges, warning devices, and driver information systems. These systems include cooling fans, electromagnetic instrument circuits (sending and receiving gauges), and electronic instrument circuits (driver information systems).

ELECTROMAGNETIC INSTRUMENT CIRCUITS

Gauges and warning lamps allow the driver to monitor a vehicle's operating conditions. These instruments differ widely from car to car, but all are analog. Digital electronic instruments are explained in the "Electronic Instrument Circuits" section in this chapter. **Warning lamps** are used in place of gauges in many cases because they are less expensive and easier to understand, although they do not transmit as much useful information as gauges do. The following paragraphs explain the general operation of analog gauges, lamps, and the sending units that control them.

Gauge Operating Principles

Common gauges use one of the following three operating principles:

- Mechanical
- Bimetallic (thermal-type)
- Electromagnetic

Mechanical gauges are operated by cables, fluid pressure, or fluid temperature. Because they do not require an electrical circuit, they do not fit into our study. The cable-driven speedometer is the most common mechanical gauge.

Bimetallic Gauges

A **bimetallic gauge** works by allowing current to flow through the bimetallic strip and heat up one of the metals faster than the other, causing the strip to bend. A typical gauge (Figure 13-1) has U-shaped bimetallic piece anchored to the gauge body at the end of one arm. The other arm has a high-resistance wire (heater coil) wound around it. Current flow through the heater coil bends the free bimetallic arm. Varying the current changes the bend in the arm. A pointer attached to the moving arm can relate the changes in current to a scale on the face of the gauge.

Ambient temperature could affect the gauge, but the U-shape of the bimetallic strip provides temperature compensation. Although ambient temperature bends the free arm in one direction, the fixed arm is bent in the other direction and the effect is cancelled.

Electromagnetic Gauges

The movement of an electromagnetic gauge depends on the interaction of magnetic fields. The following kinds of movements are commonly used:

- D'Arsonval movement
- Three-coil or two-coil movement
- Air core design

A **D'Arsonval movement** has a movable electromagnet surrounded by a permanent horseshoe magnet (Figure 13-2). The electromagnet's field opposes the permanent magnet's field, causing the electromagnet to rotate. A pointer mounted on the

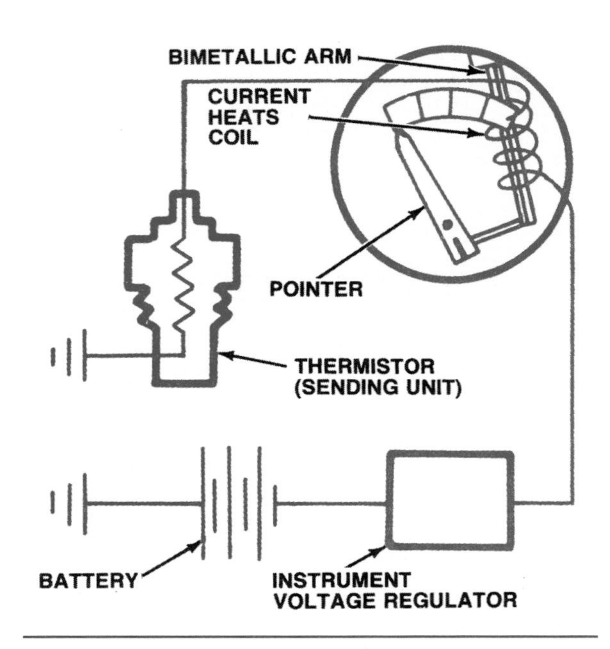

Figure 13-1. The bimetallic gauge depends upon the heat of current flow bending a bimetallic strip. (DaimlerChrysler Corporation)

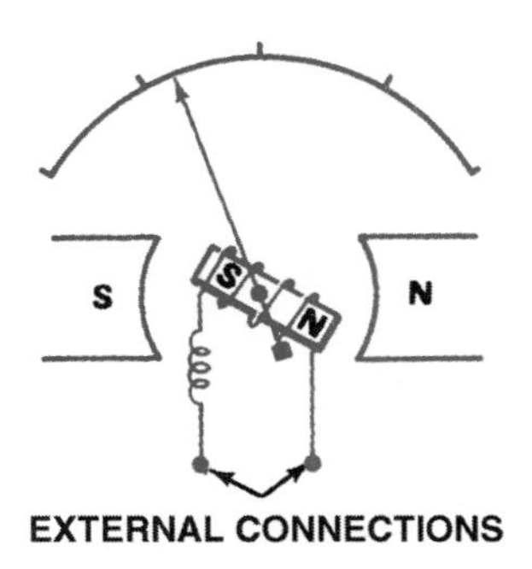

Figure 13-2. The D'Arsonval movement uses the field interaction of a permanent magnet and an electromagnet.

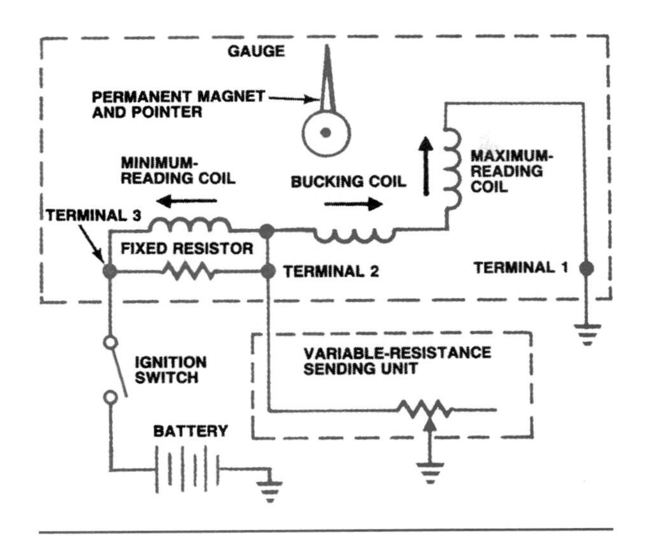

Figure 13-3. In a three-coil gauge, the variable resistance-sending unit affects current flow through three interacting electromagnets. (GM Service and Parts Operations)

electromagnet relates this movement to a scale on the face of the gauge. The amount of current flow through the electromagnet's coil determines the electromagnet's field strength, and therefore the amount of pointer movement.

A **three-coil movement** depends upon the field interaction of three electromagnets and the total field's effect on a movable permanent magnet. This type of gauge is used in GM and some late-model Ford vehicles.

The circuit diagram of a typical three-coil movement (Figure 13-3) shows that two coils are wound at right angles to each other. These are the minimum-reading coil and the maximum-reading coil. Their magnetic fields will pull the permanent magnet and pointer in opposite directions. A third coil is wound so that its magnetic field opposes that of the minimum-reading coil. This is called the bucking coil.

The three coils are connected in series from the ignition switch to ground. A fixed resistor forms a circuit branch parallel to the minimum-reading coil. The variable-resistance-sending unit forms a circuit branch to ground, parallel to the bucking and minimum-reading coils.

When sending resistance is high, current flows through all three coils to ground. Because the magnetic fields of the minimum-reading and the bucking coils cancel each other, the maximum-reading coil's field has the strongest effect on the permanent magnet and pointer. The pointer moves to the maximum-reading end of the gauge scale.

As sending unit resistance decreases, more current flows through the minimum-reading coil and the sending unit to ground than flows through the bucking and maximum-reading coils. The minimum-reading coil gains a stronger effect upon the permanent magnet and pointer, and the pointer moves to the minimum-reading end of the gauge scale.

Specific three-coil gauges may have slightly different wiring, but the basic operation remains the same. Because the circular magnet is carefully balanced, it will remain at its last position even when the ignition switch is turned off, rather than returning to the minimum-reading position, as does a bimetallic gauge.

The design of two-coil gauges varies with the purpose for which the gauge is used. In a fuel gauge, for example, the pointer is moved by the magnetic fields of the two coils positioned at right angles to each other. Battery voltage is applied to the E (empty) coil and the circuit divides at the opposite end of the coil. One path travels to ground through the F (full) coil; the other grounds through the sender's variable resistor. When sender resistance is low (low fuel), current passes through the E coil and sender resistor to move the pointer toward E on the scale. When sender resistance is high (full tank), current flows through the F coil to move the pointer toward F on the scale.

When a two-coil gauge is used to indicate coolant temperature, battery voltage is applied to both coils. One coil is grounded directly; the other grounds through the sending unit. Sender resistance causes the current through one coil to change as the temperature changes, moving the pointer.

In the **air-core gauge** design, the gauge receives a varying electrical signal from its sending unit. A pivoting permanent magnet mounted to a pointer aligns itself to a resultant field according to sending unit resistance. The sending unit resistance varies the field strength of the windings in opposition to the reference windings. The sending unit also compensates for variations in voltage.

This simple design provides several advantages beyond greater accuracy. It does not create radiofrequency interference (RFI), is unaffected by temperature, is completely noiseless, and does not require the use of a voltage limiter. Like the three- and two-coil designs, however, the air-core design remains at its last position when the ignition switch is turned off, giving a reading that should be disregarded.

■ Instrument Voltage Regulator

On early-model cars and imported vehicles, except for the air core electromagnetic design, gauges required a continuous, controlled amount of voltage. This is usually either the system voltage of 12 volts or a regulated 5–6 volts. An instrument voltage regulator (IVR) supplied that regulated voltage. The IVR can be a separate component that looks much like a circuit breaker or relay; it can also be built into a gauge. Its bimetallic strip and vibrating points act like a self-setting circuit breaker to keep the gauge voltage at a specific level. Gauges that operate on limited voltage can be damaged or give inaccurate readings if exposed to full system voltage.

Warning Lamp Operating Principles

Warning lamps alert the driver to potentially hazardous vehicle operating conditions, such as the following:

- High engine temperature
- Low oil pressure
- Charging system problems
- Low fuel level
- Unequal brake fluid pressure
- Parking brake on
- Seat belts not fastened
- Exterior lighting failure

Warning lamps can monitor many different functions but are usually activated in one of the following four ways:

- Voltage drop
- Grounding switch
- Ground sensor
- Fiber optics

The first three methods are used to light a bulb or an LED mounted on the dash panel. Fiber optics is a special application of remote light.

Voltage Drop

A bulb will light only if there is a voltage drop across its filament. Warning lamps can be installed so that equal voltage is applied to both bulb terminals under normal operating conditions. If operating conditions change, a voltage drop occurs across the filament, and the bulb will light. This method is often used to control charging system indicators, as we learned in Chapter 8.

Grounding Switch

A bulb connected to battery voltage will not light unless the current can flow to ground. Warning lamps can be installed so that a switch that reacts to operating conditions controls the ground path. Under normal conditions, the switch contacts are open and the bulb does not light. When operating conditions change, the switch contacts close. This creates a ground path for current and lights the bulb.

Ground Sensor

A **ground sensor** is the opposite of a ground switch. Here, the warning lamp remains unlit as long as the sensor is grounded. When conditions change and the sensor is no longer grounded, the bulb lights. Solid-state circuitry generally is used in this type of circuit.

Fiber Optics

Strands of a special plastic can conduct light through long, curving runs (Figure 13-4). When one end of the fiber is installed in the instrument panel and the other end is exposed to a light, the driver will be able to see that light. Changing operating conditions can cause the fiber to change from light to dark, or from one color to another. Fiber optics are usually used for accessory warning lamps, such as coolant level reminders and exterior bulb monitors.

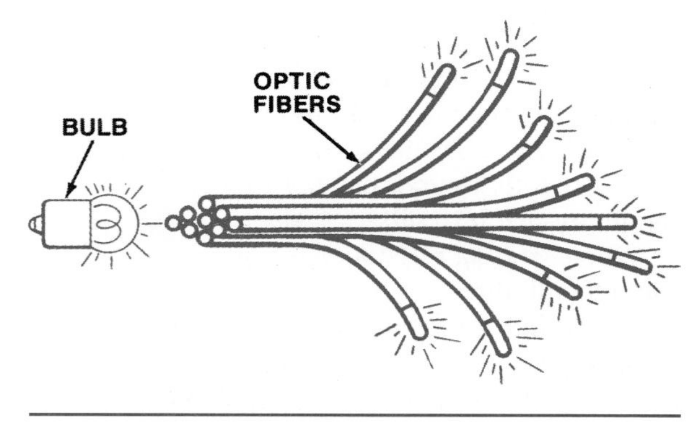

Figure 13-4. Light-carrying fibers can be used in accessory instruments.

Specific Instruments

Many different instruments have appeared in automobiles, but certain basic functions are monitored in almost all cars. Normally, a car's instrument panel will contain at least the following:

- An ammeter, a voltmeter, or an alternator warning lamp
- An oil pressure gauge or warning lamp
- A coolant temperature gauge or warning lamp
- A fuel level gauge

The following paragraphs explain how these specific instruments are constructed and installed.

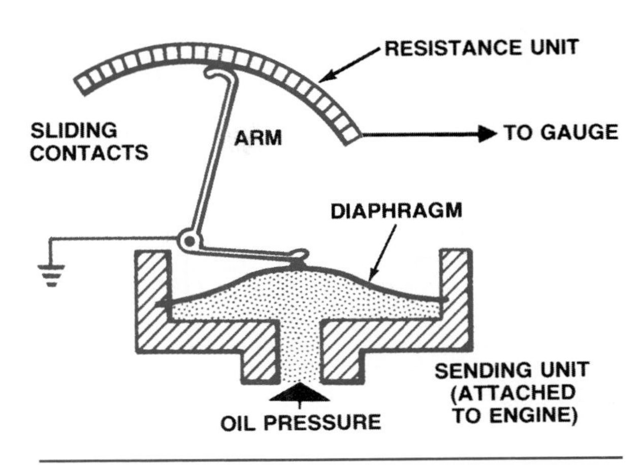

Figure 13-5. The oil pressure sending unit provides a varying amount of resistance as engine oil pressure changes.

Charging System Indicators

Ammeter, voltmeter, and warning lamp installations are covered in Chapter 8. Ammeters usually contain a D'Arsonval movement that reacts to field current flow into the alternator and charging current flow into the battery. Many late-model cars have a voltmeter instead of an ammeter. A voltmeter indicates battery condition when the engine is off, and charging system operation when the engine is running. Warning lamps light to show an undercharged battery or low voltage from the alternator. Lamps used on GM cars with a CS charging system will light when the system voltage is too high or too low.

Chrysler rear-wheel-drive (RWD) cars from 1975 on that have ammeters also have an LED mounted on the ammeter face. The LED works independently to monitor system voltage and lights when system voltage drops by about 1.2 volts. This alerts the driver to a discharge condition at idle caused by a heavy electrical load.

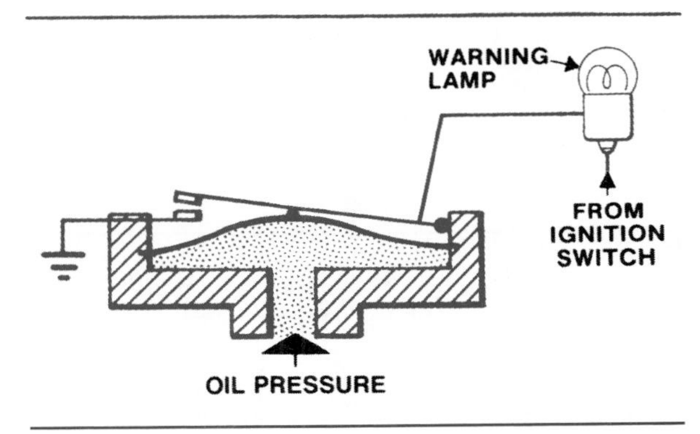

Figure 13-6. This oil pressure grounding switch has a fixed contact and a contact that is moved by the pressure-sensitive diaphragm.

Oil Pressure Gauge or Warning Lamp

The varying current signal to an oil pressure gauge is supplied through a variable-resistance sending unit that is exposed to engine oil pressure. The resistor variation is controlled by a diaphragm that moves with changes in oil pressure, as shown in Figure 13-5.

An oil pressure-warning lamp lights to indicate low oil pressure. A ground switch controls the lamp as shown in Figure 13-6. When oil pressure decreases to an unsafe level, the switch diaphragm moves far enough to ground the warning lamp circuit. Current then can flow to ground and the bulb will light. Oil pressure warning lamps can be operated by the gauge itself. When the pointer moves to the low-pressure end of the scale, it closes contact points to light a bulb or an LED.

Temperature Gauge or Warning Lamp

In most late-model cars, the temperature gauge ending unit is a thermistor exposed to engine coolant temperature, as shown in Figure 13-7. As coolant temperature increases, the resistance of the thermistor decreases and current through the gauge varies.

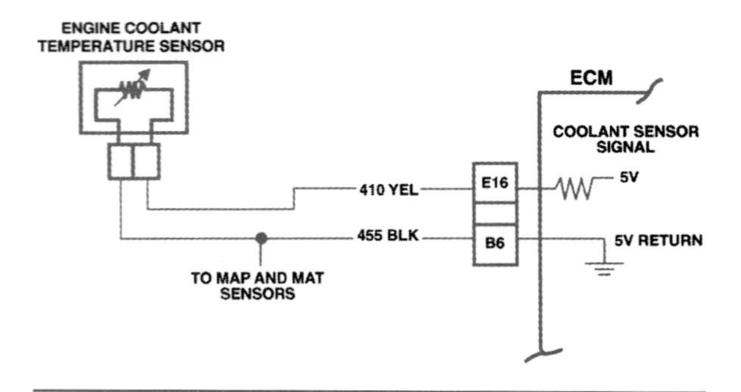

Figure 13-7. Coolant temperature gauge. (GM Service and Parts Operations)

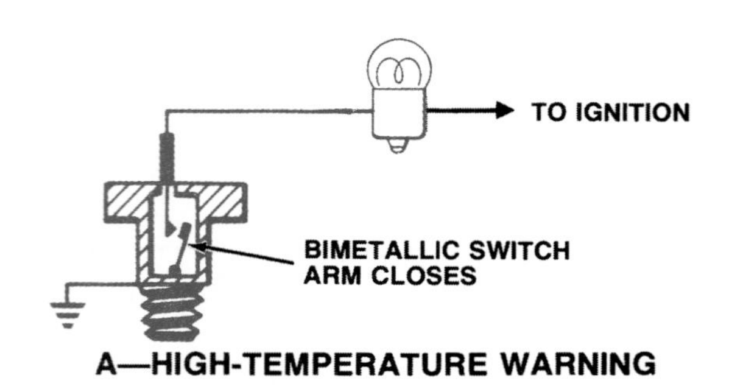

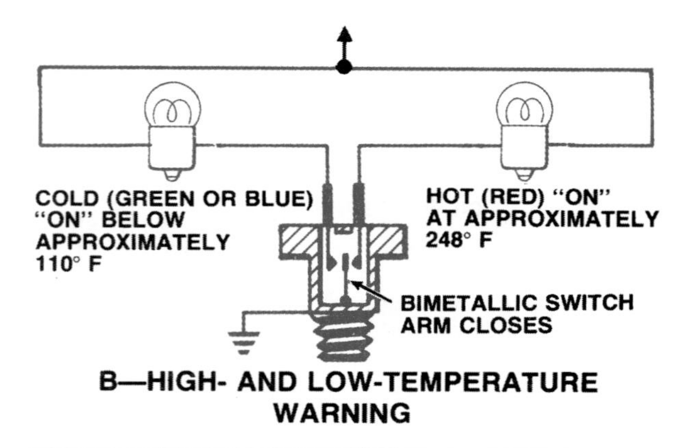

Figure 13-8. Temperature grounding switches expose a bimetallic strip to engine coolant temperature to light a high-temperature lamp or both high- and low-temperature warning lamps.

Temperature warning lamps can alert the driver to high temperature or to both low and high temperature. The most common circuit uses a bimetallic switch and reacts only to high temperature, as shown in Figure 13-8A. A ground switch has a bimetallic strip that is exposed to coolant temperature. When the temperature reaches an unsafe level, the strip bends far enough to ground the warning lamp circuit. If the circuit also reacts to low temperature, the bimetallic strip has a second set of contacts. These are closed at low temperature (Figure 13-8B) but open during normal operating temperature. The low-temperature circuit usually lights a different bulb than does the high-temperature circuit.

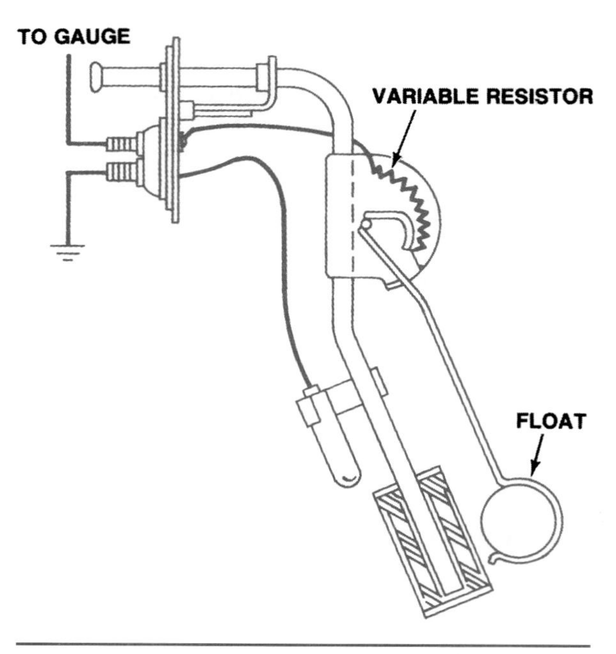

Figure 13-9. The fuel tank sending unit has a float that moves with the fuel level in the tank and affects a variable resistor.

A temperature warning lamp or an LED also can be lit by the action of the temperature gauge pointer, as explained for the oil pressure gauge.

Fuel Gauge or Warning Lamp

All modern cars have a fuel level gauge. Some have an additional warning lamp or an LED to indicate a low fuel level. A variable resistor in the fuel tank provides current control through the fuel gauge. The fuel tank sending unit has a float that moves with the fuel level, as shown in Figure 13-9. As the float rises and falls, the resistance of the sending unit changes. If a low-fuel-level indicator is used, its switch may operate through a heater or bimetallic relay to prevent flicker. Fuel-level warning lamps are operated by the action of the fuel gauge pointer, as explained for an oil pressure gauge.

Tachometer

In addition to the other instrument panel displays, some cars have tachometers to indicate engine speed in revolutions per minute (rpm). These usually have an electromagnetic movement. The engine speed signals may come from an electronic pickup at the ignition coil (Figure 13-10). Voltage pulses taken from the ignition system are processed by solid-state circuitry into signals to drive the tachometer pointer. The pointer responds to the frequency of these signals, which increase with engine speed. A filter is used to round off the pulses and remove any spikes.

Late-model vehicles with an engine control system may control the tachometer through an electronic module. This module is located on the rear of the instrument cluster printed circuit board and is the interface between the computer and tachometer in the same way the solid-state circuitry processes the ignition system-to-tachometer signals described earlier.

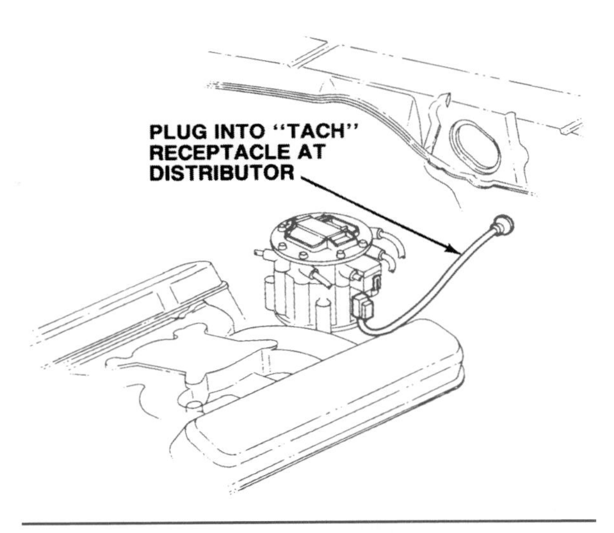

Figure 13-10. This GM HEI (high energy ignition) distributor has a special connector for a tachometer. (GM Service and Parts Operations)

MALFUNCTION INDICATOR LAMP (MIL)

Vehicles with electronic engine control systems generally have a computer-operated warning lamp on the instrument panel to indicate the need for service. In the past, this was called a Check Engine, Service Engine Soon, Power Loss, or Power Limited lamp, according to the manufacturer, as shown in the sample instrument cluster in Figure 13-11. To eliminate confusion, all domestic manufacturers now refer to it as a **malfunction indicator lamp (MIL).**

The MIL lamp alerts the driver to a malfunction in one of the monitored systems. In some vehicles built before 1995, the MIL is used to retrieve the faults or trouble codes stored in the computer memory by grounding a test terminal in the diagnostic connector. Like other warning lamps, the MIL comes on briefly as a bulb check when the ignition is turned on.

Seatbelt-Starter Interlocks

During 1974 and early 1975, U.S. federal safety standards required a seatbelt-starter interlock system on all new cars. The system required front-seat occupants to fasten their seatbelts before the car could be started. This particular standard was repealed by an act of Congress in 1975. Now, most interlock systems have been disabled so that only a warning lamp and buzzer remain.

Antilock Brake System (ABS) Warning Lamp

Vehicles with ABS have a computer-operated amber antilock warning lamp (Figure 13-12) in addition to the MIL and the standard red brake

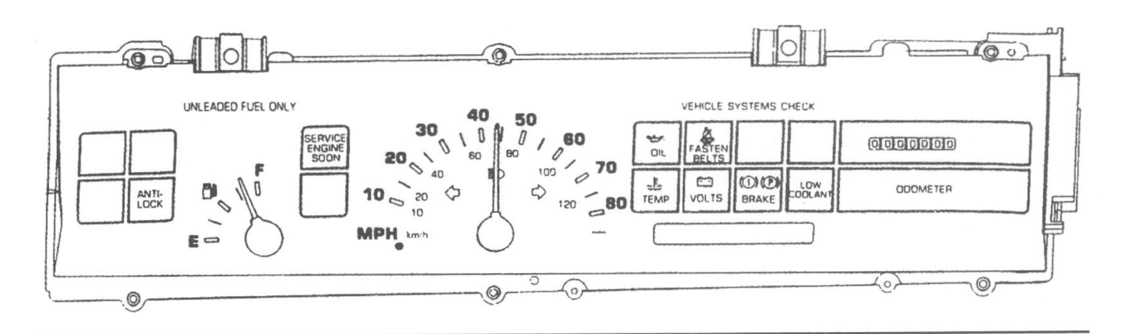

Figure 13-11. Malfunction indicator lamp in the instrument panel. (DaimlerChrysler Corporation)

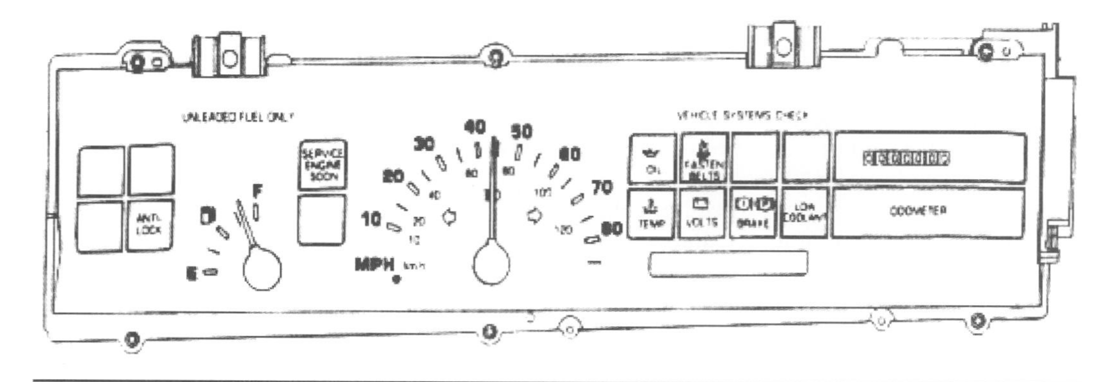

Figure 13-12. ABS lights. (DaimlerChrysler Corporation)

lamp. The antilock lamp serves the same functions for the ABS that the MIL lamp does for engine control systems as follows,

- Lights to warn of a system problem that inhibits ABS operation
- Retrieves trouble codes in the same way as the MIL lamp (specific vehicles only)
- Lights briefly at the beginning of an ignition cycle as a bulb check and to notify the driver that self-diagnostics are taking place

Buzzers, Tone Generators, Chimes, and Bells

Buzzers are a special type of warning device. They produce a loud warning sound during certain operating conditions, such as the following:

- Seatbelts not fastened
- Door open with key in ignition
- Lights left on with engine off
- Excessive vehicle speed

A typical buzzer (Figure 13-13) operates in the same way as a horn. Instead of moving a diaphragm, the vibrating armature itself creates the sound waves. In Figure 13-13, two conditions are required to sound the buzzer: The door must be open, and the key must be in the ignition. These conditions close both switches and allow current to flow through the buzzer armature and coil.

Most warning buzzers are separate units mounted on the fuse panel or behind the instrument panel. GM vehicles may have a buzzer built into the horn relay (Figure 13-14). When the ignition key is left in the switch and the door is opened, a small amount of current flows through

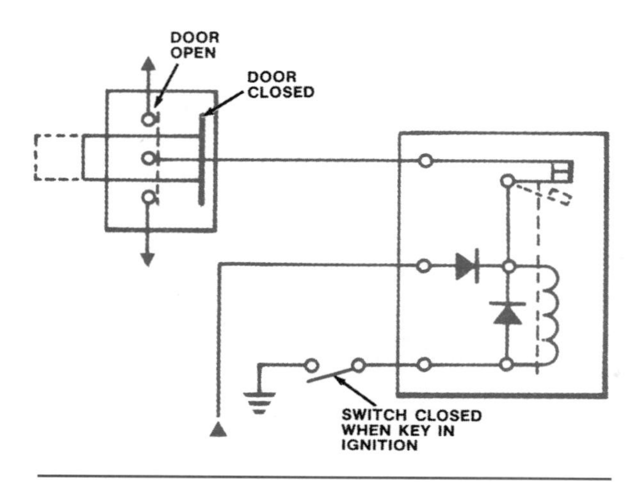

Figure 13-13. Current will flow through this warning buzzer only when both switches are closed—when the door is open and the key is in the ignition switch.

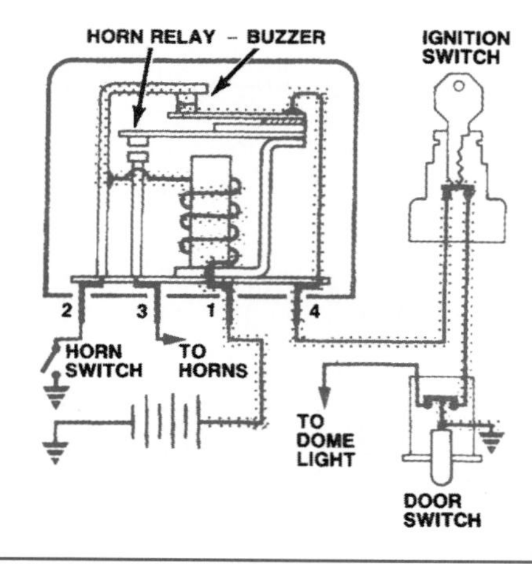

Figure 13-14. GM cars may have a buzzer built into the horn relay. Here, the buzzer is activated because both the door switch and the key switch are closed. (GM Service and Parts Operations)

the relay coil. The magnetic field is strong enough to operate the buzzer, but it is not strong enough to close the horn contacts.

Grounding switches usually activate buzzers. A timing circuit can be built into the buzzer by winding a heater coil around an internal circuit breaker (Figure 13-15) and connecting the heater coil directly to ground. When current flows to the buzzer, a small amount of current flows through the heater coil to ground. When the coil is hot enough, it will open the circuit breaker and keep it open. Current through the buzzer will stop even though the grounding switch is still closed.

Some typical buzzer warning circuits are shown in Figures 13-16 and 13-17. Figure 13-17 includes a prove-out circuit branch with a manual-grounding switch that the driver can close to check that the bulb and buzzer are still working. Some prove-out circuits operate when the ignition switch is at START, to show the driver if any bulbs or buzzers have failed.

Tone generators, chimes, and bells are mechanical devices that produce a particular sound when voltage is applied across a sound bar. Various sounds are obtained by varying the voltage. Like buzzers, they are replaced if defective.

The circuit shown in Figure 13-18 is the circuit diagram for an electronic temperature gauge. The coolant temperature sensor is an NTC thermistor with high resistance at low temperatures and low resistance at high temperatures. When a cold engine is first started, the sensor's resistance is very high, resulting in a low voltage output to the gauge display, which translates into a low

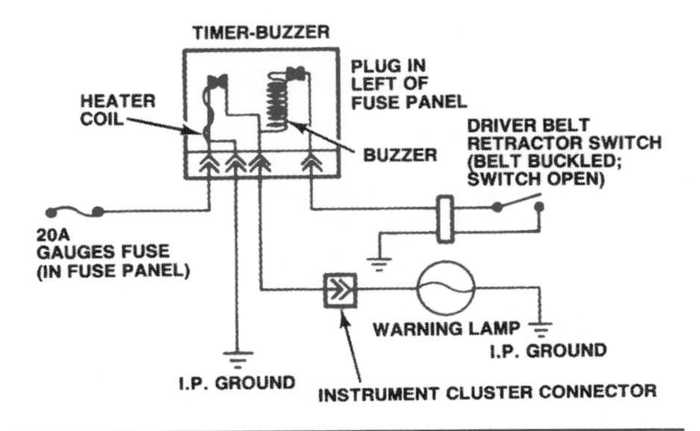

Figure 13-15. This buzzer will sound for only a few seconds each time it is activated, because of the circuit breaker and heater coil built into the unit. (GM Service and Parts Operations)

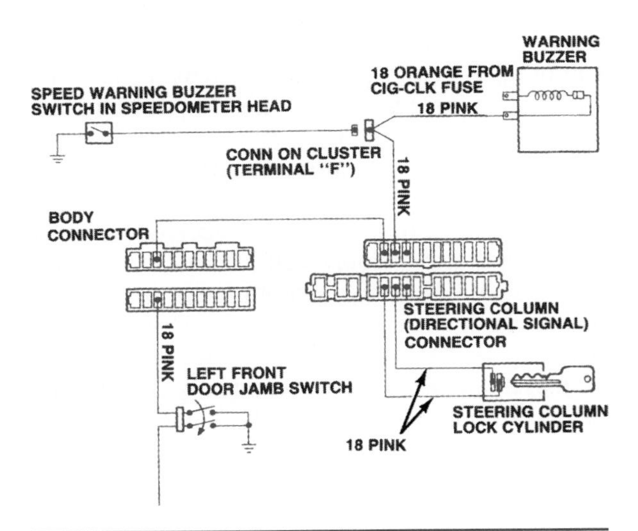

Figure 13-16. In this circuit, one buzzer responds both to excessive speed and to driver door position. (GM Service and Parts Operations)

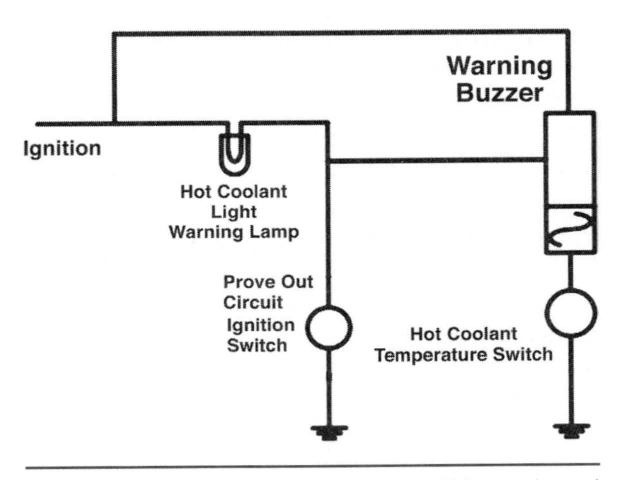

Figure 13-17. This warning buzzer will be activated if engine coolant temperature rises above a safe level.

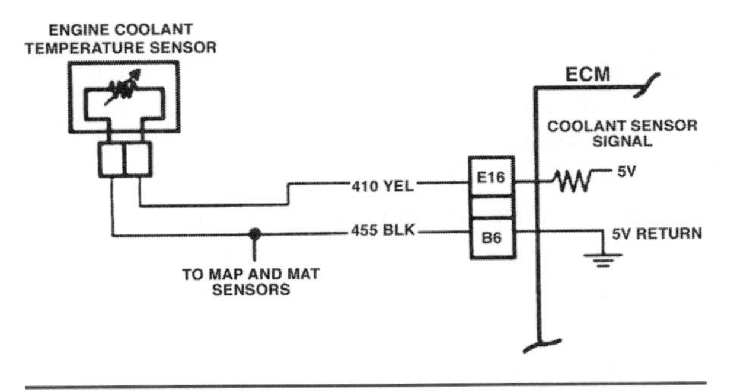

Figure 13-18. GM's electronic temperature gauge circuit is similar to that of an analog gauge. (GM Service and Parts Operations)

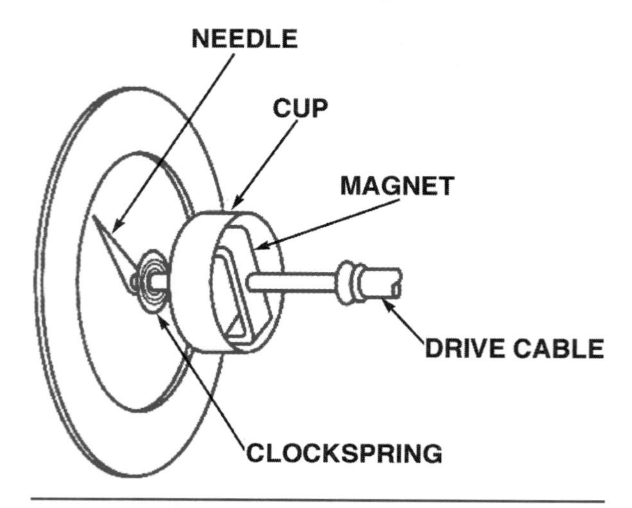

Figure 13-19. Mechanical speedometer.

coolant temperature reading on the display. As coolant temperature increases, sensor resistance decreases. This results in a higher voltage output to the gauge display, which translates into a higher coolant temperature reading.

SPEEDOMETER

A mechanical speedometer uses a flexible cable, similar to piano wire, that is encased in an outer cover (Figure 13-19). One end of the cable connects to the transmission and the other end connects to the back of the speedometer. As the vehicle moves, the cable begins to rotate. The speed that the cable rotates is proportional to the speed of the vehicle; in other words, the faster the vehicle moves, the faster the cable will turn.

The indicator needle is attached to a metal drum in the speedometer. As the cable turns, so does a magnet inside the drum. The spinning magnet creates a rotating magnetic field around the drum, causing the drum to rotate and move the indicator needle along the scale. The faster the magnet spins, the more the drum moves. The speedometer gears drive a mechanical odometer. The gears are driven by a worm gear mounted on the same shaft as the permanent magnet of the speedometer (Figure 13-20). The gears reduce the speed of the odometer cable driven by the transmission.

ODOMETER

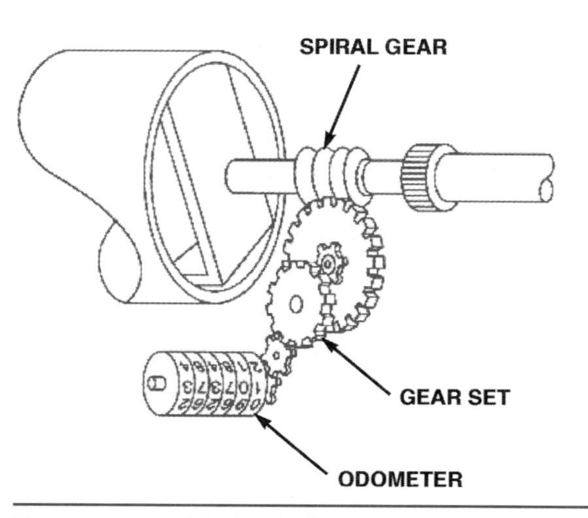

Figure 13-20. Mechanical odometer.

ELECTRONIC INSTRUMENT CIRCUITS

Electronic instruments or **driver information system (DIS)** used on late-model cars have same purpose as the traditional analog instruments: The DIS displays vehicle-operating information to the driver and includes all guage and speedometer information. The primary difference between the DIS and traditional systems is the way in which the information is displayed. The **Driver Information Center (DIC)** is a type of DIS used by many of the automotive manufacturers. The DIS and DIC are basically the same item.

Digital instrumentation is more precise than conventional analog gauges. Analog gauges display an average of the readings received from the sensor; a digital display will present exact readings. Most digital instrument panels provide for display in English or metric values. Drivers select which gauges they wish to have displayed. Most of these systems will automatically display the gauge to indicate a potentially dangerous situation. For example, if the driver has chosen the oil pressure gauge to be displayed and the engine temperature increases above set limits, the temperature gauge will automatically be displayed to warn the driver. A warning light and/or a chime will also activate, to get the driver's attention.

Most electronic instrument panels contain self-diagnostics. The tests are initiated through a scan tool or by pushing selected buttons on the instrument panel. The instrument panel cluster also initiates a self-test every time the ignition switch is turned to ACC or RUN. Usually the entire dash is illuminated and every segment of the display is lighted; ISO symbols generally flash during this test. At the completion of the test, all gauges will display current readings. A code is displayed to alert the driver if a fault is found.

Like analog instruments, electronic instruments receive inputs from a sensor or a sending unit. The information is displayed in various ways. Depending upon the gauge function and the manufacturer's design, the display may be digital numbers or a vertical, horizontal, or curved bar (Figure 13-21). The following paragraphs give three examples of how electronic instruments function.

Electronic Speedometer

Figure 13-22A shows a GM speedometer that can be either a quartz analog (swing needle) display or a digital readout. The speed signal in this system originates from a small AC voltage generator with four magnetic fields called a permanent magnet (PM) generator. This device usually is installed at the transmission or transaxle speedometer gear adapter and is driven like the speedometer cable on conventional systems.

As the PM generator rotates, it generates an AC voltage of four pulses per turn, with voltage and frequency increasing as speed increases. The unit pulses 4,004 times per mile of travel (2,488 times per kilometer) with a frequency output of 1.112 oscillations per second (hertz) per mile per hour of travel (0.691 hertz per kilometer per hour of travel).

Since the PM generator output is analog, a buffer is used to translate its signals into digital input for the processing unit. The processing unit sends a voltage back to the buffer, which switches the voltage on and off and interprets it as vehicle speed changes. If the instrument cluster has its own internal buffer as part of the cluster circuitry, the PM generator signal will go directly to the speedometer.

On some systems, the buffer may contain more than one switching function, as shown in Figure 13-22B, as it handles the ECM and cruise control systems. These secondary switching functions run at half the speed of the speedometer switching operation, or 0.556 Hz/mph (0.3456 Hz/km/h).

If the instrument cluster uses a quartz analog display (Figure 13-23), the gauge is similar to the two-coil electromagnetic gauge discussed earlier. This type of gauge is often called a swing needle or air-core gauge and does not return to zero when the ignition is turned off. When the car begins to move, the buffered speed signal is conditioned

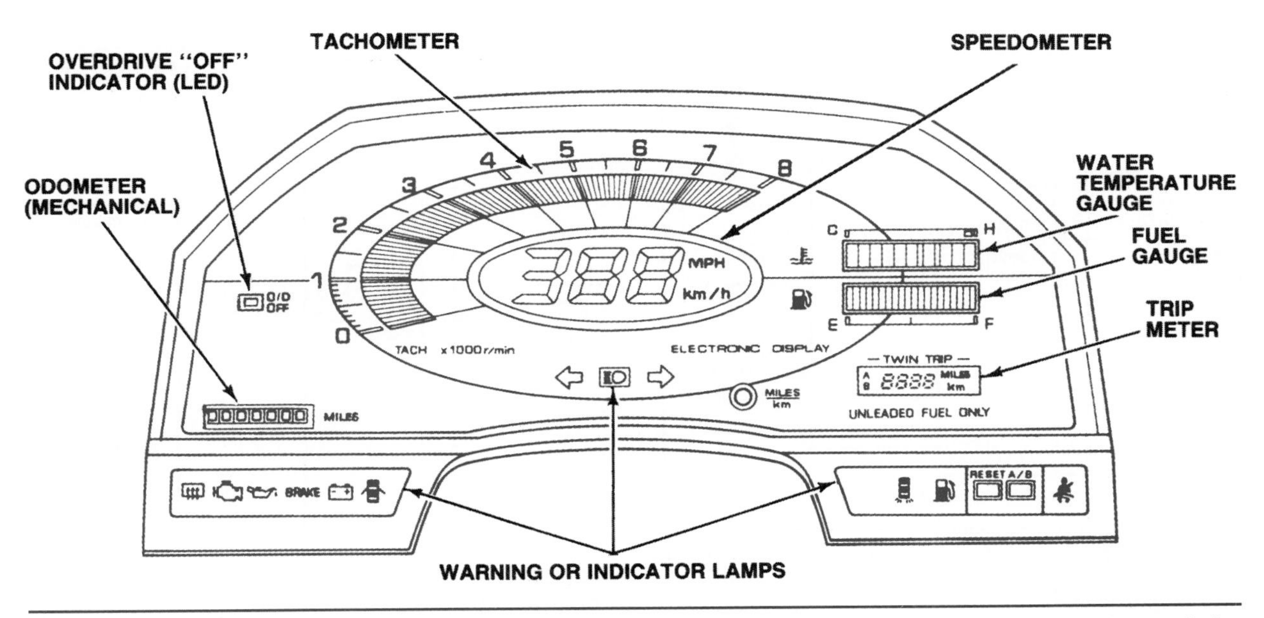

Figure 13-21. Toyota's electronic display is a digital combination meter, which uses a colored liquid crystal display (LCD) panel. (Reprinted by permission of Toyota Motor Corporation)

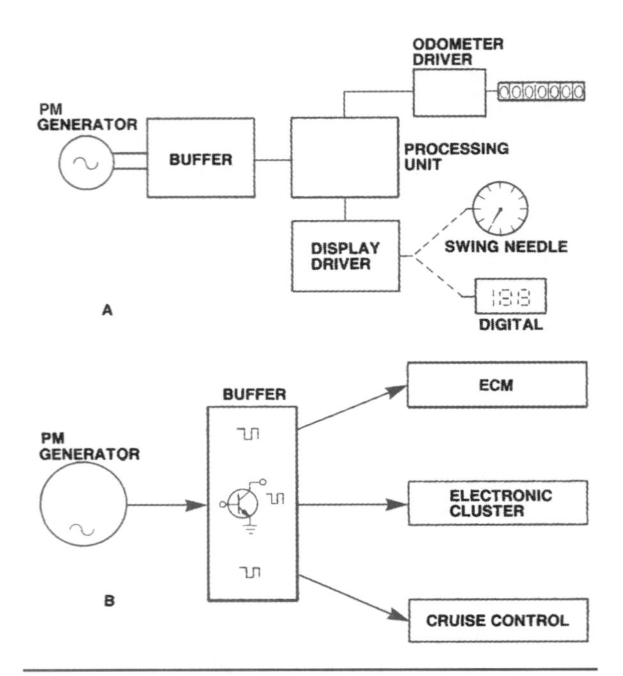

Figure 13-22. GM's electronic speedometer uses a permanent magnet (PM) generator instead of an optical sensor. In A, the buffer translates the PM generator analog signals into digital signals for the processor, which activates the display driver to operate an analog or a digital display. In B, the buffer toggles voltage on and off to interpret vehicle speed to the electronic cluster. (GM Service and Parts Operations)

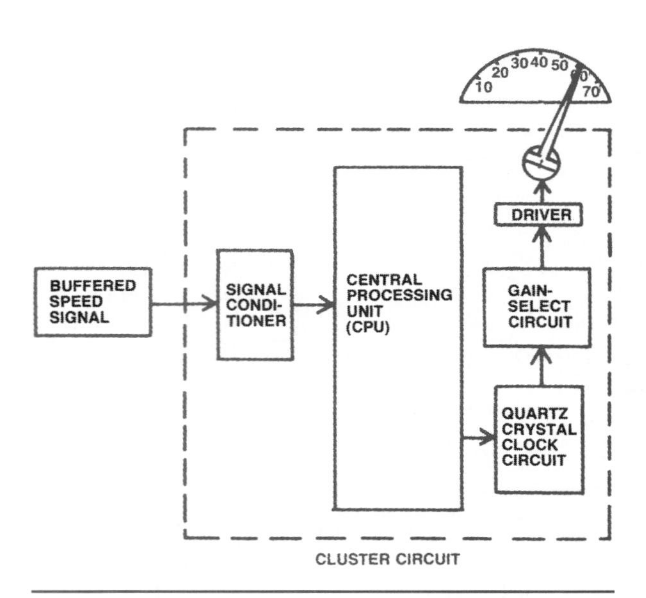

Figure 13-23. The buffered signal passes to a signal conditioner, which transmits it to the CPU where a quartz clock circuit ensures accuracy. After processing, the signal goes to a gain-select circuit, which sends it to the driver circuit for analog display.

and sent to the central processing unit (CPU). The CPU processes the digital input using a quartz-crystal clock circuit and sends it to a gain-select circuit, where it is transferred to a driver circuit. The driver circuit then sends the correct voltages to the gauge coils to move the pointer and indicate the car's speed.

Virtually the same process is followed when a digital display is used (Figure 13-24), with the following minor differences in operation:

- The CPU can be directed to display the information in either English or metric units. A select function sends the data along different circuits according to the switch position.
- The driver circuit is responsible for turning on the selected display segments at the correct intensity.

The odometer used with electronic speedometers can be electromechanical, using a stepper motor (Figure 13-25A), or an IC chip using **nonvolatile RAM** (Figure 13-25B).

The electromechanical type is similar to the conventional odometer, differing primarily in the way in which the numbers are driven. A stepper motor takes digital-pulsed voltages from the speedometer circuit board at half the buffered speed signal discussed earlier. This provides a very accurate accounting of accumulated mileage.

The IC chip retains accumulated mileage in its special nonvolatile RAM, which is not lost when

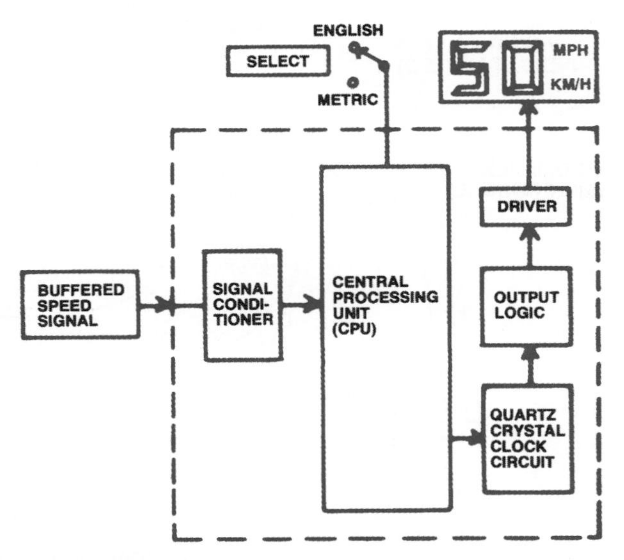

Figure 13-24. The digital cluster circuit is similar to the analog circuit, but an output-logic circuit is used instead of a gain-select circuit.

current is removed. Since its memory *cannot* be turned back, the use of this chip virtually eliminates one of the frauds often associated with the sale of used cars.

Electronic Instrument Panel

The GM electronic **instrument panel cluster (IPC)** contains the following indicators:

- The ABS indicator
- The Air Bag indicator
- The Brake indicator

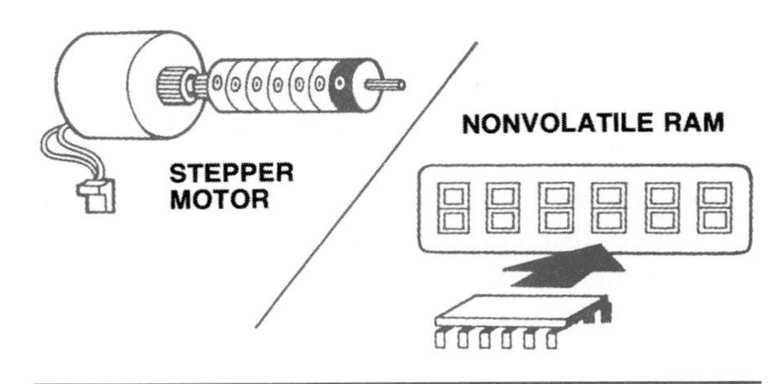

Figure 13-25. Electronic odometers may use a stepper motor or a nonvolatile RAM chip for mileage display. (GM Service and Parts Operations)

- The Cruise indicator
- The Charge indicator
- The Engine Coolant indicator
- The Engine Oil Pressure indicator
- The Fasten Belts indicator
- The Security indicator
- The Service Engine Soon indicator (MIL)

Indicators and Warning Messages

Average Fuel Economy

The average fuel economy is a function of distance traveled divided by fuel used since the parameter was last reset. When the average fuel economy is displayed in the IPC, pressing the Info Reset button on the DIC will reset the average fuel economy parameter. The average fuel economy parameter is only displayed on the DIC. The average fuel economy displays either miles or kilometers, as requested, by briefly pressing the Eng/Met button on the DIC.

Engine Coolant Temperature Gauge

The IPC displays the engine coolant temperature, as determined by the PCM. The IPC receives a Class 2

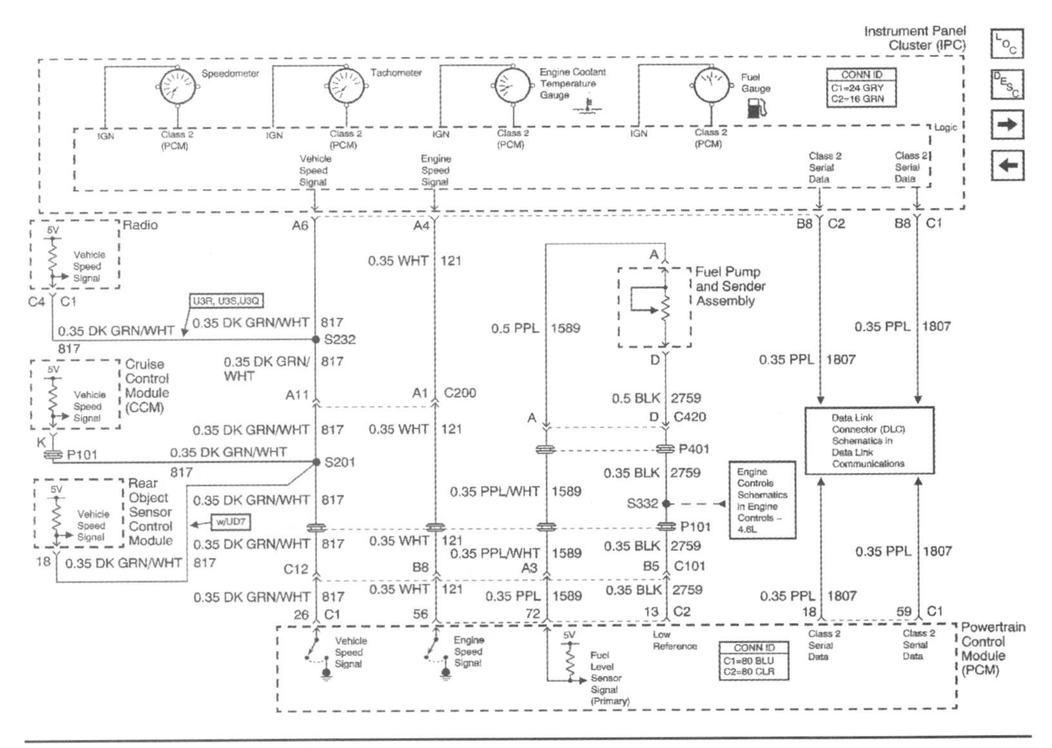

Figure 13-26. GM electronic instrument panel schematic. (Courtesy of GM Service and Parts Operations)

message from the PCM indicating the engine coolant temperature. The engine coolant temperature gauge defaults to C if the following is true:

- The PCM detects a malfunction in the engine coolant temperature sensor circuit.
- The IPC detects a loss of Class 2 communications with the PCM.

Fuel Gauge

The IPC displays the fuel level, as determined by the PCM. The IPC receives a Class 2 message from the PCM indicating the fuel level percentage. The fuel gauge defaults to empty if the following is true:

- The PCM detects a malfunction in the fuel level sensor signal circuit.
- The IPC detects a loss of Class 2 communications with the PCM.

Fuel Range

The fuel range is the estimated distance that the vehicle can travel under the current fuel economy and fuel level conditions. The driver cannot reset the fuel range parameter. "LO" is displayed in the fuel range display when the range is calculated to be less than 64 km (40 miles). The fuel range displays either miles or kilometers, as requested, by briefly pressing the Eng/Met button on the DIC.

Odometer

The IPC contains a season odometer and two trip odometers. The IPC calculates the mileage based on the vehicle-speed signal circuit from the PCM. The odometer will display "Error" if an internal IPC memory failure is detected. The odometer displays either miles or kilometers, as requested, by briefly pressing the Eng/Met button on the DIC. Pressing the Reset Trip A/B button for greater than 2 seconds will reset the trip odometer that is displayed.

PRNDL Display

The IPC displays the selected automatic transmission/transaxle gear position determined by the PCM, as sensed by the gear position selected by the driver. The IPC receives a Class 2 message from the PCM indicating the gear position. The PRNDL for the transmission gear position display blanks if the following is true:

- The PCM detects a malfunction in the transmission range switch signal circuit.
- The IPC receives a Class 2 message indicating the park position and the column park switch indicates a position other than park.
- The IPC detects a loss of Class 2 communications with the PCM.

Speedometer

The IPC displays the vehicle speed based on the information received from the PCM. The PCM converts the data from the vehicle speed sensor to a 4,000-pulses/mile signal. The IPC uses the vehicle-speed signal circuit from the PCM in order to calculate the vehicle speed. The speedometer defaults to 0 km/h (0 mph) if a malfunction in the vehicle speed signal circuit exists. The speedometer displays either miles or kilometers, as requested, by briefly pressing the Eng/Met button on the DIC.

The speedometer display configuration can be changed by using the Dspl Mode button. The speedometer can be configured in one of the following formats:

- Analog display only
- Analog and digital displays
- Digital display only

Tachometer

The IPC displays the engine speed based on information received from the PCM. The PCM converts the data from the engine to a 2-pulses-per-engine-revolution signal. The IPC uses the engine-speed signal circuit from the PCM to calculate the engine speed. The tachometer defaults to 0 rpm if a malfunction in the engine-speed signal circuit exists.

Voltmeter

The IPC displays the system voltage.

Driver Information Center (DIC)

The DIC vehicle information displays vehicle operation parameters that are available to the driver when the ignition is in the RUN position. On many applications, the driver can navigate through the various parameters displayed on the DIC using the Info Up and Info Down buttons. Some parameters can be reset by briefly pressing the Info Reset button.

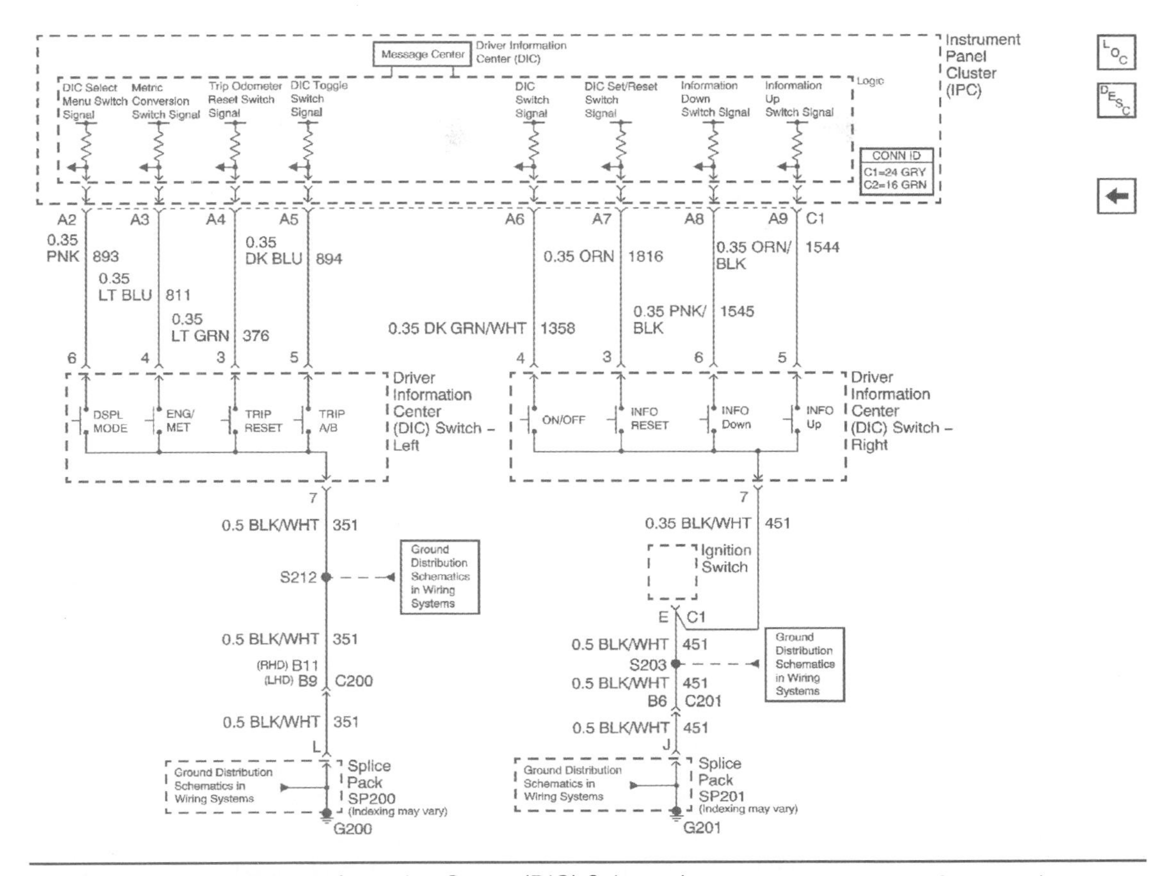

Figure 13-27. GM Driver Information Center (DIC) Schematic. (GM Service and Parts Operations)

NOTE: The Driver Information Center (DIC) is also know as a Driver Information System (DIS).

Some parameters require information from other modules. If the IPC has not received data for a parameter when the time has come to display the parameter, the IPC will blank the display on the DIC. If the IPC has determined that a Class 2 communication failure with one of the modules exists when the time has come to display the parameter, the IPC will display dashes on the DIC. The vehicle information display parameters are displayed in the following order:

1. OUTSIDE TEMP
2. RANGE
3. MPG AVG
4. MPG INST
5. FUEL USED
6. AVG MPH
7. TIMER
8. BATTERY VOLTS
9. TIRE PRESSURE
10. TACHOMETER
11. ENGINE OIL LIFE
12. TRANS FLUID LIFE
13. PHONE
14. FEATURE PROGRAMMING

When the Info Up button is pressed, the DIC will display the next parameter on the list. When the Info Down button is pressed, the DIC will display the previous parameter on the list.

Body Control Module (BCM) Computers

When more than one computer is used on a vehicle, it is often desirable to link their operations. The body control module (BCM) used on many

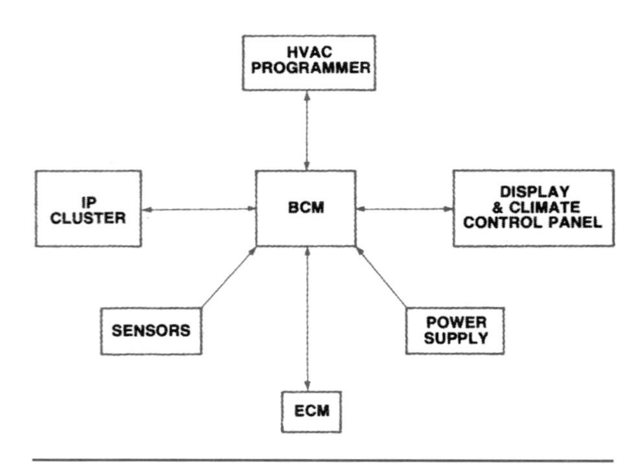

Figure 13-28. Sophisticated electronic systems are composed of several computers and use a central computer (GM calls it a body control module) to manage the system. (GM Service and Parts Operations)

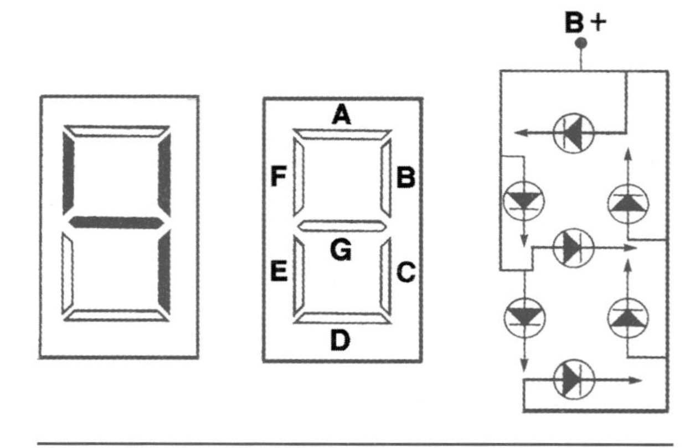

Figure 13-29. Selective application of voltage through the diodes composing a light-emitting diode (LED) results in an alphanumeric display. (GM Service and Parts Operations)

GM cars is an example. The BCM manages the communications for the multiple computer system (Figure 13-28) using a network of sensors, switches, and other microprocessors to monitor vehicle-operating conditions. Certain components also provide the BCM with feedback signals; these tell the BCM whether the components are responding to the BCM commands properly. Like the powertrain control module (PCM), which operates the engine control system, the BCM has built-in diagnostics to help locate and correct a system malfunction.

Light-Emitting Diodes (LEDs)

The light-emitting diode (LED) is a diode that transmits light when electrical current is passed through it (Figure 13-29). An LED display is composed of small dotted segments arranged to form numbers and letters when selected segments are turned on. The LED is usually red, yellow, or green. LEDs have the following major drawbacks:

- Although easily seen in the dark, they are difficult to read in direct sunlight.
- They consume considerable current relative to their brightness.

Liquid Crystal Display (LCD)

A liquid crystal display (LCD) uses sandwiches of special glass containing electrodes and polarized fluid to display numbers and characters.

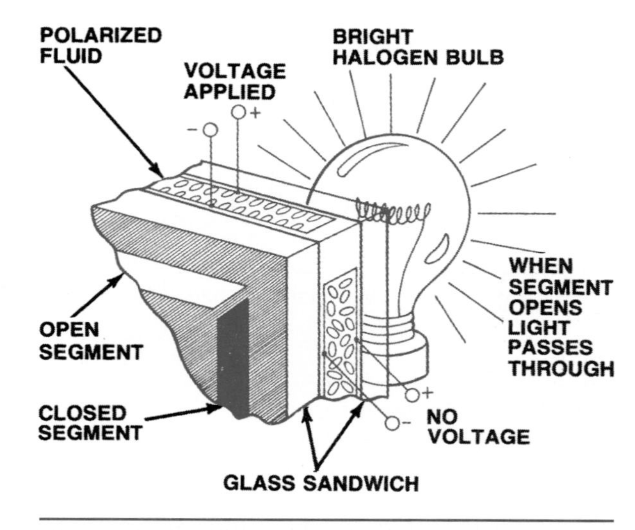

Figure 13-30. Light passes through polarized fluid to create the liquid-crystal display. (GM Service and Parts Operations)

Light cannot pass through the polarized fluid until voltage is applied. The display is very dense, however, and the various special filters used to provide colors create even more density. For this reason, halogen lights are generally placed behind the display (Figure 13-30). Although LCDs perform slowly in cold ambient temperatures, require proper alignment, and are very delicate, they have the following big advantages:

- They consume very little current relative to their brightness.
- They can be driven by a microprocessor through an interfacing output circuit.

Vacuum Fluorescent Display (VFD)

This is the most commonly used display for automotive electronic instruments, primarily because of its durability and bright display qualities. The **vacuum fluorescent display (VFD)** generates light similar to a television picture tube, with free electrons from a heated filament striking phosphor material that emits a blue-green light (Figure 13-31).

The anode segments are coated with a fluorescent material such as phosphorous. The filament is resistance wire, heated by electrical current. The filament coating produces free electrons, which are accelerated by the electric field generated by the voltage on the accelerating grid. High voltage is applied only to the anode of those segments required to form the characters to be displayed. Since the anode is at a higher voltage than the fine wire-mesh grid, the electrons pass through the grid. The phosphors on the segment anodes impressed with high voltage glow very brightly when struck by electrons; those receiving no voltage do not glow. The instrument computer determines the segments necessary to emit light for any given message and applies the correct sequences of voltage at the anodes.

VFD displays are extremely bright, and their intensity must be controlled for night viewing. This can be done by varying the voltage on the accelerating grid: the higher the voltage, the brighter the display. Intensity can also be controlled by pulse-width dimming, or turning the display on and off very rapidly while controlling the duration of on-time. This is similar to the pulse-width modulation of a carburetor mixture control solenoid or a fuel injector. The on-off action occurs so rapidly that it cannot be detected by the human eye.

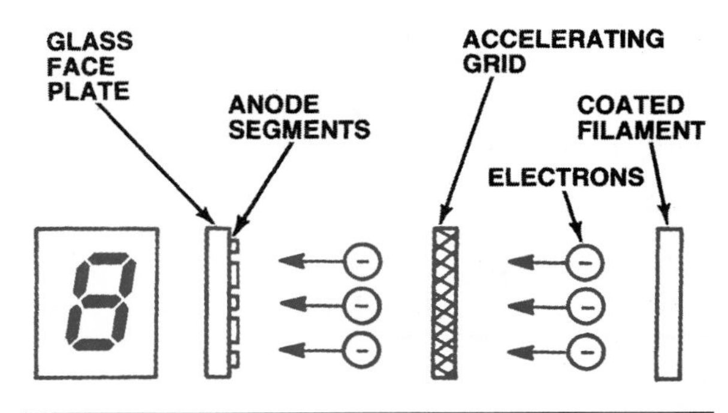

Figure 13-31. In a vacuum fluorescent display, voltage applied selectively to segment anodes makes the fluorescent material glow. (DaimlerChrysler Corporation)

Cathode Ray Tube (CRT)

Another display device used in automotive instrumentation was the *cathode-ray tube* (*CRT*), as used in the 1986–1996 Buick Riviera Buick Reatta. The CRT is essentially the same as the display used in an oscilloscope or a television set. CRTs function with an electron beam generated by an electron gun located at the rear of the tube. The CRT consists of a cathode that emits electrons and an anode that attracts them. Electrons are "shot" in a thin beam from the back of the tube. Permanent magnets around the outside neck of the tube and plates grouped around the beam on the inside of the tube shape the beam. A tube-shaped anode that surrounds the beam accelerates it as it leaves the electron gun. The beam has so much momentum that the electrons pass through the anode and strike a coating of phosphorus on the screen, causing the screen to glow at these points. The control plates are used to move the beam back and forth on the screen, causing different parts of it to illuminate. The result is a display (oscilloscope) or a picture (television).

The automotive CRT has a touch-sensitive Mylar switch panel installed over its screen. This panel contains ultra-thin wires, which are coded by row and column. Touching the screen in designated places blocks a light beam and triggers certain switches in the panel, according to the display mode desired. The switches in turn send a signal to the control circuitry, which responds by displaying the requested information on the CRT screen. In principle, this type of instrumentation combines two personal computer attributes: It has touch-screen control and is **menu driven.**

HEAD-UP DISPLAY (HUD)

The **head-up display (HUD)** is a secondary display system that projects video images onto the windshield. It is an electronic instrumentation system (Figure 13-32) that consists of a special windshield, a HUD unit containing a computer module, and a system-specific dimmer switch.

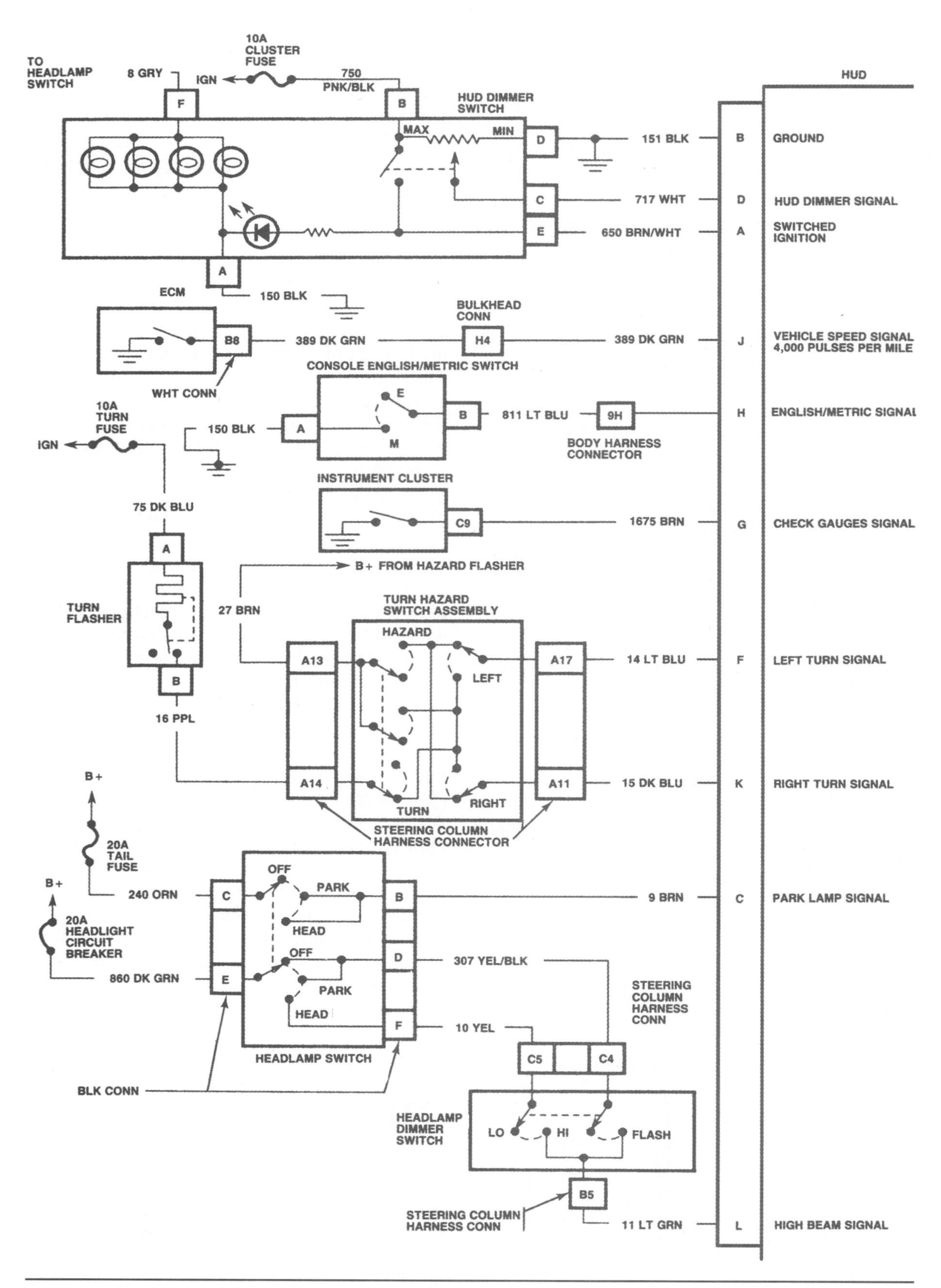

Figure 13-32. A wiring schematic of the HUD system as used by GM. (GM Service and Parts Operations)

The HUD unit processes various inputs that are part of the instrument cluster and projects frequently used driver information on the windshield area for viewing from the driver's seat. The dimmer switch provides system power for the computer module, varies the intensity of the display unit, and can change the vertical position of the display image through a mechanical cable drive system. When the ignition is turned on, the HUD unit performs a self-check routine and projects the following image (Figure 13-33) for approximately 3.5 seconds:

- Turn signal indicators
- High-beam indicator
- Check gauges indicator
- Speed (km/h or mph) indicator
- All segments of the digital speedometer

After completing the self-check, the system begins normal operation. The ECM provides vehicle speed information for HUD operation by completing a ground path to the HUD unit 4,000 times per mile. Each time the HUD unit recognizes a voltage drop at terminal J, it counts one pulse. By counting the pulses per second, the HUD unit can determine vehicle speed and project the corresponding figure on the windshield display.

Night Vision Head-Up Display (HUD)

The night vision system (Figure 13-34) used on the 2002–2003 Cadillac Deville is a monochromatic (single-color) option available to improve the vision of the driver beyond the scope of the headlamps.

Figure 13-33. The HUD system display image. (GM Service and Parts Operations)

The night vision operates only under the following conditions:

- The ignition is on.
- The front fog lamps or the headlamps are on during low light conditions. The night vision system uses the signal from the ambient light sensor to determine when low-light conditions exist.
- The night vision system is on.

The night vision system contains the following components:

- The head-up display (HUD)
- The head-up display (HUD) switches
- The night vision camera

Head-Up Display (HUD)

Night vision uses a HUD to project the night vision video images onto the windshield. The HUD projects the detected object images onto the windshield based on the video signals from the night vision camera.

Head-Up Display Switches

- **On/Off and Dimming Switch:** This turns the HUD display and the night vision system on or off. When the HUD is turned on, the system warm-up logo is displayed for a period of approximately 45 seconds. The on/off and dimming switch is also used to adjust the brightness of the video image. Moving the switch up will increase the HUD video image brightness. Moving the switch down will decrease the HUD video image brightness.
- **Up/Down Switch:** The HUD in the night vision system has an electric tilt adjust motor that adjusts the video image to the preferred windshield location of the driver. Pressing the Up/Down switch directs the tilt motor to adjust the night vision video image up or down, within a certain range.

Night Vision Camera

The night vision camera senses the heat given off by objects that are in the field of view of the camera. Warmer objects, such as pedestrians, animals, and other moving vehicles, appear whiter on the displayed image. Colder objects, such as the sky, signs and parked vehicles, appear darker on the displayed image. The night vision camera sends the detected object image information to the HUD via the high video signal circuit and the low video signal circuit.

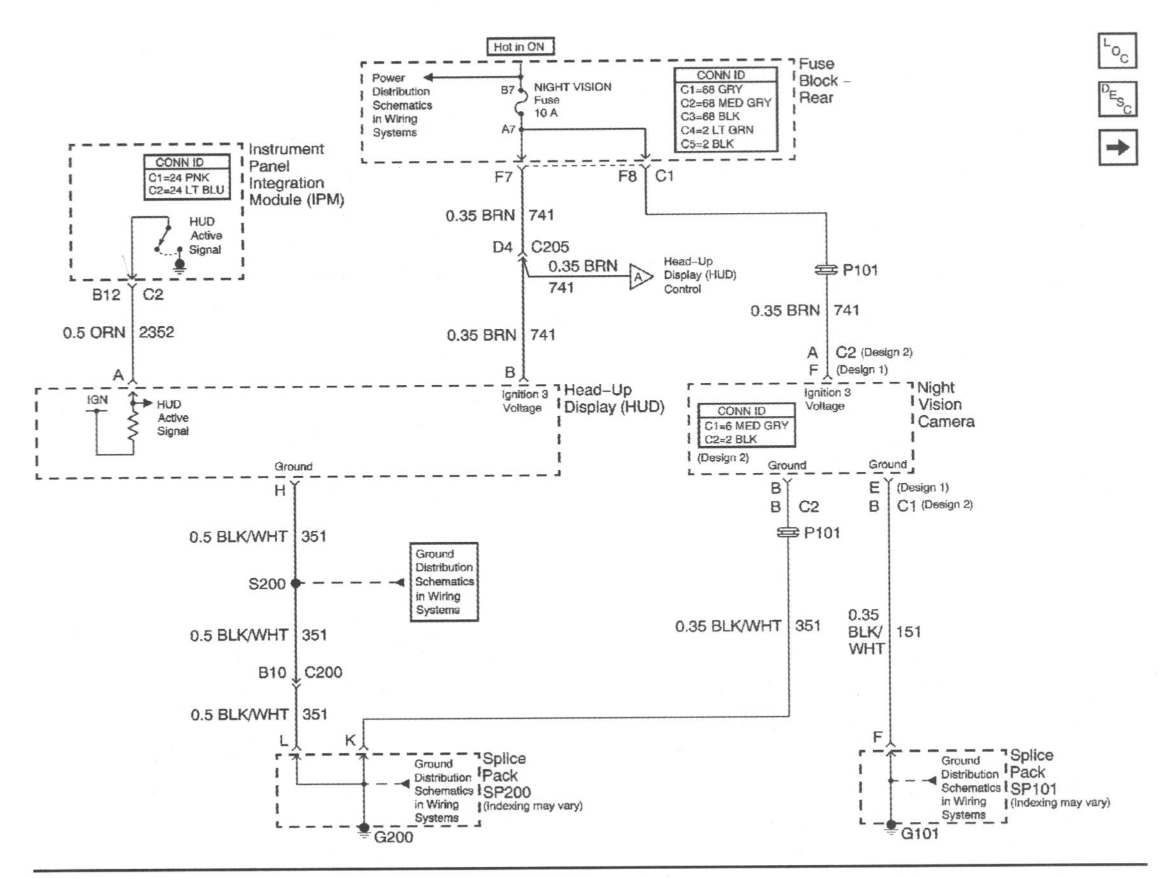

Figure 13-34. GM night vision HUD system schematic. (GM Service and Parts Operations)

SUMMARY

Instruments include gauges and warning lamps. There are various types of gauges, including mechanical, bimetallic, electromagnetic, and electronic. A voltage drop, a grounding switch, a ground sensor, or fiber optics can light warning lamps. Late-model vehicles may have a digital display instead of traditional analog gauges. Digital displays can be individual or they can be part of a more elaborate vehicle electronic system, such as a trip computer or message center. Electronic instruments or driver information systems (DISs) have the same purpose as the traditional analog instruments: The DIS displays vehicle-operating information to the driver and includes all gauge and speedometer information. The primary difference between the electronic (DIS) and traditional systems is the way in which the information is displayed. The DIS is also called a driver information center (DIC). Body control module (BCM) computers act as managers of other computers in a comprehensive vehicle system.

Electronic displays may use a light-emitting diode (LED), a liquid-crystal display (LCD), a vacuum fluorescent display (VFD), or a cathode-ray tube (CRT) to transmit information. Some instrumentation is menu-driven, giving the user an opportunity to select the information to be displayed. Touch-sensitive screens similar to those on personal computers are used instead of keyboards or pushbuttons on some late-model systems. A head-up display (HUD) is a secondary display system that projects video images onto the windshield. The GM night vision system is a head-up display (HUD) system that projects useful vehicle information in front of the driver near the windshield.

Review Questions

1. Which of the following is *not* a reason why warning lamps have replaced gauges in automobiles?
   a. Cheaper to manufacture
   b. Cheaper to install
   c. More accurate
   d. Easier to understand
2. Temperature compensation in bimetallic gauges is accomplished by:
   a. Current flow through the heater coil
   b. The shape of the bimetallic strip
   c. An external resistor in the circuit
   d. Hermetically sealing the unit
3. Gauges with three-coil movements are most often used by:
   a. General Motors
   b. Ford
   c. DaimlerChrysler
   d. Toyota
4. An oil pressure warning lamp is usually controlled by:
   a. Voltage drop
   b. Ground switch
   c. Ground sensor
   d. Manual switch
5. The sending unit in the fuel gauge uses a:
   a. Fixed resistor
   b. Zener diode
   c. Float
   d. Diaphragm
6. Buzzers are a special type of warning device that are activated by:
   a. Voltage drop
   b. Ground switches
   c. Diaphragms
   d. Optical fibers
7. Two technicians are discussing driver information systems. Technician A says the head-up display (HUD) is a secondary display system that projects video images onto the windshield. Technician B says that a liquid-crystal display (LCD) is an indicator in which electrons from a heated filament strike a phosphor material that emits light. Who is right?
   a. A only
   b. B only
   c. Both A and B
   d. Neither A nor B
8. In GM BCM computer-controlled electric fan circuits:
   a. The BCM switches the control line voltage on/off with pulse-width modulation.
   b. The fan control module switches the ground on/off.
   c. Both A and B
   d. Neither A nor B
9. Electromagnetic gauges do *not* use a:
   a. Mechanical movement
   b. D'Arsonval movement
   c. Air core movement
   d. Two- or three-coil movement
10. Which type of gauge does *not* use an instrument voltage regulator (IVR)?
    a. D'Arsonval movement
    b. Air core movement
    c. Two-coil movement
    d. Three-coil movement
11. Technician A says GM's electronic speedometer interprets vehicle speed by using a buffer to switch the voltage on/off. Technician B says GM's electronic speedometer uses a buffer to translate the analog input of the PM generator into digital signals. Who is right?
    a. A only
    b. B only
    c. Both A and B
    d. Neither A nor B
12. Technician A says an electronic speedometer cannot use a stepper motor to provide the display. Technician B says the use of nonvolatile RAM prevents the odometer from being turned back. Who is right?
    a. A only
    b. B only
    c. Both A and B
    d. Neither A nor B
13. An electronic display device using electrodes and polarized fluid to create numbers and characters is called:
    a. LCD
    b. LED

c. VFD
d. CRT

14. Which electronic display device is most frequently used because it is very bright, consumes relatively little power, and can provide a wide variety of colors through the use of filters?
    a. LCD
    b. LED
    c. VFD
    d. CRT

15. Which electronic display device is difficult to read in daylight and consumes considerable power relative to its brightness?
    a. LCD
    b. LED
    c. VFD
    d. CRT

14

Horns, Wiper, and Washer System Operation

LEARNING OBJECTIVES

Upon completion and review of this chapter, you should be able to:

- Explain the operation of an automotive horn.
- Identify the different types of wiper systems and explain their operation.
- Identify the different types of windshield washer systems and explain their operation.

KEY TERMS

Automobile Horn
Depressed Park Position
Horn Relay
Horn Switch
Washer Pumps
Wiper Switch

INTRODUCTION

This chapter introduces you to the operation of an automotive horn and washer/wiper circuits. It will further show how various original equipment manufacturers apply technology to their horn and washer/wiper systems. Chapter 14 of the *Shop Manual* shows you how to diagnose horn and wiper/washer system concerns.

HORN CIRCUITS

An **automobile horn** is a safety device operated by the driver to alert pedestrians and other motorists. Some states require two horn systems, with different sound levels for city and country use.

Circuit Diagram

Some early automobiles used a simple series horn circuit, as shown in Figure 14-1A. Battery current is supplied to the horn circuit through the fuse panel, or from a terminal on the starter relay or solenoid. The normally open horn switch is installed between the power source and the grounded horn. When the driver pushes the horn button, the horn switch closes and current flows through the circuit to sound the horn. If the car has more than one horn (Figure 14-1B), each horn will form a parallel path to ground.

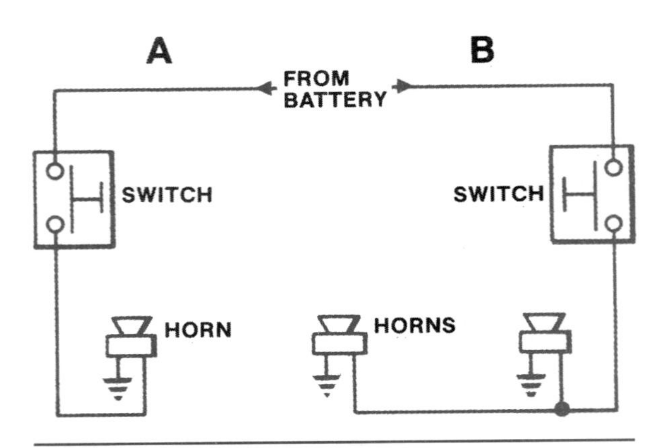

Figure 14-1. A simple circuit with a single horn in series with the switch (A), or two horns in parallel with each other and in series with the switch (B).

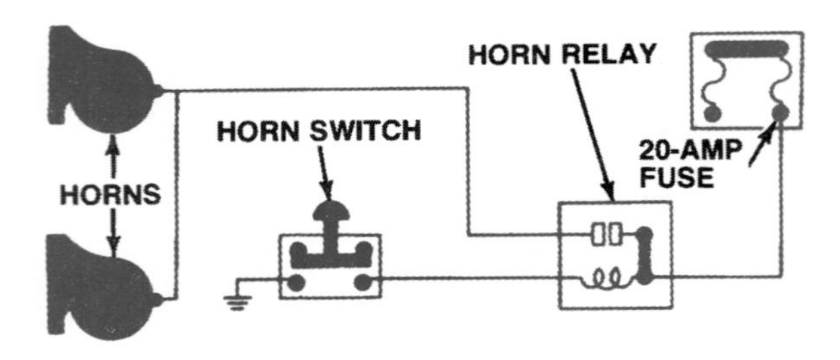

Figure 14-2. Many horn systems are controlled by a relay. (DaimlerChrysler Corporation)

Most horn circuits include a **horn relay** (Figure 14-2). The normally open relay contacts are between the power source and the grounded horn. The horn switch is between the relay coil and ground. When the horn switch is closed, a small amount of current flows through the relay coil. This closes the relay coil and allows a greater amount of current to flow through the horns.

Horn Switches, Relays, and Fuses

The **horn switch** is normally installed in the steering wheel or steering column (Figure 14-3). Contact points can be placed so that the switch will be closed by pressure at different points on the steering wheel (Figure 14-4). Some cars have a button in the center of the wheel; others have a number of buttons around the rim of the wheel, or a large separate horn ring. Many imported cars and some domestic cars have the horn button on

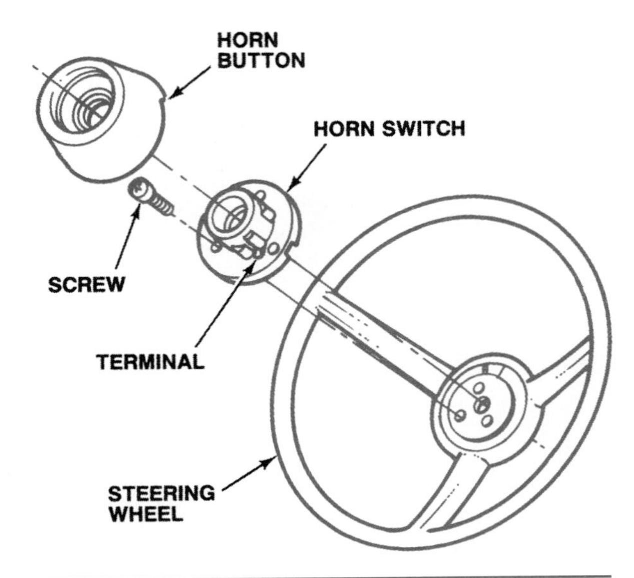

Figure 14-3. The horn switch is mounted in the steering column. (DaimlerChrysler Corporation)

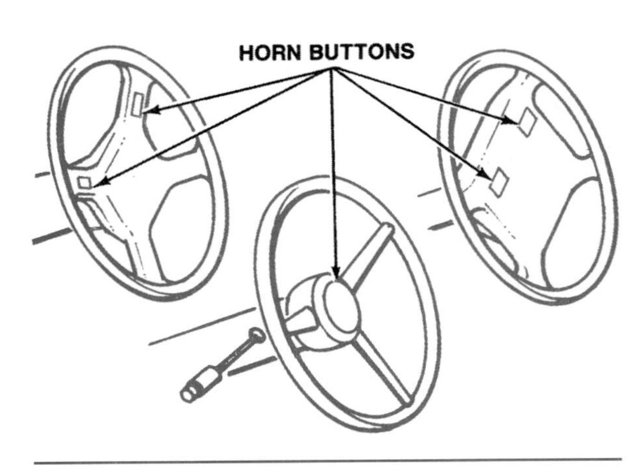

Figure 14-4. Horn buttons can be placed at various locations around the steering wheel.

the end of a multifunction lever or stalk on the steering column. All of these designs operate in the same way: Pressure on the switch causes contacts to close. When the pressure is released, spring tension opens the contacts.

Horn relays can be mounted on the fuse panel (Figure 14-5). They also can be attached to the bulkhead connector or mounted near the horns in the engine compartment. The relay is not serviceable, and must be replaced if defective. The horn circuit often shares a 15- to 20-ampere fuse with several other circuits. It may also be protected by a fusible link.

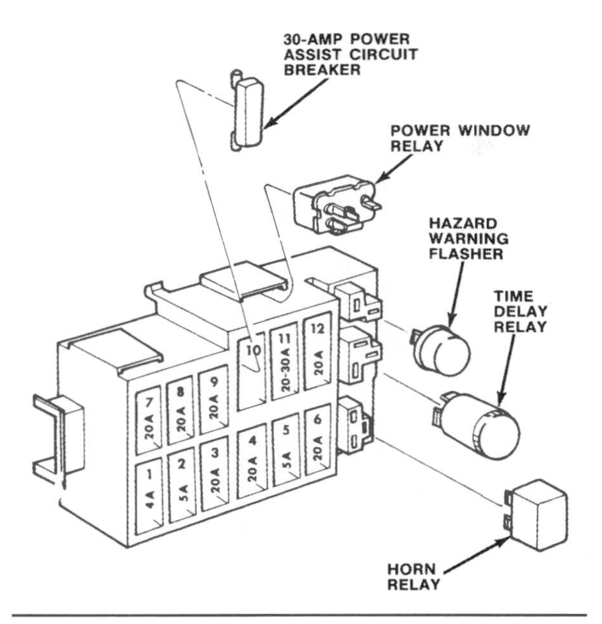

Figure 14-5. The horn relay can be mounted on the fuse panel. (DaimlerChrysler Corporation)

Horns

Except for Chrysler's air horn, which uses air pressure from the compressor, automobile horns use *electromagnetism* to vibrate a diaphragm and produce sound waves. A typical horn contains normally closed contact points in series with a coil. One of the contact points is mounted on a movable armature to which the horn diaphragm is connected.

The horn coil is in series with the horn switch or horn relay contacts. When the horn switch or horn relay contacts close, current flows through the horn coil to ground. The electromagnetic field created by the coil attracts the armature, also moving the diaphragm. The armature movement opens the contact points, which open the coil circuit. With no magnetic field to hold them, the armature and diaphragm move back to their normal positions. The points are again closed, allowing current to flow through the coil. This making and breaking of the electromagnetic circuit causes the horn diaphragm to vibrate.

Since this cycle occurs very rapidly, the resulting rapid movements of the diaphragm create sound waves. The speed or frequency of the cycling determines the pitch of the sound created. This can be adjusted by changing the spring tension on the horn armature to increase or decrease the electromagnetic pull on the diaphragm.

The History from the Bell to the Electric Horn

Many types of signal alarms have been used on cars as follows:

- Mechanical bell
- Bulb horn
- Compression whistle
- Exhaust horn
- Hand-operated horn (Klaxon)
- Electric horn

The mechanical bell was used on very early cars; the driver operated the bell with a foot pedal. The bulb horn, similar to that on a child's bicycle, proved to be inconvenient and unreliable. The compression whistle was most often used in cars with no battery or limited battery capacity; a profiled cylinder provided the whistle's power. Exhaust horns used gases from the engine exhaust; they, too, were foot-operated. The hand-operated Klaxon horn amplified a grating sound caused by a metal tooth riding over a metal gear. This did not work well, because the horn had to be near the driver rather than at the front of the car. Over the years, the electric horn has been the most popular type of signal alarm.

WINDSHIELD WIPERS AND WASHERS

Federal law requires that all cars built in, or imported into, the United States since 1968 have both a two-windshield wiper system and a windshield washer system. Wiper systems on older vehicles may be operated by engine vacuum or by the power steering hydraulic system.

Modern wiper systems are operated by *electric motors*. The washer system can be manually operated, or it can have an electric pump. Many vehicles also have a single-speed wiper and washer for the rear window. This is a completely separate system, but it operates in the same way as the windshield wiper and washer system.

Circuit Diagram

A typical *two-speed wiper system* circuit diagram is shown in Figure 14-6. The motor fields are permanent magnets. The wiper switch controls both the wiper motor speeds and the washer pump. The park switch within the wiper motor ensures that

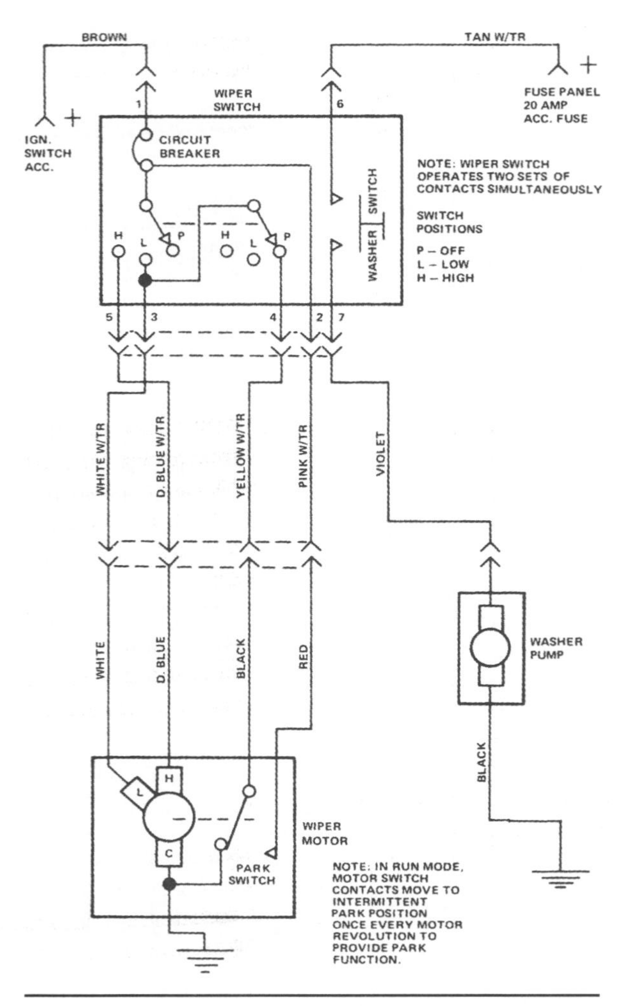

Figure 14-6. A simple two-speed wiper circuit.

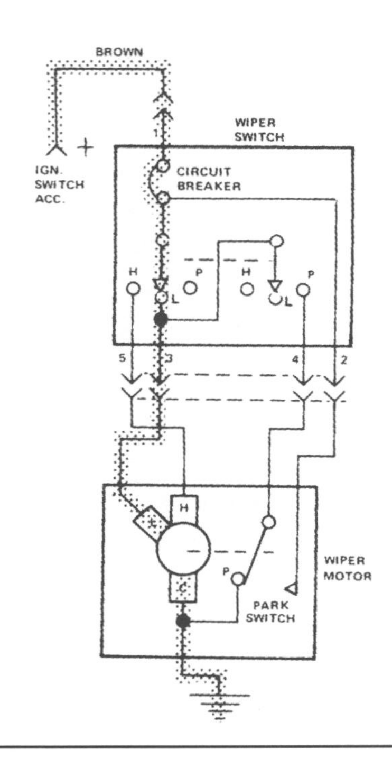

Figure 14-7. Low-speed current flow.

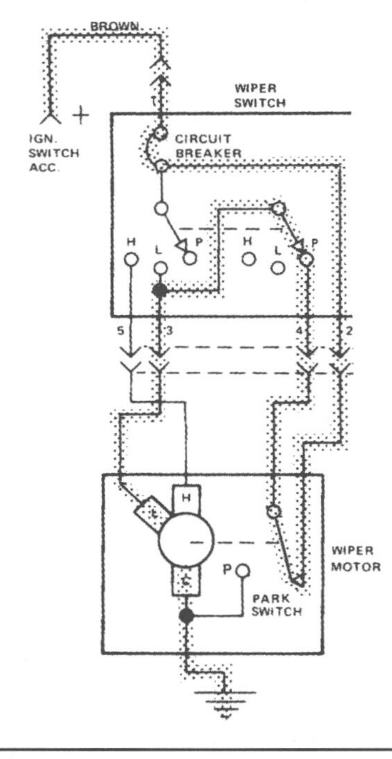

Figure 14-8. The park switch allows the motor to continue turning until the wiper arms reach their park position.

when the wiper switch is turned off, the motor will continue to turn until the wiper arms have reached the bottom edge, or park position, of the windshield. The circuit shown has a circuit breaker built into the wiper switch. The circuit breaker also can be a separate unit, or it can be mounted on the wiper motor.

Figure 14-7 shows low-speed current flow through the simple circuit. Current flows through the wiper switch contacts, the low-speed brush L, and the common (shared) brush C to ground. During high-speed operation, the current flows through the high-speed brush H and the common brush to ground. When the wiper switch is turned to park, or off, the park switch comes into the circuit.

The *park switch* is a two-position, cam-operated switch within the wiper motor. It moves from one position to the other during each motor revolution. When the wiper arms are at their park position, the park switch is at the P contact, as shown in Figure 14-8. No current flows through the park switch. At all other wiper arm positions, the park switch is held against spring tension at the other contact. If the wiper switch is turned

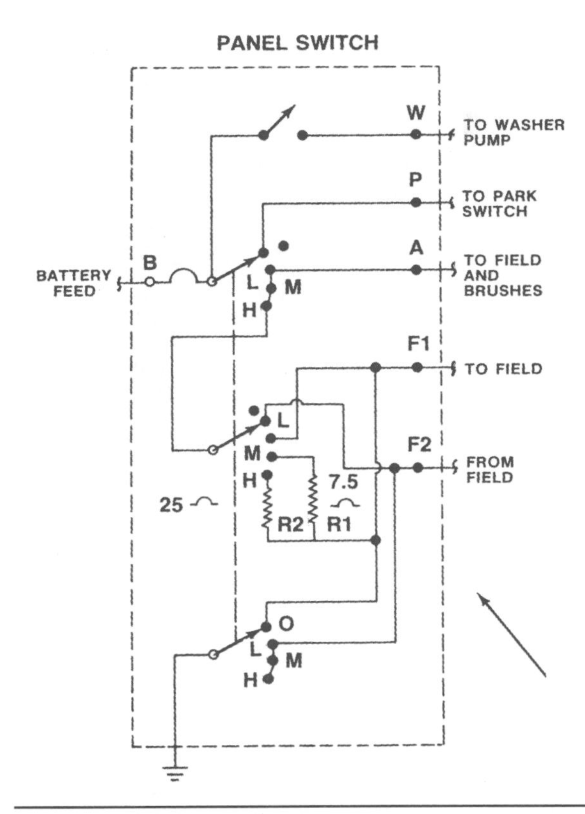

Figure 14-9. This three-speed wiper system controls motor speed by routing field current flow through various resistors. (DaimlerChrysler Corporation)

off while the wiper arms are not at their park position current will flow through the park switch to the low-speed brush. The motor will continue to turn until the wiper arms reach their park position. At that point, the park switch moves to the P contact and all current stops.

When extra features are added to the wiper system, the circuits become more complex. For example, many manufacturers offer three-speed wiper systems. These systems use electromagnetic motor fields. The switch contacts route field current through resistors of various values (Figure 14-9) to vary the wiper motor speed. Some GM two-speed wiper circuits also use this type of motor.

Many late-model vehicles have wiper arms that retreat below hood level when the switch is turned off (Figure 14-10). This is called a **depressed park position** and is controlled by the park switch. When the wiper switch is turned off, the park switch allows the motor to continue turning until the wiper arms reach the bottom edge of the windshield. The park switch then *reverses* current flow through the wiper motor, which makes a partial revolution in the opposite direction. The wiper linkage pulls the wiper arms down

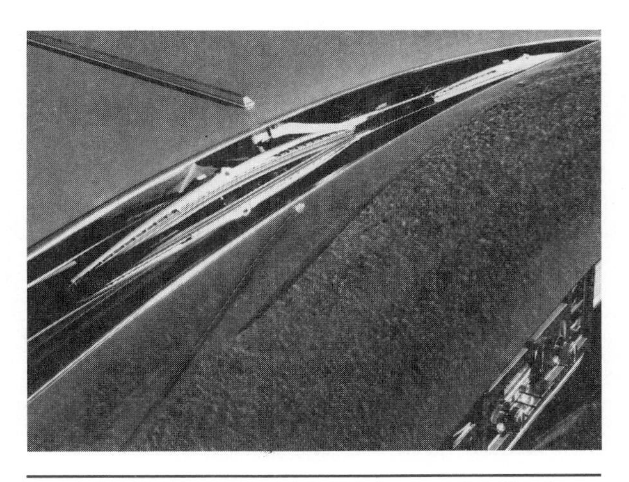

Figure 14-10. Many late-model cars have a depressed-park wiper position.

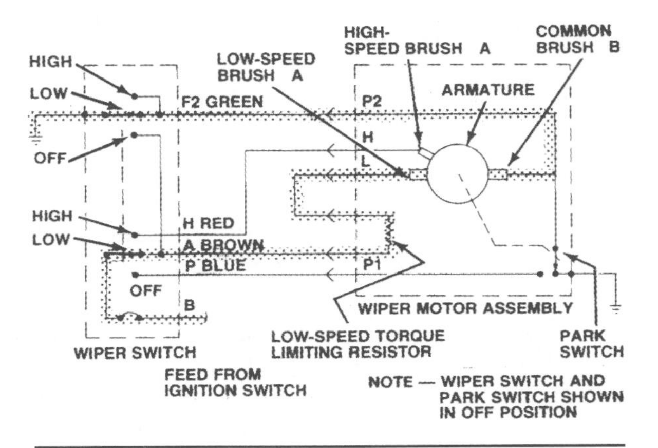

Figure 14-11. Low-speed current flow in a depressed-park system. (DaimlerChrysler Corporation)

below the level of the hood during this motor reversal. The motor reversal also opens the park switch to stop all wiper motor current flow.

A depressed-park wiper system is shown in Figure 14-11. During normal operation, current flows through either brush A or common brush B to ground. When the wiper switch is turned off (Figure 14-12), current flows through the park switch, into brush B, and through low-speed brush A to ground. This reverses the motor's rotation until the wiper arms reach the depressed park position, the park switch moves to the grounded position, and all current stops.

Many wiper systems have a low-speed intermittent or *delay mode*. This allows the wiper arms to sweep the windshield completely at intervals of three to 30 seconds. Most intermittent, or delay, wiper systems route current through a solid-state module containing a variable resistor

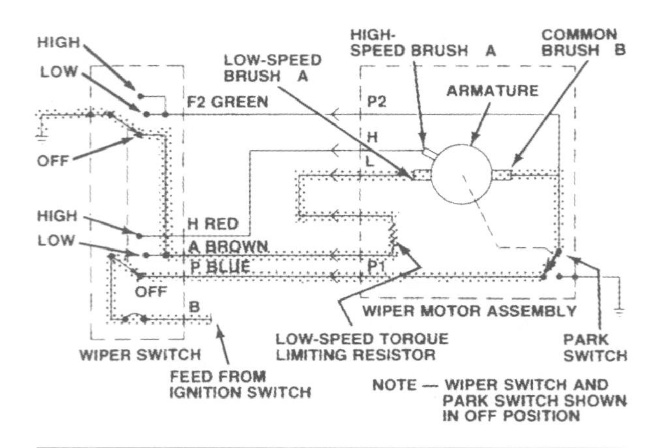

Figure 14-12. The park switch reverses current flow through the motor so that the wiper arms are pulled down into the depressed-park position. (DaimlerChrysler Corporation)

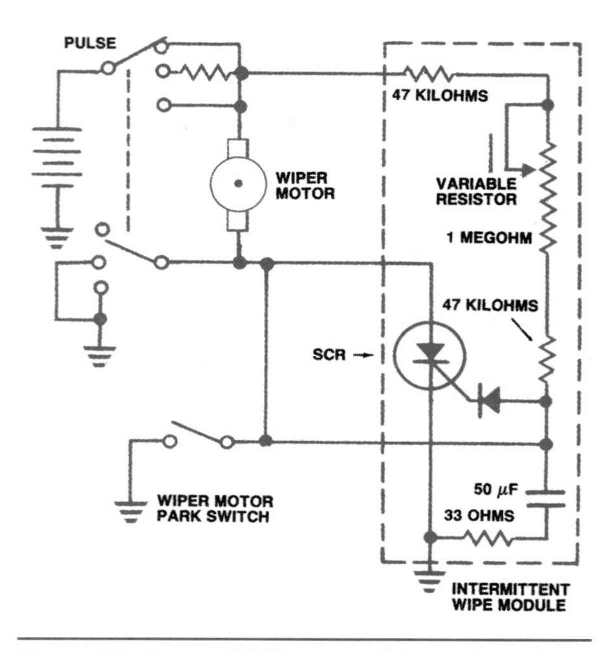

Figure 14-13. An SCR in the solid-state intermittent wiper module or control unit triggers the wiper motor for intermittent wiper arm sweeps. (DaimlerChrysler Corporation)

and a capacitor (Figure 14-13). Once the current passing through the variable resistor has fully charged the capacitor, it triggers a silicon-controlled rectifier (SCR) that allows current flow to the wiper motor. The park switch within the motor shunts the SCR circuit to ground. Current to the motor continues, however, until the wiper arms reach their park position and the park switch is opened. The driver through the variable resistor controls the capacitor rate of

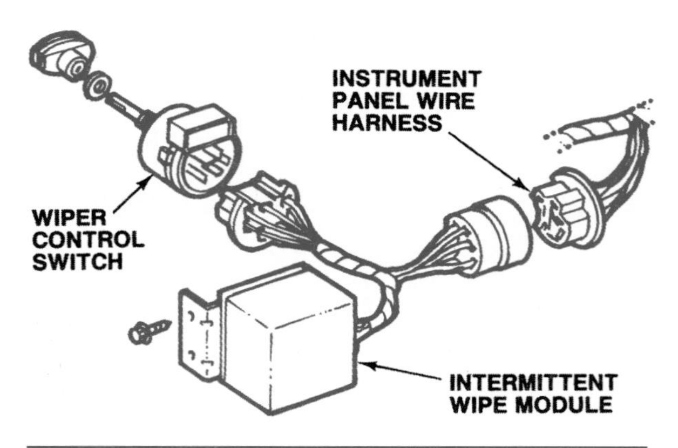

Figure 14-14. The intermittent wipe governor or module is installed between the wiper switch and the wiper motor.

charge, and therefore the interval between the wiper arm sweeps.

SCR Intermittent Wipers

On some imported cars, the intermittent or delay mode is sensitive to vehicle speed and varies from approximately 15 seconds (at low road speed) up to the wipers' normal low speed (at moderate road speed) as vehicle speed changes. The intermittent mode can be cancelled by pressing a cancel switch, and wiper speed can be set manually with the wiper switch. Intermittent wiper control circuitry on many cars is contained in a separate module that is installed between the wiper switch and the wiper motor, as shown in Figure 14-14.

Switches

The **wiper switch** is between the power source and the grounded wiper motor. The wiper switch does not receive current unless the ignition switch is turned to the Accessory or the Run position. The wiper switch may be mounted on the instrument panel, or it can be mounted in the steering column and controlled by a multifunction lever or stalk (Figure 14-15).

If the system has an electric washer pump, the pump is generally controlled by contacts within the wiper switch. The washer is usually operated by a spring-loaded pushbutton that is part of the wiper switch (Figure 14-16). Moving the switch to its wash position or pressing the pushbutton will operate the washer pump as long as the switch is held in position or is pressed.

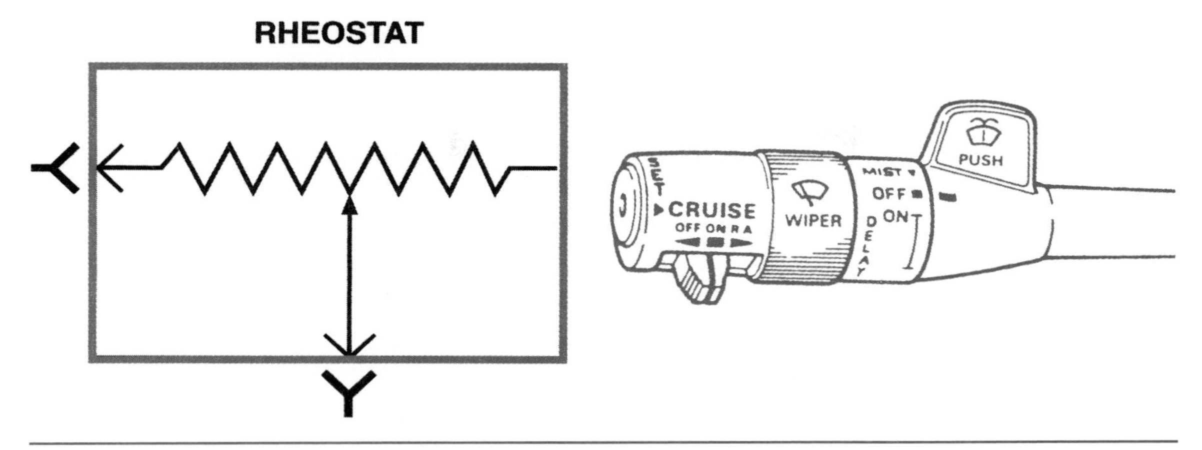

Figure 14-15. The washer switch is usually a spring-loaded pushbutton mounted on the instrument panel or on a multifunction lever. (GM Service and Parts Operations)

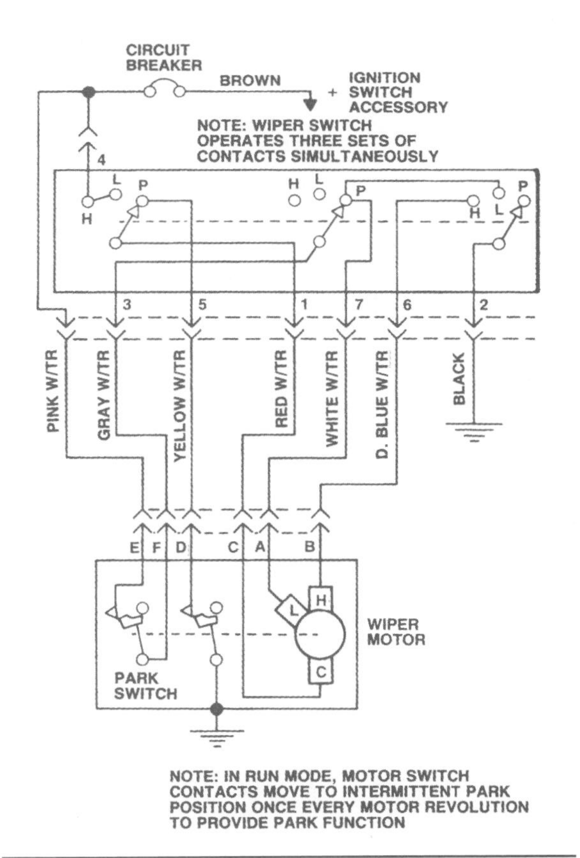

Figure 14-16. In this system, the high-speed brush is set directly opposite the common brush. The common brush is insulated, and the two speed-control brushes are grounded through the wiper switch.

Motors

Most two-speed wiper motors use permanent ceramic magnets as pole pieces. Three brushes ride on the motor's commutator. One brush is a common, or shared, brush and conducts current whenever the wiper motor is operating. The other brushes are placed at different positions relative to the motor armature. Current through one brush produces a different motor speed than current through the other brush. The wiper switch contacts route current to one of these two brushes, depending upon which wiper motor speed the driver selects.

In many wiper motors, the *high-speed brush* is placed directly opposite the common brush (Figure 14-16). The *low-speed brush* is offset to one side. This placement of the low-speed brush affects the interaction of the magnetic fields within the motor and makes the motor turn slowly. The placement of the high-speed brush causes the motor to turn rapidly. Chrysler and some GM two-speed motors vary from this pattern (Figures 14-13 and 14-17). The low-speed brush is directly opposite the common brush and the high-speed brush is offset. A resistor wired in series with the low-speed brush reduces the motor's torque at low speed. This extra resistance in the low-speed circuit results in a lower motor speed even with the reversed brush position.

The common brush can be grounded and the two speed-control brushes can be insulated, as shown in Figure 14-17. Other motors have the speed-control brushes grounded through the wiper switch contacts and the common brush insulated, as shown in Figure 14-17.

Some two-speed and all three-speed wiper motors have two electromagnetic field windings (Figure 14-18). One field coil is in series with a motor brush and is called the series field. The other field coil is a separate circuit branch directly to ground and is called the shunt field. The two coils are wound in opposite directions, so that their magnetic fields oppose each other.

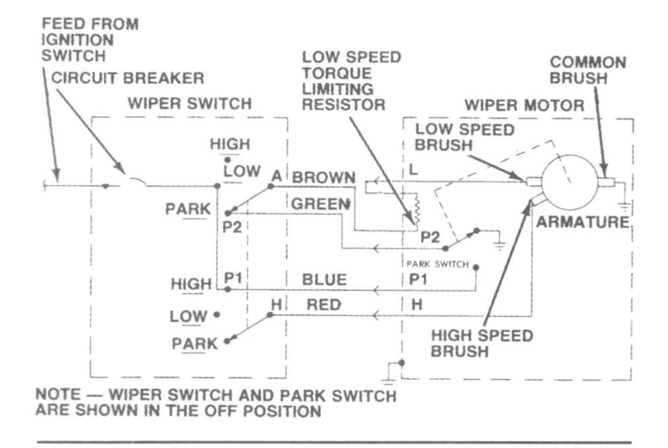

Figure 14-17. This motor has an extra resistor in the low-speed circuit, and so the low-speed brush is placed directly across from the common brush. The common brush is grounded, and the two speed-control brushes are insulated.

The wiper switch controls current through these two field coils. At low speed, about the same amount of current flows through both coils. Their opposing magnetic fields result in a weak total field, so the motor turns slowly. At medium speed (three-speed motor), current to one coil must flow through a resistor. This makes the coil's magnetic field weaker and results in a stronger total field within the motor. The motor revolves faster. At high speed, current to the coil must flow through a greater value resistor. The total magnetic field of the motor is again increased, and the motor speed increases. The resistors can act on either the shunt coil or the series coil to reduce current flow and thereby increase the motor's total field strength and speed. In Figure 14-18, the resistors act on the shunt field. Many wiper motors can be serviced to some extent, as shown in the *Shop Manual,* Chapter 14.

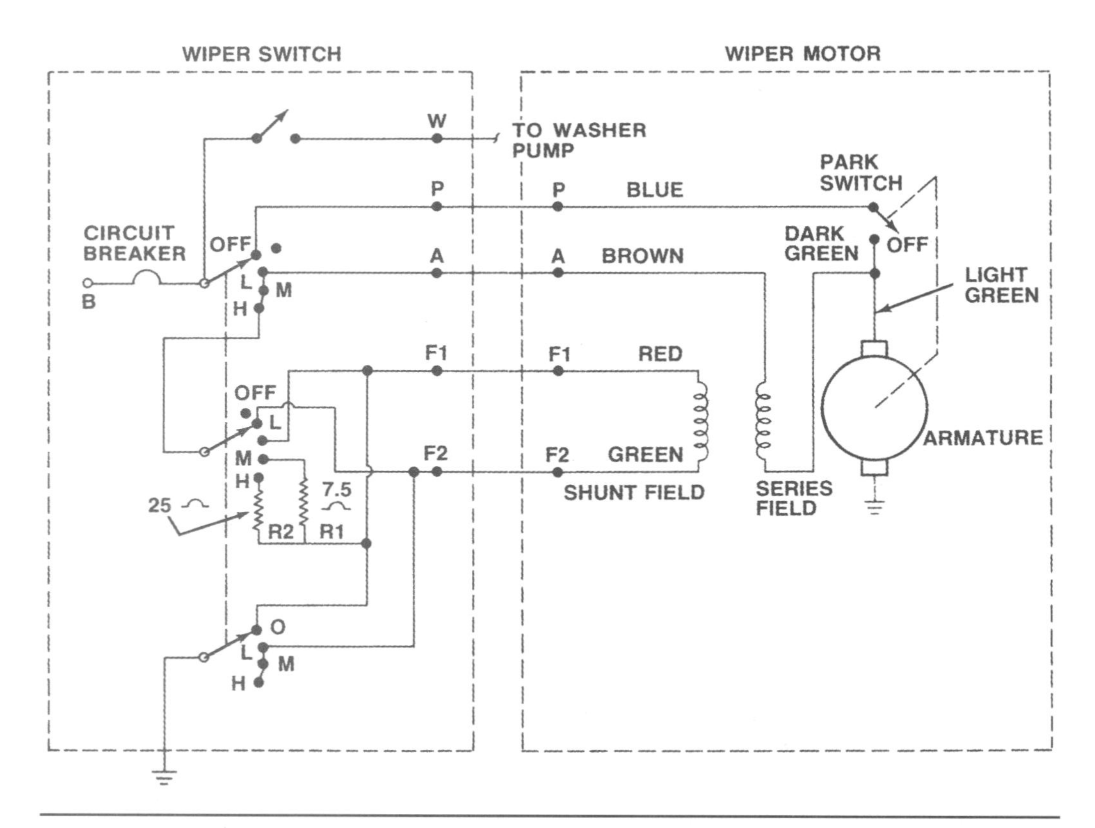

Figure 14-18. In this motor with two electromagnetic fields, the motor speed is controlled by the amount of current through one of the fields. (DaimlerChrysler Corporation)

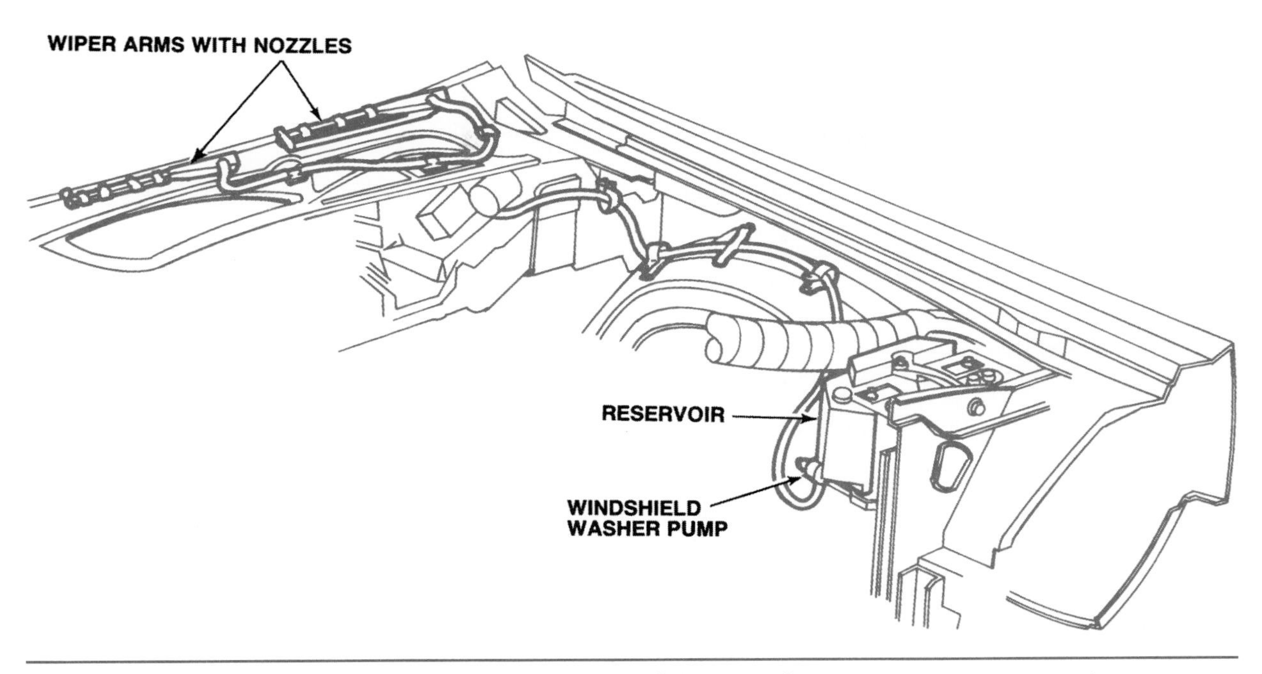

Figure 14-19. The washer pump is often mounted on the fluid reservoir. (DaimlerChrysler Corporation)

Washer Pumps

Windshield **washer pumps** (Figure 14-19) draw a cleaning solution from a reservoir and force it through nozzles onto the windshield. The unit can be a positive-displacement pump or a centrifugal pump that forces a steady stream of fluid, or it can be a pulse-type pump that operates valves with a cam to force separate spurts of fluid.

The washer pump is generally mounted in or on the fluid reservoir (Figure 14-19). Some GM pulse pumps are mounted on the wiper motor (Figure 14-20). Washer pumps are not usually serviceable, so they are replaced if they fail.

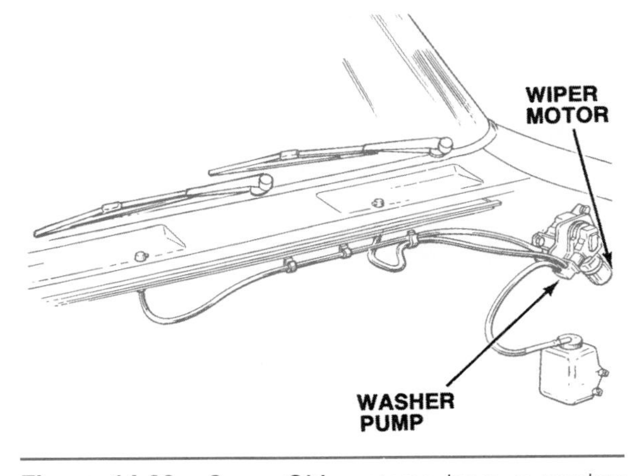

Figure 14-20. Some GM systems have a washer pump mounted on the wiper motor. (GM Service and Parts Operations)

SUMMARY

An automotive horn circuit can be a simple series circuit, or it can use a relay to control current through the horns. The horn switch is a normally open push-pull switch that is operated by the driver. Horns use electromagnetism to vibrate a diaphragm and to produce sound waves.

Windshield wiper and washer circuits have many variations. They can include a permanent magnet motor or one with electromagnetic fields. The park position can be at the bottom edge of the windshield or below the bottom edge. An intermittent wipe feature can be driver- or speed-controlled. Each of these variations requires slightly different circuitry. Washer pumps can be mounted at the cleaner reservoir or on the wiper motor. Pumps are not serviced, but are replaced.

Review Questions

1. Horn relays are sometimes included in the horn circuit to:
   a. Allow the use of two horns in the circuit
   b. Decrease the amount of current needed to activate the horn
   c. Increase the amount of current needed to activate the horn
   d. Allow the horn button to be placed on the end of a stalk on the steering column
2. Horn circuits are generally protected by a:
   a. Fuse
   b. Fusible link
   c. Either A or B
   d. Neither A nor B
3. The ________ within the wiper motor ensures that when the motor is turned off, the wiper arms will be brought to the bottom position.
   a. Wiper switch
   b. Park switch
   c. Recycle relay
   d. Park/neutral switch
4. Two-speed wiper motors generally use ________ to achieve the two speeds.
   a. Cams
   b. Reduction gears
   c. Speed-control brushes
   d. Gear reduction
5. All three-speed wiper motors have ________ fields.
   a. Electromagnetic
   b. Permanent magnet
   c. Two-series
   d. Two-shunt
6. Technician A says that when the driver pushes the horn button, electromagnetism moves an iron bar inside the horn, which opens and closes contacts in the horn circuit. Technician B says that many vehicle horn circuits include a relay. Who is right?
   a. A only
   b. B only
   c. Both A and B
   d. Neither A nor B
7. Which of the following do automotive horns use to operate?
   a. Electromagnetic induction
   b. Magnetic repulsion
   c. Magnetic resonance
   d. Electromagnetism
8. The wiper park switch is which of the following?
   a. Three-position cam-operated switch
   b. Four-position rotary switch
   c. Two-position toggle switch
   d. Two-position cam-operated switch
9. Technician A says the windshield washer pump draws a cleaning solution from a reservoir and forces it through nozzles onto the windshield. Technician B says the unit can be a positive-displacement pump or a centrifugal pump. Who is right?
   a. A only
   b. B only
   c. Both A and B
   d. Neither A nor B
10. Two technicians are discussing horn operation. Technician A says when the horn switch or horn relay contacts close, current flows through the horn coil to ground. The electromagnetic field created by the coil attracts the armature, also moving the diaphragm. Technician B says the armature movement closes the contact points, which opens the coil circuit. Who is right?
    a. A only
    b. B only
    c. Both A and B
    d. Neither A nor B

15

Body Accessory Systems Operation

LEARNING OBJECTIVES

Upon completion and review of this chapter, you should be able to:

- Identify the components of the automotive HVAC (Heater Ventilator and Air Conditioning System) and explain the operation.
- Identify radio and/or entertainment system components and explain their operation.
- Explain the operation of the rear window defroster/defogger and heated windshields.
- Explain the operation of power windows and seats.
- Explain the operation of Power Door Locks, Trunk Latches, and Seat Back Releases.
- Identify the different types of REMOTE/ Keyless Entry Systems and explain their operation.
- Identify the different types of Theft Deterrent Systems and explain their operation.
- Explain the operation of cruise control systems.
- Explain the operation of the Supplemental Restraint System (SRS).

KEY TERMS

Air Bag Module
Automatic Door Lock (ADL)
Data Link
Defroster
Heater fan
Inflator Module
Igniter Assembly
Safing sensor
Servomotor
Servo Unit
Transducer

INTRODUCTION

Electrical accessories provide driver and passenger comfort, convenience, and entertainment. New electrical accessories are introduced every year, but some systems have been common for many years. Such systems increasingly are being automated with computer control. This chapter will explain the electrical operation of some common accessory systems.

HEATING AND AIR-CONDITIONING SYSTEMS

Although heating and air-conditioning systems rely heavily on mechanical and vacuum controls, a good deal of electrical circuitry also is involved. Since the late 1970s, air-conditioning systems have become increasingly "smart," relying on solid-state modules or microprocessors for their operation. This also has complicated the job of servicing such systems.

Heater Fan

Heating systems use a **heater fan** attached to a permanent-magnet, variable-speed blower motor to force warm air into the passenger compartment (Figure 15-1). The higher the voltage applied to the motor, the faster it runs. A switch mounted on the instrument panel controls the blower operation (Figure 15-2). In most heating systems, the switch controls blower speed by directing the motor ground circuit current through or around the coils of a resistor block (Figure 15-3) mounted near the motor.

When the switch is off, the ground circuit is open and the blower motor does not run. (Some systems used in the 1970s, however, were wired so that the blower motor operated on low speed whenever the ignition was on). When the switch is turned to its low position, voltage is applied across all of the resistor coils and the motor runs at a low speed. Moving the switch to the next position bypasses one of the resistor coils. This allows more current to the blower motor, increasing its speed. When the switch is set to the highest position, all of the resistors are bypassed and

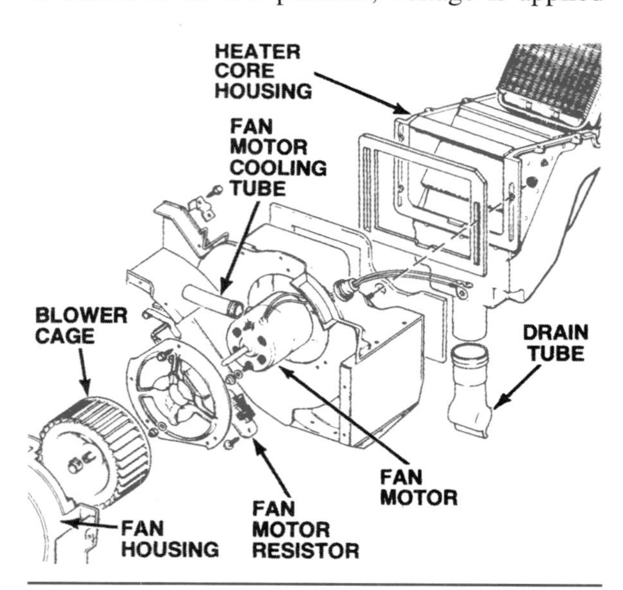

Figure 15-1. An electric motor drives the heater fan.

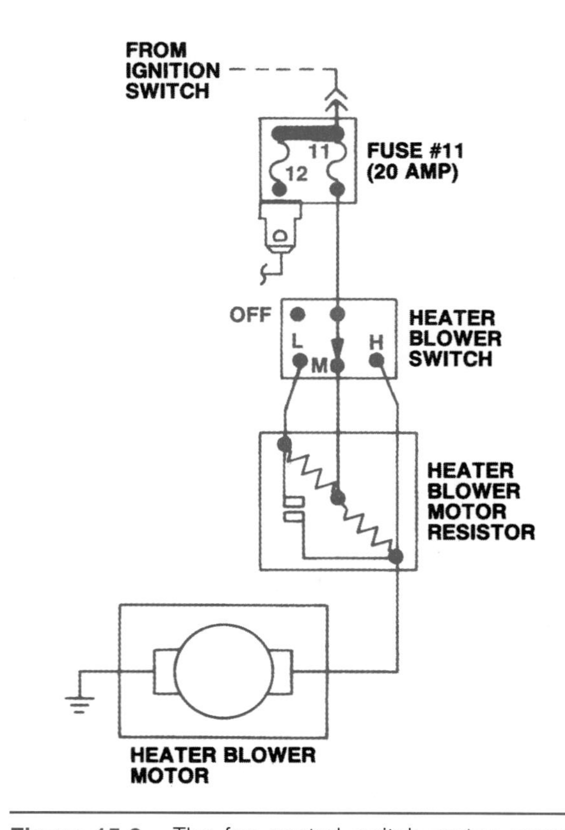

Figure 15-2. The fan control switch routes current through paths of varying resistance to control motor speed. (DaimlerChrysler Corporation)

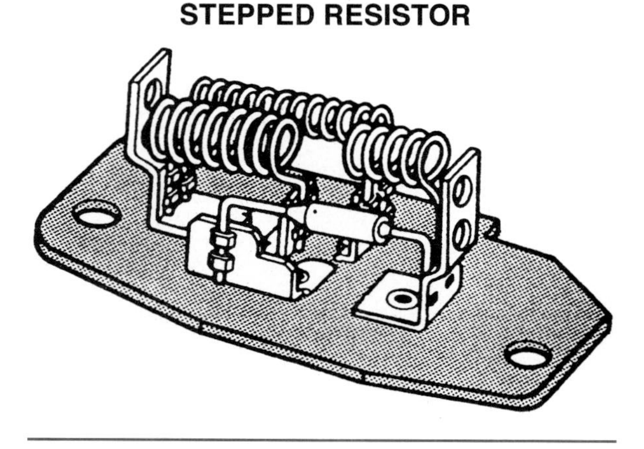

Figure 15-3. Blower motor resistors are installed on a "block" near the motor. Some resistor blocks have a thermal limiter.

full current flows to the motor, which then operates at full speed.

In some GM systems, a relay is used between the high switch position and the blower motor. Ford incorporates a thermal limiter in its resistor block, as shown in Figure 15-3. Current flows through the limiter at all blower speeds. If current passing through the limiter heats it to 212°F (100°C), the limiter opens and turns off the blower motor. When this happens, the entire resistor block must be replaced.

Air-Conditioning Fan and Compressor Clutch

Air-conditioning fan controls are similar to heater controls. In most cars that have both heating and air-conditioning systems, the same blower motor is shared by both systems (Figure 15-4). One or more switches route current through different resistors to control the blower motor speed.

In addition to the fan switch, the control assembly in the passenger compartment contains a driver-controlled air-conditioning clutch switch and an integral clutch switch activated when the function selector lever is set to the defrost position. These switches are used to operate the belt-driven compressor. A compressor that operates constantly wastes energy. To use energy more efficiently, the compressor has an electromagnetic clutch (Figure 15-5). This clutch locks and unlocks the compressor pulley with the compressor shaft. The compressor will operate only when the clutch switch is closed and the electromagnetic clutch is engaged.

Most recent air-conditioning systems use a clutch-cycling pressure switch or a pressure-cycling switch to control compressor clutch operation. This pressure-operated electric switch generally is wired in series with the clutch field

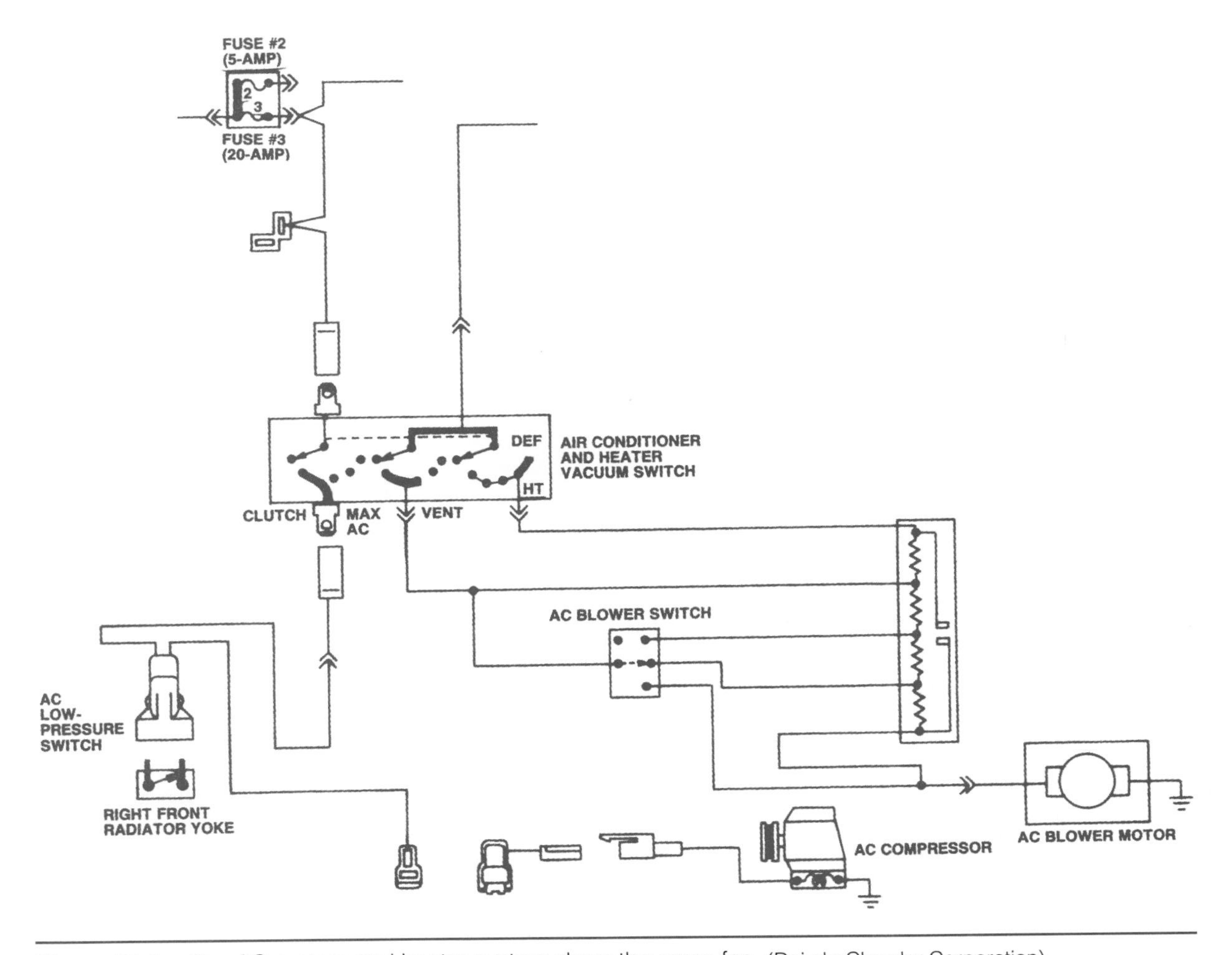

Figure 15-4. The AC system and heater system share the same fan. (DaimlerChrysler Corporation)

coil. The switch closes when the pressure on the low side of the refrigerant system rises to a specified value, engaging the clutch. When system pressure drops to a predetermined value, the switch opens to shut off the compressor. The switch operates to control evaporator core pressure and prevent icing of the evaporator cooling coils.

Air-conditioning systems may also use low- and high-pressure switches as safety devices, as follows:

- The low-pressure switch is closed during normal compressor operation and opens only when refrigerant is lost or ambient temperature is below freezing.
- The high-pressure switch is normally closed to permit compressor operation. However, if system pressure becomes excessive (generally 360–400 psi or 2,480–2,760 kPa), the switch acts as a relief valve and opens to shut off the compressor. Once pressure drops to a safe level, the switch will close again and permit the compressor to operate.

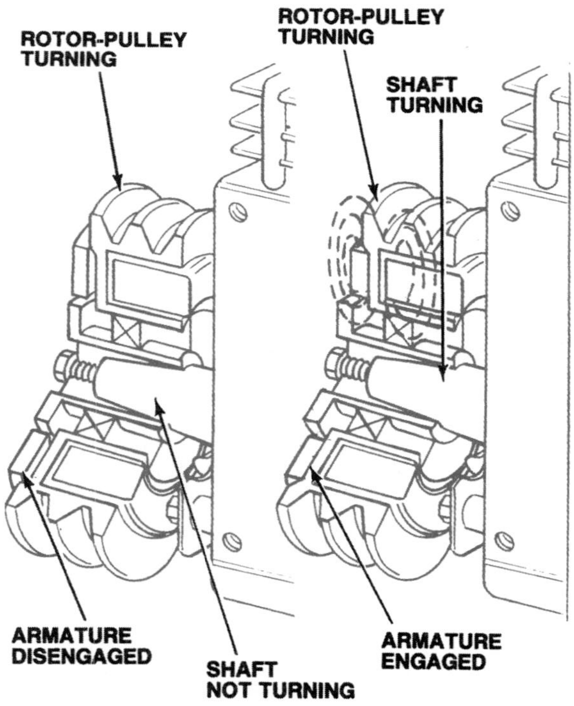

Figure 15-5. The electromagnetic clutch in this air-conditioning compressor prevents the compressor from wasting energy.

- A pressure relief valve on the compressor high-pressure side may be used instead of a high-pressure switch. Some systems have a diode installed inside the compressor clutch connector to suppress any voltage spikes that might be produced by clutch circuit interruption.

Other compressor clutch controls may include the following:

- A power-steering pressure or cutout switch to shut the compressor off whenever high power-steering loads are encountered, as during parking. The switch senses line pressure and opens or closes the circuit to the compressor clutch accordingly.
- A wide-open throttle (WOT) switch on the throttle body or accelerator pedal to open the circuit to the compressor clutch during full acceleration.
- A pressure-sensing switch in the transmission to override the WOT switch when the transmission is in high gear.

Temperature Control

The basic electrical components of most air-conditioning systems have already been described. As we have seen, they provide input to, and protection for, the refrigeration system.

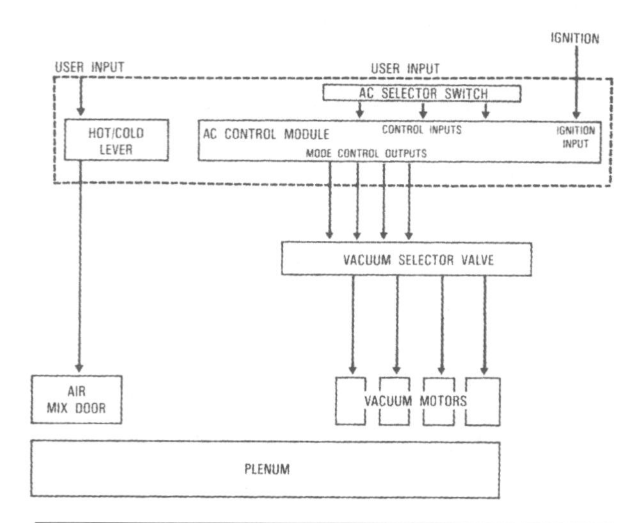

Figure 15-6. Manual air conditioning system block diagram. (GM Service and Parts Operations)

All input to the air-conditioning system begins with the control assembly mounted in the instrument panel. Temperature control can take the following forms:

- Manual control
- Semiautomatic control (programmer controlled)
- Fully automatic control (microprocessor or body computer controlled)

A manual temperature control system does not provide a method by which the system can function on its own to maintain a preset temperature. The user, through the mechanical control assembly, must make system input. Once the air-conditioning switch is turned on, the temperature selection made, and the blower speed set, the system functions with vacuum-operated mode door actuators and a cable-actuated air-mix door. Figure 15-6 is a block diagram of such a system.

Automatic Temperature Control (ATC)

With a semiautomatic temperature control system, the user still selects the mode but the actuators are electrically operated. Selecting the mode does not directly control the actuator; it creates an electrical input to an independent module or programmer (Figure 15-7). On Chrysler vehicles, the electronic **servomotor** performs the programmer function (Figure 15-8). Two sensors are added to inform the programmer of ambient temperature and in-car temperature (Figure 15-9). The programmer calculates the resistance values provided by the temperature dial setting and the two additional sensors

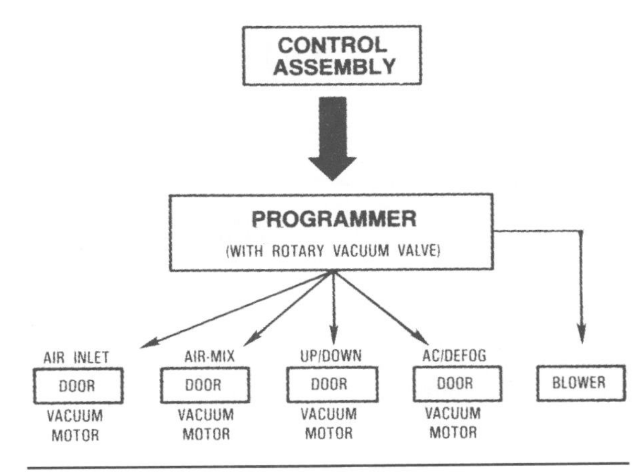

Figure 15-7. Semiautomatic AC systems use an electronic programmer to translate mechanical control movement into actuator signals. (GM Service and Parts Operations)

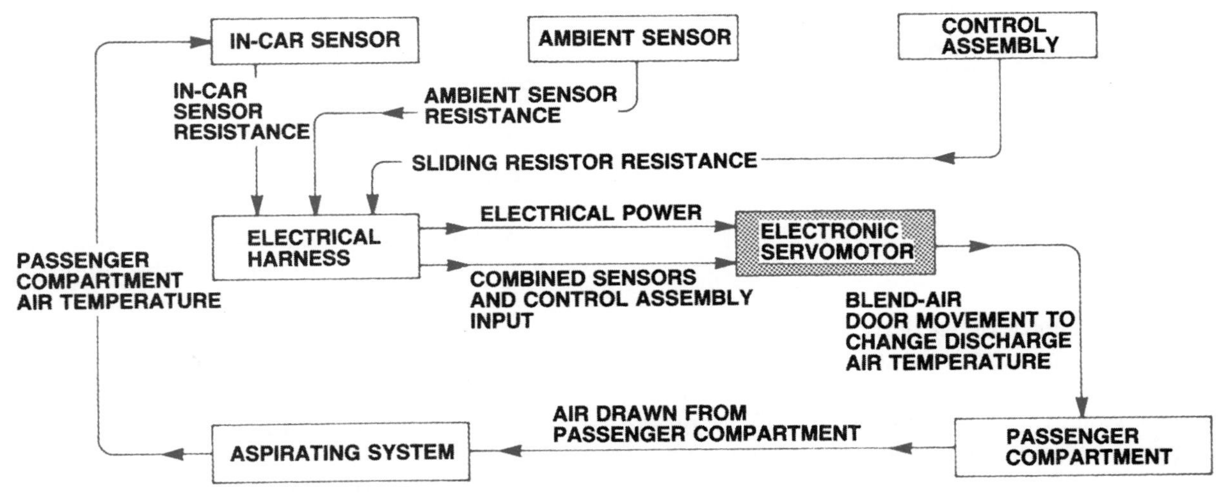

Figure 15-8. An electronic servomotor takes the place of a programmer in DaimlerChrysler's semiautomatic AC system. (DaimlerChrysler Corporation)

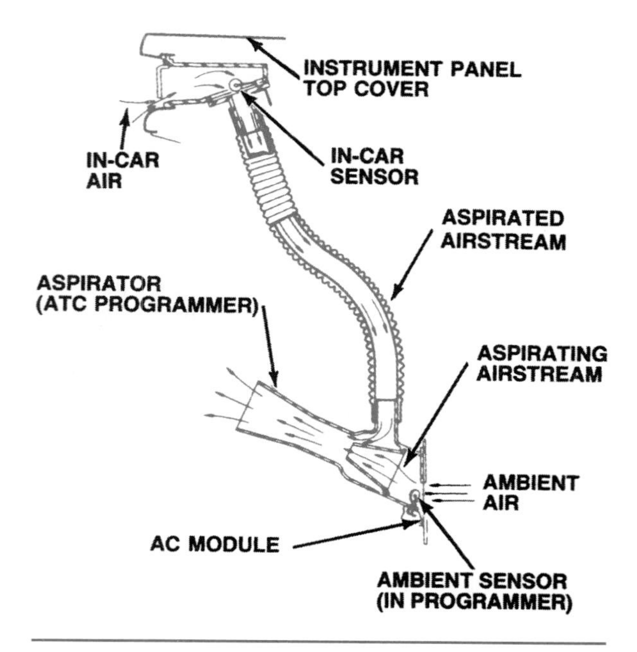

Figure 15-9. Resistance values from in-car and ambient temperature sensors are coupled with the resistance provided by the control assembly temperature dial to direct the programmer. (GM Service and Parts Operations)

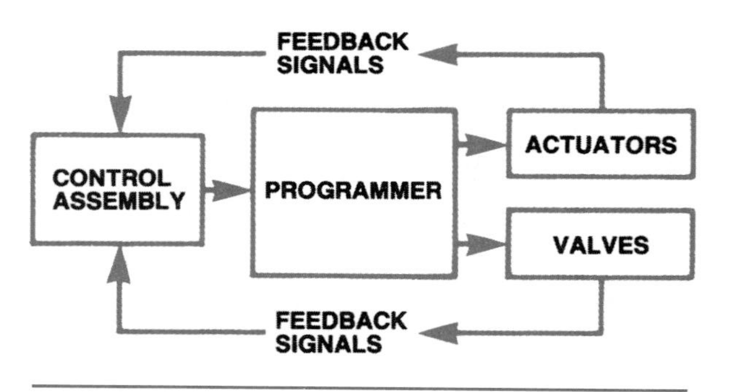

Figure 15-10. The actuators used in fully automatic AC systems provide feedback signals that allow the control assembly to monitor system operation. (GM Service and Parts Operations)

and adjusts cables or vacuum selector valves to maintain the preset temperature. The semiautomatic temperature control system differs from a manual system primarily in the use of the programmer; actuators and doors are still moved by mechanical linkage and cables.

In a fully automatic temperature control system, the control assembly is electronic instead of manual. The user selects the mode and the temperature. The control assembly microprocessor sends the appropriate signals to the programmer to operate the system. Electric servomotors are used as actuators to send a feedback signal to the electronic control assembly (Figure 15-10). This lets the control assembly monitor the system and make whatever adjustments are required to maintain the desired system temperature.

Since the control assembly is constantly monitoring the system, it knows when a malfunction occurs and can transmit this information to the service technician.

Semiautomatic Control (Programmer Controlled)

The GM C61 system is representative of a semiautomatic control system. Once the user has selected the mode and temperature, the system automatically controls blower speed, air temperature, air delivery, system turn-on, and compressor operation. It does this with a programmer inserted between the control assembly and the actuators (Figure 15-7), as well as two temperature sensors. The ambient sensor installed in the programmer is exposed to ambient airflow through a hole in the module wall; the in-car sensor is located under the instrument panel top cover. Figure 15-9 shows the sensor locations. Both sensors are disc-type thermistors that provide a return voltage signal to the programmer based on variable resistance. The programmer is built into the air-conditioning control assembly (Figure 15-11) and contains the following:

- A DC amplifier that receives a weak electrical signal from the sensors and control assembly and sends a strong output signal proportional to the input signal it receives
- A transducer that converts the amplifier signal to a vacuum signal that actuates the vacuum motor
- A vacuum checking relay that has a check valve to maintain a constant vacuum signal to the vacuum motor and the rotary vacuum valve
- A vacuum motor to actuate the rotary shaft that drives the air-mix door link
- A rotary vacuum valve to route vacuum to control the mode doors and operate the heater water valve; this valve does the same job as a vacuum selector valve in a manual system

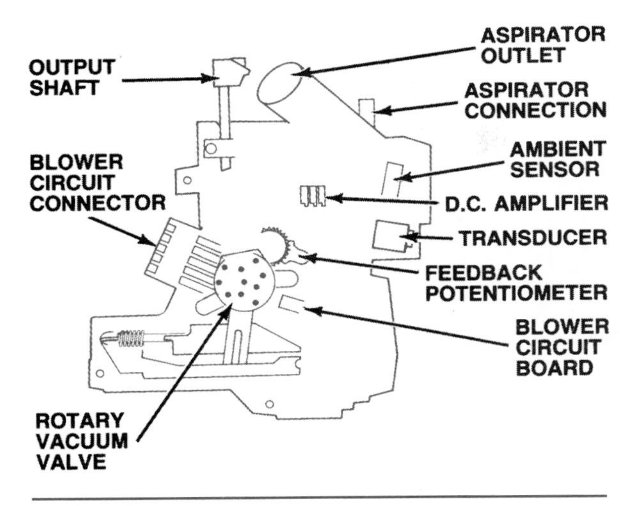

Figure 15-11. Components of the programmer used in GM's C61 AC system. (GM Service and Parts Operations)

- A feedback potentiometer to inform the programmer of system corrections required by changing temperature demands

A circuit board electrical switch is mounted on the base of the control assembly. The rotary switch contacts are positioned by the mode-select lever to provide the correct electrical path to the compressor clutch. The temperature dial varies the resistance of a wire-wound rheostat installed directly above it. The programmer uses the total resistance provided by the temperature dial and temperature sensors to calculate how the system should function.

To use this type of system, the driver need only set the control assembly in the auto mode and select a temperature. From this point on, the programmer controls the system operation by automatically setting the mode and blower speed and adjusting the air-mix doors to maintain the desired air temperature. Note that we have added nothing to the underhood portion of the air-conditioning system; we have only modified the operation of the control system by adding a device to maintain temperature within a selected narrow range.

Fully Automatic Control (BCM Controlled)

The electronic climate control (ECC) system used by Cadillac (Figure 15-12) is similar to the ETCC system just described. When used with a body control computer (BCM), the control assembly contains an electronic circuit board, but the BCM acts as the microprocessor. The BCM is constantly in touch with the climate control panel on the control assembly through a **data link,** or digital signal path (serial data line) provided for communication. The panel transfers user requests to the BCM, which sends the correct data to the panel for display.

Like the other semiautomatic and fully automatic systems we have looked at, the user selects the mode and temperature. The system automatically controls blower speed, air temperature, air delivery, system turn-on, and compressor operation. Although the ECC system functions similarly to the ETCC system, there are differences in compressor cycling methods. In a system without BCM control, the compressor clutch is grounded through the low-pressure switch. The power module thus cycles power to the compressor clutch (Figure 15-14A). In a BCM-controlled system, the compressor clutch current is received through a fuse and the power steering cutout switch or diode; the power module cycles the ground circuit for the compressor, Figure 15-14B.

The electronic comfort control (ECC) system used by Oldsmobile is BCM controlled and can be used as either a fully automatic or a manual system. When used manually, the driver can control blower speed and air delivery mode, but the system will continue to control temperature automatically. In addition to the BCM, power module, programmer, and control panel assembly used in other BCM-controlled systems, the ECC system uses inputs from the engine (electronic) control module (ECM). This allows the BCM to check several engine and compressor conditions before it turns the compressor on.

The BCM communicates with the ECM, the ECC panel, and the programmer on the serial data line to transmit data serially (one piece after another). The serial data line acts like a party telephone line; while the BCM is communicating with the ECM, the programmer and the ECC panel can "hear" and understand the conversation. They also can process and use the information communicated, but they cannot cut in on the transmission. For example, suppose the ECM is sending engine data to the BCM. The ECC panel computer, which needs to display engine rpm to the driver, "listens" in on the conversation, picks

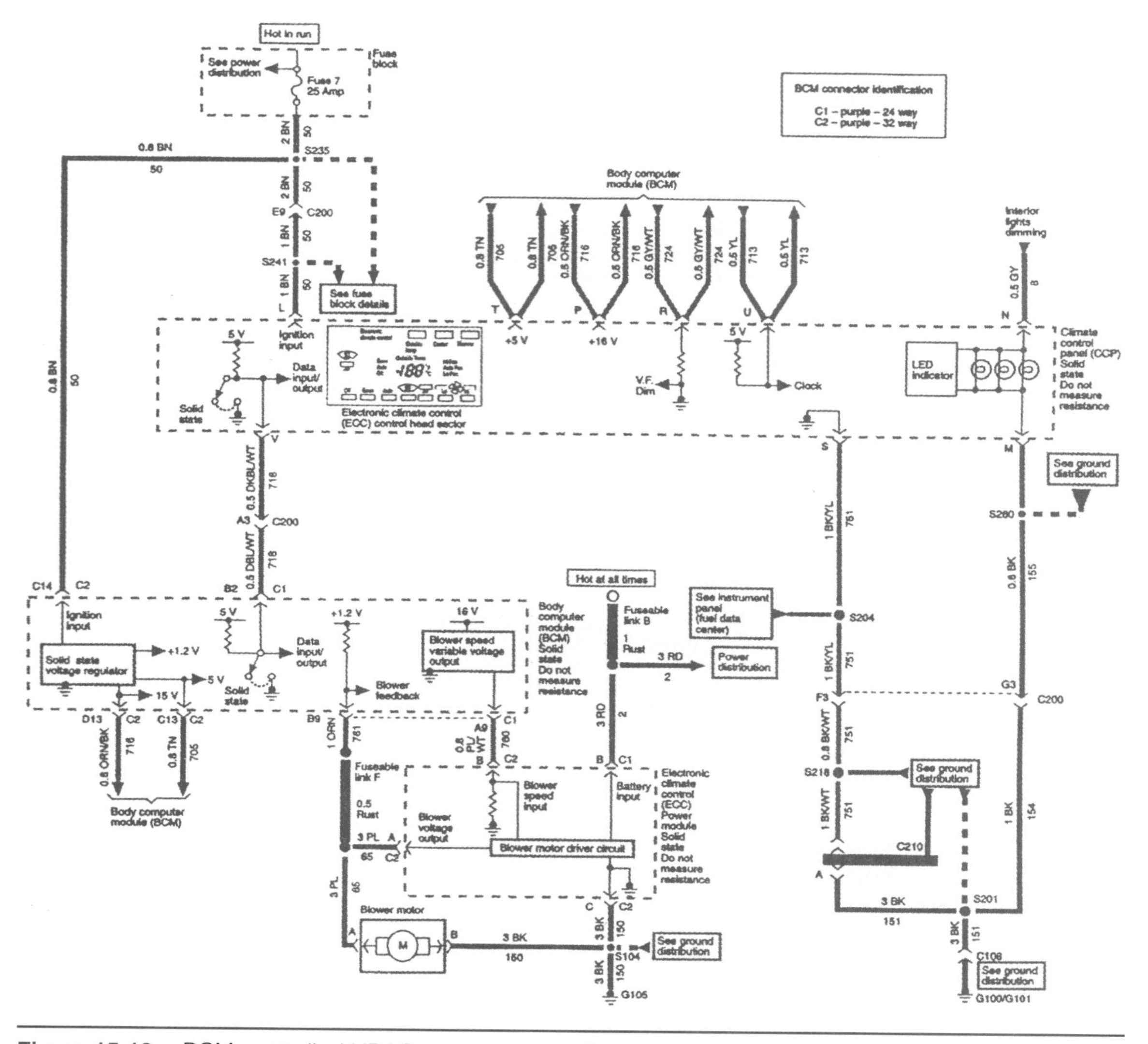

Figure 15-12. BCM-controlled HBAC system schematic. (GM Service and Parts Operations)

up the data it needs, and displays it on its panel. When the ECM is finished, it momentarily transmits a 5-volt signal to declare the line idle and the BCM opens a conversation with the next device it needs to talk with.

The programmer controls air delivery and temperature on instructions from the BCM, using a series of vacuum solenoids that control the mode door operation. The programmer also has a motor that controls the air-mix door position to regulate temperature. When directed to change blower speed by the BCM, the programmer sends a variable voltage signal to the power module, which sends the required voltage to the blower motor.

The ECC system is the most complex of the ones we've discussed, and this is reflected in the diagnostic sequence designed into the overall system network. When any subsystem exceeds its programmed limits, the system sets a trouble code and in some cases provides a backup function. The instrument panel cluster and the ECC panel (Figure 15-13) are used to access and control the self-diagnostic features. When the technician accesses the diagnostic mode, any stored BCM and ECM codes are displayed, along with various BCM and ECM parameters, discrete inputs and outputs, and any BCM output-override information.

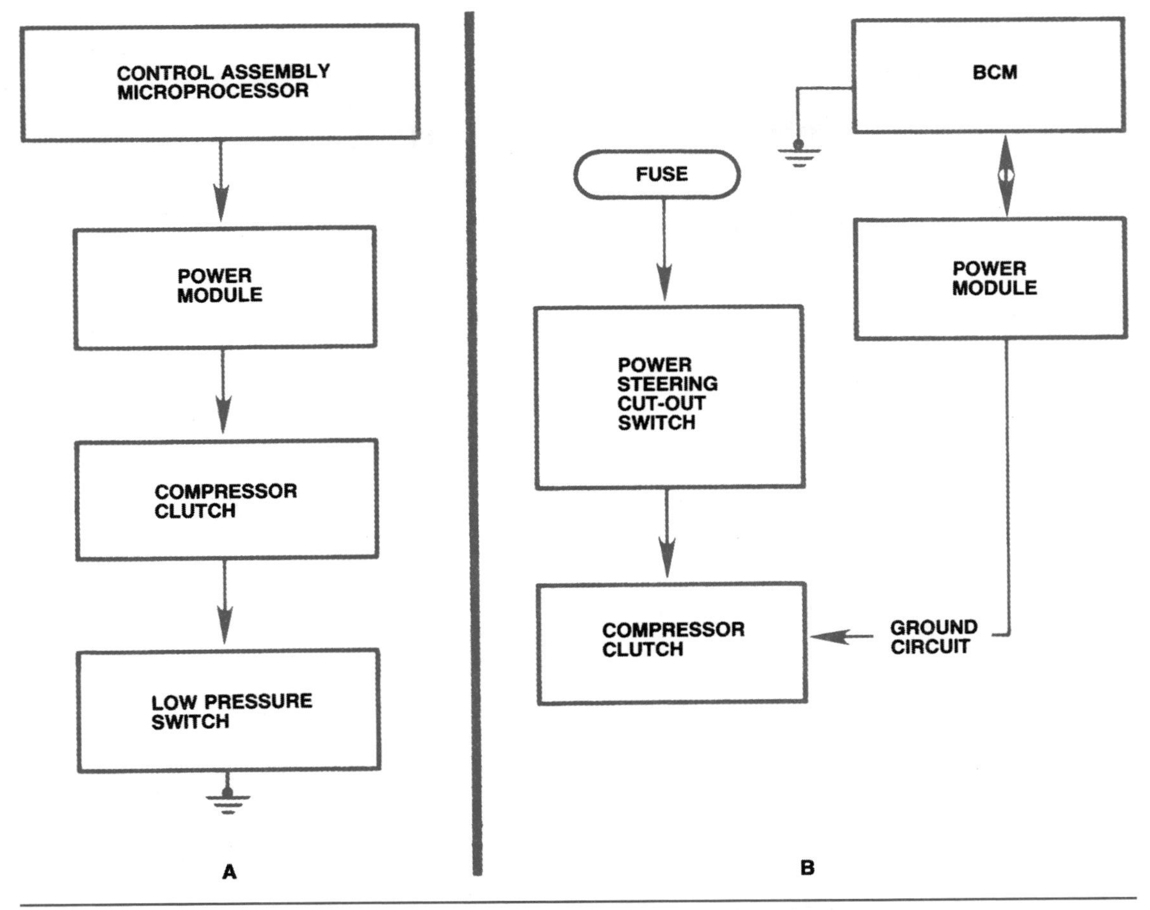

Figure 15-13. Non-BCM-controlled ECC systems ground the compressor clutch through the low-pressure switch and provide power through the power module (A). BCM-controlled ECC systems send power through a fuse and a power steering cutout switch, and cycle the ground through the power module (B). (GM Service and Parts Operations)

CLASS 2 IPM-CONTROLLED HVAC SYSTEMS

GM Electronically Controlled Blower Motor

HVAC Module

Most of the luxury model cars, including GM, have automatic *heating, ventilating, and air conditioning (HVAC)* systems that are computer controlled (Figure 15-14). The HVAC control module is a computer device that interfaces between the operator and the HVAC system to maintain air temperature and distribution settings. The control module sends switch input data to the instrument panel module (IPM) and receives display data from the IPM through signal and clock circuits. The control module does not retain any HVAC DTCs (diagnostic trouble codes) or settings.

Instrument Panel Module (IPM)

A function of the IPM operation is to process HVAC system inputs and outputs. Also, the IPM acts as the HVAC control module's Class 2 interface. The battery positive voltage circuit provides power that the IPM uses for Keep-Alive Memory (KAM). If the battery positive voltage circuit loses power, then all HVAC DTCs and settings will be erased from KAM. The ignition voltage circuit provides a device on signal. The IPM supports the following features:

- Driver set temperature
- Passenger set temperature

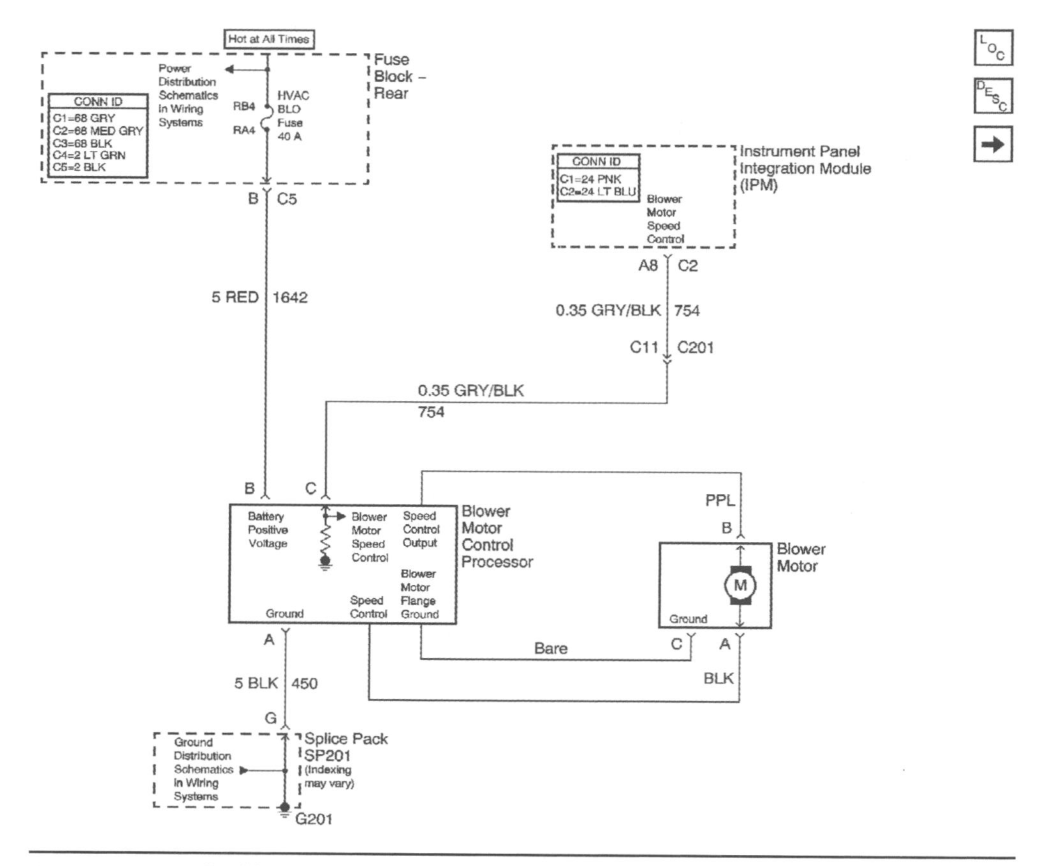

Figure 15-14. Cadillac Deville electronically controlled blower motor schematic. (GM Service and Parts Operations)

- Mode
- Blower motor speed
- A/C compressor request, auto ON or A/C OFF

This information will be stored inside the HVAC control module (IPM) memory. When a different driver identification button is selected, the HVAC control module will recall the appropriate driver settings. When the HVAC control module (IPM) is first turned on, the last stored settings for the current driver will be activated, except for the rear defrost and heated seat settings.

In automatic operation, the HVAC control module will maintain the comfort level inside of the vehicle by controlling the A/C compressor clutch, the blower motor, the air temperature actuators, the mode actuator, and recirculation.

To place the HVAC system in automatic mode, the following is required:

- The blower motor switch must be in the AUTO position.
- The air temperature switch must be in any other position besides 60 or 90 degrees.
- The mode switch must be in the AUTO position.

Once the desired temperature is reached, the blower motor, mode, recirculation, and temperature actuators will automatically adjust to maintain the temperature selected (except in the extreme temperature positions). The HVAC control module performs the following functions to maintain the desired air temperature:

- Regulate blower motor speed
- Position the air temperature actuator
- Position the mode actuator
- Position the recirculation actuator
- Request A/C operation

When the warmest position is selected in automatic operation, the blower speed will increase gradually until the vehicle reaches normal operating temperature. When normal operating temperature is reached, the blower will stay on high speed and the air temperature actuators will stay in the full heat position. When the coldest position is selected in automatic operation, the blower will stay on high and the air temperature actuators will stay in the full cold position.

In cold temperatures, the automatic HVAC system will provide heat in the most efficient manner. The vehicle operator can select an extreme temperature setting but the system will not warm the vehicle any faster. In warm temperatures, the automatic HVAC system will also provide air conditioning in the most efficient manner. Selecting an extreme cool temperature will not cool the vehicle any faster.

RADIOS AND ENTERTAINMENT SYSTEMS

Entertainment radios (Figure 15-15) are available in a wide variety of models. The complexity of systems varies from the basic AM radio to the compact disc (CD) player with high power amplifiers and multiple speakers. However, the overall operation of the radio itself, electrically, is basically the same. The major components in a basic AM system are a radio receiver and speaker. In the more complex stereo systems, the major components are an AM/FM radio receiver, a stereo amplifier, a sound amplifier switch, several speakers, and possibly a power antenna system.

In addition, many of the newer designs utilize a control module to aid in system diagnostics, memory presets, and other advanced features. The use of a scan tool and the appropriate service manual will enable the technician to determine the correct repair.

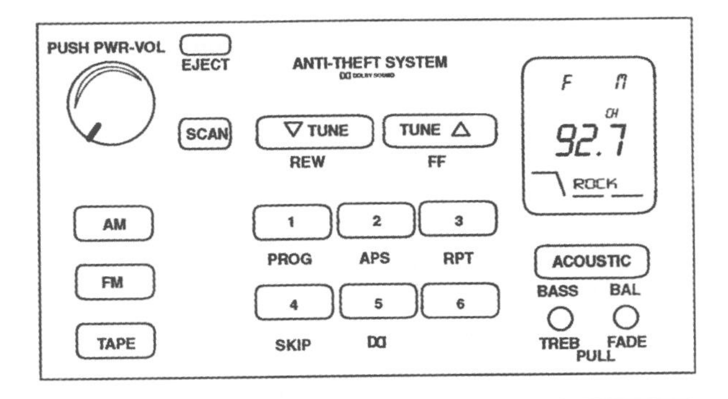

Figure 15-15. Radio face panel.

The inner circuitry of radios, tape players, CD players, power amplifiers, and graphic equalizers is beyond the scope of this text. However, a technician must understand the external circuitry of sound systems in order to troubleshoot them. Most sound units and speakers are grounded. In a few four-speaker systems, the speakers are insulated from their mountings. Current flows from the sound unit, through all of the speakers, and back to ground.

Entertainment System Diagnostics

Internal diagnostic examination of the radio should be left to the authorized radio service center. However, the automotive technician should be able to analyze and isolate radio reception conditions to the area of the component causing the condition. All radio conditions can be isolated to one of five general areas:

- Antenna system
- Radio chassis (receiver)
- Speaker system
- Radio noise suppression equipment
- Sound system

Radio Operation

Operation of the AM radio requires only that power from the fuse panel be available at the radio. The radio intercepts the broadcast signals with its antenna and produces a corresponding input to the system speaker. In addition, some radios have built-in memory circuits to ensure that the radio returns to the previously selected station when the radio or ignition switch is turned off and back on again. Some of these memory circuits require an additional power input from the fuse panel that remains hot at all times. The current draw is very small and requires no more power than a clock. However, if battery power is removed, the memory circuit has to be reset.

The service manual and owner's guide for the vehicle contain detailed information concerning radio operation. If the radio system is not working, check the fuse. If the fuses are okay, refer to the service manual. Remember, the radio chassis (receiver) itself should only be serviced by a qualified radio technician or specialty radio service shop. If you determine that the radio itself is the problem, remove the radio and send it to a qualified radio technician.

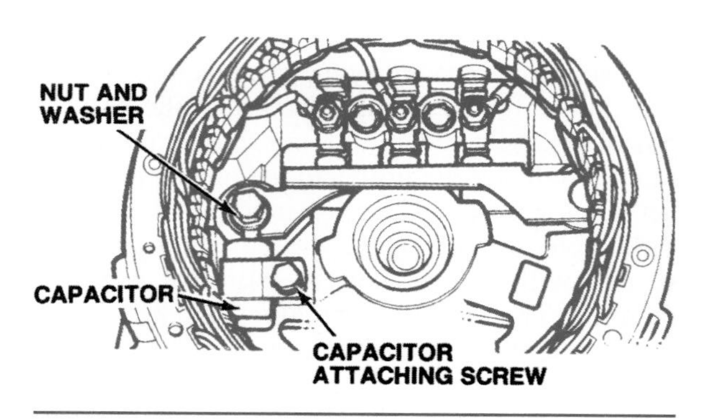

Figure 15-16. RFI capacitors can be installed inside the alternator. (DaimlerChrysler Corporation)

Antitheft Audio Systems

Most radio systems have built-in devices that make the audio system soundless if stolen. If the power source for the audio system is cut, the antitheft system operates so that even if the power source is reconnected, the audio system will not produce any sound. Some systems require an ID number selected by the customer to be entered. When performing repairs on vehicles equipped with this system, the customer should be asked for the ID number prior to disconnecting the battery terminals or removing the audio system. After the repairs, the technician or customer must input the ID number to regain audio system operation.

Other systems sense a specific code from the control module to allow the audio system to operate. This means the radio will not operate unless it is installed in the correct vehicle. Always refer to the vehicle service manual before removing a stereo to determine if it is equipped with any antitheft devices and the procedures for removal.

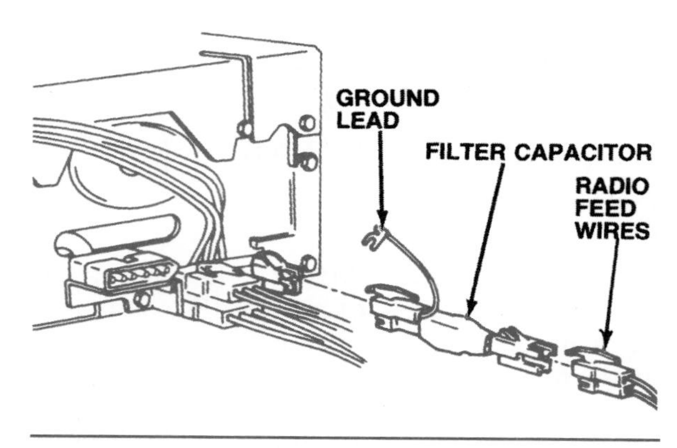

Figure 15-17. An RFI capacitor may be installed near the radio. (GM Service and Parts Operations)

Noise Suppression

The vehicle's ignition system is a source of radio interference. This high-voltage switching system produces a radio frequency electromagnetic field that radiates at AM, FM, and CB frequencies. Although components have been designed into the vehicle to minimize this concern, the noise is more noticeable if the radio is turned slightly off channel when listening to FM programs. Vehicle electrical accessories and owner add-on accessories may also contribute to radio interference. Furthermore, many noise sources are external to the vehicle, such as power lines, communication systems, ignition systems of other vehicles, and neon signs.

In addition to resistance-type spark plugs and cables, automobiles use capacitors and ground straps to suppress radio static or interference caused by the ignition and charging systems. Capacitors may be mounted as follows:

- Inside the alternator (Figure 15-16)
- Behind the instrument panel, near the radio (Figure 15-17)
- At the ignition coil with the lead connected to the coil primary positive terminal (Figure 15-18A)
- In a module mounted at the wiper motor and connected in series between the motor and wiring harness (Figure 15-18B)

Ground straps are installed to conduct small, high-frequency electrical signals to ground. They require a large, clean, surface-contact area. Such ground straps are installed in various locations depending upon the vehicle. Some common locations are as follows:

- Radio chassis to cowl
- Engine to cowl
- Across the engine mounts
- From air-conditioning evaporator valve to cowl

The small bulb that lights the sound unit controls may be part of the instrument panel

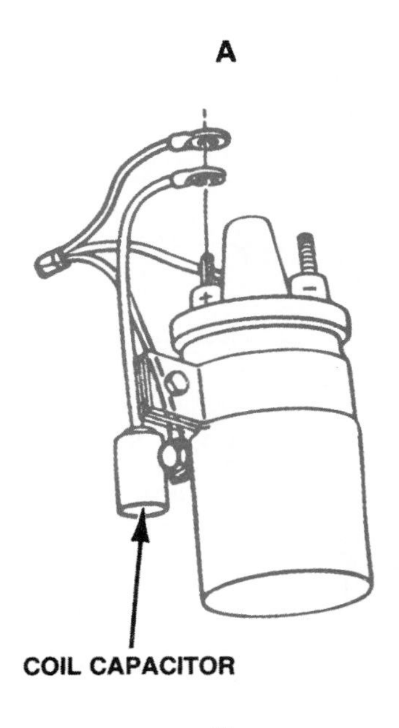

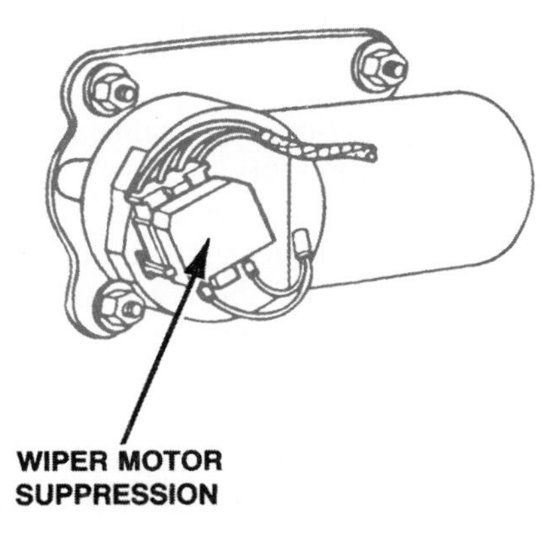

Figure 15-18. RFI capacitors may be installed on the ignition coil or wiper motor. (DaimlerChrysler Corporation)

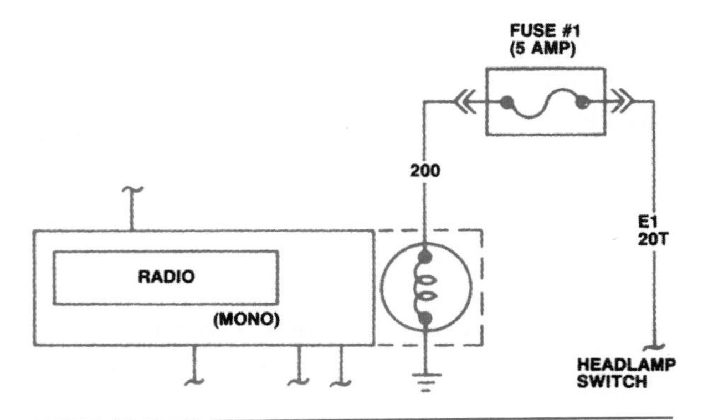

Figure 15-19. The radio illumination bulb is controlled by the IP light circuit. (DaimlerChrysler Corporation)

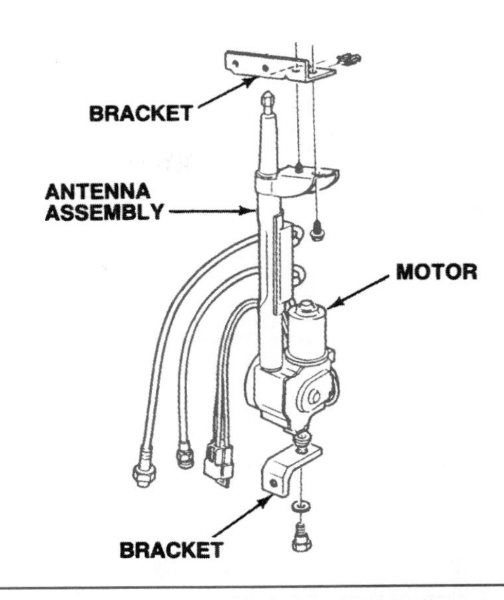

Figure 15-20. The motor of an electrically extended radio antenna is usually installed inside the wheel well under a protective cover. (GM Service and Parts Operations)

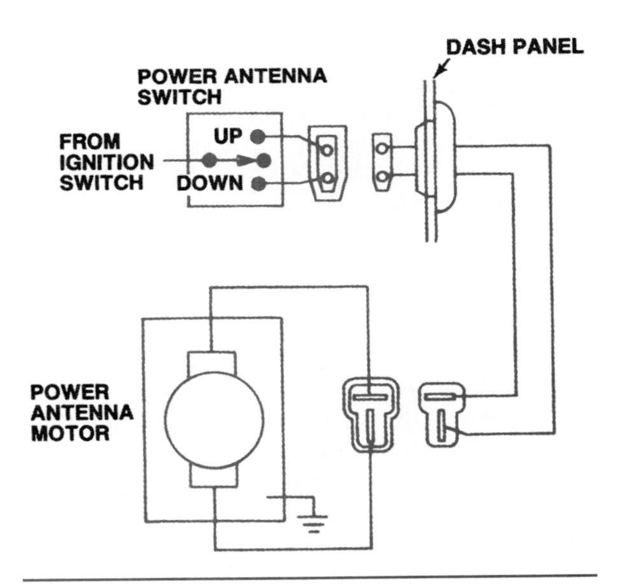

Figure 15-21. A separate switch can control power antennas. (DaimlerChrysler Corporation)

circuitry (Figure 15-19) or part of the sound unit's internal circuitry. Some cars use electrically extended radio antennas (Figure 15-20). The antenna motor may be automatically activated when the radio is turned on, or a separate switch (Figure 15-21) may control it. A relay may control current to the antenna motor.

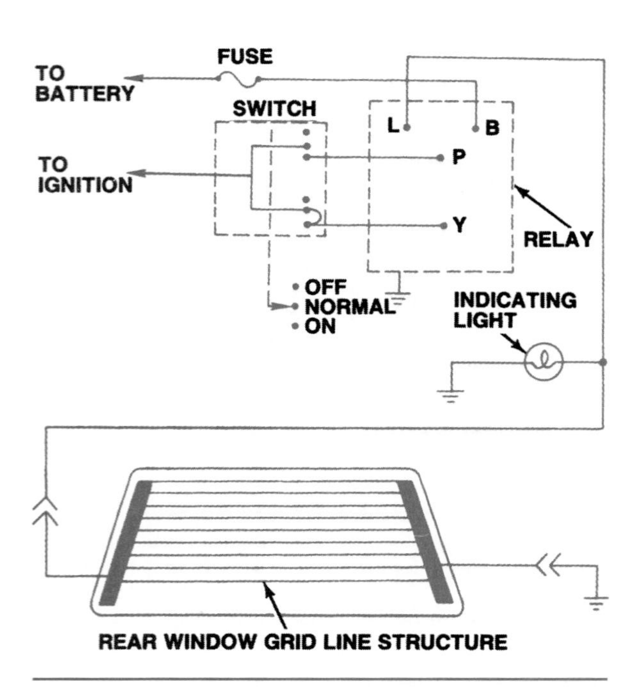

Figure 15-22. Defroster circuit with grid. (DaimlerChrysler Corporation)

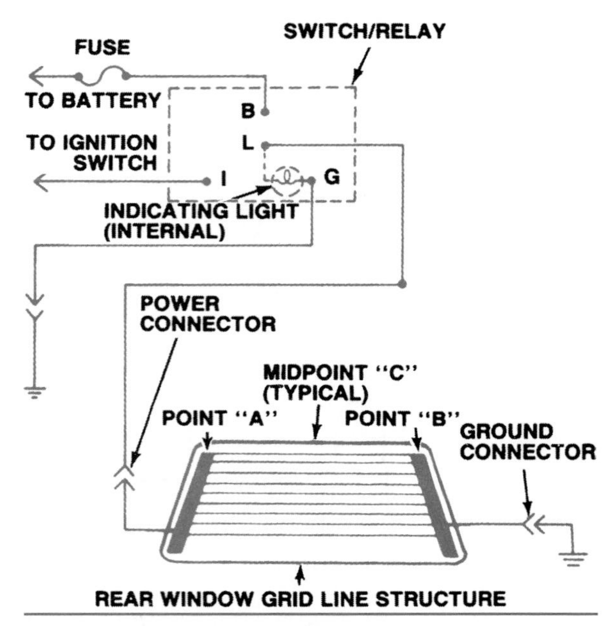

Figure 15-23. Late-model system with a solid-state timing module that turns off the defroster current automatically. (DaimlerChrysler Corporation)

REAR-WINDOW DEFOGGER AND DEFROSTER

Some older vehicles have a rear-window defogger, which is a motor-driven fan similar to that used in the heating system but mounted behind the rear seat near the rear window. It is controlled by a separate switch that routes current through circuits of varying resistance (like a heater fan) to change motor speed. Heat is provided electrically by a length of resistance wire in the defogger unit. The resistance heater is connected in parallel with the motor so that it heats when the motor is running at either high or low speed.

Rear Window Defroster

A **defroster** is a grid of electrical heating conductors that is bonded to the rear window glass (Figure 15-22). The defroster grid is sometimes called a defogger. Current through the grid may be controlled by a separate switch and a relay (Figure 15-22) or by a switch-relay combination (Figure 15-23). In both designs, when the switch is closed, the relay is energized and an indicator lamp is lit. The relay contact points conduct current to the rear window grid.

Most late-model systems have a solid-state timing module that turns off the defroster current automatically. In the system shown in Figure 15-23, the switch ON position energizes the relay's pull-in and hold-in coils. The switch NORMAL position keeps the hold-in coil energized so that the relay points remain closed. Cleaning the inside rear glass should be done carefully to avoid scratching the grid material and causing an open in the circuit.

POWER WINDOWS

Car doors can contain motors to raise and lower the window glass (Figure 15-24). The motors usually are the permanent-magnet type and are insulated at their mounting and grounded through the control switch (Figure 15-25) or the master switch. Each control switch operates one motor, except for the driver's door switch. This is a master switch that can control any of the motors. Some systems have a mechanical locking device that allows only the driver's switch to control any of the motors.

The single-motor control switches each have one terminal that is connected to battery voltage. Each of the other two switch terminals is con-

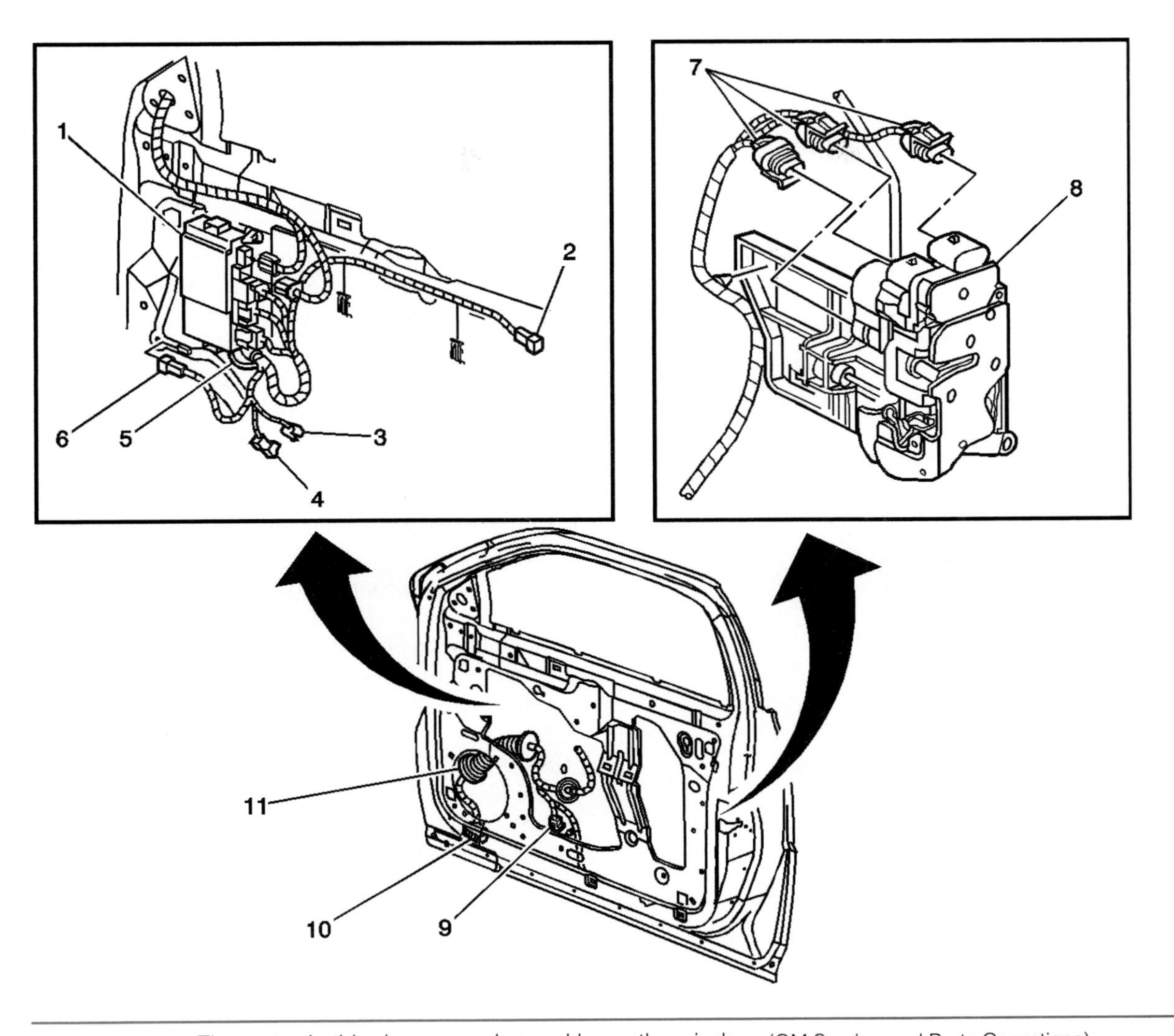

Figure 15-24. The motor in this door can raise and lower the window. (GM Service and Parts Operations)

nected to one of the two motor brushes. The window is moved up or down by reversing the direction of motor rotation. Motor rotation is controlled by routing current into one brush or the other. Each individual window switch is connected in series with the driver's master switch. Current from the motor must travel through the master switch to reach ground.

POWER SEATS

Electrically adjustable seats can be designed to move in several ways, as follows:

- Two-way systems move forward and backward.
- Four-way systems move forward, backward, and front edge up and down.
- Six-way systems, used in most late-model applications, move the entire seat forward, backward, up, and down; tilt the upper cushion forward and backward, and move the lower cushion front edge up and down, and rear edge up and down.

GM makes a typical two-way power seat system, as shown in Figure 15-26. The series-connected motor has two electromagnetic field windings that are wound in opposite directions. One winding receives current from the forward switch position. The second winding receives current from the rear switch position. Current through one winding will make the motor turn in

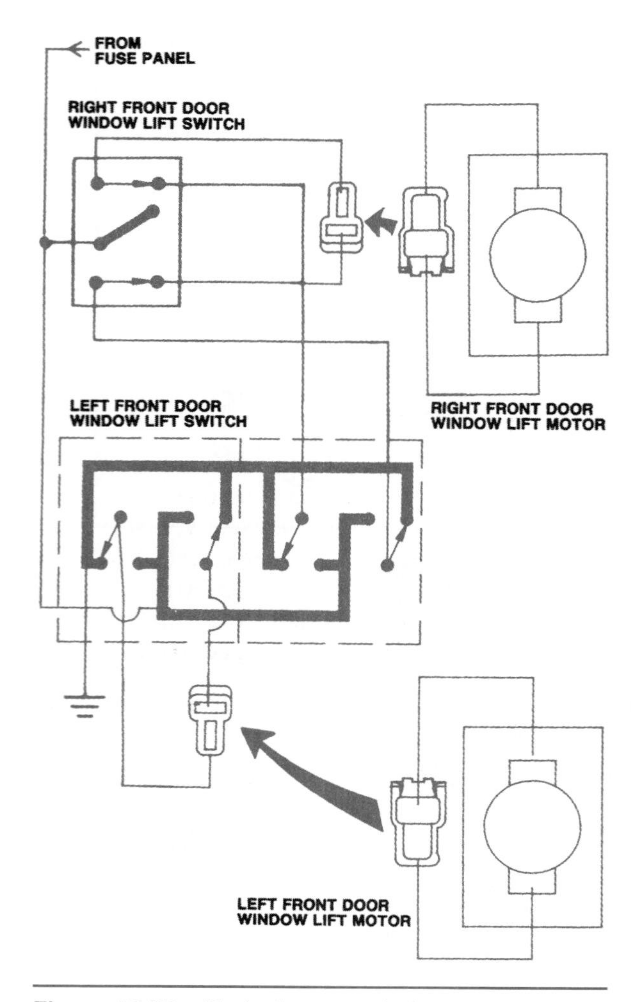

Figure 15-25. Typical power window control circuit. (DaimlerChrysler Corporation)

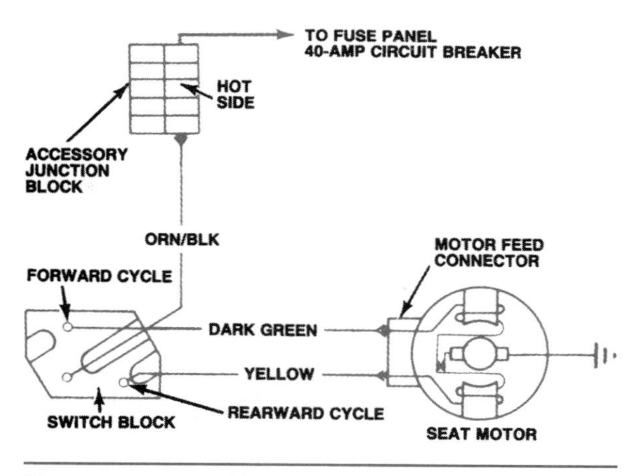

Figure 15-26. A GM two-way power seat has a motor with electromagnetic fields; current through the fields determines the direction of the motor and thus the movement of the seat. (GM Service and Parts Operations)

one direction; current through the opposite winding will make the motor turn in the opposite direction. The motor armature is linked to the seat mounting by a transmission that translates this rotary motion into seat motion.

Ford and GM have made four-way power seat systems that contain two reversible motor armatures in one housing. Ford's motors have permanent-magnet fields, while GM's motors have series-connected electromagnetic fields. One motor is linked to a transmission that moves the seat forward and backward. The other motor's transmission tilts the front edge of the seat. A single four-position switch controls both motors. The switch contacts shift current to different motor brushes (Ford) or to different field windings (GM) to control motor reversal.

Early GM six-way power seat systems use one reversible motor that can be connected to one of three transmissions. Transmission hookup is controlled by three solenoids (Figure 15-27). The control switch is similar to that used by Ford and Chrysler, but the circuitry differs. Current must flow through one of the solenoids to engage a transmission, then through a relay to ground. The relay points conduct current to the motor brushes. Additional switch contacts conduct current to the electromagnetic motor windings.

Chrysler, Ford, and late-model GM six-way power seat systems use three reversible motor

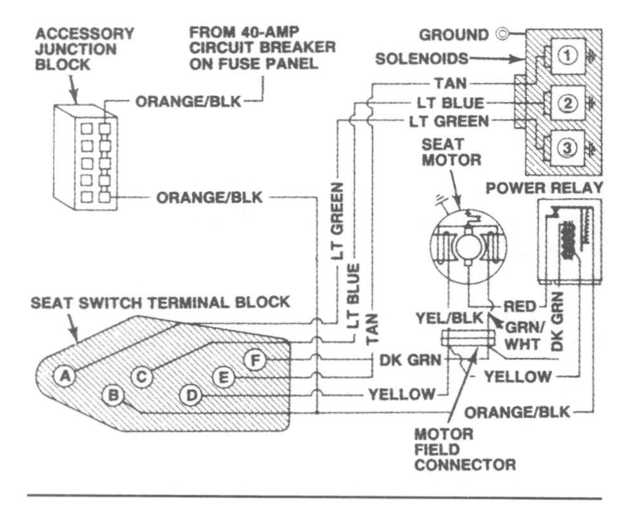

Figure 15-27. GM's early six-way power seat systems use one reversible motor that can be connected to one of three transmissions. Transmission hookup is controlled by three solenoids. (GM Service and Parts Operations)

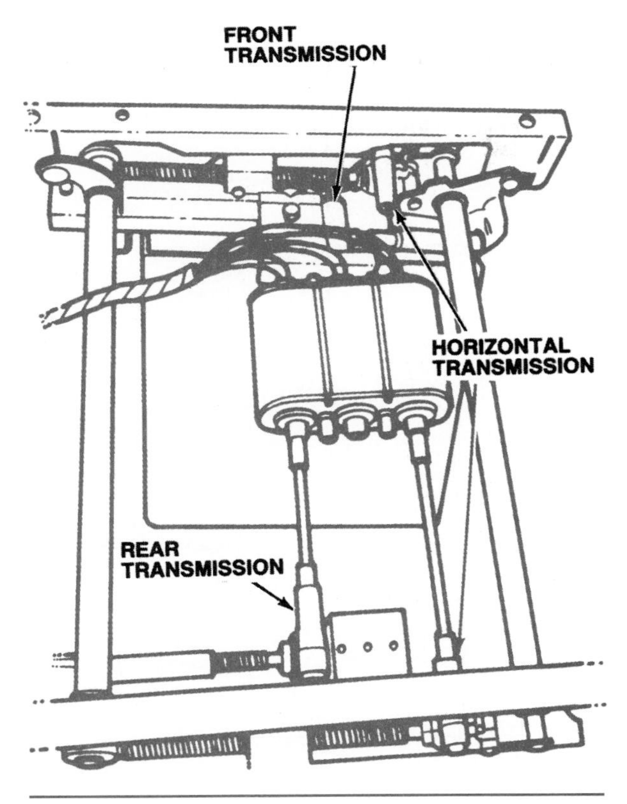

Figure 15-28. Chrysler, Ford, and late-model GM six-way power seat systems use three reversible motor armatures in one housing. (DaimlerChrysler Corporation)

armatures in one housing (Figure 15-28). The control switches have two two-position knobs that control edge tilt and a four-position knob that controls forward, backward, up and down seat movement. The switch contacts shift the current to different motor brushes to control motor reversal. The permanent-magnet motors are grounded through the switch and may contain an internal circuit breaker.

HEATED SEATS

Most manufacturers of premium cars and SUVs offer heated front seats, and in some cases back seats as well (Figure 15-29). Most vehicle heated seat systems consist of four heated seats: two in the front and two in the rear. Figure 15-30 shows a typical heated seat system schematic. Most heated seat systems consist of the following components:

- Heated seat module or controller
- Heated seat switch
- Seat back heating element
- Seat cushion heating element

The rear integration module (RIM), driver door module (DDM), left rear door module

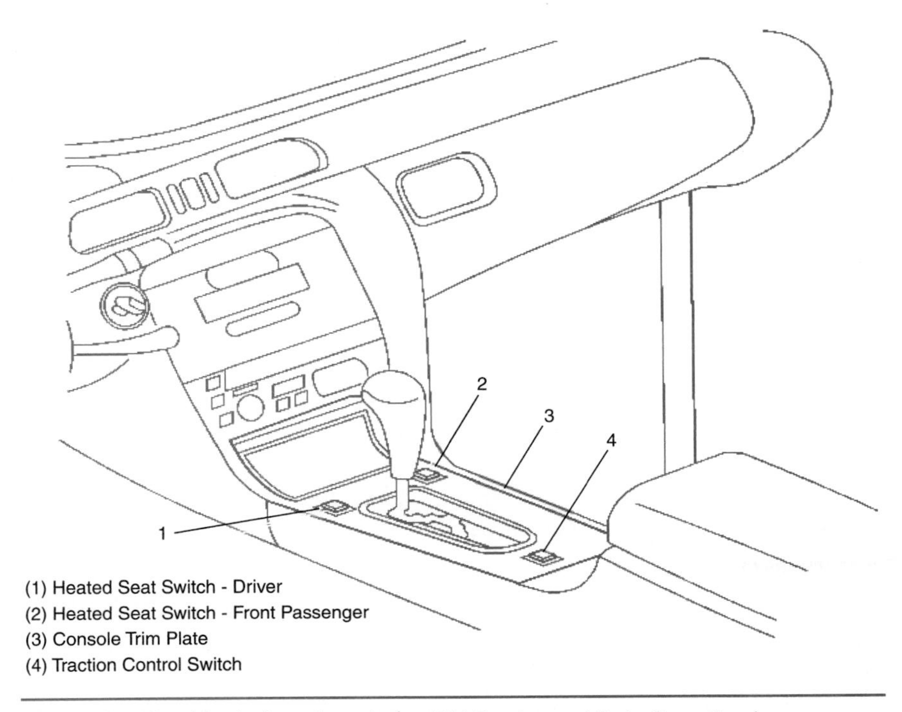

Figure 15-29. Heated seat controls. (GM Service and Parts Operations)

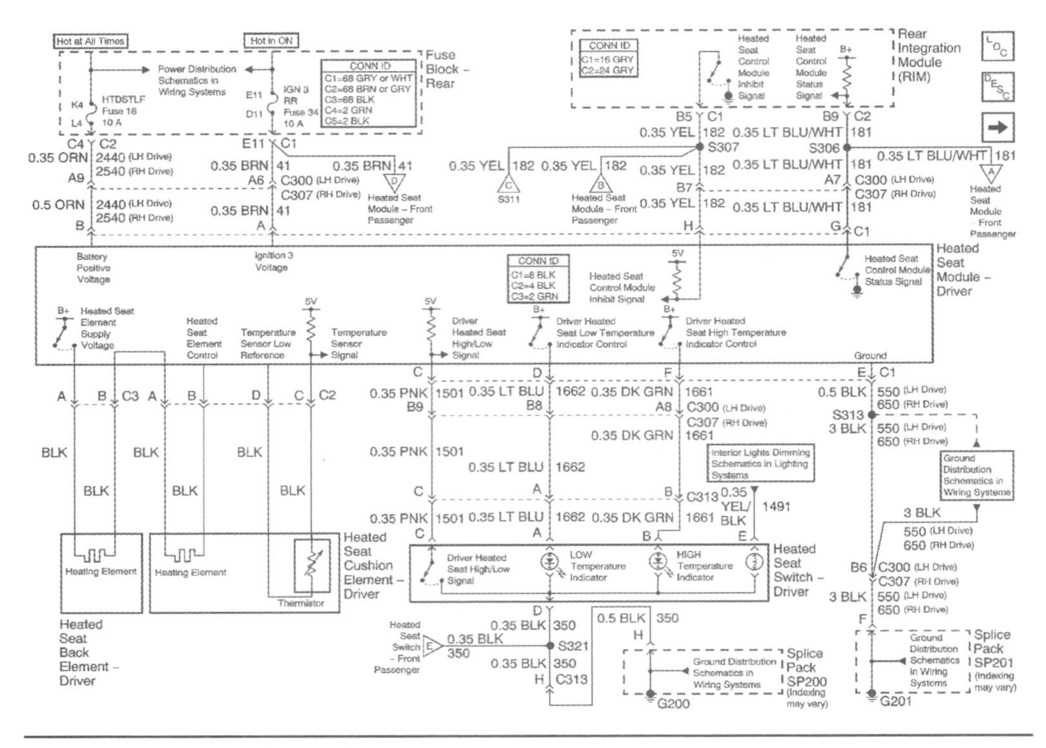

Figure 15-30. Heated seat circuit. (GM Service and Parts Operations)

(LRDM), and right rear door module (RRDM) are also involved in the operation of the heated seats. The system is functional only with the ignition switch in the ON position.

Power and Grounds

Battery positive voltage is supplied to the front and rear heated seat module through the ignition 3 voltage circuit and the IGN 3 fuse located in the rear fuse block. This voltage is used to power up the module. Battery positive voltage is also supplied at all times to all four heated seat modules from the fuses located in the rear fuse block.

The modules to apply voltage to the seat heating elements use this battery voltage. The left and right front heated seat modules are grounded through the module ground circuit and G302. The left and right rear heated seat modules are grounded through the module ground circuit and G301. The left and right front heated seat switches are grounded through the switch ground circuit and G200. The left and right rear heated seat switches are grounded through the switch ground circuit provided by the associated door module.

Temperature Regulation

The heated seat system is designed to warm the seat cushion and seat back to approximately 42°C (107.6°F) when in the high position, and 37°C (98.6°F) when in the low position. The heated seat module monitors the seat temperature through the temperature sensor signal circuit and the temperature sensor (thermistor) that is located in the seat cushion. The temperature sensor is a variable resistor: its resistance changes as the temperature of the seat changes. When the temperature sensor resistance indicates to the heated seat module that the seat has reached the desired temperature, the module opens the ground path of the seat heating elements through the heated seat element control circuit. The module will then cycle the element control circuit open and closed in order maintain the desired temperature.

Front Heated Seat Operation

When the heated seat switch is first pressed, the heated seat high/low signal circuit of the heated seat module is momentarily grounded through the HI/LO switch contacts, indicating a high heat command. In response to this signal, the heated seat module applies battery positive voltage to the seat cushion/back heating elements, setting the temperature level to high heat. When the heated seat switch is pressed a second time, the heated seat high/low signal circuit of the heated seat module is again momentarily grounded through the HI/LO switch contacts, indicating a low heat command. In response to this second signal, the heated seat module then sets the temperature level to low heat. When the heated seat switch is pressed a third time, the heated seat high/low signal circuit of the heated seat module is again momentarily grounded through the HI/LO switch contacts, indicating a heat off command. In response to this signal, the heated seat module removes battery voltage from the seat heating elements.

Front Heated Seat Switch Indicators

When the heated seat is off and the front heated seat switch is pressed once, the heated seat temperature is set to high heat. The heated seat module applies 5 volts through the heated seat high temperature indicator control circuit to the heated seat switch, illuminating the high temperature indicator. When the switch is pressed a second time the heated seat temperature is set to low heat. The heated seat module applies 5 volts through the heated seat low temperature indicator control circuit to the heated seat switch, illuminating the low temperature indicator. After the switch is pressed a third time, the heated seat is turned off, and the front heated seat module removes the voltage from the low temperature indicator.

Rear Heated Seat Operation

When the heated seat switch is first pressed, the heated seat switch signal circuit of the rear door module is momentarily grounded through the switch contacts, indicating a high heat command. The rear door module then sends a simple buss interface (SBI) message to the driver's door module (DDM), indicating the high heat command. The DDM then sends out a Class 2 message to the rear integration module (RIM), indicating the high heat command. The RIM momentarily sends a 35-millisecond one-shot pulse signal that is pulled low through the heated seat switch signal circuit of the rear heated seat module, indicating the high heat command. In response to this signal, the heated seat module will then apply battery positive voltage to the seat cushion/back heating elements, setting the temperature level to high heat.

When the switch is pressed a second time, the heated seat switch signal circuit of the rear door module is again momentarily grounded, indicating a low heat command. The rear door module then sends out a SBI message to the DDM, indicating the low heat command. The DDM then sends out a Class 2 message to the RIM, indicating the low heat command. The RIM again momentarily sends a 35-millisecond one-shot pulse signal that is pulled low through the heated seat switch signal circuit of the heated seat module, indicating the low heat command. In response to this signal, the heated seat module then sets the temperature level to low heat.

After the switch is pressed a third time, the heated seat switch signal circuit of the rear door module is again momentarily grounded, indicating a heat off command. The rear door module then sends out a SBI message to the DDM, indicating the heat off command. The DDM then sends out a Class 2 message to the RIM, indicating the heat off command. The RIM again momentarily sends a 35-millisecond one-shot pulse signal that is pulled low through the heated seat switch signal circuit of the heated seat module, indicating the heat off command. In response to this signal, the heated seat module then removes the battery voltage from the seat heating elements, turning off the heated seats.

Rear Heated Seat Switch Indicators

When the heated seat is off and the rear heated seat switch is pressed once, the heated seat temperature is set to high heat. The rear door module applies battery positive voltage through the heated seat high temperature indicator control circuit to the heated seat switch, illuminating the high temperature indicator. When pressed a second time, the heated seat temperature is set to low heat. The rear door module the applies battery positive voltage through the heated seat low temperature indicator control circuit to the heated seat switch, illuminating the low temperature indicator. After the switch is pressed a third time, the heated seat is turned off, and the rear door module removes the battery voltage from the low temperature indicator.

Load Management

Three levels of load management are controlled by the DIM. The DIM sends the status of the load management to the RIM via a Class 2 message. The ON/OFF status of the heated seats is reported to the RIM through the status-signal circuit of each heated seat module. The RIM inhibits the heated seat function for the heated seats through the heated seat module inhibit-signal circuit, according to the level of load management. During load shed level 00, the RIM leaves the heated seat inhibit-signal circuit open so that each heated seat module is in the normal mode of operation. During load shed level 01, the RIM will cycle the signal from High to Low every 0.25 second to set the heat level to the low setting. During load shed level 02, the RIM will supply a constant ground through the heated seat module inhibit-signal circuit to the heated seat modules. In response to this signal, the heated seat module then removes the battery voltage from the seat heating elements. The instrument cluster will display a Battery Saver Active message.

POWER DOOR LOCKS, TRUNK LATCHES, AND SEAT-BACK RELEASES

Solenoids and motors are used to control door, trunk, and seat-back latches, and locks. Door and trunk systems are usually controlled by separate switches mounted near the driver. Door-jamb switches usually control seat-back latches.

In some GM door lock systems, current flows through a solenoid winding to ground (Figure 15-31) when the driver closes the switch. The solenoid core movement either locks or unlocks the door, depending upon which switch position is selected. Some Ford and Chrysler electric door locks use a relay-controlled circuit as shown in Figure 15-32.

Current from the control switch flows through the relay coil, closing the relay contacts. The contacts route current directly from the fuse panel to the solenoid windings. Other Ford, Chrysler, and GM power door locks use an electric motor to move the locking mechanism. The electric motor receives current through a relay (Figure 15-32)

Power trunk latches use an insulated switch and a grounded solenoid coil (Figure 15-33). Power seat-back releases can be automatically controlled by grounding door-jamb switches (Figure 15-34). Opening one of the front doors energizes a relay; the relay contacts conduct current to solenoids, which unlatch both seat backs.

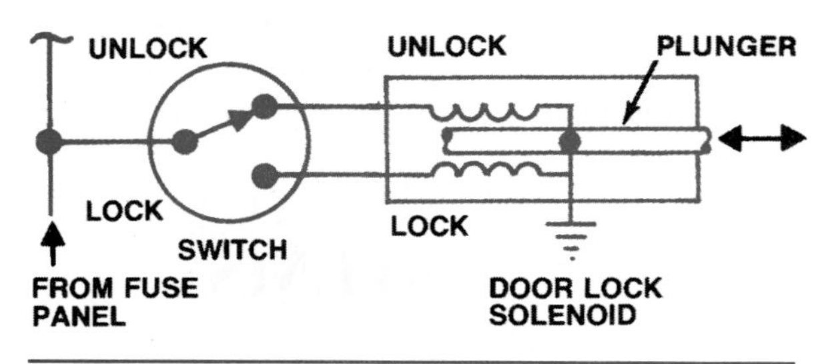

Figure 15-31. GM power door lock system. (GM Service and Parts Operations)

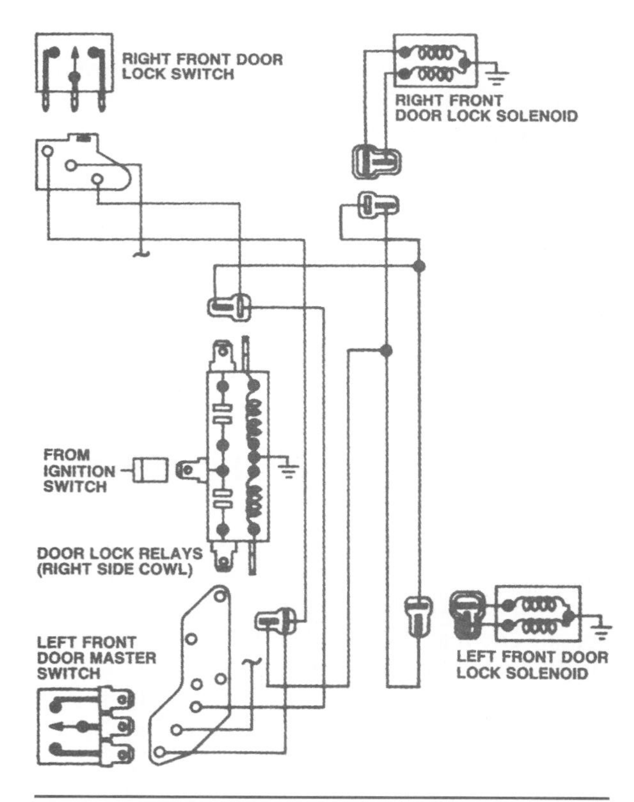

Figure 15-32. DaimlerChrysler power door lock system. (DaimlerChrysler Corporation)

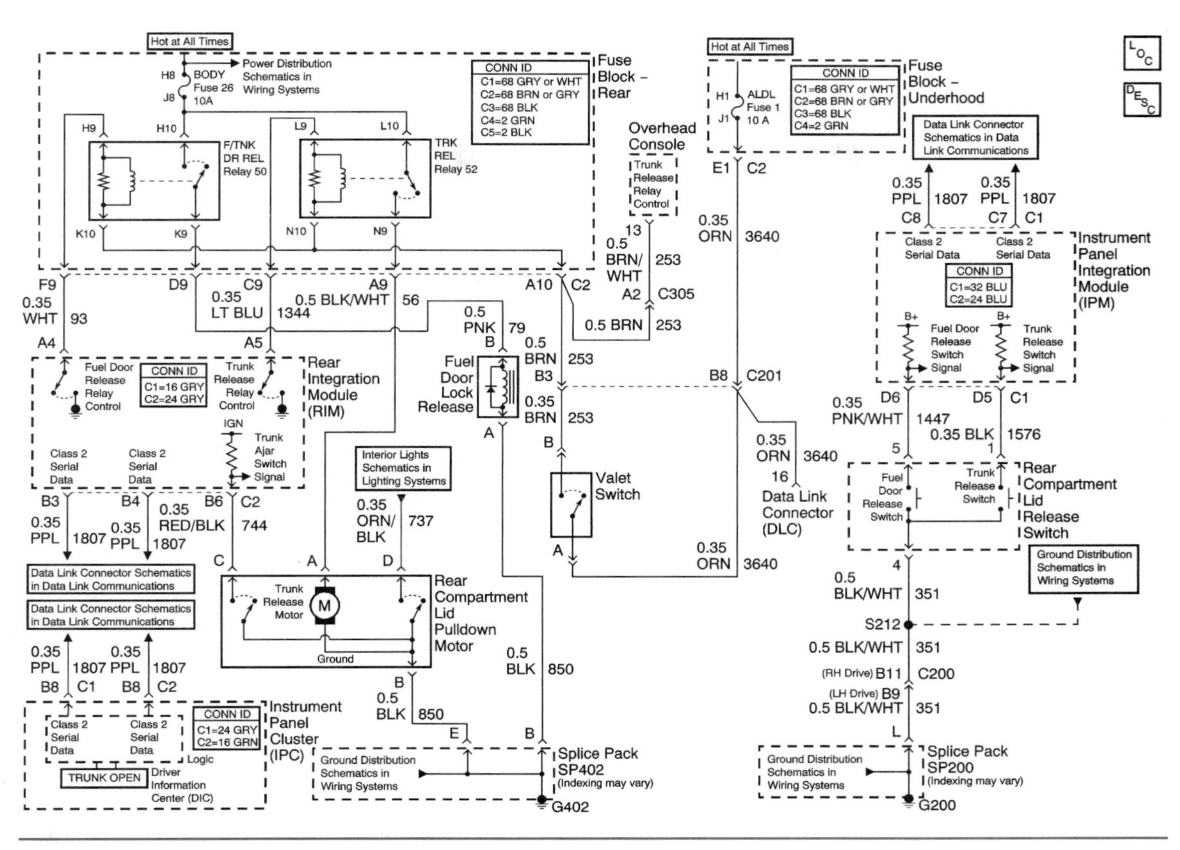

Figure 15-33. Typical power trunk lid latch system. (GM Service and Parts Operations)

AUTOMATIC DOOR LOCK (ADL) SYSTEM

General Motors and Ford both use an **automatic door lock (ADL)** system in the power door lock system on some of their models, as a safety and convenience feature. Ford ADL systems are an integral part of the keyless entry system, while General Motors ADL systems are available on vehicles regardless of whether they have keyless entry.

On GM vehicles with automatic transaxles, placing the gear selector in Drive automatically locks all vehicle doors when the ignition is ON. All doors unlock automatically when the gear selector is returned to the Park position. Individual doors can be unlocked manually from the inside, the front doors can be unlocked with the key from outside, or all the doors can be unlocked electrically while in Drive.

System Operation

The ADL feature may be a function of the chime module, an ADL controller, or a multifunction alarm module, depending on the vehicle model. In a typical General Motors ADL circuit, voltage is applied to the chime module, ADL controller, or alarm module. When the doors are closed, the ignition is in the Run position, and the gear selector is placed in Drive, the module or controller sends current to ground through the lock relay coil in the ADL relay (Figure 15-35). Current passes through the relay, door lock motors, and unlock relay to ground, locking the doors. The module or controller then removes current from the relay coil to prevent damage to the lock motors. When the vehicle stops and the gear selector is returned to Park, voltage is sent to the unlock relay coil in the ADL relay. The doors are unlocked by current passing through the relay, door lock motors, and lock relay to ground.

REMOTE/KEYLESS ENTRY SYSTEMS

In the late 1970s, Ford developed the first keyless entry system used on domestic vehicles. Chrysler and GM both offer keyless entry options on some

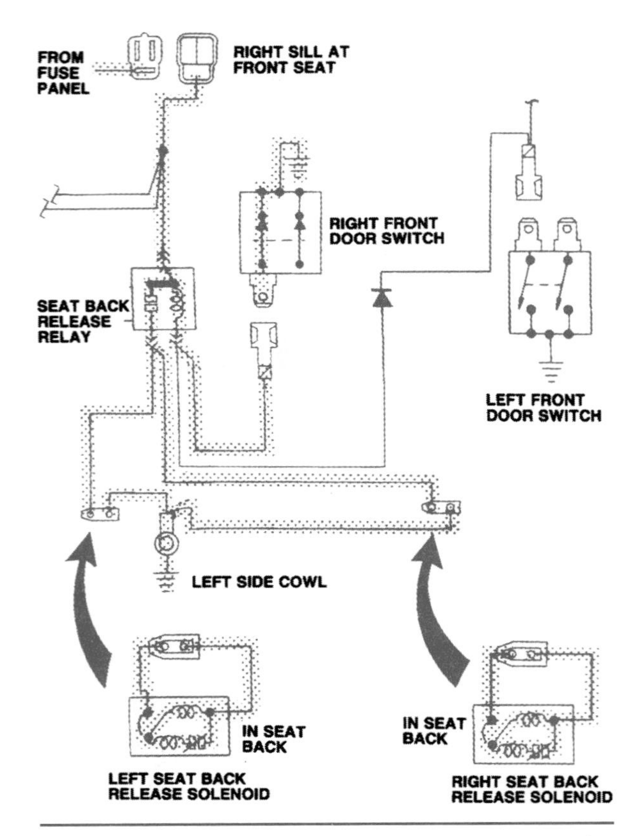

Figure 15-34. Power seat back releases can be automatically controlled by grounding door-jamb switches. (DaimlerChrysler Corporation)

current models. Since their applications differ substantially in design, concept, and operation, we will look at the Ford version first.

Ford

Ford's keyless entry system has remained substantially unchanged since its introduction. It provides a convenient entry method when the vehicle keys have been forgotten, or accidentally locked inside. The system consists of a five-button keypad secured to the outer panel on the driver's door, a microprocessor-relay control module, and connecting wiring.

The keyless entry system incorporates two additional subsystems: one for illuminated entry and the other for automatic door locks. Operating as a single system, it performs the following functions:

- Unlocks the driver's door
- Unlocks other doors or the deck lid when a specific keypad button is depressed within five seconds after unlocking the driver's door
- Locks all doors from outside the vehicle when the required keypad buttons are depressed simultaneously
- Turns on the interior lamps and the illuminated keyhole in the driver's door
- Automatically locks all doors when they are closed, the driver's seat is occupied, the ignition switch is on, and the gear selector is moved through the reverse position.

A linear keypad using calculator-type buttons is installed in the driver's door and used to input a numerical code to the control module. The five-keypad buttons are numbered 1-2, 3-4, 5-6, 7-8, and 9-0 from left to right. The numerical code used to open the door, however, is a derivative of a five-digit keypad code stamped on the control module and printed on a sticker attached to the inside of the deck lid. This code refers to the location of the five buttons on the keypad, not the keypad button number. For example, if the module number is 23145, the doors will unlock only if the keys are depressed in that order. If the module requires replacement, a sticker bearing the new module number is applied over the old sticker on the deck lid.

The control module's program operates the keyless entry, illuminated entry, and ADL systems. Two 14-pin connectors (one brown and one gray) connect the wiring harness to the control module; the brown connector also connects the keypad harness to the module. The following components provide inputs to the control module:

- Keypad buttons
- Door handles
- Courtesy lamp switch
- Driver's seat sensor
- Transmission backup lamp switch
- Ignition switch
- Door lock and unlock switches
- Door-ajar switch

The following components receive output signals from the control module:

- Keypad lamps
- Interior courtesy lamps
- Door lock LEDs
- Deck-lid-release solenoid
- Door lock solenoids

Ford added a remote keyless entry feature on some models, which uses a handheld radio transmitter with three buttons for door lock control from outside the vehicle. If the vehicle is equipped

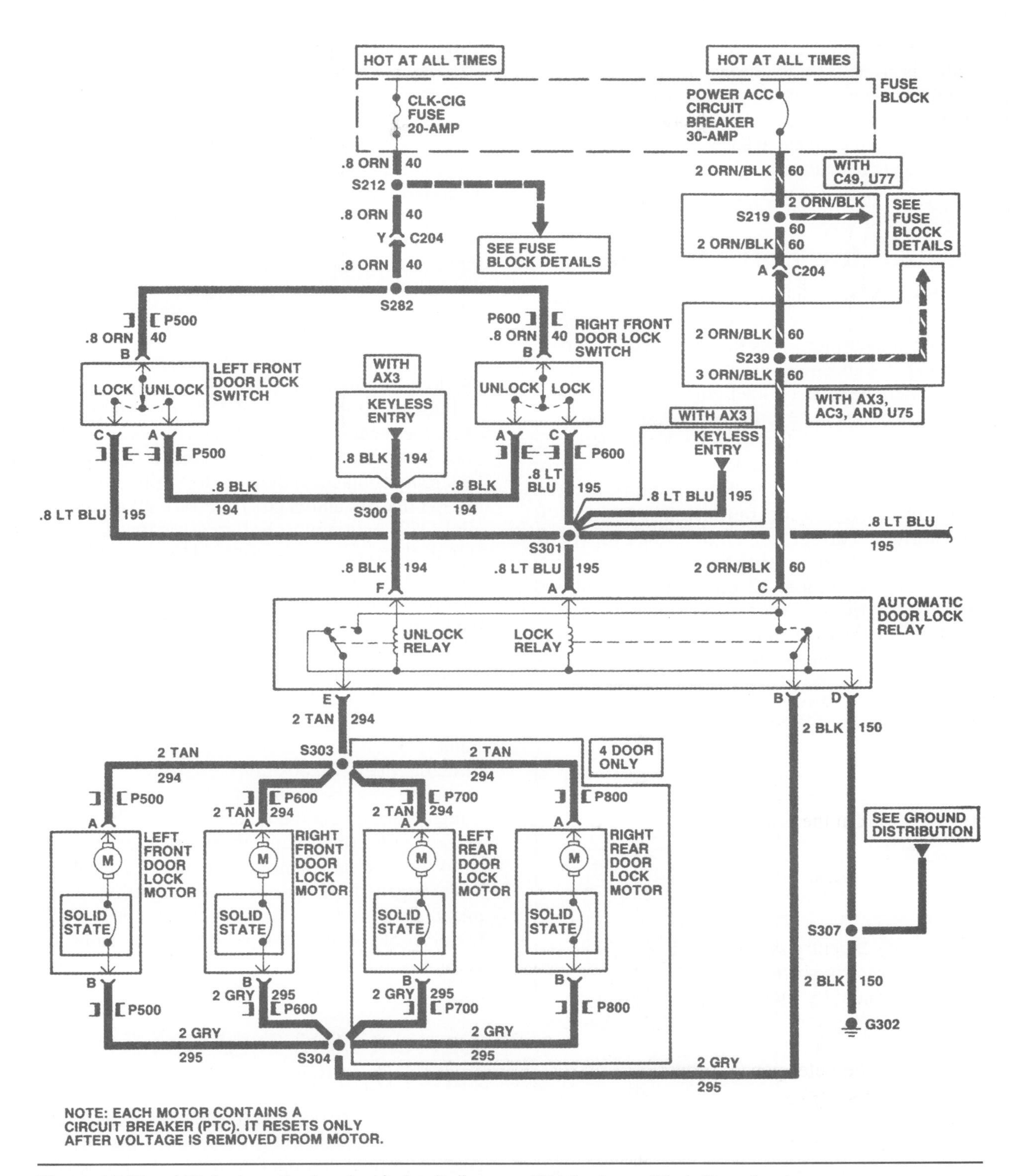

Figure 15-35. ADL (automatic door lock) circuit diagram. (GM Service and Parts Operations)

with the Ford antitheft system, a four-button transmitter is used. The additional button is marked "Panic" and allows the driver to activate the alarm in an emergency. The system operates essentially the same as the Delco RKE system described in the following section.

GM Keyless Entry Systems

In 1993, GM introduced a Passive Keyless Entry (PKE) system on the Corvette. In this system, a key-fob transmitter locks the doors as the person carrying it walks away from the car, and unlocks them when the carrier comes close to the car again. The owner does not even need to push a button.

Other GM vehicles use the Delco Remote Keyless Entry (RKE) or Remote Lock Control (RLC) system. The key fob contains a radio transmitter with three buttons (Figure 15-36) that allow the driver to lock or unlock the doors and trunk lid from outside the car. The transmitter contains a random 32-bit access code stored in a PROM; the same code is stored in a receiver module located in the trunk. This receiver detects and decodes UHF signals from the transmitter within a range of approximately 33 feet (10 meters).

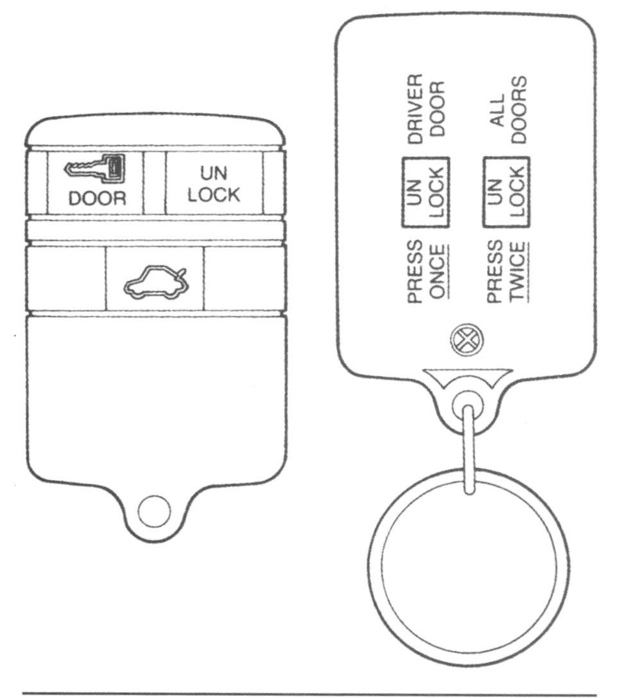

Figure 15-36. Delco Remote Keyless Entry (RKE) or Remote Lock Control (RLC) system radio transmitter. (Delphi Automotive Systems)

Depressing the DOOR button on the transmitter sends a signal to the receiver. If the signal contains a valid access code (VAC), the receiver supplies battery voltage to the lock relay coil. This energizes the lock relay, which sends current to the door lock motors. The LH door lock motor is grounded through receiver terminal B and internal contacts; all other motors are grounded through the unlock relay contacts in the door lock relay (Figure 15-37). When the lock function is used, the receiver grounds circuit 156, turning on the interior lights for two seconds to indicate that the doors are locked.

If the transmitter UNLOCK button is depressed once, the receiver sends battery voltage to the LH door lock motor, which is grounded through the lock relay contacts in the door lock relay, and only the LH door is unlocked. To unlock all doors, the UNLOCK button must be depressed twice. At this signal, the receiver also sends battery voltage to the unlock relay coil in the door lock relay. This energizes the unlock relay, which sends current to the other three door lock motors. The lock relay contacts in the door lock relay provide ground for the motors, which unlock the doors. When the unlock function is used, the receiver also grounds circuit 156 to turn on the interior lights for approximately 40 seconds, or until the ignition switch is turned to the RUN position.

Depressing the trunk lid release button on the transmitter signals the receiver to supply battery voltage at terminal H. This current energizes the trunk lid release solenoid, allowing the trunk lid to be opened. However, if the ignition is on and the transaxle position switch is not in Park, battery voltage is not supplied and the trunk lid cannot be opened.

Current 2002-2003 GM Keyless Entry Systems

The keyless entry system (Figure 15-38) is a supplementary vehicle entry device; use the keyless entry system in conjunction with a door lock key. Radio frequency interference (RFI) or discharged batteries may disable the system.

Keyless entry allows you to operate the following components:

- The door locks
- The rear compartment lid release

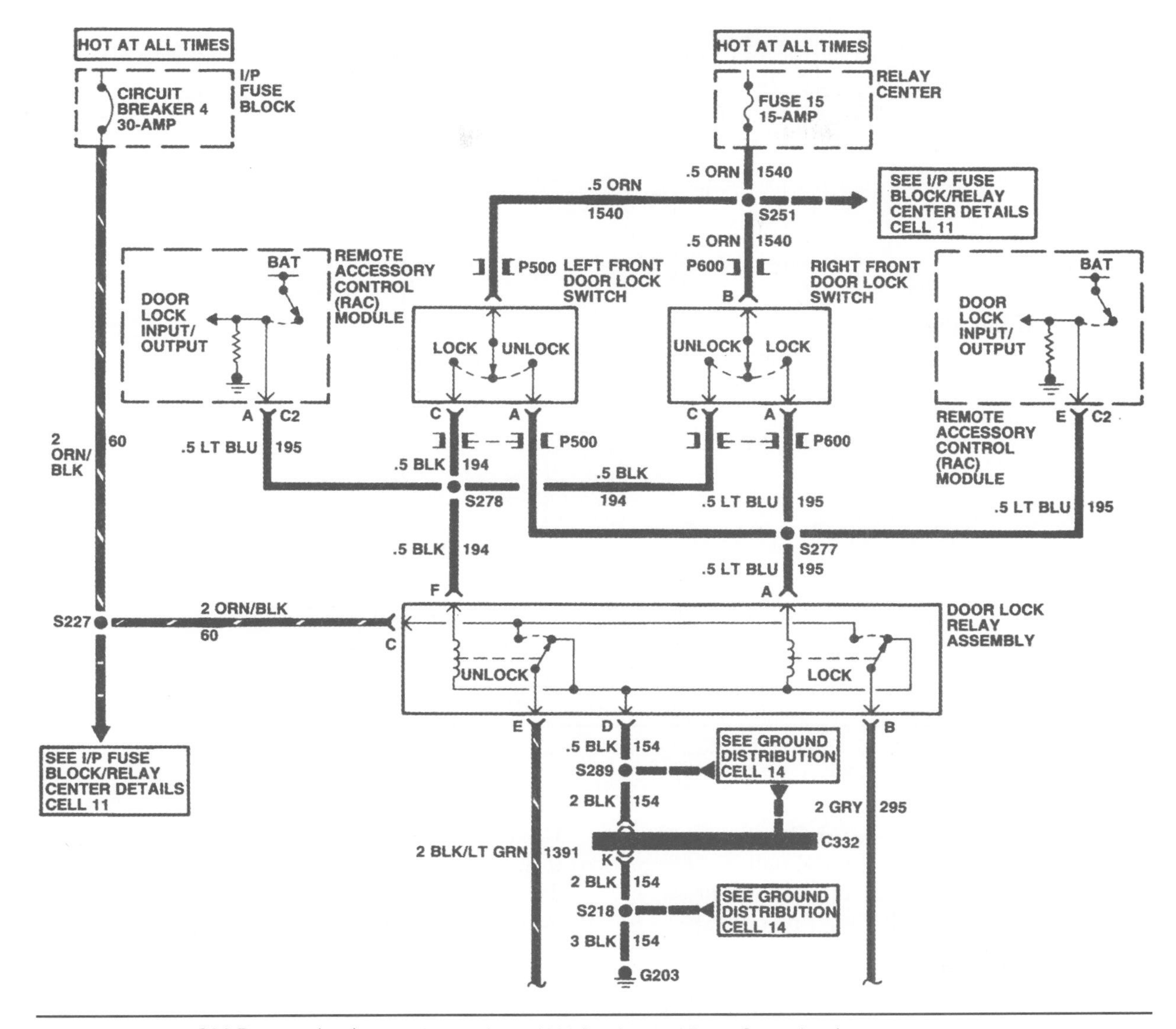

Figure 15-37. GM Remote keyless entry system. (GM Service and Parts Operations)

- The illuminated entry lamps
- The fuel door release

The keyless entry system has the following main components:

- The transmitters
- The remote control door lock receiver (RCDLR)

When you press a button on a transmitter, the transmitter sends a signal to the RCDLR. The RCDLR interprets the signal and activates the requested function via a Class 2 message over the serial data line.

Unlock Driver's Door Only

Momentarily press the UNLOCK button in order to perform the following functions:

- Unlock the driver's door only.
- Illuminate the interior lamps for approximately 40 seconds or until the ignition is turned ON.
- Flash the exterior lights, if selected ON in personalization.
- Disarm the content theft deterrent (CTD) system, if equipped.
- Deactivate the CTD system when in the Alarm Mode.

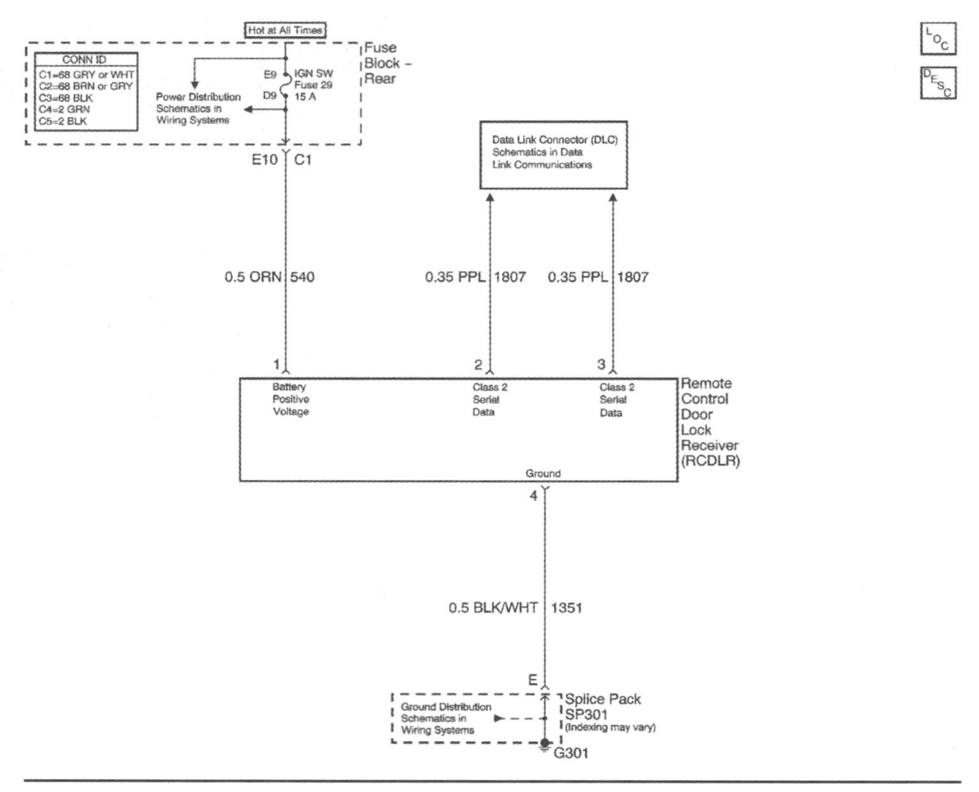

Figure 15-38. 2002 GM Cadillac Seville keyless entry system schematic. (GM Service and Parts Operations)

Unlock All Doors—Second Operation

Momentarily press the UNLOCK button a second time, within four seconds of the first press, in order to perform the following functions:

- Unlock the remaining doors.
- Illuminate the interior lamps for approximately 40 seconds or until the ignition is turned ON.
- Flash the exterior lights, if selected ON in personalization.
- Chirp the horn, if selected ON in personalization.

Lock All Doors

Press the LOCK button in order to perform the following functions:

- Lock all of the doors and immediately turn off the interior lamps.
- Flash the exterior lights, if selected ON in personalization.
- Chirp the horn, if selected ON in personalization.
- Arm the content theft deterrent (CTD) system.

Rear Compartment Lid Release

If the vehicle transaxle is in Park or Neutral and the ignition is in the OFF position, a single press of the rear compartment release button will open the rear compartment lid. The interior lamps will not illuminate.

Fuel Door Release

If the vehicle transaxle is in PARK or NEUTRAL and the ignition is in the OFF position, a single press of the fuel-door release button will open the fuel door.

Keyless Entry Personalization

The exterior lamps and horn chirp may be personalized for two separate drivers as part of the remote activation verification feature.

Rolling Code

The keyless entry system uses rolling code technology. Rolling code technology prevents anyone from recording the message sent from the transmitter and using the message in order to gain entry to the vehicle. The term *rolling code* refers to the way that the keyless entry system sends and receives the signals. The transmitter sends the signal in a different order each time. The transmitter and the remote control door lock receiver (RCDLR) are synchronized to the appropriate order. If a programmed transmitter is out of synchronization, it sends a signal that is not in the order that the RCDLR expects. This will occur after 256 presses of any transmitter button that is out of range of the vehicle.

Automatic Synchronization

The keyless entry transmitters do not require a manual synchronization procedure. If needed, the transmitters automatically resynchronize when any button on the transmitter is pressed within range of the vehicle. The transmitter will operate normally after the automatic synchronization.

DaimlerChrysler Keyless Entry Systems

The DaimlerChrysler keyless entry system also uses a key fob-style radio transmitter, (Figure 15-39) to unlock and lock the vehicle doors and deck lid. This multipurpose system is similar to many aftermarket theft-deterrent systems, since it turns on the interior lamps, disarms the factory-installed antitheft system, and chirps the horn whenever it is used.

The transmitter attaches to the key ring and has three buttons for operation within 23 feet (7 meters) of the vehicle module receiver. The transmitter has its own code stored in the module memory. If the transmitter is lost or stolen, or an additional one is required, a new code must be stored in module memory. Figure 15-40 shows the integration of the keyless entry, illuminated entry, vehicle theft security, and power door lock systems with the BCM.

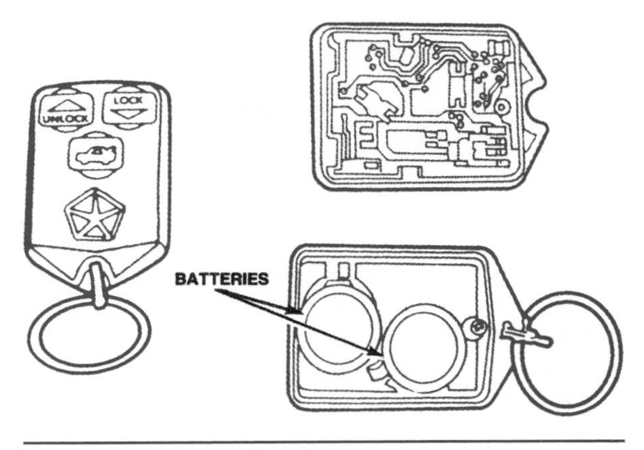

Figure 15-39. DaimlerChrysler keyless entry system key fob-style radio transmitter. (DaimlerChrysler Corporation)

THEFT DETERRENT SYSTEMS

Antitheft systems are usually aftermarket installations, although in recent years, some manufacturers have offered factory-installed systems on the luxury vehicles in their model line. Basic antitheft systems provide a warning when a forced entry is attempted through the car doors or the trunk lid. A starter interlock feature is incorporated on some models.

System functioning relies on strategically located switches installed in the door jambs, the door lock cylinders, and the trunk lock cylinder (Figure 15-41). After the system is armed, any tampering with the lock cylinders or an attempt to open any door or the trunk lid without a key causes the alarm controller to trigger the system.

Once a driver has closed the doors and armed the system, an indicator lamp in the instrument cluster comes on for several seconds, and then goes out. The system is disarmed by unlocking a front door from the outside with the key or turning the ignition on within a specified time. If the alarm has been set off, the system can be disarmed by unlocking a front door with the key.

Delphi (Delco) UTD System

The Delphi universal theft deterrent (UTD) system was introduced on some 1980 GM models and offered as an option until it was superseded by the personal automotive security system (PASS). The circuitry, logic, and power relays that operate the system are contained within a controller module.

When the system is armed by the driver, a security system warning lamp in the instrument panel glows for four to eight seconds after the doors have been closed, then shuts off. The system can be disarmed without sounding the alarm by unlocking a front door from the outside with a key, or by turning the ignition switch on. If the alarm has sounded, it can be shut off by unlocking one of the front doors with a key.

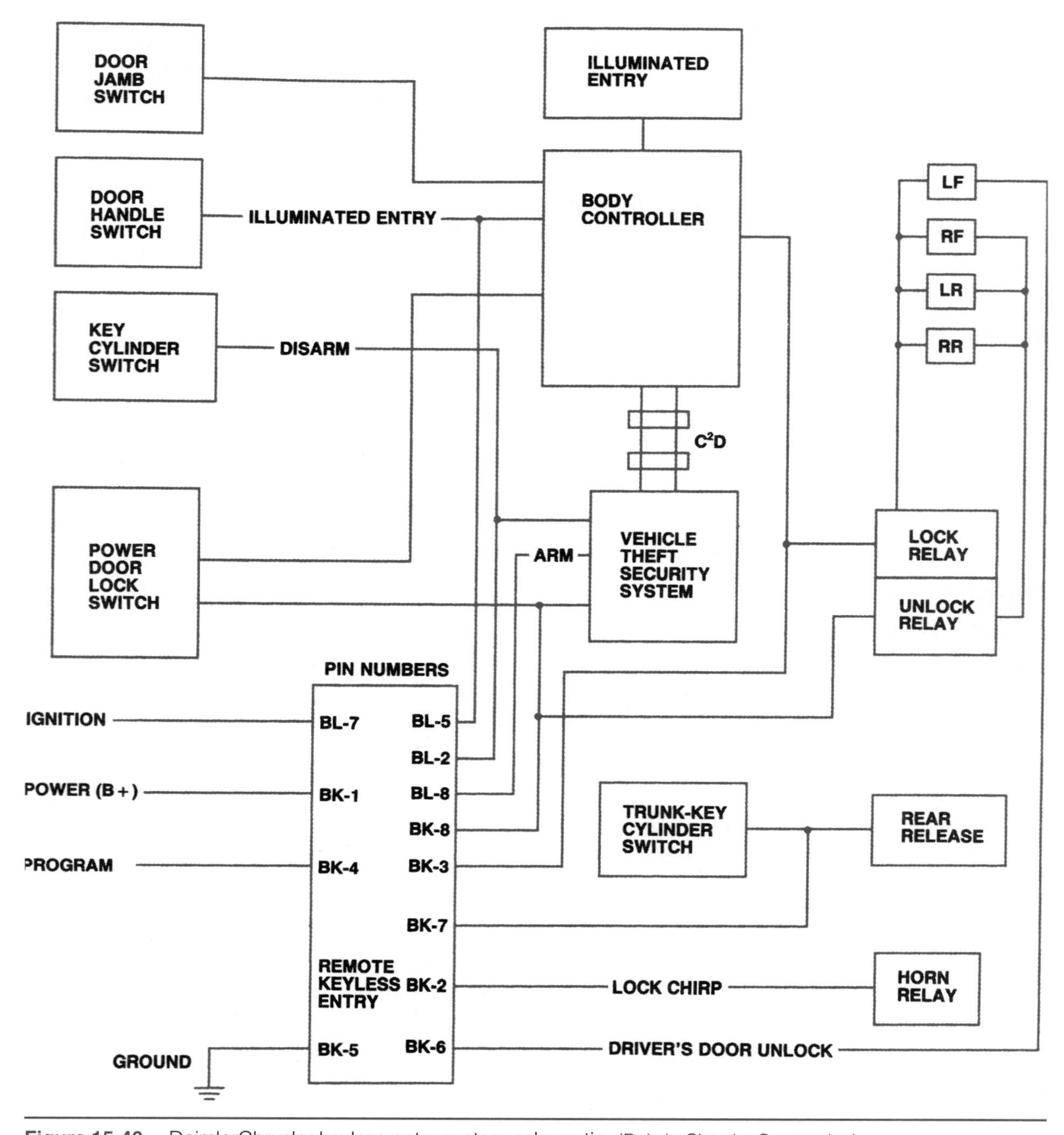

Figure 15-40. DaimlerChrysler keyless entry system schematic. (DaimlerChrysler Corporation)

If the system is armed and a door is opened forcibly, a two-terminal doorjamb switch activates the alarm through one terminal. The other switch terminal operates the interior lights. On vehicles with power door locks, the circuits are separated by a diode. Tamper switches are installed in all door locks and the trunk lid lock (Figure 15-42). The switches are activated by any rotation or in-and-out movement of the lock cylinders during a forced entry. A disarm switch in the LH door cylinder (Figure 15-43) allows the owner to deactivate the system without sounding the alarm before entering the vehicle. All tamper switches should be kept clean, as corrosion can cause the system to activate without apparent reason.

Exact wiring of the UTD system depends on the particular vehicle and how it is equipped. To understand just how the system works on a given vehicle, you must have the proper wiring diagram.

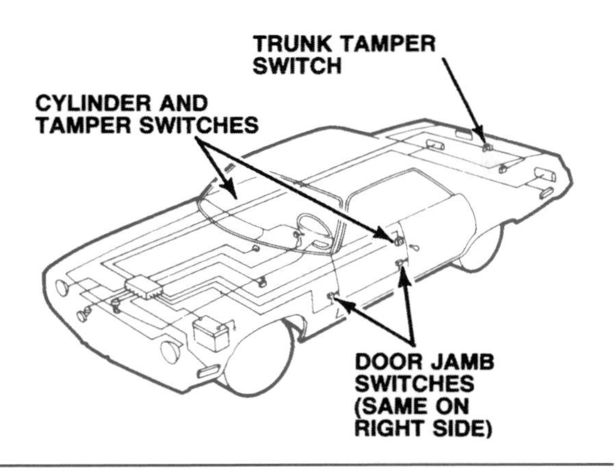

Figure 15-41. System functioning relies on strategically located switches installed in the door jambs, the door lock cylinders, and the trunk lock cylinder. (GM Service and Parts Operations)

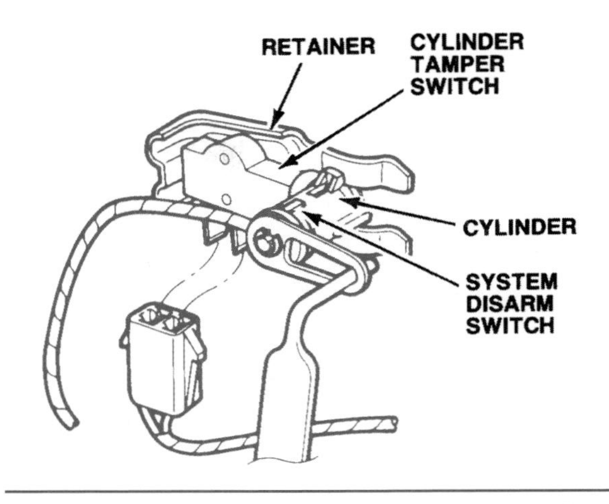

Figure 15-43. The UTD disarm switch is part of the LH door lock cylinder. (GM Service and Parts Operations)

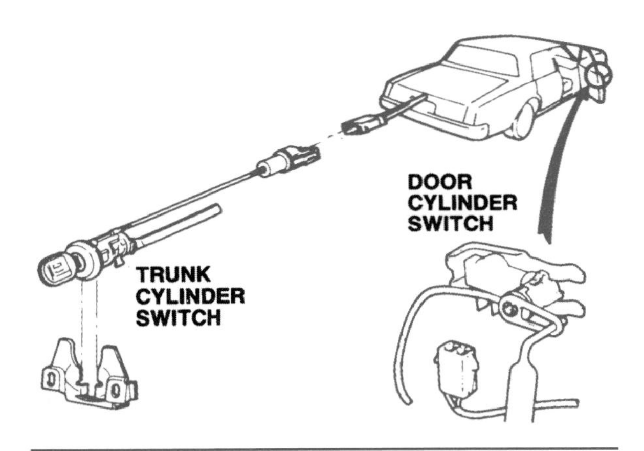

Figure 15-42. Tamper switches are installed in all door locks and the trunk lid lock. (GM Service and Parts Operations)

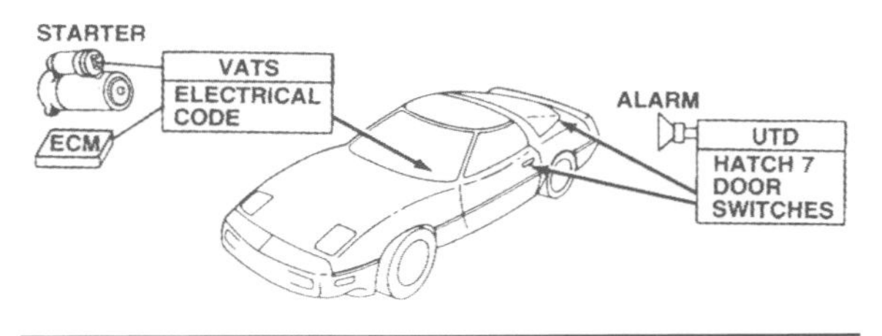

Figure 15-44. Delphi (Delco) VATS system. (GM Service and Parts Operations)

Delphi (Delco) VATS/PASS-Key II™ System

The Delco (Delphi) vehicle antitheft system (VATS), introduced as standard equipment on the 1986 Corvette (Figure 15-44) functions as an ignition-disable system. It is not designed to prevent a forced entry, but to protect the steering column lock if an intruder breaks into the vehicle. When used on Corvettes with the UTD system, the combination is called the forced entry alarm system (FES). When used on other GM vehicles, VATS is called PASS-Key II™. The system (Figure 15-45) consists of the following components:

- Resistor ignition key
- Steering column lock cylinder with resistor-sensing contact
- VATS or PASS-Key II™ decoder module
- Starter enable relay
- PCM
- Wiring harness

A small resistor pellet embedded in the ignition key contains one of 15 different resistance values. The key is coded with a number that indicates which resistor pellet it contains. Resistor pellet resistance values vary according to key code and model year. To operate the lock, the key must have the proper mechanical code (1 of 2,000); to close the starter circuit, it must also have the correct electrical code (1 of 15).

Inserting the key in the ignition lock cylinder brings the resistor pellet in contact with the resistor sensing contact. Rotating the lock applies battery power to the decoder module (Figure 15-45). The sensing contact sends the resistance value of the key pellet to the decoder module, where it is compared to a fixed resistance value stored in memory. If the resistor code and the fixed value are the same, the decoder module energizes the starter enable relay, which closes the circuit to the starter solenoid and allows the engine to crank. At the same time, the module sends a pulse-width modulated (PWM) cranking fuel-enable signal to the PCM.

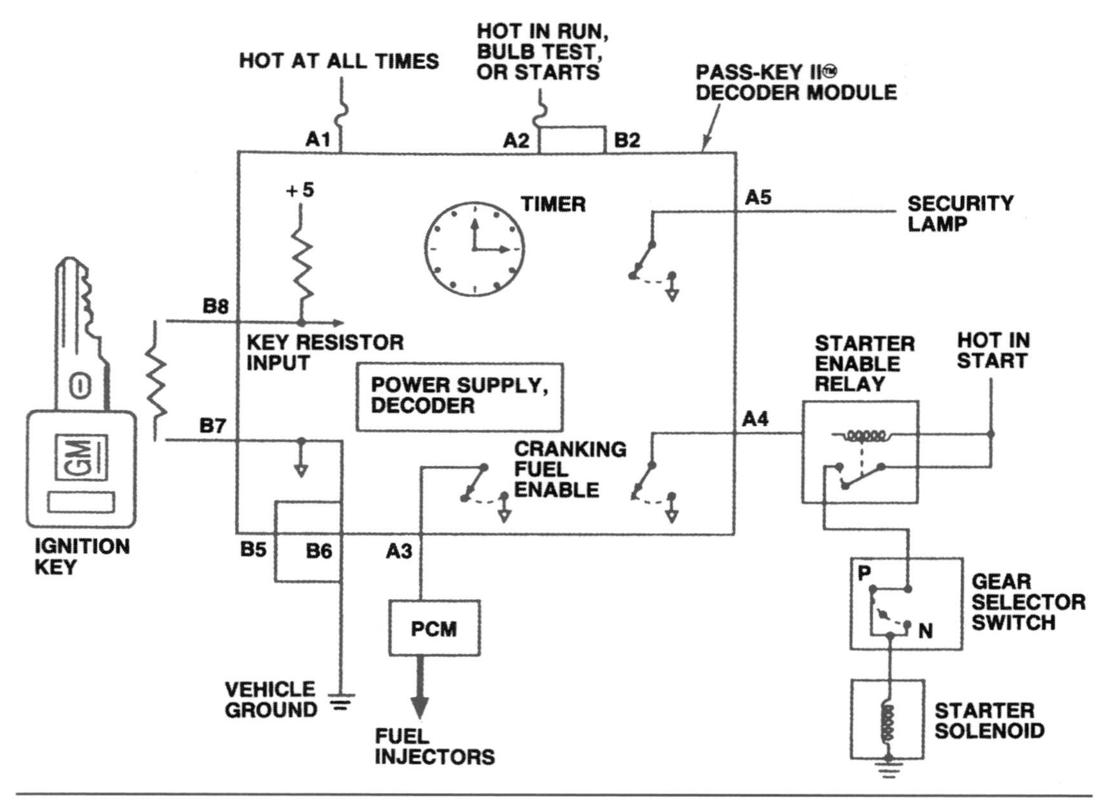

Figure 15-45. PASS-Key circuit diagram. (GM Service and Parts Operations)

If the key resistor code and the module's fixed resistance value do not match, the module shuts down for two to four minutes. Repeating the attempt to start the vehicle with the wrong key will result in continued module shutdowns. During vehicle operation, the key resistor pellet inputs are continually read. If the module sees an open, short, or incorrect resistance value for 60 consecutive seconds, a Security indicator lamp comes on and remains lighted until the fault is corrected. The lamp also comes on for five seconds when the ignition is first turned on. This serves as a bulb check and indicates that the system is functioning properly.

DaimlerChrysler Antitheft Security System

This passive-arming theft-deterrent system (Figure 15-46) is factory installed on high-line Chrysler models and functions like many aftermarket alarm installations. When combined with the Remote Keyless Entry (Figure 15-40), the system becomes an active arming system. Once armed, the doors, hood, and trunk lid all are monitored for unauthorized entry.

The system is passively armed by activating the power door locks before closing the driver's door; it will not arm if the doors are locked manually. The system is actively armed if the doors are locked with the RKE transmitter. A SET lamp in the instrument cluster flashes for 15 seconds during the arming period. If a forcible entry is attempted while the system is armed, it responds by sounding the horn, flashing the park and taillamps, and activating an engine kill feature.

The system is passively disarmed by unlocking either front door with the key, or actively disarmed by using the RKE transmitter. If the alarm has been activated during the driver's absence, the horn will blow three times when the vehicle is disarmed as a way of informing the driver of an attempted entry or tampering.

Ford Antitheft System

This antitheft system bears many similarities to the Delco UTD theft-deterrent system. It is installed on luxury models, uses many of the

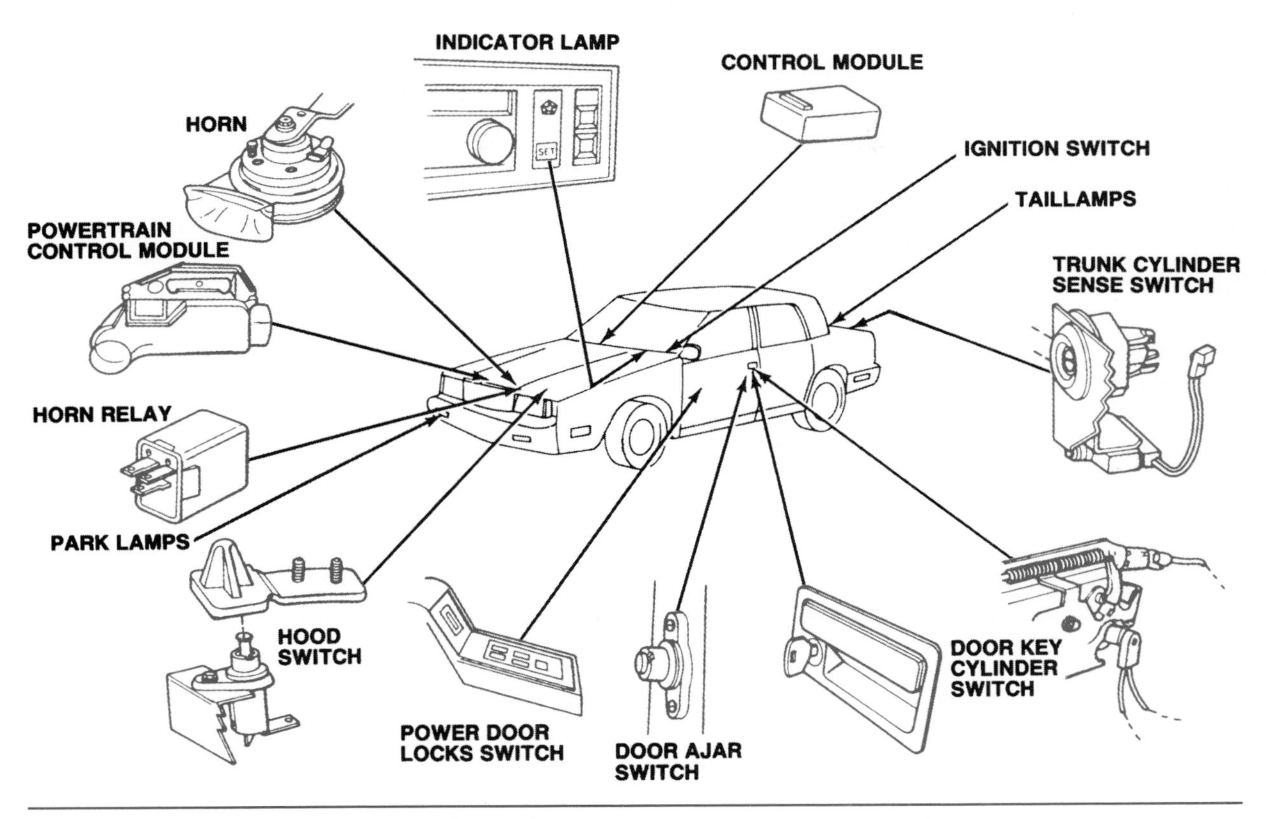

Figure 15-46. DaimlerChrysler antitheft security system. (DaimlerChrysler Corporation)

same components, and functions in essentially the same way. Once the system is armed, any tampering with the doors, hood, or trunk lid signals the control module. Once triggered, the system flashes the low-beam headlamps, the parking lamps, and alarm indicator lamp on and off; sounds the horn; and interrupts the starter circuit. The system is composed of the following components:

- Antitheft control module
- Antitheft warning indicator
- Door-key unlock switches
- Hood switch
- Trunk-lid lock-cylinder tamper switch
- Ignition-key lock-cylinder sensor

It also incorporates the following components from other systems:

- Power door lock switches
- Door-ajar switches
- Horn relay
- Low-beam headlamps
- Parking lamps
- Keyless entry module
- Starter relay

CRUISE CONTROL SYSTEMS

The *cruise control system* is one of the most popular electronic accessories installed on today's vehicles. During open-road driving it can maintain a constant vehicle speed without the continued effort of driver. This helps reduce driver fatigue and increases fuel economy. Several override features built into the cruise control system allow the vehicle to be accelerated, slowed, or stopped.

Problems with the system can vary from no operation, to intermittent operation, to not disengaging. To diagnose these system complaints, today's technicians must rely on their knowledge and ability to perform an accurate diagnosis. Most of the system is tested using familiar diagnostic procedures; build on this knowledge and ability to diagnose cruise control problems. Use system schematics, troubleshooting diagnostics, and switch continuity charts to assist in isolating the cause of the fault.

Most vehicle manufacturers have incorporated self-diagnostics into their cruise control systems. This allows some means of retrieving trouble codes to assist the technician in locating system faults.

On any vehicle, perform a visual inspection of the system. Check the vacuum hoses for disconnects, pinches, loose connections, etc. Inspect all wiring for tight, clean connections. Also, look for good insulation and proper wire routing. Check the fuses for opens and replace as needed. Check and adjust linkage cables or chains, if needed. Some manufacturers require additional preliminary checks before entering diagnostics. In addition, perform a road test (or simulated road test) in compliance with the service manual to confirm the complaint.

CAUTION: When servicing the cruise control system, you will be working close to the air bag and antilock brake systems. The service manual will instruct you when to disarm and/or depressurize these systems. Failure to follow these procedures can result in injury and additional costly repairs to the vehicle.

When engaged, the cruise control components set the throttle position to the desired speed. The speed is maintained unless heavy loads and steep hills interfere. The cruise control is disengaged whenever the brake pedal is depressed. The common speed or cruise control system components function in the following manner.

Cruise Control Switch

The cruise control switch (Figure 15-47) is located on the end of the turn signal lever or near the center or sides of the steering wheel. There are usually several functions on the switch, including off-on, resume, and engage buttons. The switch is different for resume and non-resume systems.

The **transducer** is a device that controls the speed of the vehicle. When the transducer is engaged, it senses vehicle speed and controls a vacuum source (usually the intake manifold). The vacuum source is used to maintain a certain position on a servo. The speed control is sensed from the lower cable and casing assembly attached to the transmission.

The **servo unit** is connected to the throttle by a rod or linkage, a bead chain, or a Bowden cable. The servo unit maintains the desired car speed by receiving a controlled amount of vacuum from the transducer. The variation in vacuum changes the position of the throttle. When a vacuum is applied, the servo spring is compressed and the throttle is positioned correctly. When the vacuum is released, the servo spring is relaxed and the system is not operating.

Two switches are activated by the position of the brake pedal. When the pedal is depressed, the brake-release switch disengages the system.

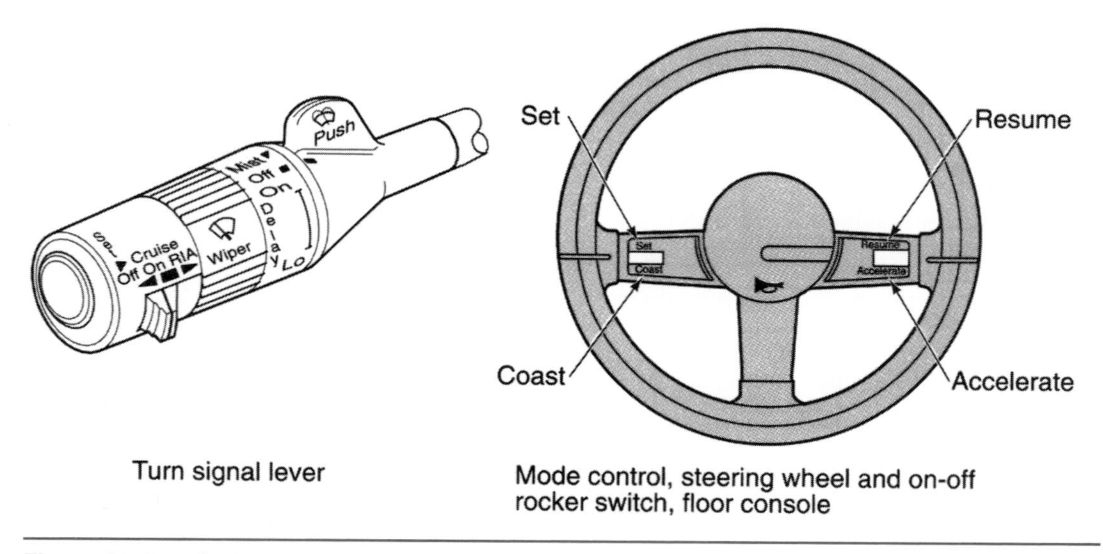

Figure 15-47. Cruise control switch.

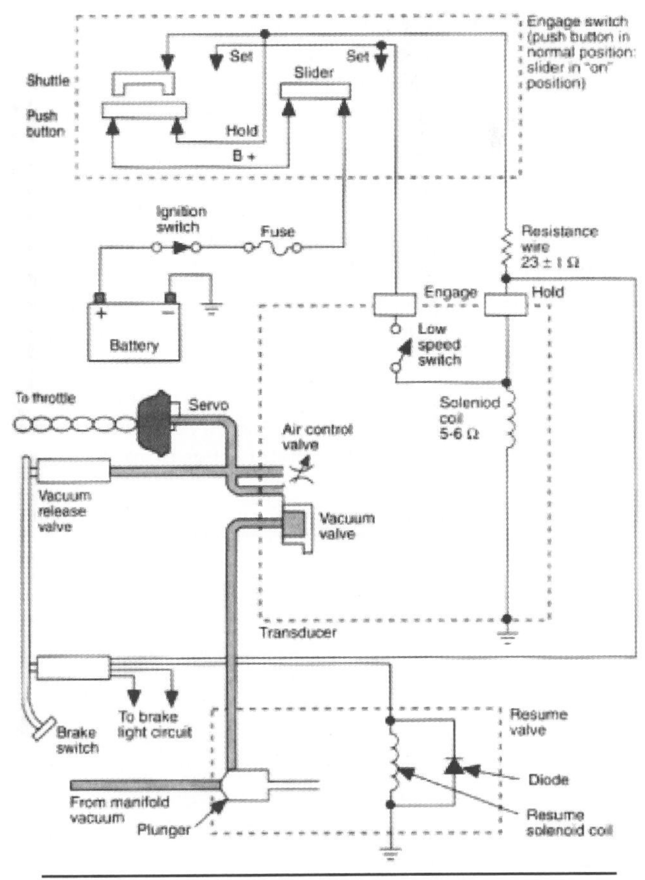

Figure 15-48. Cruise control system schematic. (GM Service and Parts Operations)

A vacuum-release valve is also used to disengage the system when the brake pedal is depressed.

Electrical and Vacuum Circuits

Figure 15-48 shows an electrical and vacuum circuit diagram. The system operates by controlling vacuum to the servo through various solenoids and switches.

Electronic Cruise Control Components

Cruise control can also be obtained by using electronic components rather than mechanical components. Depending on the vehicle manufacturer, several additional components may be used.

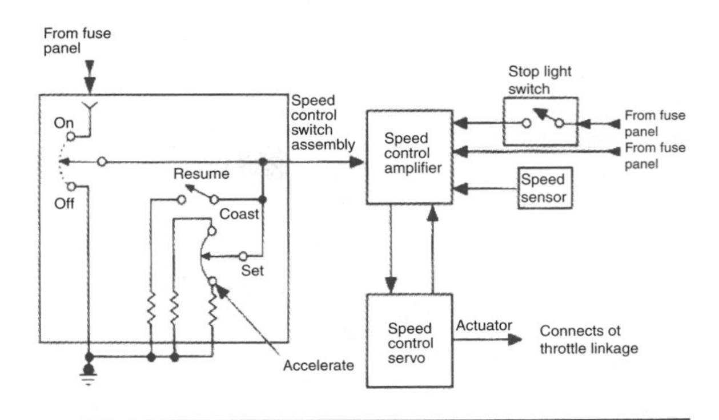

Figure 15-49. Cruise control system component circuit. (GM Service and Parts Operations)

The electronic control unit is used to control the servo unit. The servo unit is again used to control the vacuum, which in turn controls the throttle. The vehicle speed sensor (VSS) *buffer amplifier* is used to monitor or sense vehicle speed. The signal created is sent to the electronic control module. A generator speed sensor may also be used in conjunction with the VSS. The clutch switch is used on vehicles with manual transmissions to disengage the cruise control when the clutch is depressed. The accumulator is used as a vacuum storage tank on vehicles that have low vacuum during heavy load and high road speed.

Figure 15-49 shows how electronic cruise control components work together. The servo unit controls the throttle position, using a vacuum working against spring pressure to operate an internal diaphragm. The controller controls the servo unit vacuum circuit electronically. The controller has several inputs that help determine how it will affect the servo, including a brake-release switch (clutch-release switch), a speedometer, a buffer amplifier (generator speed sensor), a turn signal lever mode switch, and speed-control switches on the steering wheel.

SUPPLEMENTAL RESTRAINT SYSTEMS

A typical *supplemental inflatable restraint (SIR)* or air bag system (Figure 15-50) includes three important elements: the electrical system, air bag module, and knee diverter. The electrical system includes the impact sensors and the electronic control module. Its main functions are

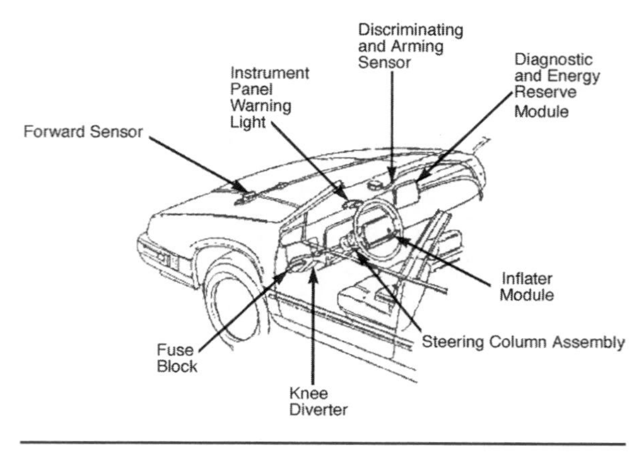

Figure 15-50. SRS components. (DaimlerChrysler Corporation)

to conduct a system self-check to let the driver know that it is functioning properly, to detect an impact, and to send a signal that inflates the air bag. The air bag module is located in the steering wheel for the driver and in the dash panel for passengers; it contains the air bag and the parts that cause it to inflate. The knee diverter cushions the driver's knee from impact and helps prevent the driver from sliding under the air bag during a collision. It is located underneath the steering column and behind the steering column trim.

Electrical System Components

The electrical system generally has the following parts:

Diagnostic Monitor Assembly

The diagnostic monitor contains a microcomputer that monitors the electrical system components and connections. The monitor performs a self-check of the microcomputer internal circuits and energizes the system readiness indicator during prove-out and whenever a fault occurs. System electrical faults can be detected and translated into coded indicator displays. If a certain fault occurs, the microcomputer disables the system by opening a thermal fuse built into the monitor. If a system fault exists and the indicator is malfunctioning, an audible tone signals the need for service. If certain faults occur, the system is disarmed by a firing circuit disarm device incorporated within the monitor or diagnostic module.

Trouble codes can be retrieved through the use of a scan tool or flash codes, and on some models through the digital panel cluster (if equipped). As with all diagnostics, consult the appropriate service manual for the correct procedures.

CAUTION: When servicing the air bag system, the service manual will instruct you when and how to disarm the system. Failure to follow these procedures can result in injury and additional costly repairs to the vehicle.

An air-bag-system backup power supply is included in the diagnostic monitor to provide air bag deployment power if the battery or battery cables are damaged in an accident before the crash sensors close. The power supply depletes its stored energy approximately one minute after the positive battery cable is disconnected.

Sensors

The sensors detect impact (Figure 15-51) and signal the air bag to inflate. At least two sensors must be activated for the air bag to inflate. There are usually five sensors: two at the radiator support, one at the right-hand fender apron, one at the left-hand fender apron, and one at the cowl in the passenger compartment. However, a few systems use only two sensors—one in front of the radiator and another in the passenger compartment. There is an **interlock** between the sensors, so two or more must work together to trigger the system. Keep in mind that air bag systems are designed to deploy in case of frontal collisions

CAUTION: The backup power supply energy must be depleted before any air bag component service is performed. To deplete the backup power supply energy, disconnect the positive battery cable and wait one minute.

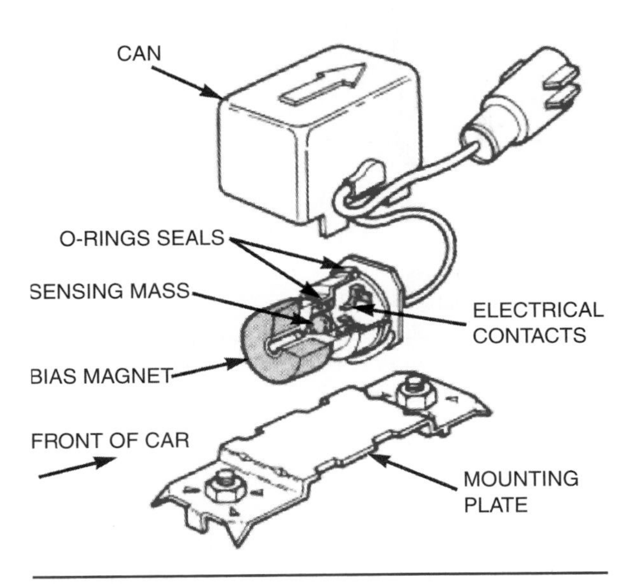

Figure 15-51. SRS sensors. (DaimlerChrysler Corporation)

only. Although the design of individual systems varies, the vehicle must be traveling a minimum of 12–28 mph before the system is armed and ready for deployment.

All the sensors use some type of *inertia switching mechanism* that provides for the breakaway of a metal ball from its captive magnet. This function causes a signal to activate a portion of the deployment program set up in the control processor. The system is still capable of directly applying battery power to the squib or detonator. At least two sensors, one safing sensor and one front crash sensor, must be activated to inflate the air bag.

Safing Sensors

An integrated version of this network includes a **safing sensor** (Figure 15-52), sometimes attached to the original crash sensor. This device confirms the attitude and magnitude of the frontal deceleration forces and offers the microprocessor a second opinion before actual deployment. This is all it takes to complete the firing sequence, and the bag will deploy.

Wiring Harness

The wiring harness connects all system components into a complete unit. The wires carry the electricity that signals the air bag to inflate. The harness also passes the signals during the self-diagnosis sequence.

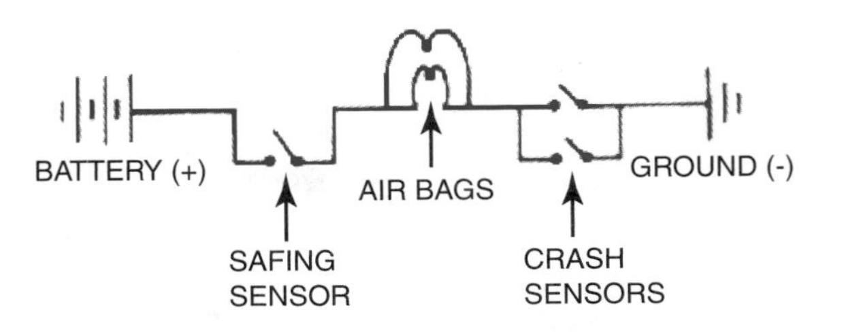

Figure 15-52. Safing sensor. (GM Service and Parts Operations)

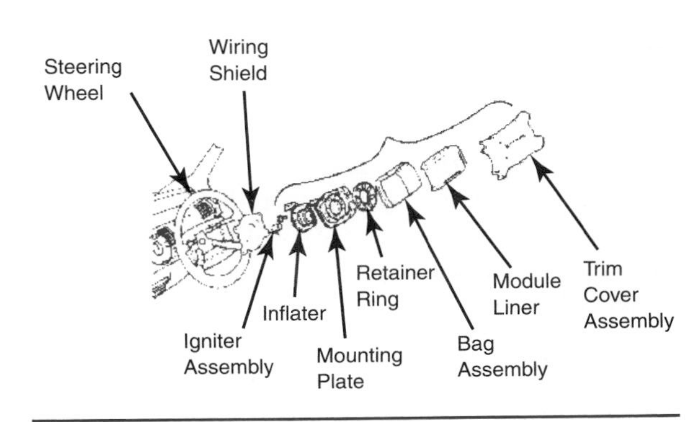

Figure 15-53. Air bag module. (DaimlerChrysler Corporation)

SIR or Air Bag Readiness Light

This light lets the driver know the air bag system is working and ready to do its job. The readiness lamp lights briefly when the driver turns the ignition key from OFF to RUN. A malfunction in the air bag system causes the light to stay on continuously or to flash, or the light might not come on at all. Some systems have a tone generator that sounds if there is a problem in the system or if the readiness light is not functioning.

Air Bag Module

The bag itself is composed of nylon and is sometimes coated internally with neoprene. All the **air bag module** (Figure 15-53) components are packaged in a single container, which is mounted in the center of the steering wheel or in the dash panel on the passenger side. The entire assembly must be serviced as one unit when repair of the air bag system is required. The air bag module is made up of the following components.

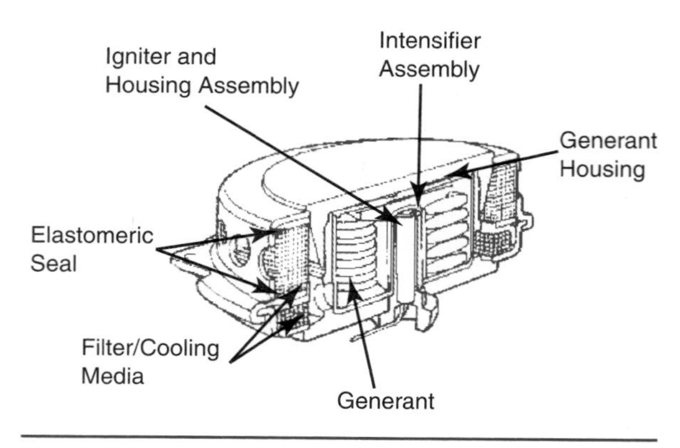

Figure 15-54. Igniter assembly. (GM Service and Parts Operations)

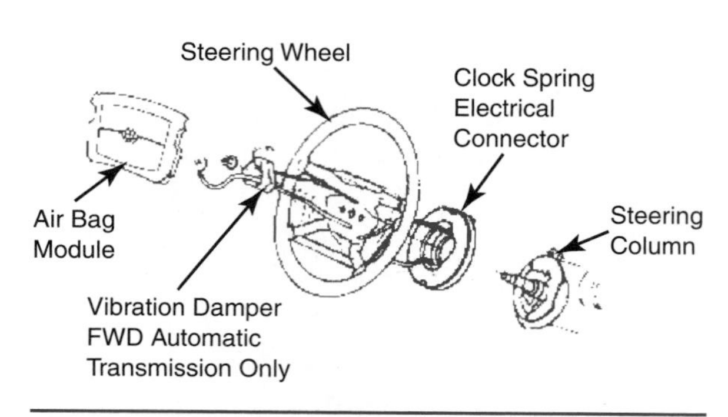

Figure 15-55. Liner and steering wheel trim cover. (GM Service and Parts Operations)

Igniter Assembly (Figure 15-55)

Inflation of the air bag is caused by an explosive release of gas. For the explosion to occur, a chemical reaction must be started. The **igniter assembly** does this when it receives a signal from the air bag monitor. Actually, the igniter is a two-pin bridge device: When the electrical current is applied, it arcs across the two pins, creating a spark that ignites a squib (canister of gas) that generates zirconic potassium perchlorate (ZPP). This material ignites the propellant. Some newer model air bags now use solid propellant and argon. This gas has a stable structure, cools more quickly, and is inert as well as non-toxic.

Inflator Module

The *inflator module* contains the ZPP. Once it triggers the igniter, the propellant charge is progressive, burning sodium azide, which converts to nitrogen gas as it burns. It is the nitrogen gas that fills the air bag.

Almost as soon as the bag is filled, the gas is cooled and vented, deflating the assembly as the collision energy is absorbed. The driver is cradled in the envelope of the supplemental restraint bag instead of being propelled forward to strike the steering wheel or be otherwise injured by follow-up inertia energy from seat belt restraint systems. In addition, a certain degree of facial protection against flying objects is obtained just when it is needed.

It is important to remember that only the tandem action of at least one main sensor and a safing sensor initiates safety restraint system activation. The micro-controller also provides failure data and trouble codes for use in servicing various aspects of most systems.

Mounting Plate and Retainer Ring

The *mounting plate and retainer ring* attach the air bag assembly to the inflator. They also keep the entire air bag module connected to the steering wheel.

Liner and Steering Wheel Trim Cover (Figure 15-56)

The liner houses the air bag; the trim cover goes over the exterior of the steering wheel hub. Passenger-side air bags are very similar in design to the driver's unit. The actual capacity of gas required to inflate the bag is much greater because the bag must span the extra distance between the occupant and the dashboard at the passenger seating location. The steering wheel and column make up this difference on the driver's side.

WARNING: When the air bag is deployed, a great deal of heat is generated. Although the heat is not harmful to passengers, it may damage the clock spring electrical connector. When replacing a deployed air bag module, examine all of the electrical connections for signs of scorching or damage. If damage exists, it must be repaired.

SUMMARY

Heating and air-conditioning systems share a motor-driven fan. The motor speeds usually are controlled by varying the resistance in the motor circuit. Air-conditioning systems also have an electromagnetic clutch on the compressor. The clutch is energized by the air-conditioning system control switch. A clutch-cycling pressure switch turns the compressor clutch on and off as needed to maintain desired evaporator pressure and temperature. Other switches are used in the clutch circuit to protect the system from high or low pressure, or to shut the clutch off under certain conditions, such as wide-open throttle.

Older vehicles may use a fan and motor as a rear-window defogger; this system is similar to the heating system. A rear-window defroster or defogger on late-model vehicles is a grid of conductors attached to the rear window. A relay usually controls current to the conductors. Ford's heated windshield system uses current directly from the alternator.

The parts of a sound system that concern most service technicians include the way the sound unit and the speakers are mounted, interference capacitors, panel illumination bulbs, and power antennas.

Power windows are moved by a reversible motor. Motor direction is controlled by current through different brushes. The driver's-side master switch controls all of the windows, because each individual switch is grounded through the master switch.

Power seats can be moved by one, two, or three motors and various transmission units. Permanent-magnet motors or electromagnetic field motors can be used. Current to the motors sometimes is controlled by a relay.

Power door locks, trunk latches, and seat-back releases can be moved by solenoids or motors. Relays are often used to control current to the solenoid or motor.

ADL systems are a safety feature integrated with the power door locks. They automatically lock all doors before the vehicle is driven. Keyless entry systems are both convenience and safety features: They allow a driver access to a vehicle by entering a code through a keypad on the driver's door, or by depressing a button on a key chain transmitter.

Theft-deterrent systems such as the Delco UTD and Ford Antitheft System are factory installed on luxury cars. Such systems use other vehicle systems to sound an alarm when the car is tampered with. The Delco VATS/PASS-Key II™ system uses a resistor ignition key that is "read" by the lock cylinder when inserted. If the resistance and the memory value do not match, the system shuts down the ignition and the starter.

When engaged, the cruise control components set the throttle position to the desired speed. The speed is maintained unless heavy loads and steep hills interfere. The cruise control is disengaged whenever the brake pedal is depressed. The cruise control switch is located on the end of the turn signal lever or near the center or sides of the steering wheel. There are usually several functions on the switch, including off-on, resume, and engage buttons. The switch is different for resume and non-resume systems.

The transducer is a device that controls the speed of the vehicle. When the transducer is engaged, it senses vehicle speed and controls a vacuum source (usually the intake manifold). The vacuum source is used to maintain a certain position on a servo. The speed control is sensed from the lower cable and casing assembly attached to the transmission. The servo unit is connected to the throttle by a rod or linkage, a bead chain, or a Bowden cable. The servo unit maintains the desired car speed by receiving a controlled amount of vacuum from the transducer. The variation in vacuum changes the position of the throttle. When a vacuum is applied, the servo spring is compressed and the throttle is positioned correctly. When the vacuum is released, the servo spring is relaxed and the system does not operate.

Two switches are activated by the position of the brake pedal. When the pedal is depressed, the brake release switch disengages the system. A vacuum release valve is also used to disengage the system when the brake pedal is depressed.

The Supplemental Inflatable Restraint (SIR) or air bag system includes three important elements. The electrical system includes the impact sensors and the electronic control module. Its main functions are to conduct a system self-check to let the driver know that it is functioning properly, to detect an impact, and to send a signal that inflates the air bag. The air bag module is located in the steering wheel for the driver and in the dash panel for passengers. It contains the air bag and the parts that cause it to inflate. The knee diverter cushions the driver's knee from impact and helps prevent the driver from sliding under the air bag during a collision. It is located underneath the steering column and behind the steering column trim. Newer vehicles contain SIR systems in the side panels and headliner or curtains.

Review Questions

1. Technician A says an electronic climate control (ECC) system with BCM control cycles the power to the compressor clutch. Technician B says an electronic climate control (ECC) system without BCM control cycles the ground circuit to the compressor clutch. Who is right?
   a. A only
   b. B only
   c. Both A and B
   d. Neither A nor B
2. Technician A says Ford's heated windshield system uses a conductive grid bonded to the outside surface of the glass. Technician B says the conductive grid is applied to the back of the outer glass layer before it is laminated to the inner glass layer. Who is right?
   a. A only
   b. B only
   c. Both A and B
   d. Neither A nor B
3. Constant operation of the compressor in automotive air-conditioning systems is prevented by:
   a. A solenoid
   b. A servomagnet
   c. An electromagnetic clutch
   d. A one-way clutch
4. Individual switches on automobile power window circuits must be connected in ______ with the driver's side master switch.
   a. Series
   b. Shunt
   c. Parallel
   d. Series-parallel
5. The air-conditioning compressor clutch can be controlled by:
   a. A power steering cutout switch
   b. A pressure cycling switch
   c. Both A and B
   d. Neither A nor B
6. Technician A says the user selects the mode in a semiautomatic temperature control system, but the actuators are electrically operated. Technician B says a semiautomatic temperature control system uses in-car and ambient temperature sensors. Who is right?
   a. A only
   b. B only
   c. Both A and B
   d. Neither A nor B
7. Technician A says the body control module (BCM) talks to other computers in an ECC system on a serial data line. Technician B says the ECC system programmer activates the actuators on a serial data line. Who is right?
   a. A only
   b. B only
   c. Both A and B
   d. Neither A nor B
8. GM power seat system motors use:
   a. Permanent-magnet fields
   b. Electromagnetic fields
   c. Both A and B
   d. Neither A nor B
9. The Delphi (Delco) vehicle antitheft system (VATS) uses a ________ in the ignition key.
   a. Thermistor
   b. Potentiometer
   c. Magnet
   d. Resistor
10. Power window systems use:
    a. Unidirectional motors
    b. Reversible motors
    c. Stepper motors
    d. Servomotors
11. Technician A says Ford's heated windshield system will work only if the in-car temperature is above 40°F (4°C). Technician B says the heated windshield module can control the EEC-IV module under certain circumstances. Who is right?
    a. A only
    b. B only
    c. Both A and B
    d. Neither A nor B
12. Technician A says a keyless entry system incorporates the function of an illuminated entry system. Technician B says moving the gear selector into Drive with the ignition on activates an ADL system. Who is right?
    a. A only
    b. B only

c. Both A and B
d. Neither A nor B

13. Technician A says the Delco Remote Keyless Entry (RKE) and the Remote Lock Control (RLC) are different keyless entry systems. Technician B says the RKE and RLC systems are subsystems of the VAT and PASS systems. Who is right?
    a. A only
    b. B only
    c. Both A and B
    d. Neither A nor B

14. Technician A says if the resistance value of the ignition key in the PASS system does not match the UHF value stored in the receiver's memory, the vehicle will not start. Technician B says a factory-installed theft-deterrent system is a complex multiple-circuit system. Who is right?
    a. A only
    b. B only
    c. Both A and B
    d. Neither A nor B

15. In a cruise control system, which component controls the amount of vacuum that is applied to the servo unit?
    a. The throttle body
    b. The cruise control switch
    c. The transducer
    d. The brake pedal position switch

16. Which of the following actions will deactivate cruise control operation?
    a. Applying the accelerator pedal
    b. Applying the brake
    c. Driving up a steep hill
    d. Releasing the accelerator pedal

17. In an electronically controlled cruise control system, what component is used to monitor vehicle speed?
    a. Wheel speed sensor
    b. Generator speed sensor
    c. Turbine speed sensor
    d. Vehicle speed sensor

18. What enables an air bag to deploy in an accident, even if the battery becomes disconnected during the crash?
    a. A mechanical push-arm igniter
    b. A backup power supply
    c. The velocity versus solid object sensor
    d. A "sudden stop" signal received from the vehicle speed sensor

19. Though different manufacturers have different specifications, how fast must a vehicle be moving before the air bag system is armed and ready to deploy if needed?
    a. Any speed above zero
    b. Between 3 and 10 mph
    c. Between 12 and 28 mph
    d. At least 45 mph

20. When servicing an air bag, what is the first step you should take?
    a. Disconnect the battery negative cable.
    b. Disconnect the clockspring electrical connector.
    c. Take resistance measurements at all electrical connectors.
    d. Disconnect the impact sensors.

Glossary

AC Generator (Alternator): A device in which the movement of magnetic lines through a stationary conductor generates voltage. A magnet called a rotor is turned inside a stationary looped conductor called a stator.

Air Core Gauge: A gauge design in which there is no magnetic core. A field created by the sending unit resistance moves a pivoting permanent magnet.

Amber: A translucent, yellowish resin, derived from fossilized trees. When you rub it against other materials, it became charged with an unseen force that had the ability to attract other lightweight objects such as feathers, somewhat like a magnet picks up metal objects.

Ampere: This unit expresses how many electrons move through a circuit in one second. A current flow of 6.25 pico (billion/billion) or 6.25×10^{18} electrons per second is equal to one ampere.

Asymmetrical: Different on both sides of center. In an asymmetrical low-beam headlamp, the light beam is spread farther to one side of center than to the other.

Batteries: A chemical device that produces direct current (DC) from a chemical reaction. It converts electrical energy into chemical energy, which is stored until the battery is connected to an external circuit.

Bipolar: A transistor that uses both holes and electrons as current carriers.

CCA Rating: CCA or Cold Cranking Amps is an important measurement of battery capacity because it measures the discharge load, in amps, that a battery can supply for 30 seconds at 0°F while maintaining a voltage of 1.2 volts per cell (7.2 volts per battery) or higher. The CCA rating generally falls between 300 and 970 for most passenger cars.

Capacitor: A device that opposes a change in voltage.

Cell: A case enclosing one element in an electrolyte. Each cell produces approximately 2.1 to 2.2 volts. Cells are connected in series.

Circuit Number: The number, or number and letter that manufacturers use to identify a specific electrical circuit in a diagram.

Clutch Start Switch: A starting safety switch that is operated by the clutch pedal.

Color Coding: The use of colored insulation on wire to identify an electrical circuit.

Combustible Materials: Materials such as gasoline, paint, and oily rags that will burn when set on fire.

Component Symbols: Symbols on a wiring diagram that identify automotive components.

Compound Motor: A motor that has both series and shunt field windings. Often used as a starter motor.

Connectors: Devices used to provide a strong permanent connection of two or more wires and to protect terminals from the elements and wear.

Corrosive: If a material burns the skin, or dissolves metals and other materials, a technician should consider it hazardous.

Cranking Performance Rating: A battery rating based on the amperes of current that a battery can supply for 30 seconds at 0°F, with no battery cell falling below 1.2 volts.

Current: This is the flow of electrons.

Current Flow Diagram: Developed SPX Valley Forge Technical Information Systems. Electrical circuit diagrams are printed in color so the lines match the color of the wires. The name of the color is printed beside the wire. The metric wire gauge may also be printed immediately before the color name. The circuit flow from positive to negative or top to bottom.

Cycling: Battery electrochemical action and operation from charged to discharged and back. One complete cycle is operation from fully charged to discharged and back to fully charged.

D'Arsonval Movement: A small, current-carrying coil mounted within the field of a permanent horseshoe magnet. Interaction of the magnetic fields causes the coil to rotate. Used as a measuring device within electrical gauges and test meters.

Data Link: A digital signal path for communications between two or more components of an electronic system.

Delta-Type Stator: An AC Generator (alternator) stator design in which the three windings of a three-phase AC Generator (alternator) are connected end-to-end. The beginning of one winding is attached to the end of another winding. Delta-type stators are used in AC Generators (alternator) that must provide high-current output.

Detented: Positions in a switch that allow the switch to stay in that position. In an ignition switch, the On, Off, Lock, and Accessory positions are detented.

Drain: The field-effect transistor (PET) layer, which collects current carriers (similar to the collector of a bipolar transistor).

Electricity: When an atom is not balanced, it becomes an ion. Ions try to regain their balance of equal protons and electrons by exchanging electrons with nearby atoms. This is known as the flow of electric current or electricity.

Electrolyte: The chemical solution in a battery that conducts electricity and reacts with the plate materials. A reactive solution that when in the presence of two dissimilar metals causes a chemical action that creates electricity in an automotive wet-cell battery.

Electromagnetism: Any magnetic field created by electrical current flow.

Electrostatic Discharge or ESD: An electrostatic charge can build up on the surface of your body. ESD occurs when you touch something with your charge that can be discharged to the other surface.

Element: A complete assembly of positive plates, negative plates, and separators making up one cell of a battery.

Eyewash Fountain: An eyewash fountain is similar to a drinking water fountain, but the eyewash fountain has water jets placed throughout the fountain top to wash out your eyes after they have contaminated with a dangerous liquid.

Field Circuit: The charging system circuit that delivers current to the AC Generator (alternator) field.

Fire Extinguishers: Device used to extinguish a fire.

Full-Wave Rectification: A process by which all of an ac sine wave voltage is rectified and allowed to flow as dc.

Ground: The ground path is that side of the circuit that carries one half of the current in the circuit; i.e., automobile engine, frame, and body.

Ground Cable: The battery cable that provides a round connection from the vehicle chassis to the battery.

Group Number: A battery identification number that indicates battery dimensions, terminal design, hold-down location, and other physical features.

Half-Wave Rectification: A process by which only one-half of an ac sine wave voltage is rectified to dc.

Hazard Communication Standard: Published by the Occupational Safety and Health Administration (OSHA) in 1983. Originally, this document provides information on how to handle hazardous materials in a work situation.

Hazardous Waste Materials: Chemicals, or components, that the shop no longer needs, and these materials pose a danger to the environment and people if they are disposed of in ordinary garbage cans or sewers.

Head-Up Display (HUD) System: A secondary display system that projects video images onto the windshield.

Hold-In Winding: The coil of small-diameter wire in a solenoid that is used to create a magnetic field to hold the solenoid plunger in position inside the coil solenoid, allowing full battery current to flow to the starter motor. This type of solenoid can be designed to provide an alternate path to the ignition coil during starting. This bypasses the resistance wire normally used to lower coil voltage during engine operation and provides a hotter spark during starting.

Ignitable: A liquid is hazardous if it has a flash point below 1400°F (600°C), and a solid is hazardous if it ignites spontaneously.

Installation Diagram: A drawing that shows where the wires, switches, loads, attachment hardware, and other parts of an electrical circuit are located.

Insulated, or Hot Cable: The battery cable that conducts battery current to the automotive electrical system.

Kirchhoff's Law of Current (Kirchhoff's 1st Law): Kirchhoff's Law of Current: States that the current flowing into a junction or point in an electrical circuit must equal the current flowing out.

Kirchhoff's Law of Voltage Drops (Kirchhoff's 2nd Law): Kirchhoff's Law of Voltage Drops states that voltage will drop in exact proportion to the resistance and that the sum of the voltage drops must equal the voltage applied to the circuit.

Lap Winding: A method of wiring a motor armature. The two ends of a conductor are attached to two commutator bars that are next to each other.

Lift Pads (Points): Areas on the vehicle that you place the lift arms against, to lift the vehicle.

Light-Emitting Diode (LED): A gallium-arsenide diode that emits energy as light. Often used in automotive indicators.

Lines of Force (Lines of Flux): A magnetic field is made up of many invisible lines of force. These lines of force can be compared to *current* used in electricity.

Liquid Crystal Display (LCD): An indicator consisting of a sandwich of glass containing electrodes and polarized fluid. Voltage applied to the fluid allows light to pass through it.

Lockout/tagout: This OSHA procedure is designed to prevent electrical equipment from being started while being repaired or maintained.

Logic Symbol: A symbol identifying the type of electronic gate in a digital or logic circuit.

Magnetic Field Intensity: Refers to the magnetic filed strength (force) exerted by the magnetic filed in magnetism and can be compared to *voltage* in electricity.

Material Safety Data Sheets (MSDS): The MSDS sheets provide the following information about the hazardous material: chemical name, physical characteristics, protective handling equipment, explosion/fire hazards, incompatible materials, health hazards, medical conditions aggravated by exposure, emergency and first-aid procedures, safe handling, and spill/leak procedures.

Menu-Driven: A computer program that allows the user to select choices from a list or "menu". As each choice is made, another menu allows the user to make another choice to achieve the desired end result.

Metric Wire Sizes: The wire sizes listed as 0.5, 1.0, 1.5, 4.0, or 6.0 are actual sizes in millimeters. These numbers are the cross section area of the conductor in square millimeters (mm^2).

Multiplex Wiring System: An electrical circuit in which several devices share signals on a common conductor. Signals may be transmitted in parallel form by a solid-state switching device or in serial form over a peripheral data bus or fiber-optic cable.

Multiplexing: is defined as a means of sending two or more messages simultaneously over the same channel.

Nonvolatile RAM: Random access memory (RAM) that retains its information when current to the chip is removed.

Neutral Junction: The center connection of the three windings in a Y-type stator.

Occupational Safety and Health Act (OSHA): The purposes of this legislation is to assist and encourage the citizens of United States in their efforts to assure safe and healthful working conditions by providing research, information, education, and training in the field of occupational safety and health.

Ohm's Law: States that voltage equals current times resistance and is expressed as $E = IR$. Ohm's Law is based on the fact that it takes 1 volt of electrical potential to push 1 ampere of current through 1 ohm of resistance.

Output Circuit: The charging system circuit that sends voltage and current to the battery and other electrical systems and devices.

Plates: Built on grids of conductive materials, which act as a framework for the dissimilar metals.

Primary Battery: A battery in which chemical processes destroy one of the metals necessary to create electrical energy. Primary batteries cannot be recharged.

Primary Wiring: The low-voltage wiring in an automobile electrical system.

Pull-In Winding: The coil of large-diameter wire in a solenoid that is used to create a magnetic field to pull the solenoid plunger into the coil.

Radio Choke: A coil of extremely fine wire used to absorb oscillations created by the making and breaking of an electrical circuit.

Radioactive: Any substance that emits measurable levels of radiation. When individuals bring containers of highly radioactive substance into the shop environment, qualified personnel with the appropriate equipment must test them.

Reactive: Any material, which reacts violently with water or other chemical, is considered hazardous. When exposed to low pH acid solutions, if a material releases cyanide gas, hydrogen sulfide gas, or similar gases, it is considered hazardous.

Reserve Capacity Rating: A battery rating based on the number of minutes a battery at 80°F can supply 25 amperes, with no battery cell falling below 1.75 volts.

Rheostat: A variable resistor used to control current.

Right-to-Know Laws: These laws state that employees have a right to know when the materials they use at work are hazardous.

Safety Glasses: Eye glasses that protect the eyes.

Safety Stands: Metal stands that you place under a vehicle to support it once you have lifted it with a hydraulic jack or lift.

Schematic Diagram: A drawing that shows all of the different circuits in a complete electrical system.

Secondary Battery: A battery in which chemical processes can be reversed. A secondary battery can be recharged so that it will continue to supply voltage.

Series Contacts: The normally closed set of contacts in a double-contact regulator. When they open, field current must flow through a resistor.

Series Motor: A motor that has only one path for current flow through the field and armature windings. Commonly used for starter motors.

Servomotor: An electric motor that is part of a feedback system used for automatic control of a mechanical device, such as in a temperature-control system.

Shorting Contacts: The normally open set of contacts in a double-contact regulator. When closed, they short-circuit the field to ground.

Shunt Motor: A motor that has its field windings wired in parallel with its armature. Not used as a starter motor, but often used to power vehicle accessories.

Sine Wave Voltage: The constant change, from zero volt to a positive peak, and then to a negative peak, and back to zero, of an induced alternating voltage in a conductor.

Single-Phase Current: Alternating current created by a single-phase voltage.

Single-Phase Voltage: The sine wave voltage induced within one conductor by one revolution of an AC Generator (alternator) rotor.

SLA: Sealed Lead Acid Battery. These new batteries do not require -and do not have the small gas vent used on previous maintenance-free batteries.

Solenoid: Similar to a relay in the way it operates. The major difference is that the solenoid core moves instead of the armature, as in a relay. This allows the solenoid to change current flow into mechanical movement.

Solenoid-Actuated Starter: A starter that uses a solenoid both to control current flow in the starter circuit and to engage the starter motor with the engine flywheel.

Solvents: Chemicals that are used to clean parts, tools, and other items essential in vehicle maintenance and collision repair operations.

Source: The field-effect transistor (PET) layer that supplies current-carrying holes or electrons (similar to the emitter of a bipolar transistor).

Specific Gravity: The weight of a volume of liquid divided by the weight of the same volume of water at a given temperature and pressure. Water has a specific gravity of 1.000. Produces the brief high-current flow required of a starting battery.

Starting Safety Switch: A neutral start switch. It keeps the starting system from operating when a car's transmission is in gear.

State-of-Charge Indicator or Built-in Hydrometer: Installed in the battery top. The indicator shows whether the electrolyte has fallen below a minimum level, and it also functions as a *go/no-go gauge.*

Sulfation: The crystallization of lead sulfate on the plates of a constantly discharged battery.

Symmetrical: The same on both sides of center. In a symmetrical high-beam headlamp, the light beam is spread the same distance to both sides of center.

Thermistor: A resistor specially constructed so that its resistance changes as its temperature changes. Also called a thermal resistor.

Thermocouple: A small device that gives off a low voltage when two dissimilar metals are heated.

Three-Coil Movement: A gauge design that depends upon the field interaction of three electromagnets and the total field effect on a movable permanent magnet.

Three-Phase Current: Three overlapping, evenly spaced, single-phase currents that make up the total ac output of an AC Generator (alternator).

Torque: Twisting or rotating force; usually expressed in foot-pounds, inch-pounds, or Newton-meters.

Toxic: Materials are hazardous if they leak one or more of eight different heavy metals in concentrations greater than 100 times the primary drinking water standard.

Transformers: Electrical devices that work on the principle of mutual induction. Transformers are typically constructed of a primary winding (coil), secondary winding (coil) and a common core. The principle of a transformer is essentially that of flowing current through a primary coil and inducing current flow in a secondary or output coil.

Used Oil: This is engine oil that has been used.

Vacuum Fluorescent Display (VFD): An indicator in which electrons from a heated filament strike a phosphor material that emits light.

Valence: This means the ability to combine.

Valley Forge Diagram: A type of current flow system diagram in which current travels from top to bottom and only relevant information is given.

Voltage Creep: Excessive voltage at high speeds due to excessive field current flows through a single-contact regulator. Also called voltage drift.

Weatherproof Connectors: Used in the engine compartment and body harnesses of late-model GM cars. This type of connector has a rubber seal on the wire ends of the terminals, with secondary sealing covers on the rear of each connector half. Such connectors are particularly useful in electronic systems where moisture or corrosion in the connector can cause a voltage drop.

Wire Gauge: Wire size numbers based on the cross-section area of the conductor. Larger wires have lower gauge numbers.

Wire Gauge Number: An expression of the cross section area of the conductor.

Wiring Harness: A bundle of wires enclosed in a plastic cover and routed to various areas of the vehicle. Most harnesses end in plug-in connectors. Harnesses are also called looms.

Y-Type Stator: An AC Generator (alternator) stator design in which one end of each of the three windings in a three-phase AC Generator (alternator) is connected at a neutral junction. This design is used in AC Generators (alternator) that require high voltage at low speed.

Index